Edited by
Achille G. Gravanis and
Synthia H. Mellon

**Hormones in Neurodegeneration,
Neuroprotection, and Neurogenesis**

Related Titles

Kvetnansky, R., Aguilera, G., Goldstein, D., Jezova, D., Krizanova, O., Sabban, E., Pacak, K.

Stress, Neurotransmitters, and Hormones

Neuroendocrine and Genetic Mechanisms

2009
ISBN: 978-1-57331-692-7

Nothwang, H., Pfeiffer, E (eds.)

Proteomics of the Nervous System

2008
ISBN: 978-3-527-31716-5

Bader, M. (ed.)

Cardiovascular Hormone Systems

From Molecular Mechanisms to Novel Therapeutics

2008
ISBN: 978-3-527-31920-6

von Bohlen und Halbach, O., Dermietzel, R.

Neurotransmitters and Neuromodulators

Handbook of Receptors and Biological Effects

2006
ISBN: 978-3-527-31307-5

Cutolo, M., Bijlsma, J. W. J., Straub, R. H., Masi, A. T. (eds.)

Neuroendocrine Immunology in Rheumatic Diseases

Translation from Basics to Clinics

2010
ISBN: 978-1-57331-769-6

Edited by Achille G. Gravanis and Synthia H. Mellon

Hormones in Neurodegeneration, Neuroprotection, and Neurogenesis

WILEY-VCH Verlag GmbH & Co. KGaA

The Editors

Prof. Dr. Achille G. Gravanis
Dept. of Pharmacology
Univ. of Crete Med. School
71110 Heraklion
Greece

Prof. Dr. Synthia H. Mellon
Ctr. Reproductive Sciences
University of California
P.O. Box 0556
San Francisco, CA 94143-0556
USA

Cover
The cover photo depicts sympathetic neurons in culture, isolated from superior cervical ganglia (SCG) of E17 mouse embryos (With kind permission from the Dept. of Pharmacology, School of Medicine, University of Crete).

■ **Limit of Liability/Disclaimer of Warranty:**
While the publisher and author have used their best efforts in preparing this book, they make no representations or warranties with respect to the accuracy or completeness of the contents of this book and specifically disclaim any implied warranties of merchantability or fitness for a particular purpose. No warranty can be created or extended by sales representatives or written sales materials. The Advice and strategies contained herein may not be suitable for your situation. You should consult with a professional where appropriate. Neither the publisher nor authors shall be liable for any loss of profit or any other commercial damages, including but not limited to special, incidental, consequential, or other damages.

Library of Congress Card No.: applied for

British Library Cataloguing-in-Publication Data
A catalogue record for this book is available from the British Library.

Bibliographic information published by the Deutsche Nationalbibliothek
The Deutsche Nationalbibliothek lists this publication in the Deutsche Nationalbibliografie; detailed bibliographic data are available on the Internet at <http://dnb.d-nb.de>.

© 2011 WILEY-VCH Verlag & Co. KGaA, Boschstr. 12, 69469 Weinheim, Germany

Wiley-Blackwell is an imprint of John Wiley & Sons, formed by the merger of Wiley's global Scientific, Technical, and Medical business with Blackwell Publishing.

All rights reserved (including those of translation into other languages). No part of this book may be reproduced in any form – by photoprinting, microfilm, or any other means – nor transmitted or translated into a machine language without written permission from the publishers. Registered names, trademarks, etc. used in this book, even when not specifically marked as such, are not to be considered unprotected by law.

Cover Design   Formgeber, Eppelheim
Typesetting   Laserwords Private Limited, Chennai, India
Printing and Binding   Fabulous Printers Pte Ltd, Singapore

Printed in Singapore
Printed on acid-free paper

ISBN: 978-3-527-32627-3

Foreword

To the scientific keywords of this book, I certainly like to add "ageing" and "hope." Clearly, the main risk of neurodegeneration is increased lifespan, which characterizes the current evolution of mankind and, at the same time, today we have hope and even some remedies for improved understanding and means of maintaining active brain function until we die.

The CNS receives, modulates, and transfers to the body information from the environment; it controls our mental and affective life and is therefore largely in command of our behavior and activities. The hormonal system (which includes more than 50 distinct important molecules) is the main vector of the CNS, which gives orders to the rest of the body. Conversely, many hormones synthesized at the "periphery" have access to the brain and can modify its (re)actions. Hormones interact between them at levels of regulated synthesis, metabolism and activities. Age-dependent changes in secretion, distribution, receptors, and metabolism of hormones are important to study; they may be responsible for opposite effects according to the "age." If the core of the receptors' structure is the same in most target cells, the receptor-associated proteins vary according to the tissue, physiological state, and so on, including the age of the person.

The CNS is certainly the most complex system of our body. With reference to hormones, not only does it control many glandular productions and therefore their activities, but it itself also synthesizes some of them already produced by peripheral glands, such as in the case of neurosteroids. For example [1], progesterone, made in Schwann cells and oligodendrocytes, is obviously the same as the progesterone coming from ovaries; it is, however, understandable that it works preferentially (probably or even uniquely) in the neighboring nerves and myelin, we even know that "classical" glandular progesterone may have the same activities. Does this double origin influence an important secondary sex characteristic explaining the pathological differences between men and women? Up to now, we believe that "neuroprogesterone" does not act through receptors different from those found at the periphery (nuclear receptors).

Among neurosteroids (synthesized in the CNS by definition) [2], pregnenolone is globally the most abundant, and it itself displays activities in the CNS, which have not yet been demonstrated elsewhere in the organism. MAP2, an important

Hormones in Neurodegeneration, Neuroprotection, and Neurogenesis.
Edited by Achille G. Gravanis and Synthia H. Mellon
Copyright © 2011 WILEY-VCH Verlag GmbH & Co. KGaA, Weinheim
ISBN: 978-3-527-32627-3

microtubule-associated protein abundant in the brain, is a pregnenolone receptor, binding the steroid specifically, thus stimulating tubulin polymerization [3]. Its derivative "MAP4343," which is not metabolized to progesterone, is active to treat spinal cord injuries (European patent EP15831538, 2008). In addition, pregnenolone, when sulfated at the 3-position, becomes a ligand of the $GABA_A$ receptor, antagonist to allopregnanolone, and also an active ligand of an NMDA receptor [4].

The steroid receptors bind reversibly to the heat shock protein HSP90, and while working on the hetero-oligomeric forms of these receptors, an immunophilin (binding immunosuppressants) protein FKBP52 was discovered. This protein, abundant in the brain, is able to bind to Tau (a microtubule-associated protein) and display an "anti Tau effect" on Tau activity on tubulin polymerization [5]. These results deriving from hormonal studies (even without direct action of steroids) have led to unveil novel mechanisms applicable to neurodegenerative diseases (whether hormones will or will not additionally interfere has not been detected so far).

The previously cited examples make logical that this book includes a remarkable variety of functions for hormones acting, from and/or for, on the CNS. An incomplete list includes the interesting actions of estrogens; progesterone (and progestins); androgens themselves or in association with peptides of the insulin-growth hormone – IGF family; the role of leptin; the subtle and varied effects of glucocorticosteroids and dehydroepiandrosterone; and the effects of somatostatin, CRH, PACAP, and erythropoietin. No doubt therapeutic issues may arise from these studies. Hormonal activities can be manipulated at the level of production (an example with etifoxine stimulating the neurosteroid pregnenolone production) [6], at the level of receptors (hormone antagonists), and at the level of their metabolism (via transport protein function and hepatic and renal functions).

I take the liberty to advise the readers of the book to keep it near their desk: they will have the privilege to conveniently obtain information on "new neuroendocrinology" as well as on appropriate references of talented authors of the reports.

Kremlin–Bicêtre, November 2010   *Etienne Emile Baulieu*

References

1. Koenig, H.L., Schumacher, M., Ferzaz, B., Do Thi, A.N., Ressouches, A., Guennoun, R., Jung-Testas, I., Robel, P., Akwa, Y., and Baulieu, E.E. (1995) Progesterone synthesis and myelin formation by Schwann cells. *Science* **268**, 1500–1503.
2. Baulieu, E.E. (1997) in *Recent Progress in Hormone Research* vol. 52 (ed. P.M. Conn), The Endocrine Society Press, Bethesda, pp. 1–32.
3. Fontaine-Lenoir V., Chambraud B., Fellous A., David S., Duchossoy Y., Baulieu E.E., and Robel P. (2006) Microtubule-associated protein 2 (MAP2) is a neurosteroid receptor. *Proc. Natl. Acad. Sci. USA.*, **103**, 4711–4716.
4. Baulieu, E.E., Robel, P., and Schumacher, M. (eds) (1999) *Neurosteroids. A new regulatory function in the nervous system.* Humana Press, Totowa, New Jersey.

5. Chambraud B., Sardin E., Giustiniani J., Dounane O., Schumacher M., Goedert M., and Baulieu E.E. (2010) A role for FKBP52 in Tau protein function. *Proc. Natl. Acad. Sci. USA.*, **107**, 2658–2663.
6. Girard C., Liu S., Cadepond F., Adams D., Lacroix C., Verleye M., Gillardin J.M., Baulieu E.E., Schumacher M., and Schweizer-Groyer G. (2008) Etifoxine improves peripheral nerve regeneration and functional recovery. *Proc. Natl. Acad. Sci. USA.*, **105**, 20505–20510.

Contents

List of Contributors XIX

Part I Estrogens, Progestins, Allopregnanolone and Neuroprotection 1

1 Interactions of Estradiol and Insulin-like Growth Factor-I in Neuroprotection: Implications for Brain Aging and Neurodegeneration 3
María-Angeles Arévalo, Luis M. Garcia-Segura, and Iñigo Azcoitia
1.1 Introduction: Hormones, Brain Aging, and Neurodegeneration 3
1.2 Estradiol, IGF-I, Brain Aging, and Neuroprotection 4
1.3 Molecular Interactions of Estrogen Receptors and IGF-I Receptor in the Brain 5
1.4 Regulation of IGF-I Receptor Signaling by Estradiol in the Brain 5
1.5 Regulation of Estrogen Receptor Transcriptional Activity by IGF-I in Neural Cells 6
1.6 Implications of the Cross Talk between Estrogen Receptors and IGF-I Receptors for Brain Aging, and Neurodegeneration 6
Acknowledgment 8
References 8

2 Structure–Nongenomic Neuroprotection Relationship of Estrogens and Estrogen-Derived Compounds 13
James W. Simpkins, Kun Don Yi, Evelyn Perez, and Douglas Covey
2.1 Introduction 13
2.2 *In vitro* Assessments of Structure–Neuroprotective Activity Relationships 14
2.2.1 Estradiol and Other Known Estratrienes 15
2.2.2 A-Ring Derivatives 15
2.2.3 B- and C-Ring Derivatives 16
2.2.4 D-Ring Derivatives 17
2.2.5 Correlation between Inhibition of TBARs and Protection against Glutamate and IAA 17
2.2.6 Estrogen Receptor Binding 17

Hormones in Neurodegeneration, Neuroprotection, and Neurogenesis.
Edited by Achille G. Gravanis and Synthia H. Mellon
Copyright © 2011 WILEY-VCH Verlag GmbH & Co. KGaA, Weinheim
ISBN: 978-3-527-32627-3

2.2.7	Correlation between Inhibition of TBARs or Neuroprotection and ER Binding 18
2.2.8	Interpretation of *In vitro* Findings 19
2.3	*In vivo* Assessment of Structure–Neuroprotective Activity Relationships 20
2.4	*In vitro* Assessment of Structure–Cell Signaling Relationships 20
2.5	Summary 24
	Acknowledgment 24
	References 24

3	**Progestins and Neuroprotection: Why the Choice of Progestin Matters** 29
	Meharvan Singh
3.1	Introduction 29
3.2	The Biology of Progesterone 30
3.3	Membrane-Associated Progesterone Receptors 31
3.4	Progesterone-Induced Protection 32
3.5	Mechanisms Underlying Progesterone's Protective Effects 33
3.6	Medroxyprogesterone Acetate 34
	Acknowledgments 37
	References 37

4	**Endogenous and Synthetic Neurosteroids in the Treatment of Niemann–Pick Type C Disease** 41
	Synthia H. Mellon, Wenhui Gong, and Marcus D. Schonemann
4.1	Introduction 41
4.2	Niemann–Pick Type C Disease as a Model of Disrupted Neurosteroidogenesis 43
4.3	Steroidogenesis and Neurosteroidogenesis in NP-C 44
4.4	Treatment of NP-C Mice with Allopregnanolone 45
4.5	Mechanism of Allopregnanolone Action: $GABA_A$ Receptor 46
4.6	Mechanism of Allopregnanolone Action: Pregnane-X Receptor 48
4.7	Mechanism of Allopregnanolone Action: Reduction of Cellular Oxidative Stress 48
4.8	Conclusions – Mechanisms of Allopregnanolone Action in Treatment of NP-C and Other Neurodegenerative Diseases 49
	Acknowledgments 51
	References 51

Part II	**Glucocorticoids, Dehydroepiandrosterone, Neuroprotection and Neuropathy** 61

5	**Glucocorticoids, Developmental "Programming," and the Risk of Affective Dysfunction** 63
	Bayanne Olabi and Jonathan Seckl
5.1	Introduction to Programming 64

5.2	Programming 65
5.2.1	Epidemiology 65
5.2.2	Birth Weight and Neuropsychiatric Disorders 66
5.3	Glucocorticoids and Fetal Development 66
5.4	Glucocorticoids: the Endocrine Programming Factor 68
5.4.1	Placental 11β-HSD2: a Barrier to Maternal Glucocorticoids 69
5.4.2	Glucocorticoid Programming 70
5.4.3	Transgenerational Effects 71
5.4.4	The Placenta 71
5.4.5	A Common Mechanism? 71
5.5	Fetal Tissue Glucocorticoid Sensitivity 72
5.6	Stress and Glucocorticoids: Key Programmers of the Brain 74
5.6.1	Programming the HPA Axis 74
5.6.2	Sex-Specific Effects 75
5.6.3	Programming Behavior 75
5.7	CNS Programming Mechanisms 76
5.7.1	The GR Gene: a Common Programming Target? 77
5.7.2	Epigenetics 78
5.8	Glucocorticoid Programming in Humans 78
5.8.1	Clinical Use of Prenatal Glucocorticoid Therapy 78
5.8.2	Consequences of Human Fetal Glucocorticoid Overexposure 79
5.8.3	Programming and Posttraumatic Stress Disorder (PTSD) 80
5.8.4	Programming Other Glucocorticoid Metabolizing Enzymes 81
5.9	Future Perspectives and Therapeutic Opportunities 82
5.10	Overview 83
	References 84
6	**Regulation of Structural Plasticity and Neurogenesis during Stress and Diabetes; Protective Effects of Glucocorticoid Receptor Antagonists** 103
	Paul J. Lucassen, Carlos P. Fitzsimons, Erno Vreugdenhil, Pu Hu, Charlotte Oomen, Yanina Revsin, Marian Joëls, and Edo Ronald de Kloet
6.1	The Stress Response 103
6.2	HPA Axis and Glucocorticoids 104
6.3	Glucocorticoid Actions 104
6.4	Feedback Regulation 104
6.5	Stress and Depression 105
6.6	Stress-Induced Viability Changes in the Hippocampus: Effect on Function, Volume, Cell Number, and Apoptosis 106
6.7	Effects of Stress on Dendritic Atrophy, Spine, and Synaptic Changes 107
6.8	Adult Hippocampal Neurogenesis 107
6.9	Effect of Stress on Adult Hippocampal Neurogenesis 109
6.10	Normalization of the Effects of Stress on the Hippocampus by Means of GR Blockade 110

6.11	Normalization of Hippocampal Alterations during Diabetes Mellitus Using the GR Antagonist Mifepristone *113*	
6.12	Concluding Remarks *115*	
	Acknowledgments *115*	
	Disclosure *115*	
	References *116*	

7	**Neuroactive Steroids and Peripheral Neuropathy** *121*	
	Roberto C. Melcangi, Silvia Giatti, Marzia Pesaresi, Donatella Caruso, and Marc J. Tetel	
7.1	Introduction *121*	
7.2	Regulation of Neuroactive Steroid Responsiveness in Peripheral Nerves *122*	
7.2.1	Synthesis and Metabolism of Neuroactive Steroids *122*	
7.2.2	Classical and Nonclassical Steroid Receptors are Expressed in Peripheral Nerves *123*	
7.3	Schwann Cell Responses to Neuroactive Steroids *123*	
7.4	Sexually Dimorphic Changes of Neuroactive Steroid Levels Induced by Pathology in Peripheral Nerves *126*	
7.5	Neuroactive Steroids as Protective Agents in PNS *126*	
7.5.1	Aging Process *126*	
7.5.2	Physical Injury *127*	
7.5.3	Diabetic Neuropathy *128*	
7.6	Chemotherapy-Induced Peripheral Neuropathy *128*	
7.7	Concluding Remarks *129*	
	Acknowledgments *129*	
	References *129*	

8	**Neuroprotective and Neurogenic Properties of Dehydroepiandrosterone and its Synthetic Analogs** *137*	
	Ioannis Charalampopoulos, Iakovos Lazaridis, and Achille Gravanis	
8.1	Introduction *137*	
8.2	Neuroprotective and Neurogenic Effects of DHEA in Hippocampal Neurons *138*	
8.3	Neuroprotective Effects of DHEA in Nigrostriatal Dopaminergic Neurons *140*	
8.4	Neuroprotective Effects of DHEA in Autoimmune Neurodegenerative Processes *141*	
8.5	Neuroprotective Effects of DHEA against Brain Ischemia and Trauma *142*	
8.6	Signaling Pathways Involved in the Effects of DHEA on Neuronal Cell Fate *144*	
8.7	Therapeutic Perspectives of DHEA and its Synthetic Analogs in Neurodegenerative Diseases *146*	
	References *147*	

9	**Neurosteroids and Pain** *155*	
	Christine Patte-Mensah, Laurence Meyer, Véronique Schaeffer, Cherkaouia Kibaly, and Ayikoe G. Mensah-Nyagan	
9.1	Introduction *155*	
9.2	General Background on Neurosteroids *155*	
9.3	Overview on Pain *156*	
9.4	Involvement of Endogenous Neurosteroids in the Control of Pain *157*	
9.4.1	Evidence for the Local Production of Neurosteroids in the Spinal Circuit *157*	
9.4.2	Endogenous Neurosteroids and Pain Modulation *162*	
9.5	Conclusion *164*	
	Acknowledgments *164*	
	References *164*	

Part III Polypeptide Hormones and Neuroprotection *171*

10	**The Insulin/IGF-1 System in Neurodegeneration and Neurovascular Disease** *173*	
	Przemyslaw (Mike) Sapieha and Lois Smith	
10.1	Introduction *173*	
10.2	Insulin and Insulin Growth Factors *174*	
10.3	Local versus Systemic Actions *174*	
10.4	Insulin/IGF Signaling Pathway *175*	
10.5	The Insulin/IGF Axis in the Brain *176*	
10.6	Insulin/IGF and Neuroprotection *176*	
10.7	Alzheimer's Disease *178*	
10.8	Parkinson's Disease *179*	
10.9	Vascular Dementia *179*	
10.10	Neurovascular Degeneration *180*	
10.11	Conclusion *182*	
	References *182*	

11	**Leptin Neuroprotection in the Central Nervous System** *189*	
	Feng Zhang, Suping Wang, Armando P. Signore, Zhongfang Weng, and Jun Chen	
11.1	Introduction *189*	
11.1.1	Origin, Source, and Structure of Leptin *189*	
11.1.2	Functions of Leptin *189*	
11.1.3	Leptin Receptors *190*	
11.1.4	Leptin Transport across the Blood–Brain Barrier *191*	
11.2	Mutation of Leptin or Leptin Receptors *192*	
11.3	Neurotrophic Role of Leptin *193*	
11.4	Leptin Neuroprotection against Disorders of the Central Nervous System *193*	
11.4.1	Acute Neurological Disorders *193*	

11.4.2	Neurodegenerative Diseases and Other Disorders	196
11.4.3	Leptin Neuroprotective Mechanisms	199
11.5	Significance	199
	References	199

12 Somatostatin and Neuroprotection in Retina 205
Kyriaki Thermos

12.1	Introduction	205
12.2	Somatostatin and Related Peptides	206
12.3	Somatostatin Receptors and Signaling	206
12.4	Somatostatin and its Receptors in Retina	206
12.5	Localization of Somatostatin Receptors in Retinal Neurons	207
12.5.1	Sst_1	208
12.5.2	Sst_2	208
12.5.3	Sst_3	209
12.5.4	Sst_4	209
12.5.5	Sst_5	209
12.6	Somatostatin Receptor Function in Retinal Circuitry	209
12.6.1	Effects on Glutamate Release	210
12.6.2	Effects on Dopamine Release	210
12.6.3	Effects on Nitric Oxide/GMP	210
12.6.4	Effects on Somatostatin Release	211
12.7	Neuroprotection by Somatostatin Analogs	212
12.7.1	Retinal Ischemia and Excitotoxicity	212
12.7.2	Anti-Ischemic Actions of SRIF	212
12.7.2.1	*Ex vivo* Studies	212
12.7.2.2	*In vivo* Studies	213
12.8	Mechanisms of SRIF's Neuroprotection	213
12.8.1	Involvement of NO/cGMP	213
12.8.2	NO/cGMP Mediates SRIF's Neuroprotective Effects	214
12.9	Therapeutic Potential of Somatostatin Agents	216
12.10	Conclusions	217
	Acknowledgments	217
	Abbreviations	218
	References	218

13 Neurotrophic Effects of PACAP in the Cerebellar Cortex 227
Anthony Falluel-Morel, Hubert Vaudry, Hitoshi Komuro, Dariusz C. Gorecki, Ludovic Galas, and David Vaudry

13.1	Expression of PACAP and its Receptors in the Developing Cerebellum	227
13.2	Effects of PACAP on Granule Cell Proliferation	229
13.3	Effects of PACAP on Granule Cell Migration	229
13.4	Effects of PACAP on Granule Cell Survival	231

13.5	Effects of PACAP on Granule Cell Differentiation	*231*
13.6	Functional Relevance	*233*
	Acknowledgments	*234*
	References	*234*

14 The Corticotropin-Releasing Hormone in Neuroprotection *237*
Christian Behl and Angela Clement

14.1	Introduction	*237*
14.2	The CRH Family of Proteins and Molecular Signal Transduction	*238*
14.3	From the Physiology to the Pathophysiology of CRH	*239*
14.4	CRH and Neurodegenerative Conditions	*240*
14.5	Protective Activities of CRH	*240*
14.6	Lessons from the Heart	*244*
14.7	Outlook	*245*
	References	*245*

15 Neuroprotective and Neurogenic Effects of Erythropoietin *251*
Helmar C. Lehmann and Ahmet Höke

15.1	Introduction	*251*
15.2	EPO in Models of Neonatal Hypoxic-Ischemic Brain Injury	*251*
15.3	EPO in Models of Ischemic Stroke in Adults	*253*
15.4	EPO in Models of Traumatic Brain Injury and Spinal Cord Trauma	*257*
15.5	EPO in Experimental Autoimmune Encephalomyelitis	*257*
15.6	EPO in Models of Peripheral Neuropathy	*258*
15.7	Summary	*260*
	References	*260*

Part IV Hormones and Neurogenesis *265*

16 Thyroid Hormone Actions on Glioma Cells *267*
Min Zhou, Harold K. Kimelberg, Faith B. Davis, and Paul J. Davis

16.1	Introduction	*267*
16.2	Origins of Glioma	*267*
16.3	Glioma Cell Biology	*268*
16.4	Thyroid Hormone Analogs, Transport, and Metabolism	*270*
16.5	Thyroid Hormones and Brain Development	*270*
16.6	Nongenomic Actions of Thyroid Hormones	*271*
16.7	Hypothyroidism Suppresses Growth of Glioma in Patients	*273*
16.8	Molecular Mechanisms of Hypothyroidism-Induced Clinical Suppression of Glioma Progression	*273*
16.8.1	Thyroid Hormone and Proliferation of Tumor Cells	*274*

16.8.2	Angiogenic Action of Thyroid Hormones	274
16.8.3	Antiapoptotic Action of Thyroid Hormones	274
16.8.4	Tumor Suppression Actions of Tetrac	275
16.9	Future Perspectives	275
	References	276

17 Gonadal Hormones, Neurosteroids, and Clinical Progestins as Neurogenic Regenerative Agents: Therapeutic Implications *281*
Lifei Liu and Roberta Diaz Brinton

17.1	Introduction	281
17.2	Gonadal Hormones, Neurosteroids, and Neurogenesis	282
17.2.1	Ovarian Hormone Regulation of Adult Neurogenesis	284
17.2.2	Estrogen Regulation of Adult Neurogenesis	284
17.2.3	Progestagen Regulation of Adult Neurogenesis	286
17.2.4	Progesterone Regulation of Adult Neurogenesis	287
17.2.5	Clinical Progestin Regulation of Neurogenesis	287
17.2.6	Androgen Regulation of Adult Neurogenesis	288
17.2.7	Testosterone and DHT Regulation of Adult Neurogenesis	288
17.2.8	DHEA Regulation of Adult Neurogenesis	289
17.3	Neurosteroid Regulation of Adult Neurogenesis	290
17.3.1	Pregnenolone and Pregnenolone Sulfate Regulation of Adult Neurogenesis	290
17.3.2	Allopregnanolone Regulation of Adult Neurogenesis	291
17.4	Gonadal Steroids, Clinical Progestins, and Neurosteroids as Neuroregenerative Therapeutics: Challenges and Strategies	292
17.4.1	Targeting Neurogenesis as a Treatment for Neurodegenerative Disease	293
17.4.2	APα as a Regenerative Factor to Promote Functional Neurogenesis and Diminish Alzheimer's Pathology	294
	References	295

18 Gonadotropins and Progestogens: Obligatory Developmental Functions during Early Embryogenesis and their Role in Adult Neurogenesis, Neuroregeneration and Neurodegeneration *305*
Craig S. Atwood and Sivan Vadakkadath Meethal

18.1	Introduction	305
18.2	Hormonal Regulation of Human Embryogenesis	305
18.2.1	The Missing Links	305
18.2.2	Trophoblastic Hormone Secretion	306
18.2.3	Human Embryonic Stem Cells: A Complete Model System for Understanding the Cellular and Molecular Mechanisms Regulating Early Human Embryogenesis	308
18.2.4	Progesterone, Human Chorionic Gonadotropin, and Early Human Embryogenesis	308

18.2.4.1	Regulation of Blastulation by Human Chorionic Gonadotropin and Progesterone *308*	
18.2.4.2	Regulation of Neurulation by Human Chorionic Gonadotropin and Progesterone *310*	
18.2.4.3	Regulation of Organogenesis by Human Chorionic Gonadotropin and Progesterone *312*	
18.2.5	Opioid Signaling and Early Human Embryogenesis *312*	
18.3	Progesterone: an Essential Neurotrophic Hormone during All Phases of Life *313*	
18.4	Age-Related Loss of Progesterone: Implications in the Pathophysiology of Neurodegenerative Diseases *314*	
18.4.1	Alzheimer's Disease *315*	
18.4.1.1	Amyloid-β Precursor Protein and Neurogenesis *315*	
18.4.1.2	Hormonal Regulation of Neurogenesis via Modulation of AβPP Metabolism *316*	
18.4.1.3	Progesterone in the Treatment of AD *316*	
18.4.2	Stroke *317*	
18.5	Conclusion *318*	
	References *319*	

19 **Human Neural Progenitor Cells: Mitotic and Neurogenic Effects of Growth Factors, Neurosteroids, and Excitatory Amino Acids** *331*
Masatoshi Suzuki, Jacalyn McHugh, and Narisorn Kitiyanant

19.1	Introduction *331*	
19.2	Neural Stem/Progenitor Cells as a Model of Human Cortical Development *331*	
19.3	Mitotic and Neurogenic Effects of a Neurosteroid: Dehydroepiandrosterone (DHEA) *332*	
19.4	Glutamate Enhances Proliferation and Neurogenesis in hNPCs *336*	
19.5	Increased Neurogenic "Radial Glial"-like Cells within Human Neurosphere Cultures *338*	
19.6	Conclusions *341*	
	Acknowledgments *342*	
	References *342*	

20 **Corticosterone, Dehydroepiandrosterone, and Neurogenesis in the Adult Hippocampus** *347*
Joe Herbert and Scarlet Bella Pinnock

20.1	Background *347*	
20.2	Glucocorticoids and Neurogenesis in the Adult Hippocampus *348*	
20.2.1	Regulation by Corticoid Levels *348*	
20.2.2	Regulation by the Corticoid Diurnal Rhythm *350*	
20.2.3	Dehydroepiandrosterone (DHEA) *354*	
20.2.4	Downstream Actions: pCREB and Wnt3a *357*	
20.2.5	Relevance to Depression *358*	

20.3 Conclusion 360
Acknowledgments 360
References 360

Index *367*

List of Contributors

María-Angeles Arévalo
Instituto Cajal
Consejo Superior de
Investigaciones Científicas
(C.S.I.C.)
Avenida Doctor Arce 37
28002 Madrid
Spain

Craig S. Atwood
Department of Medicine
University of Wisconsin-Madison
School of Medicine and Public
Health and Geriatric Research
Education and Clinical Center
Veterans administration Hospital
2500 Overlook Terrace
Madison, WI 53705
USA

and

Case Western Reserve University
Department of Pathology
2103 Cornell Road
Cleveland, OH 44106
USA

and

School of Exercise
Biomedical and Health Sciences
Edith Cowan University
270 Joondalup Drive
Joondalup, 6027 WA
Australia

Iñigo Azcoitia
Universidad Complutense
Departamento de Biología Celular
Facultad de Biología
José Antonio Novais 2
28040 Madrid
Spain

Etienne Emile Baulieu
INSERM UMR 788
80 rue du Général Leclerc
94276 Kremlin-Bicêtre
France

Christian Behl
University Medical Center of the
Johannes Gutenberg University
Mainz
Institute for Pathobiochemistry
55099 Mainz
Germany

List of Contributors

Roberta Diaz Brinton
University of Southern California
School of Pharmacy
Department of Pharmacology and
Pharmaceutical Sciences
Los Angeles, CA 90033
USA

Donatella Caruso
University of Milan
Department of Pharmacological
Sciences
Via Balzaretti 9
20133 Milano
Italy

Ioannis Charalampopoulos
University of Crete
Department of Pharmacology
School of Medicine
Stavrakia
71110 Heraklion
Greece

Jun Chen
University of Pittsburgh
Department of Neurology and
Center of Cerebrovascular
Disease Research
3550 Terrace St.
Pittsburgh
PA 15213
USA

Angela Clement
University Medical Center of the
Johannes Gutenberg University
Mainz
Institute for Pathobiochemistry
55099 Mainz
Germany

Douglas Covey
Washington University School of
Medicine
Department of Developmental
Biology
St. Louis, MO
USA

Faith B. Davis
Ordway Research Institute
Signal Transduction Laboratory
Albany, NY 12208
USA

Paul J. Davis
Ordway Research Institute
Signal Transduction Laboratory
Albany, NY 12208
USA

Anthony Falluel-Morel
University of Rouen
INSERM U982
Place E. Blondel
76821 Mont-Saint-Aignan
France

and

European Institute for Peptide
Research (IFRMP 23)
Place E. Blondel
76821 Mont-Saint-Aignan
France

and

International Laboratory
Samuel de Champlain
76821 Mont-Saint-Aignan
France

Carlos P. Fitzsimons
University of Amsterdam
Centre for Neuroscience
Swammerdam Institute of Life
Sciences, Amsterdam
Science Park 904, 1098 XH
The Netherlands

and

Leiden University Division of
Medical Pharmacology
Leiden Amsterdam Center for
Drug Research
Einsteinweg 55
2300RA Leiden
The Netherlands

Ludovic Galas
European Institute for Peptide
Research (IFRMP 23)
Place E. Blondel
76821 Mont-Saint-Aignan
France

and

International Laboratory
Samuel de Champlain
76821 Mont-Saint-Aignan
France
and

Regional Platform for Cell
Imaging (PRIMACEN)
IFRMP 23
76821 Mont-Saint-Aignan
France

Luis M. Garcia-Segura
Instituto Cajal
Consejo Superior de
Investigaciones Científicas
(C.S.I.C.)
Avenida Doctor Arce 37
28002 Madrid
Spain

Silvia Giatti
University of Milan
Department of Endocrinology
Pathophysiology and applied
Biology
Via Balzaretti 9
20133 Milano
Italy

Wenhui Gong
University of California
Department of Obstetrics
Gynecology, and Reproductive
Sciences, The Center for
Reproductive Sciences
513 Parnassus Avenue
San Francisco, CA 94143
USA

Dariusz C. Gorecki
University of Portsmouth
School of Pharmacy and
Biomedical Sciences
St. Michael Boulevard
Portsmouth PO1 2DT
UK

Achille Gravanis
University of Crete
Department of Pharmacology
School of Medicine
Stavrakia
71110 Heraklion
Greece

Joe Herbert
University of Cambridge
Cambridge Centre for Brain Repair
Forvie Site, Robinson Way
Cambridge CB20PY
UK

Pu Hu
University of Amsterdam
Centre for Neuroscience
Swammerdam Institute of Life Sciences
Science Park 904
1098 XH Amsterdam
The Netherlands

and

University of Science and Technology of China
Hefei National Laboratory for Physical Sciences at Microscale
and Department of Neurobiology and Biophysics
Hefei, Anhui
PR China

Ahmet Höke
Johns Hopkins University School of Medicine
Department of Neurology and Neuroscience
Pathology Building Room 509
600 N. Wolfe Street
Baltimore, MD 21287
USA

Marian Joëls
University of Amsterdam
Centre for Neuroscience
Swammerdam Institute of Life Sciences
Science Park 904
1098 XH Amsterdam
The Netherlands

and

University Medical Center Utrecht
Department Neuroscience and Pharmacology
The Netherlands

Cherkaouia Kibaly
Université de Strasbourg
Equipe Stéroïdes Neuromodulateurs et Neuropathologies
Unité de Physiopathologie et Médecine Translationnelle
Faculté de Médecine
11 rue Humann
67000 Strasbourg
France

Harold K. Kimelberg
Ordway Research Institute
Signal Transduction Laboratory
Albany, NY 12208
USA

Narisorn Kitiyanant
Mahidol University
Institute of Molecular Biosciences
Phutthamonthon 4 Rd
Nakhonpathom 73170
Thailand

List of Contributors

Edo Ronald de Kloet
Leiden University
Division of Medical
Pharmacology
Leiden Amsterdam Center for
Drug Research
Einsteinweg 55
2300RA Leiden
The Netherlands

Hitoshi Komuro
The Cleveland Clinic Foundation
Lerner Research Institute
Department of Neurosciences
Cleveland, Ohio 44195
USA

Iakovos Lazaridis
University of Crete
Department of Pharmacology
School of Medicine
Stavrakia
71110 Heraklion
Greece

Helmar C. Lehmann
Heinrich Heine University
Düsseldorf
Department of Neurology
Moorenstrasse 5
40225 Düsseldorf
Germany

Lifei Liu
University of Southern California
School of Pharmacy
Department of Pharmacology and
Pharmaceutical Sciences
Los Angeles, CA 90033
USA

Paul J. Lucassen
University of Amsterdam
Centre for Neuroscience
Swammerdam Institute of Life
Sciences, Science Park 904
1098 XH
Amsterdam
The Netherlands

Jacalyn McHugh
The Cedars-Sinai Regenerative
Medicine Institute
8700 Beverly Blvd.
Los Angeles, CA 90048
USA

Sivan Vadakkadath Meethal
University of Wisconsin-Madison
Department of Neurological
Surgery
School of Medicine and Public
Health
600 Highland Avenue
Madison, WI 53792
USA

Roberto C. Melcangi
University of Milan
Department of Endocrinology
Pathophysiology and applied
Biology
Via Balzaretti 9
20133 Milano
Italy

Synthia H. Mellon
University of California
Department of Obstetrics
Gynecology, and Reproductive
Sciences, The Center for
Reproductive Sciences
513 Parnassus Avenue
San Francisco, CA 94143
USA

Ayikoe G. Mensah-Nyagan
Université de Strasbourg
Equipe Stéroïdes
Neuromodulateurs et
Neuropathologies
Unité de Physiopathologie et
Médecine Translationnelle
Faculté de Médecine
11 rue Humann
67000 Strasbourg
France

Laurence Meyer
Université de Strasbourg
Equipe Stéroïdes
Neuromodulateurs et
Neuropathologies
Unité de Physiopathologie et
Médecine Translationnelle
Faculté de Médecine
11 rue Humann
67000 Strasbourg
France

Bayanne Olabi
The Queen's Medical Research
Institute, Endocrinology Unit
Centre for Cardiovascular Science
47 Little France Crescent
Edinburgh EH16 4TJ
UK

Charlotte Oomen
University of Amsterdam
Centre for Neuroscience
Swammerdam Institute of Life
Sciences
Science Park 904, 1098 XF
Amsterdam
The Netherlands

Christine Patte-Mensah
Université de Strasbourg
Equipe Stéroïdes
Neuromodulateurs et
Neuropathologies
Unité de Physiopathologie et
Médecine Translationnelle
Faculté de Médecine
11 rue Humann
67000 Strasbourg
France

Evelyn Perez
Laboratory of Experimental
Gerontology
Neurocognitive Aging Section
National Institute on Aging
Baltimore, MD
USA

Marzia Pesaresi
University of Milan
Department of Endocrinology
Pathophysiology and applied
Biology
Via Balzaretti 9
20133 Milano
Italy

Scarlet Bella Pinnock
University of Cambridge
Cambridge Centre for Brain
Repair
Forvie Site, Robinson Way
Cambridge CB2 OPY
UK

Yanina Revsin
Leiden University
Division of Medical
Pharmacology
Leiden Amsterdam Center for
Drug Research
Einsteinweg 55
2300RA Leiden
The Netherlands

Przemyslaw (Mike) Sapieha
Harvard Medical School
Department of Ophthalmology
300 Longwood Avenue
Boston, MA
USA

and

University of Montreal
Faculty of Medicine
5415 Assomption Boulevard
Montreal, Quebec
Canada

Véronique Schaeffer
Université de Strasbourg
Equipe Stéroïdes
Neuromodulateurs et
Neuropathologies
Unité de Physiopathologie et
Médecine Translationnelle
Faculté de Médecine
11 rue Humann
67000 Strasbourg
France

Marcus D. Schonemann
University of California
Department of Obstetrics
Gynecology and Reproductive
Sciences, The Center for
Reproductive Sciences
513 Parnassus Avenue
San Francisco, CA 94143
USA

Jonathan Seckl
The Queen's Medical Research
Institute, Endocrinology Unit
Centre for Cardiovascular Science
47 Little France Crescent
Edinburgh EH16 4TJ
UK

Armando P. Signore
University of Pittsburgh
Department of Neurology and
Center of Cerebrovascular
Disease Research
3550 Terrace St.
Pittsburgh
PA 15213
USA

James W. Simpkins
University of North Texas Health
Science Center, Department of
Pharmacology & Neuroscience
Institute for Aging and
Alzheimer's Disease Research
Fort Worth, TX
USA

Meharvan Singh
University of North Texas Health
Science Center, Department of
Pharmacology & Neuroscience
Institute for Aging and
Alzheimer's Disease Research
Center FOR HER
3500 Camp Bowie Blvd.
Fort Worth, TX, 76107
USA

Lois Smith
Harvard Medical School
Department of Ophthalmology
300 Longwood Avenue
Boston, MA
USA

Masatoshi Suzuki
University of Wisconsin Madison
Department of Comparative
Biosciences
School of Veterinary Medicine
2015 Linden Dr. Madison
WI 53706
USA

Marc J. Tetel
Wellesley College
Neuroscience Program
106 Central St.
Wellesley, MA 02481
USA

Kyriaki Thermos
University of Crete
Department of Pharmacology
School of Medicine
71 110 Heraklion
Greece

David Vaudry
University of Rouen
INSERM U982
Place E. Blondel
76821 Mont-Saint-Aignan
France

and

European Institute for Peptide
Research (IFRMP 23)
Place E. Blondel
76821 Mont-Saint-Aignan
France

and

International Laboratory
Samuel de Champlain
76821 Mont-Saint-Aignan
France

and

Regional Platform for Cell
Imaging (PRIMACEN)
IFRMP23
76821 Mont-Saint-Aignan
France

Hubert Vaudry
University of Rouen
INSERM U982
Place E. Blandel
76821 Mont-Saint-Aignan
France

and

European Institute for Peptide
Research (IFRMP 23)
Place E. Blondel
76821 Mont-Saint-Aignan
France

and

International Laboratory
Samuel de Champlain
76821 Mont-Saint-Aignan
France

and

Regional Platform for Cell
Imaging (PRIMACEN)
IFRMP 23
76821 Mont-Saint-Aignan
France

Erno Vreugdenhil
Leiden University, Division of
Medical Pharmacology
Leiden Amsterdam Center for
Drug Research
Einsteinweg 55
2300RA Leiden
The Netherlands

Suping Wang
University of Pittsburgh
Department of Neurology and
Center of Cerebrovascular
Disease Research
3550 Terrace St.
Pittsburgh, PA 15213
USA

Zhongfang Weng
University of Pittsburgh
Department of Neurology and
Center of Cerebrovascular
Disease Research
3550 Terrace St.
Pittsburgh, PA 15213
USA

Kun Don Yi
University of North Texas Health
Science Center, Department of
Pharmacology and Neuroscience,
Institute for Aging and
Alzheimer's Disease Research
Fort Worth, TX
USA

Feng Zhang
University of Pittsburgh
Department of Neurology and
Center of Cerebrovascular
Disease Research
3550 Terrace St.
Pittsburgh, PA 15213
USA

Min Zhou
Ordway Research Institute
Signal Transduction Laboratory
Albany, NY 12208
USA

Part I
Estrogens, Progestins, Allopregnanolone and Neuroprotection

1
Interactions of Estradiol and Insulin-like Growth Factor-I in Neuroprotection: Implications for Brain Aging and Neurodegeneration

María-Angeles Arévalo, Luis M. Garcia-Segura, and Iñigo Azcoitia

Estradiol and insulin-like growth factor-I (IGF-I) interact in the nervous system to promote neuroprotection. This interaction is mediated by multiple intracellular signaling mechanisms in which several molecules such as estrogen receptor (ER) α, phosphatidylinositol 3-kinase (PI3K), glycogen synthase kinase 3β (GSK3β), and β-catenin play a central role. Decreased hormonal levels in plasma with aging and changes in the synthesis of the ligands, and the expression of the receptors in the brain under neurodegenerative conditions may alter the neuroprotective interactions of estradiol and IGF-I.

1.1
Introduction: Hormones, Brain Aging, and Neurodegeneration

The aging process affects all tissues and organs, including the brain. Individual variations in decline of cognitive skills, development of affective disorders, and neurodegenerative diseases with aging suggest that brain deterioration is not only the result of age *per se* but probably also represents a failure to adapt to age-associated homeostatic changes [1]. Hormones are involved somewhat in the aging process since, with aging, the levels of many of them change in the plasma. Several hormones such as the growth hormone, IGF-I, dehydroepiandrosterone, and sex hormones decrease with aging in mammals [1]. In humans, their change is associated in time with the progression of neurodegenerative disorders, increased depressive symptoms, and other psychological disturbances [1]. This suggests that the modification in hormone levels, due to aging, may have a negative impact on brain function. Alternatively, since the brain is an important center for endocrine control, brain aging may be involved in the hormonal changes. These hormonal changes may, in part, represent a positive adaptive response to the aging process [1].

Even if the hormonal changes are a general positive adaptation to aging, they may have a negative impact on the brain. The decrease in the levels of neuroprotective hormones in older people may result in reduced protection against the environmental and genetic factors that promote neurodegeneration. However, not all hormones exert neuroprotective effects. There is considerable evidence to

Hormones in Neurodegeneration, Neuroprotection, and Neurogenesis.
Edited by Achille G. Gravanis and Synthia H. Mellon
Copyright © 2011 WILEY-VCH Verlag GmbH & Co. KGaA, Weinheim
ISBN: 978-3-527-32627-3

show a link between stress, hypothalamo-pituitary-adrenal (HPA) axis dysfunction, memory disorders, and aging [2]. The hippocampus is vulnerable to both stress and aging. Stress and glucocorticoids alter hippocampal neurogenesis [3] and may lead to hippocampal damage. Not only may stress in adulthood increase brain damage and brain aging, but also stress during brain development may result in permanent brain abnormalities in adult life. Prenatal stress increases anxiety-like behavior and induces dysfunction of the negative feedback of the HPA axis [4]. With aging, rats subjected to prenatal stress exhibited hyperactivity of the HPA axis associated with spatial learning impairments [5, 6].

1.2
Estradiol, IGF-I, Brain Aging, and Neuroprotection

Some hormones, such as estradiol and IGF-I, may antagonize the damaging effects of adrenal steroids. IGF-I attenuates spatial learning deficits in aged rats and promotes neurogenesis in the hippocampus [6]. Chronic IGF-I infusion in the brain restores the spatial learning abilities of aged rats that were stressed during prenatal life. IGF-I also up-regulates neurogenesis in the hippocampus of these animals and reduces their HPA axis dysfunction [6]. Interestingly, IGF-I increases estradiol levels in the plasma of aged rats that are submitted to prenatal stress [6]. Estradiol, in turn, stimulates neurogenesis in the hippocampus of young and old rats and prevents hippocampal damage induced by excitotoxic injuries [1]. In addition, the signaling of estradiol and IGF-I interacts to promote neuroprotection [7]. Consequently, low levels of protective hormones, such as IGF-I and estradiol, in older individuals may increase the risk of neural damage induced by stress hormones or by previous stressful experiences. To understand how these hormones affect the aging process in the brain and to develop adequate protocols for possible hormone therapies to prevent brain deterioration, we need to know their neuroprotective mechanisms and how these are modulated in the aged brain.

IGF-I and estradiol prevent neuronal cell death in different experimental models of neurodegenerative diseases. The interaction of IGF-I and estradiol in neuroprotection has been assessed in ovariectomized rats *in vivo*, using systemic administration of kainic acid to induce degeneration of hippocampal hilar neurons [8], an experimental model of excitotoxic cell death. Both the systemic administration of estradiol and the intracerebroventricular infusion of IGF-I prevent hilar neuronal loss induced by kainic acid. The neuroprotective effect of estradiol is blocked by the intracerebroventricular infusion of an IGF-I receptor antagonist, while the neuroprotective effect of IGF-I is blocked by the intracerebroventricular infusion of the ER antagonist ICI 182, 780 [8]. Similar results have been obtained in ovariectomized rats after the unilateral infusion of 6-hydroxydopamine into the medial forebrain bundle to injure the nigrostriatal dopaminergic pathway [9], a model of Parkinson's disease. Pretreatment with estradiol or IGF-I significantly prevents the loss of substantia nigra compacta neurons and the related motor disturbances.

Blockage of IGF-I receptor by the intracerebroventricular administration of an IGF-I receptor antagonist attenuates the neuroprotective effects of both estrogen and IGF-I [9]. Furthermore, the neuroprotective action of estradiol against 1-methyl-4-phenyl-1,2,3,6-tetrahydropyridine (MPTP) toxicity in the nigrostriatal system of male mice is associated with the regulation of IGF-I receptor signaling [10]. These findings suggest that the neuroprotective actions of estradiol and IGF-I after brain injury depend on the coactivation of both ERs and IGF-I receptor in neural cells.

1.3
Molecular Interactions of Estrogen Receptors and IGF-I Receptor in the Brain

The interaction of the neuroprotective actions of estradiol and IGF-I may be the consequence of an interaction of ERα and the IGF-I receptor in a macromolecular complex associated with components of the PI3K signaling pathway [11]. Immunohistochemical analyses have shown that ERs and IGF-I receptor colocalize in different neuronal and glial populations in the rat central nervous system *in vivo* [12, 13]. In addition, immunoprecipitation studies have shown that estradiol administration to adult ovariectomized rats results in a transient increase in the association between IGF-I receptor and ERα in the brain [14]. The interaction is coincident in time with the increase in tyrosine phosphorylation of IGF-I receptor, suggesting a possible causal relationship. Estradiol also increases the interaction between p85 and insulin receptor substrate (IRS)-1 [14], one of the first events in the signal transduction of the IGF-I receptor, further suggesting that the increase in IGF-I receptor phosphorylation induced by estradiol reflects functional activation of this receptor. In addition, ERα interacts with other components of the IGF-I receptor signaling pathway, such as the p85 subunit of the PI3K [14]. This interaction is present in control ovariectomized animals and increases after estradiol treatment. A similar estradiol-induced association of ERα and p85 occurs in the mammary cancer cell line MCF-7 [15] and in human vascular endothelial cells [16]. Interestingly, the interaction between ERα and the IGF-I receptor is also increased by the intracerebroventricular administration of IGF-I [14]. These findings suggest that the interaction of ERα with the IGF-I receptor is part of the mechanisms involved in the signaling of both IGF-I and estradiol in the brain.

1.4
Regulation of IGF-I Receptor Signaling by Estradiol in the Brain

In vitro and *in vivo* studies have shown that, in the brain, estradiol may rapidly activate the PI3K and the mitogen-activated protein kinase (MAPK) signaling pathways. These are the two main signal transduction cascades coupled to the IGF-I receptor. Estradiol also induces the phosphorylation of Akt, one of the main effectors of the PI3K pathway in the brain [7, 17–20]. Both MAPK activation and PI3K/Akt activation are involved in the neuroprotective effects of estradiol in different experimental models of neurodegeneration [21–27]. Akt activation may mediate the effects of estradiol on the activity and expression of antiapoptotic molecules, such as Bcl-2 [10, 28]. The activation of Akt by estradiol also has implications for neuroprotection via the modulation of GSK3β activity [29]. Under pathological conditions, GSK3β may be responsible for the hyperphosphorylation of Tau in Alzheimer's disease [30] and its inhibition is associated with the activation of survival pathways in neurons [31]. Interestingly, estradiol increases the amount of inactive GSK3β (ser-9 phosphorylated) and decreases the phosphorylation of Tau in the hippocampus [29]. In addition, the inhibition of GSK3β by estradiol in neuroblastoma cells and in primary cortical neurons is associated with the stabilization of β-catenin and its translocation to the cell nucleus, where it acts as a cotranscriptional modulator using canonical T cell factor (TCF)/lymphoid enhancer binding factor-1 (LEF-1)-mediated transcription, in a manner similar to that produced by Wnt3a [32].

1.5
Regulation of Estrogen Receptor Transcriptional Activity by IGF-I in Neural Cells

In addition to classical activation of ER by estradiol binding, ER transcriptional activity can be regulated by ligand-independent mechanisms. Intracellular kinase signaling pathways, activated by extracellular growth or trophic factors, regulate the ability of ERs to promote changes in gene expression. IGF-I is one of the extracellular regulators of these kinase pathways that have been shown to promote ER-dependent transcription. In different cell lines, including neuroblastoma cells, IGF-I may activate ERs in the absence of estradiol and regulate ER-mediated gene expression [33–37]. In neuroblastoma cells, IGF-I may have a different regulation of the activity of ERα depending on whether the ER ligand is present or not. In the absence of estradiol, IGF-I increases ERα activity by using the Ras/MAPK signaling pathway [34]. In contrast, IGF-I negatively regulates ERα transcriptional activity in the presence of estradiol. This effect of IGF-I is mediated by the PI3K/Akt/GSK3β pathway, which induces the translocation of β-catenin to the cell nucleus. In turn, β-catenin binds to ERα in the nucleus and inhibits its transcriptional activity [38]. Thus, by using different components of its signaling system, IGF-I may regulate ligand-independent and -dependent ER transcriptional activity in neuronal cells.

1.6
Implications of the Cross Talk between Estrogen Receptors and IGF-I Receptors for Brain Aging, and Neurodegeneration

The cross talk between ERα and IGF-I receptor in neural cells discussed in the previous sections is summarized in Figure 1.1. From the analysis of this figure, it is evident that, depending on the levels of estradiol and IGF-I, the transcriptional

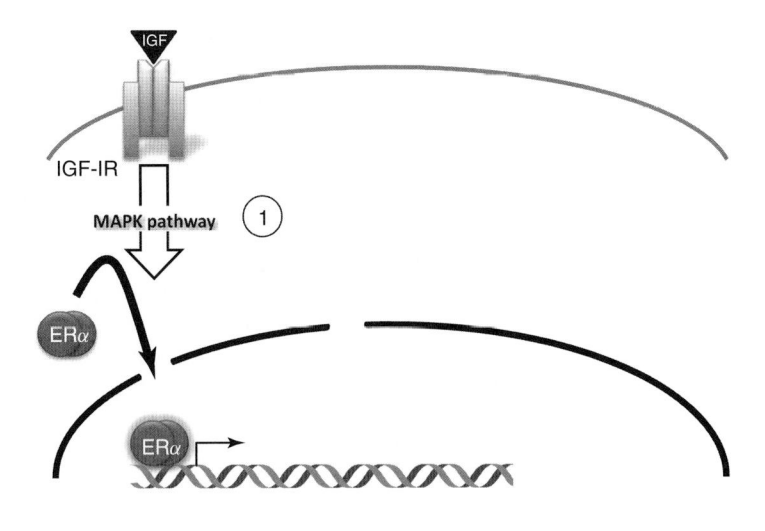

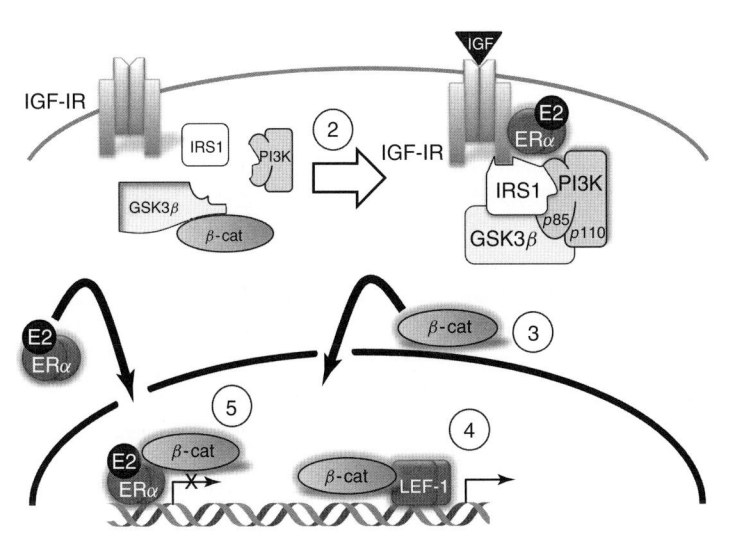

Figure 1.1 Summary of the cross talk between ERα and IGF-I receptor in neural cells. When estradiol is present at a subthreshold concentration, IGF-I enhances ERα-mediated transcriptional regulation through the MAPK pathway (1). When estradiol concentration reaches the threshold, both IGF-I and estradiol induce the association of ERα and IGF-I receptor in a macromolecular complex in which several components of the PI3K/Akt/GSK3β signaling are present (2). In this scenario, GSK3β is detached from β-catenin, which enters the cell nucleus (3), binds LEF family transcription factors and activates transcription (4). β-catenin can also bind ERα and inhibit ERα-mediated transcriptional regulation (5).

regulation mediated by this cross talk may have a very different outcome. In addition, we should also remember that estradiol directly regulates ER-dependent and ER-independent transcription and rapid membrane-initiated ER signaling by alternative mechanisms [39–43]. The coupling of the signaling pathways of IGF-I and estradiol in neural cells may act as a coincidence signal detector for timely coordination of the endocrine and paracrine effects of both factors. The function of this coincidence signal detector may be highly relevant during aging, when both estradiol and IGF-I levels decrease in the plasma [1]. Furthermore, if aging is associated with neurodegeneration, the situation becomes more complex since local estrogen synthesis, local IGF-I synthesis, and expression of ERs and the IGF-I receptor are increased in the injured brain [44–48]. Therefore, the coupling between ER and IGF-I receptor neuroprotective signaling may be altered in the brain of older people with neurodegenerative diseases.

Acknowledgment

The authors acknowledge financial support from the Spanish Ministerio de Ciencia e Innovación (grants BFU 2008-02950-C03-01 and BFU 2008-02950-C03-02).

References

1. Garcia-Segura, L.M. (2009) *Hormones and Brain Plasticity*, Oxford University Press, New York.
2. McEwen, B.S. (2002) Sex, stress and the hippocampus: allostasis, allostatic load and the aging process. *Neurobiol. Aging*, **23**, 921–939.
3. Mirescu, C. and Gould, E. (2006) Stress and adult neurogenesis. *Hippocampus*, **16**, 233–238.
4. Maccari, S., Darnaudéry, M., Morley-Fletcher, S., Zuena, A.R., Cinque, C., and VanReeth, O. (2003) Prenatal stress and long-term consequences: implications of glucocorticoid hormones. *Neurosci. Biobehav. Rev.*, **27**, 119–127.
5. Vallée, M., Maccari, S., Dellu, F., Simon, H., Le Moal, M., and Mayo, W. (1999) Long-term effects of prenatal stress and postnatal handling on age-related glucocorticoid secretion and cognitive performance: a longitudinal study in the rat. *Eur. J. Neurosci.*, **11**, 2906–2916.
6. Darnaudéry, M., Perez-Martin, M., Belizaire, G., Maccari, S., and Garcia-Segura, L.M. (2006) Insulin-like growth factor 1 reduces age-related disorders induced by prenatal stress in female rats. *Neurobiol. Aging*, **27**, 119–127.
7. Mendez, P., Azcoitia, I., and Garcia-Segura, L.M. (2005) Interdependence of oestrogen and insulin-like growth factor-I in the brain: potential for analysing neuroprotective mechanisms. *J. Endocrinol.*, **185**, 11–17.
8. Azcoitia, I., Sierra, A., and Garcia-Segura, L.M. (1999) Neuroprotective effects of estradiol in the adult rat hippocampus: interaction with insulin-like growth factor-I signalling. *J. Neurosci. Res.*, **58**, 815–822.
9. Quesada, A. and Micevych, P.E. (2004) Estrogen interacts with the IGF-1 system to protect nigrostriatal dopamine and maintain motoric behavior after 6-hydroxdopamine lesions. *J. Neurosci. Res.*, **75**, 107–116.
10. D'Astous, M., Mendez, P., Morissette, M., Garcia-Segura, L.M., and Di Paolo, T. (2006) Implication of the phosphatidylinositol-3 kinase/ protein kinase B signaling pathway

in the neuroprotective effect of estradiol in the striatum of 1-methyl-4-phenyl-1,2,3,6-tetrahydropyridine mice. *Mol. Pharmacol.*, **69**, 1492–1498.

11. Marin, R., Díaz, M., Alonso, R., Sanz, A., Arévalo, M.A., and Garcia-Segura, L.M. (2009) Role of estrogen receptor alpha in membrane-initiated signaling in neural cells: interaction with IGF-1 receptor. *J. Steroid Biochem. Mol. Biol.*, **114**, 2–7.

12. Cardona-Gomez, G.P., DonCarlos, L., and Garcia-Segura, L.M. (2000) Insulin-like growth factor I receptors and estrogen receptors colocalize in female rat brain. *Neuroscience*, **99**, 751–760.

13. Quesada, A., Romeo, H.E., and Micevych, P. (2007) Distribution and localization patterns of estrogen receptor-beta and insulin-like growth factor-1 receptors in neurons and glial cells of the female rat substantia nigra: localization of ERbeta and IGF-1R in substantia nigra. *J. Comp. Neurol.*, **503**, 198–208.

14. Mendez, P., Azcoitia, I., and Garcia-Segura, L.M. (2003) Estrogen receptor alpha forms estrogen-dependent multimolecular complexes with insulin-like growth factor receptor and phosphatidylinositol 3-kinase in the adult rat brain. *Brain Res. Mol. Brain Res.*, **112**, 170–176.

15. Castoria, G., Migliaccio, A., Bilancio, A., Di Domenico, M., de Falco, A., Lombardi, M., Fiorentino, R., Varricchio, L., Barone, M.V., and Auricchio, F. (2001) PI3-kinase in concert with Src promotes the S-phase entry of oestradiol-stimulated MCF-7 cells. *EMBO J.*, **20**, 6050–6059.

16. Simoncini, T., Hafezi-Moghadam, A., Brazil, D.P., Ley, K., Chin, W.W., and Liao, J.K. (2000) Interaction of oestrogen receptor with the regulatory subunit of phosphatidylinositol-3-OH kinase. *Nature*, **407**, 538–541.

17. Toran-Allerand, C.D., Singh, M., and Sétáló, G. (1999) Novel mechanisms of estrogen action in the brain: new players in an old story. *Front. Neuroendocrinol.*, **20**, 97–121.

18. Garcia-Segura, L.M., Sanz, A., and Mendez, P. (2006) Cross-talk between IGF-I and estradiol in the brain: focus on neuroprotection. *Neuroendocrinology*, **84**, 275–279.

19. Bourque, M., Dluzen, D.E., and Di Paolo, T. (2009) Neuroprotective actions of sex steroids in Parkinson's disease. *Front Neuroendocrinol.*, **30**, 142–157.

20. Cardona-Gomez, G.P., Mendez, P., and Garcia-Segura, L.M. (2002) Synergistic interaction of estradiol and insulin-like growth factor-I in the activation of PI3K/Akt signaling in the adult rat hypothalamus. *Brain Res. Mol. Brain Res.*, **107**, 80–88.

21. Guerra, B., Diaz, M., Alonso, R., and Marin, R. (2004) Plasma membrane oestrogen receptor mediates neuroprotection against beta-amyloid toxicity through activation of Raf-1/MEK/ERK cascade in septal-derived cholinergic SN56 cells. *J. Neurochem.*, **91**, 99–109.

22. Kuroki, Y., Fukushima, K., Kanda, Y., Mizuno, K., and Watanabe, Y. (2001) Neuroprotection by estrogen via extracellular signal-regulated kinase against quinolinic acid-induced cell death in the rat hippocampus. *Eur. J. Neurosci.*, **13**, 472–476.

23. Singer, C.A., Figueroa-Masot, X.A., Batchelor, R.H., and Dorsa, D.M. (1999) The mitogen-activated protein kinase pathway mediates estrogen neuroprotection after glutamate toxicity in primary cortical neurons. *J. Neurosci.*, **19**, 2455–2463.

24. Honda, K., Sawada, H., Kihara, T., Urushitani, M., Nakamizo, T., Akaile, A., and Shimohama, S. (2000) Phosphatidylinositol 3-kinase mediates neuroprotection by estrogen in cultured cortical neurons. *J. Neurosci. Res.*, **60**, 321–327.

25. Yu, X., Rajala, R.V., McGinnis, J.F., Li, F., Anderson, R.E., Yan, X., Li, S., Elias, R.V., Knapp, R.R., Zhou, X., and Cao, W. (2004) Involvement of insulin/phosphoinositide 3-kinase/Akt signal pathway in 17 beta-estradiol-mediated neuroprotection. *J. Biol. Chem.*, **279**, 13086–13094.

26. Zhang, L., Rubinow, D.R., Xaing, G., Li, B.S., Chang, Y.H., Maric, D., Barker, J.L., and Ma, W. (2001) Estrogen protects against beta-amyloid-induced neurotoxicity in rat hippocampal neurons by activation of Akt. *NeuroReport*, **12**, 1919–1923.
27. Quesada, A., Lee, B.Y., and Micevych, P.E. (2008) PI3 kinase/Akt activation mediates estrogen and IGF-1 nigral DA neuronal neuroprotection against a unilateral rat model of Parkinson's disease. *Dev. Neurobiol.*, **68**, 632–644.
28. Cardona-Gomez, G.P., Mendez, P., DonCarlos, L.L., Azcoitia, I., and Garcia-Segura, L.M. (2001) Interactions of estrogens and insulin-like growth factor-I in the brain: implications for neuroprotection. *Brain Res. Brain Res. Rev.*, **37**, 320–334.
29. Cardona-Gomez, G.P., Perez, M., Avila, J., Garcia-Segura, L.M., and Wandosell, F. (2004) Estradiol inhibits GSK3 and regulates interaction of estrogen receptors, GSK3 and beta-catenin, in the hippocampus. *Mol. Cell. Neurosci.*, **25**, 363–373.
30. Lovestone, S., Reynolds, C.H., Latimer, D., Davis, D.R., Anderton, B.H., Gallo, J.M., Hanger, D., Mulot, S., Marquardt, B., Stabel, S., Woodgett, J.R., and Miller, C.C.J. (1994) Alzheimer's disease-like phosphorylation of the microtubule-associated protein tau by glycogen synthase kinase-3 in transfected mammalian cells. *Curr. Biol.*, **4**, 1077–1086.
31. Cross, D.A., Alessi, D.R., Cohen, P., Andjelkovich, M., and Hemmings, B.A. (1995) Inhibition of glycogen synthase kinase-3 by insulin mediated by protein kinase B. *Nature*, **378**, 785–789.
32. Varea, O., Garrido, J.J., Dopazo, A., Mendez, P., Garcia-Segura, L.M., and Wandosell, F. (2009) Estradiol activates beta-catenin dependent transcription in neurons. *PLoS ONE*, **4**, e5153.
33. Ma, Z.Q., Santagati, S., Patrone, C., Pollio, G., Vegeto, E., and Maggi, A. (1994) Insulin-like growth factors activate estrogen receptor to control the growth and differentiation of the human neuroblastoma cell line SK-ER3. *Mol. Endocrinol.*, **8**, 910–918.
34. Patrone, C., Gianazza, E., Santagati, S., Agrati, P., and Maggi, A. (1998) Divergent pathways regulate ligand-independent activation of ERα in SK-N-BE neuroblastoma and COS-1 renal carcinoma cells. *Mol. Endocrinol.*, **12**, 835–841.
35. Klotz, D.M., Hewitt, S.C., Ciana, P., Raviscioni, M., Lindzey, S.C., Foley, J., Maggi, A., DiAugustine, R.P., and Korach, K.S. (2002) Requirement of estrogen receptor-alpha in insulin-like growth factor-1 (IGF-1)-induced uterine responses and in vivo evidence for IGF-1/estrogen receptor cross-talk. *J. Biol. Chem.*, **277**, 8531–8537.
36. Martin, M.B., Franke, T.F., Stoica, G.E., Chambon, P., Katzenellenbogen, B.S., Stoica, B.A., McLemore, M.S., Olivo, S.E., and Stoica, A. (2000) A role for Akt in mediating the estrogenic functions of epidermal growth factor and insulin-like growth factor I. *Endocrinology*, **141**, 4503–4511.
37. Font de Mora, J. and Brown, M. (2000) AIB1 is a conduit for kinase-mediated growth factor signaling to the estrogen receptor. *Mol. Cell. Biol.*, **20**, 5041–5047.
38. Mendez, P. and Garcia-Segura, L.M. (2006) Phosphatidyl inositol 3 kinase (PI3K) and glycogen synthase kinase 3 (GSK3) regulate estrogen receptor mediated transcription in neuronal cells. *Endocrinology*, **147**, 3027–3039.
39. Hammes, S.R. and Levin, E.R. (2007) Extra-nuclear steroid receptors: nature and actions. *Endocr. Rev.*, **28**, 726–741.
40. Vasudevan, N. and Pfaff, D.W. (2008) Non-genomic actions of estrogens and their interaction with genomic actions in the brain. *Front. Neuroendocrinol.*, **29**, 238–257.
41. Kelly, M.J. and Rønnekleiv, O.K. (2008) Membrane-initiated estrogen signaling in hypothalamic neurons. *Mol. Cell. Endocrinol.*, **290**, 14–23.
42. Levin, E.R. (2005) Integration of the extranuclear and nuclear actions of estrogen. *Mol. Endocrinol.*, **19**, 1951–1959.
43. Simpkins, J.W., Yi, K.D., and Yang, S.H. (2009) Role of protein phosphatases and mitochondria in the neuroprotective effects of estrogens. *Front. Neuroendocrinol*, **30** (2), 93–105.

44. Garcia-Ovejero, D., Azcoitia, I., DonCarlos, L.L., Melcangi, R.C., and Garcia-Segura, L.M. (2005) Glia-neuron crosstalk in the neuroprotective mechanisms of sex steroid hormones. *Brain Res. Brain Res. Rev.*, **48**, 273–286.
45. Garcia-Segura, L.M. (2008) Aromatase in the brain: not just for reproduction anymore. *J. Neuroendocrinol.*, **20**, 705–712.
46. Saldanha, C.J., Duncan, K.A., and Walters, B.J. (2009) Neuroprotective actions of brain aromatase. *Front. Neuroendocrinol.*, **30**, 106–118.
47. Beilharz, E.J., Russo, V.C., Butler, G., Baker, N.L., Connor, B., Sirimanne, E.S., Dragunow, M., Werther, G.A., Gluckman, P.D., Williams, C.E., and Scheepens, A. (1998) Co-ordinated and cellular specific induction of the components of the IGF/IGFBP axis in the rat brain following hypoxic-ischemic injury. *Brain Res. Mol. Brain Res.*, **59**, 119–134.
48. Guan, J., Bennet, L., Gluckman, P.D., and Gunn, A.J. (2003) Insulin-like growth factor-1 and post-ischemic brain injury. *Prog. Neurobiol.*, **70**, 443–462.

2
Structure–Nongenomic Neuroprotection Relationship of Estrogens and Estrogen-Derived Compounds

James W. Simpkins, Kun Don Yi, Evelyn Perez, and Douglas Covey

2.1
Introduction

Estrogens are potent, centrally active molecules. One important aspect of the estrogens' effects on neurons is their neuroprotective activity, which is seen against a variety of different toxicities including serum deprivation, oxidative stress, amyloid β (Aβ) peptide-induced toxicity, and glutamate-induced excitotoxicity [1]. Further, both endogenous and exogenous estrogens as well as estrogen analogs have been shown to have neuroprotective effects against the pathological events seen in experimental animal models of cerebral ischemia [2–12]. The pathological mechanisms that are activated during stroke include oxidative stress, free radical activity, excitotoxicity, inflammatory response, mitochondrial dysfunction, and apoptosis, all of which are antagonized by estrogens. However, the mechanisms by which estrogens are protective remain elusive. It is clear that the effects of estrogens are complex and multifaceted. The mechanisms of the neuroprotective actions of estrogen are characterized by classical estrogen receptor (ER)-dependent genomic and nongenomic actions, the latter of which is expressed by rapid interactions with neuronal membranes and intracellular signal transduction pathways. In addition, estrogens have also been shown to have an intrinsic antioxidant structure that lies in the phenolic ring of the compounds, which provide the antioxidant/redox cycling activity in neurons [13, 14].

As a result of the early cessation of the Women's Health Initiative (WHI) study due to increased risks of cardiovascular disease, stroke, blood clots, breast cancer, and dementia in women on either combined estrogen–progestin or estrogen-alone therapy, the scientific community has struggled to reconcile the numerous epidemiological, experimental, and clinical studies that have shown hormone therapy (HT) to be protective against a wide variety of pathological diseased states such as Parkinson's disease (PD), Alzheimer's disease (AD), stroke, and cardiovascular diseases. However, recent reevaluations of the WHI, other clinical trials, and cohort studies, such as the Nurses' Health Study, suggest that effects of HT are dependent

---

Hormones in Neurodegeneration, Neuroprotection, and Neurogenesis.
Edited by Achille G. Gravanis and Synthia H. Mellon
Copyright © 2011 WILEY-VCH Verlag GmbH & Co. KGaA, Weinheim
ISBN: 978-3-527-32627-3

on reproductive stage and the extent of preexisting neurodegenerative or cardiovascular disease at the onset of HT [15–19]. Specifically, emerging evidence indicates that younger (perimenopausal or early postmenopausal) women derive neuro- and cardiovascular protection from HT, whereas the initiation of such therapy in older individuals who are likely to have extensive preexisting conditions may be ineffective or even detrimental [20–25].

Inasmuch as many of the side effects of chronic HT are related to peripheral effects of orally administered estrogen preparations acting through known ERs [20–25], several strategies have been undertaken to capitalize on the potent neuroprotective effects of estrogens, while avoiding the side effects of chronic peripheral ER activation. First, selective estrogen receptor modulators (SERMs) are compounds that act as agonists on one ER and antagonists on the other ER [26]. This strategy has attempted to enhance ERβ agonist activity while antagonizing ERα [27, 28], given that the ERα is highly expressed in the breast, uterus, and liver and its chronic activation may account for the now known side effects of chronic estrogen administration to postmenopausal women.

An alternative strategy that we have used is based on our observations that nonfeminizing estrogens, like 17α-estradiol (αE2) [4] and the enantiomer of 17β-estradiol (*ent*-E2) [8] are potently neuroprotective. We reasoned that a substantial portion of the neuroprotective activity of estrogens was independent of their ability to activate ERs.

2.2
In vitro Assessments of Structure–Neuroprotective Activity Relationships

We undertook a program of drug discovery that had the following objectives:

1) To synthesize and test compounds that are more potent than known, naturally occurring estrogens.
2) To synthesize and test compounds that did not bind to ERs *in vitro* and do not stimulate estrogen-responsive tissues *in vivo*.

To achieve these goals, to date we have synthesized more than a hundred compounds using a variety of strategies to achieve both, enhanced potency and reduced ER interaction. To assess these compounds, we used a three-pronged approach. First, we tested all compounds for neuroprotective potency *in vitro* against two levels of two insults, glutamate and iodoacetic acid (IAA), and assessed binding to ERα and ERβ. Second, the most potent compounds were tested in a transient middle cerebral artery occlusion model for their ability to protect brain tissue form the effect of cerebral ischemia. Finally, we assessed the effect of some of these nonfeminizing estrogens on intracellular signaling pathways that are known to mediate the rapid neuroprotective response to estrogens.

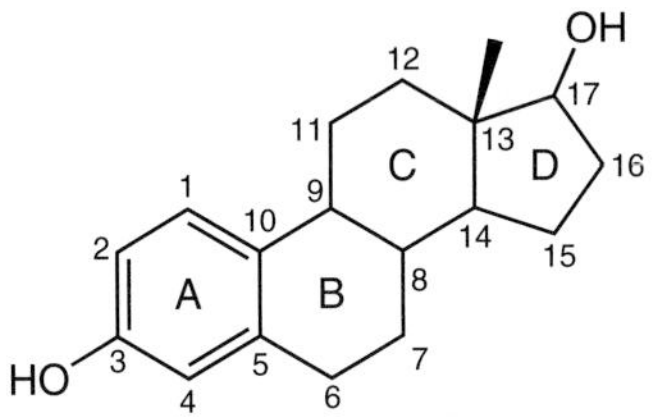

Figure 2.1 Structure of 17β-estradiol with carbon numbers and ring designations indicated.

2.2.1
Estradiol and other known Estratrienes

17β-estradiol (βE2, Figure 2.1) was able to protect HT-22 cells against glutamate toxicity with an ED_{50} in the range of 1.4–2.0 µM while βE2 was less effective in protecting HT-22 cells against IAA toxicity and required 2.9–4.7 µM concentrations to inhibit 50% of IAA-induced cell death. Other known estrogenic compounds, estrone (E1), diethylstilbesterol [11], αE2, and ent-E2, were also able to inhibit oxidative stress-induced toxicity [29].

2.2.2
A-Ring Derivatives

We evaluated 35 compounds with modifications to the phenolic A-ring [29]. A-ring derivatives with electron-donating constituents that stabilized the phenoxy radical were more potent than βE2 in protecting these cells from glutamate and IAA toxicity. Our primary synthetic strategy was to replace the hydrogen with a bulky alkyl group at the 2 or 4 positions of the A-ring. For those compounds that included the addition of a single group to the 2-carbon position of the A-ring or 4-carbon position of the A-ring, the average EC_{50} value for protection against 10 mM glutamate and 20 mM IAA was 0.82 ± 0.48 and 0.49 ± 0.14 µM, respectively (Figure 2.2). When two groups flanked the 3-OH position, neuroprotection was enhanced with EC_{50} values for protection against 10 mM glutamate and 20 mM IAA being 0.09 ± 0.05 and 0.14 ± 0.02 µM, respectively (Figure 2.2).

C3 methoxy ether analogs of the phenolic A-ring that lack an additional phenol substituent did not protect cells from death induced by glutamate or IAA. Further negative controls included replacing the hydroxyl group at the 3-position of the A-ring with a CH_2OH or carboxylic acid group. These compounds were also ineffective in protecting HT-22 cells from cell death. Extending the length of the steroid by adding a (4-hydroxyphenyl)ethyl group to the 2-position of the A-ring increased neuroprotective properties. Removing one of the two hydroxyl groups did not affect neuroprotection against glutamate, but decreased potency in the IAA model by approximately 300-fold. Displacing the phenolic hydroxyl group from the steroid backbone, thus extending the length of the molecule, was more protective against toxicity than βE2. Relocating the hydroxyl group of the phenolic A-ring from the 3-position to the 2-position

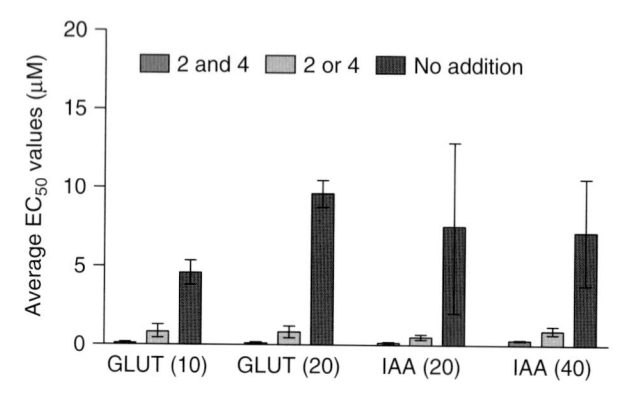

Figure 2.2 Neuroprotective EC_{50} values of A-ring-substituted estratrienes. EC_{50} values were grouped for those compounds that had one substitution added to either the 2- or 4-position, compounds that had additions made to both the 2- and 4-positions, and compounds that did not fit into either of these two categories (no addition). Depicted are mean ± SEM ($n = 7$). When SEM bars are not depicted, they are too small for presentation. The significance of differences among groups was determined by one-way analysis of variance (ANOVA) with a Tukey's multiple-comparisons test for planned comparisons between groups when significance was detected. $P < 0.05$ versus the other two groups. (Reproduced with permission from Perez et al., 2006 [65].)

decreased the protection against glutamate cytotoxicity, but not against IAA toxicity. Switching the hydroxyl group to the 2-position as well as adding bulky groups (1-adamantyl or *tert*-butyl) to the 3-position greatly improved neuroprotective potencies [29].

2.2.3
B- and C-Ring Derivatives

We evaluated 15 compounds with modifications to the B- or C-ring of the estratriene backbone. Hydroxyl additions to the B- or C-ring make the normally lipophilic estrogens more hydrophilic. In all cases, whether above or below the plane of the ring, when hydroxyl groups were added to the B- and C-rings, neuroprotection was completely abolished. Opening of the B-ring results in decreased rigidity of the molecule. Opening the ring structure of estratrienes did not further prevent IAA-induced cell death. Producing a nonplanar conformation so that the A- and B-rings lie perpendicular to the C- and D-rings did not enhance or hinder neuroprotection. Introducing conjugated double bonds into the B- or C-rings of estratrienes is yet another way to increase the stability of the phenoxy radical. As a group, they were 179- and 488-fold more potent than $\beta E2$ against glutamate and IAA toxicity, respectively. Steroids having an aromatic B-ring, as those that exist in conjugated equine estrogens, with intact phenolic hydroxyl groups were more potent against IAA toxicity but performed equally well against glutamate toxicity when compared to $\beta E2$.

2.2.4
D-Ring Derivatives

Esterification of the 17-hydroxyl with benzoic acid, introduction of the side chain found in 25-hydroxycholesterol, a 2-hydroxy-1-methylethyl side chain, or a spirocyclopentyl group at the C16 position of the D-ring, did not enhance the neuroprotection potency. Indeed, the modifications to the 17-position decreased neuroprotective potency against glutamate-induced cell killing. Complete removal of a 17-substituent, however, enhanced neuroprotection.

The combination of the 17β-hydroxyl and 13α-methyl groups induces a boat conformation in the C-ring. The combination of the 17α-hydroxyl and 13α-methyl groups does not change the chair configuration of the C-ring, but the D-ring is orthogonal and above the A-, B-, and C-rings. These changes reduced potencies against glutamate and IAA-induced cell death.

2.2.5
Correlation between Inhibition of TBARs and Protection against Glutamate and IAA

The potency of the estratrienes for protection against glutamate and IAA toxicity was compared with the antioxidant capacity of these compounds to inhibit peroxidation events. There was a positive correlation between inhibition of iron-induced lipid peroxidation and neuroprotective potency in the glutamate- and IAA-induced cell death models. Inhibition of TBAR levels correlated with potencies for neuroprotection against glutamate and IAA. When graphically depicted as categories of low, medium, and high, EC_{50} values for TBARs inhibition (a surrogate for lipid peroxidation inhibition) correlated with neuroprotection (Figure 2.3), although higher concentrations of estratrienes were required to inhibit peroxidation than were needed for neuroprotection. This latter observation indicates that inhibition of lipid peroxidation is not the only mechanism involved in the neuroprotective activity of these compounds.

2.2.6
Estrogen Receptor Binding

Competition binding experiments revealed that βE2 bound to ERα and ERβ with EC_{50} values of 3.0 and 4.5 nM, respectively. DES, as expected, bound slightly better and had EC_{50} values of 2.6 nM for ERα and 1.8 nM for ERβ. E1, αE2, and *ent*-estradiol bound to these receptors with less affinity. Additional double bonds in the steroid structure did not drastically change the affinity of the estratriene for the ERs. Additions to the A-ring drastically affected binding to the ERs. Additions to the 2- and 4-positions completely abolished the affinity of estratriene for both ERs. Adamantyl groups added to the 2-position, likewise, abolished binding to the ERs. The 1-methyl-allyl, *sec*-butyl, *tert*-butyl additions to the 2-position of the A-ring greatly reduced the binding affinity for the receptors.

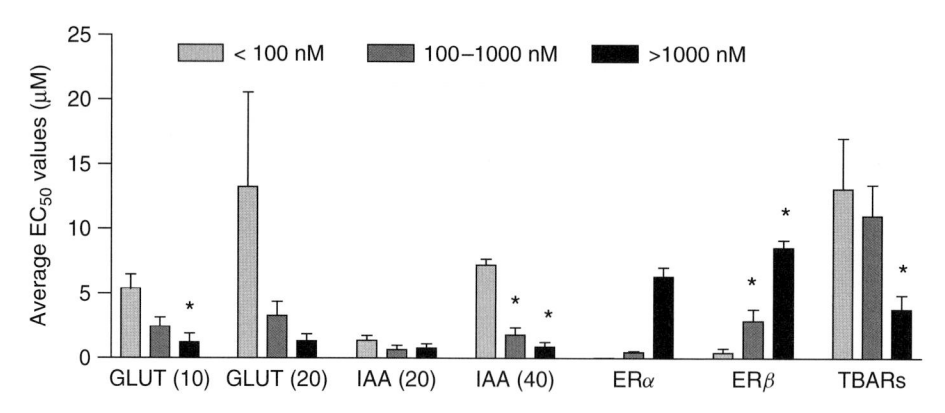

Figure 2.3 Graphical representation of the relationship between lipid peroxidation and neuroprotection. Data were categorized into three groups: those compounds that had EC$_{50}$ values for TBAR inhibition of 2 mM, those compounds with EC$_{50}$ of 2–10 mM, and those compounds requiring 410 mM to inhibit lipid peroxidation by 50%. The significance of differences among groups was determined by one-way analysis of variance (ANOVA) with a Tukey's multiple-comparisons test for planned comparisons between groups when significance was detected. $P < 0.05$ versus the other two groups. (Reproduced with permission from Perez et al., 2006.)

Midsize additions (*tert*-butyl) to the 2-position and 2-hydroxy-1-methyl-ethyl additions to the C17 position of the estratriene further reduced the binding capacity. While bulkier additions (1-methyl-allyl) and *sec*-butyl to the 4-position also completely inhibited ER binding, smaller additions (methyl) were approximately 69- and 58-fold less effective in binding to the ERs. Hydroxyl additions to the B- and C-rings in the α- or β-positions inhibited binding activity. Hydroxyl groups in the β configuration at the C11 position and of E1 and βE2 bound 10-fold less to ERα than βE2. Changes to the planarity of the steroid ring structure also produced changes to the affinity for ERs. Opening the ring structure at the 9-position diminished the affinity of the steroid for both ERs.

2.2.7
Correlation between Inhibition of TBARs or Neuroprotection and ER Binding

The potency of the estratrienes for ER binding was compared to protection against glutamate and IAA toxicity as well as the antioxidant capacity of these compounds to inhibit peroxidation events. In all cases, ERα or ERβ binding presented a slight negative correlation with neuroprotective protection and lipid peroxidation inhibition (Figure 2.3). When graphically depicted as categories of low, medium, and high EC$_{50}$ values for ERα binding, there was a negative correlation with ERα binding and lipid peroxidation inhibition as well as neuroprotection. A similar relationship was observed for ERβ binding (Figure 2.3).

2.2.8
Interpretation of *In vitro* Findings

This study expands the findings of Green [5] and Behl [30, 31], showing that estratrienes are neuroprotective agents. It also demonstrates that compounds that have electron-donating substituents, which increase the stability of the phenoxy radical, provide better neuroprotective abilities. The donated hydrogen radical can quench free radicals formed in oxidative stress conditions. Conjugated double bonds in the B- and C-rings increase phenoxy radical stability by resonance stabilization. Polar groups added to the B- and C-rings abolished estrogen's ability to protect cells from oxidative stress. These hydroxyl groups, which introduce hydrophilicity into the middle of the scaffold, may impact the way these steroids fit into the hydrophobic lipid bilayer and, thus, their ability to affect lipid oxidation events. D-ring substituents that increase lipophilicity such as the 17-benzoate and the spirocyclopentyl group at C16 decrease neuroprotection. Other 17-substituents such as the 2-hydroxy-1-methyl-ethyl and 5-hydroxy-1,5-dimethyl-hexyl groups, which contain a hydrophilic group in a hydrophic side chain, are also less neuroprotective. Solid-state NMR studies have shown that the 17-benzoate group orients the phenolic A-ring close to the polar head groups of membrane lipids [32]. This orientation would locate the phenolic A-ring away from the unsaturated double bonds in the fatty acids of the membrane lipids and could contribute to the decreased neuroprotective activity of this compound. Protection required a phenolic hydroxyl group as the 3-*O*-methyl congener proved ineffective. The phenol group did not have to be part of the A-ring, as addition of a phenol group to the 3-position was just as, if not more, protective against both toxicities. Repositioning the hydroxyl in the 2-carbon of the A-ring was still, albeit less, protective against glutamate toxicity and was a better protector against IAA cell toxicity. This result shows that the position of the hydroxyl group is not restricted to the 3-position of the A-ring.

This study demonstrates that estratrienes are potent lipid antioxidants. This is a well-known property as previous studies have determined the radical scavenging, iron chelating, and total antioxidant properties of various estratrienes [33–35]. We have confirmed other reports that show 2- and 4-additions to the phenolic A-ring are potent antioxidants [36]. Further, their antioxidant potency correlates with their ability to protect HT-22 cells from oxidative stress-induced toxicity. In a cell-free system, we have shown that estratrienes are able to inhibit iron-induced lipid peroxidation. Estratrienes performed better in the lipid-based antioxidant model, a TBARs assay, than in scavenging ROS from an aqueous phase. This trend occurs in spite of a study showing that antioxidant capacities of various estrogenic structures were found to be dependent on the aqueous or lipophilic nature of the assay system [37, 38].

Redox iron cycling is complex and a Fe(II)/Fe(III) 5 : 1 ratio is optimal in initiating peroxidation events. We added ferric ions at a 50 mM concentration, whereas estratrienes were able to inhibit lipid peroxidation events at doses below this iron load. While chelation has been shown to occur with steroidal molecules at a 3 : 1 stoichiometry [39], many of our compounds were effective at low micromolar concentrations, indicating that alternative mechanisms are involved, perhaps indicating the redox cycling of estrogens [13, 14]. The lipophilicity of estratrienes leads to their accumulation in the hydrophobic plasma membranes and affects membrane fluidity [40, 41] and increased lipid ordering [41]. Estrogen localization to this lipid environment places them at the site of key peroxidation events, and in conjunction with their antioxidant properties decreases oxidative damage. The lipid membrane is also the site of various signal transduction processes including PI3K signaling and phosphatidylserine flipping in apoptosis.

More than 95% of the binding affinity of estratrienes for the ER is dependent on the aromaticity of the A-ring. One way to prevent ER binding is to hinder the 3-OH group with bulky groups in the 2- and 4-positions. This masking is important for three reasons: (i) it increases the stability of the phenoxy radical, (ii) it imparts hydrophobicity to the A-ring and thus orients it toward the center of the lipid bilayer where lipid peroxidation events occur [32], and (iii) it inhibits estratriene binding to ERs [42]. We found the third condition to hold in our competition binding experiments. The most potent neuroprotective estratrienes, compounds with 3-hydroxyl groups flanked by two groups (such as adamantyl moieties), did not bind to ERs. We also obtained a slight negative correlation between neuroprotection and ERα and ERβ binding.

In conclusion, we found that placement of large bulky groups on the A-ring of estratrienes, while retaining its phenolic nature, produced compounds that are neuroprotective antioxidants but that were devoid of ER binding. These compounds may be useful in the protection of the brain against neurodegenerative diseases as well as more acute insults. Lacking ER binding activity, these compounds have the additional feature of avoiding many of the side effects of estrogens.

2.3
In vivo Assessment of Structure–Neuroprotective Activity Relationships

Till date, eight of the compounds described herein have been tested for brain protection in a rat model of middle cerebral artery occlusion and all eight compounds have proven to be neuroprotective against this model of stroke [29, 43–48]. This result suggests that the neuroprotective actions of estrogens are related to their antioxidant properties and not to their ability to bind ERs both, *in vitro* and *in vivo*. Further, uterotrophic assays showing an increase in weight with E1, but not with nonfeminizing estrogen analogs, substantiate this finding [48].

2.4
In vitro Assessment of Structure–Cell Signaling Relationships

Finally, we assessed the effect of some of these nonfeminizing estrogens on intracellular signaling pathways that are known to mediate the rapid neuroprotective response to estrogens. Our strategy was to determine if nonfeminizing estrogens were as (or more) effective in affecting neuroprotective signaling.

In view of the relative lack of binding of these analogs to ERα or ERβ [29, 48], intracellular signaling pathways may be involved in mediating their neuroprotective effects. Indeed, experimental studies have shown that the neuroprotective effects of estrogens are mediated at least in part through the rapid but acute phosphorylation of signaling proteins, such as adenylyl cyclase, Akt, PKA, PKC, and MAPK [45, 49–54]. Changes in the activity of these enzymes can regulate the phosphorylation of numerous cellular substrates such as intermediary signaling proteins Rsk, p38, JNK, and the nuclear transcriptional factors CREB and cfos/c-Jun, which may ultimately mediate cell survival changes (for a review, see [55]). However, ERK1/2 is rapidly and persistently phosphorylated in response to oxidative and excitotoxic stresses caused by glutamate and the presence of estrogens prevents this persistent phosphorylation (Figure 2.4); [56]. These data suggest that prolonged phosphorylation of ERK1/2 is detrimental to cell survival; this is supported by the observations by Smith and colleagues who noted aberrant neuronal expression of phosphorylated ERK1/2 and other MAPKs in AD brains in association with markers of oxidative stress [57]. Others have also shown that MAPK phosphorylation is associated with a variety of sporadic and familial neurodegenerative diseases characterized by tau deposits [58]. Patients with PD and other Lewy body diseases have increased phosphorylated ERK1/2 in neurons of substantia nigra and midbrain [59]. In addition,

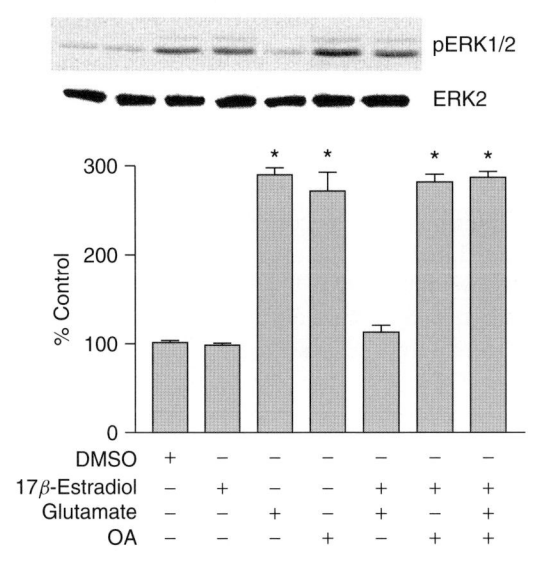

Figure 2.4 Phosphorylation of ERK following glutamate treatment in the presence and absence of various estrogens and/or okadaic acid. HT-22 cells were treated with 100 nM okadaic acid, 10 mM glutamate, and/or 10 µM 17β-estradiol. Cells were harvested after 24 h of treatment for western blot analysis of pERK. Depicted graphs are mean ± SEM for $n = 5$ with a representative blot. *$P < 0.05$ versus control.

increased ERK1/2 phosphorylation has been noted in the vulnerable pneumbra following acute ischemic stroke in humans [60].

Because this chronic phosphorylation of ERK1/2 is thought to send a neurotoxic or neurodegenerative signal, it has been speculated that estrogens attenuate the persistent ERK1/2 phosphorylation to mediate neuroprotection. Indeed, it has been shown that estrogens alone cause a rapid but transient increase in ERK1/2 phosphorylation [52, 61]. In addition, it has also demonstrated that estrogens attenuate the persistent ERK1/2 phosphorylation caused by cytotoxic insults [56–60]. Glutamate-induced toxicity causes an increase in phosphorylation of ERK1/2 ([56]; Figure 2.4). Estrogens alone did not alter the phosphorylation state of ERK1/2 compared with vehicle control following a 24 h treatment period. However, the presence of estrogens attenuated the elevated phosphorylation of ERK1/2 induced by glutamate. However, estrogens had no effect on okadaic acid-induced ERK1/2 phosphorylation, and the presence of this broad-spectrum protein phosphatase inhibitor abolished the estrogen-mediated reduction in phosphorylation ERK1/2 caused by glutamate (Figure 2.4). Phosphorylation of ERK1/2 in response to glutamate toxicity corresponds to decreases in serine/threonine phosphatase expression, and inhibitors of ERK1/2 phosphorylation attenuate the glutamate-induced cytotoxicity [56, 62]; Figure 2.4). Therefore, one of the mechanisms of estrogen-mediated neuroprotection is by preservation of serine/threonine phosphatase protein expression and activity via an ER-independent pathway that ultimately prevents persistent ERK1/2 hyperphosphorylation as evidenced by the similar effects of the estrogen analogs as compared to βE2 [56, 62–64].

That these effects of βE2 and nonfeminizing estrogens are important to the viability of neurons is demonstrated in Figure 2.5. HT-22 cells treated with glutamate show a profound increase in ROS production, lipid peroxidation, and protein oxidation (Figure 2.5). Simultaneous treatment with βE2, αE2, ent-E2, or 2-adamantyl-estrogens effectively prevented these effects of glutamate toxicity (Figure 2.5). In contrast, when protein phosphatases were inhibited, all these markers of oxidative stress increased and all of the estrogens were ineffective in preventing these increases. Collectively, these data indicate that nonfeminizing estrogens prevent glutamate toxicity in part by preventing the prooxidant response to this substance, in a manner that requires the presence of protein phosphatase activity.

Figure 2.5 Effects of glutamate, okadaic acid, 17β-estradiol, and E2 analogs on oxidative stress markers in HT-22. (a) ROS production, (b) lipid peroxidation, and (c) protein carbonylation were measured in neurons treated simultaneously with 100 nM okadaic acid, 10 mM glutamate, and/or 10 mM 17β-estradiol, 10 mM 17α-estradiol, 10 mM ENT E2, and/or 1 mM ZYC3. All data were normalized to percentage survival of nontreated control. Depicted are mean $\pm$ SEM for six independent experiments with two replicates per experiment for ROS production and six independent experiments each for lipid peroxidation and protein carbonylation experiments. *$P < 0.05$ versus vehicle control.

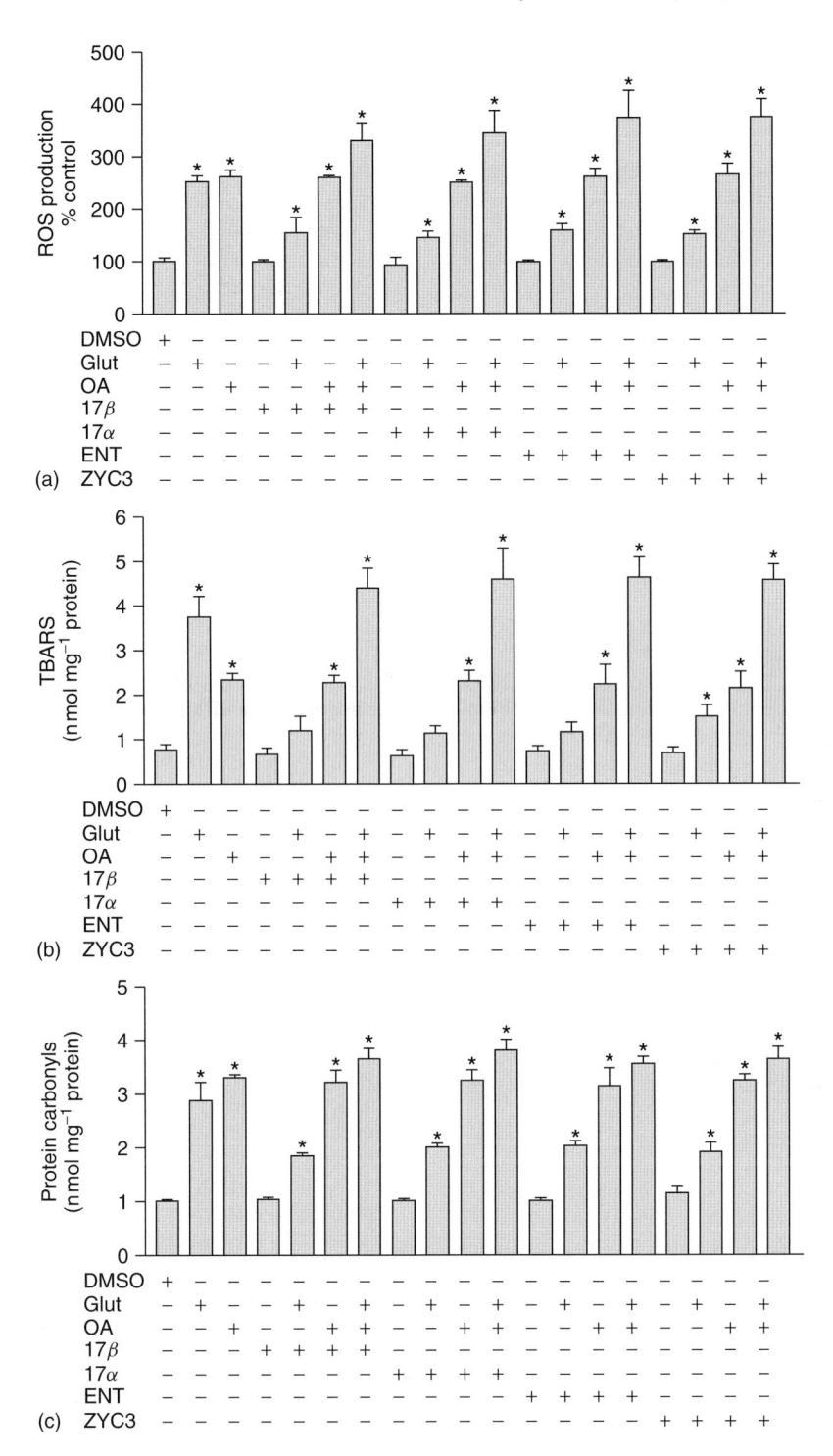

2.4 In vitro Assessment of Structure–Cell Signaling Relationships

2.5
Summary

In summary, we have shown that the phenolic A-ring of estrogens is essential for neuroprotective activity. Also, chemical modifications of the estrogen scaffold that enhance stability of the A-ring phenol radical increase neuroprotective potency of estrogens. This was achieved by the placement of substituents on the 2- and/or 4-position of the phenolic A-ring of estradiol and/or by introducing conjugated double bonds into the B- or C-rings of estratrienes. The resulting, potent estratrienes were active in an animal model for stroke and in *in vitro* models for neuroprotective signaling. Collectively, this synthetic approach to the discovery of potent, nonfeminizing estrogens could provide a new means for achieving neuroprotection without the unwanted side effects of naturally occurring estrogens.

Acknowledgment

Supported by NIH Grants P01 AG10485, P01 AG22550, and P01 AG27956.

References

1. Green, P.S. and Simpkins, J.W. (2000) Neuroprotective effects of estrogens: potential mechanisms of action. *Int. J. Dev. Neurosci.*, **18**, 347–358.
2. Alkayed, N.J., Harukuni, I., Kimes, A.S., London, E.D., Traystman, R.J., and Hurn, P.D. (1998) Gender-linked brain injury in experimental stroke. *Stroke*, **29**, 159–165; discussion 166.
3. Dubal, D.B., Kashon, M.L., Pettigrew, L.C., Ren, J.M., Finklestein, S.P., Rau, S.W., and Wise, P.M. (1998) Estradiol protects against ischemic injury. *J. Cereb. Blood Flow Metab.*, **18**, 1253–1258.
4. Green, P.S., Bishop, J., and Simpkins, J.W. (1997) 17 alpha-estradiol exerts neuroprotective effects on SK-N-SH cells. *J. Neurosci.*, **17**, 511–515.
5. Green, P.S., Gordon, K., and Simpkins, J.W. (1997) Phenolic A ring requirement for the neuroprotective effects of steroids. *J. Steroid Biochem. Mol. Biol.*, **63**, 229–235.
6. Green, P.S., Gridley, K.E., and Simpkins, J.W. (1996) Estradiol protects against beta-amyloid (25-35)-induced toxicity in SK-N-SH human neuroblastoma cells. *Neurosci. Lett.*, **218**, 165–168.
7. Green, P.S., Gridley, K.E., and Simpkins, J.W. (1998) Nuclear ER-independent neuroprotection by estratrienes: a novel interaction with glutathione. *Neuroscience*, **84**, 7–10.
8. Green, P.S., Yang, S.H., Nilsson, K.R., Kumar, A.S., Covey, D.F., and Simpkins, J.W. (2001) The nonfeminizing enantiomer of 17beta-estradiol exerts protective effects in neuronal cultures and a rat model of cerebral ischemia. *Endocrinology*, **142**, 400–406.
9. Gridley, K.E., Green, P.S., and Simpkins, J.W. (1997) Low concentrations of estradiol reduce beta-amyloid (25-35)-induced toxicity, lipid peroxidation and glucose utilization in human SK-N-SH neuroblastoma cells. *Brain Res.*, **778**, 158–165.
10. Simpkins, J.W., Wang, J., Wang, X., Perez, E., Prokai, L., and Dykens, J.A. (2005) Mitochondria play a central role in estrogen-induced neuroprotection. *Curr. Drug Targets CNS Neurol. Disord.*, **4**, 69–83.

11. Sudo, S., Wen, T.C., Desaki, J., Matsuda, S., Tanaka, J., Arai, T., Maeda, N., and Sakanaka, M. (1997) Beta-estradiol protects hippocampal CA1 neurons against transient forebrain ischemia in gerbil. *Neurosci. Res.*, **29**, 345–354.
12. Yang, S.H., Shi, J., Day, A.L., and Simpkins, J.W. (2000) Estradiol exerts neuroprotective effects when administered after ischemic insult. *Stroke*, **31**, 745–749; discussion 749–750.
13. Prokai, L., Prokai-Tatrai, K., Perjesi, P., Zharikova, A.D., Perez, E.J., Liu, R., and Simpkins, J.W. (2003) Quinol-based cyclic antioxidant mechanism in estrogen neuroprotection. *Proc. Natl. Acad. Sci. U.S.A.*, **100**, 11741–11746.
14. Prokai, L., Prokai-Tatrai, K., Perjesi, P., Zharikova, A.D., and Simpkins, J.W. (2003) Quinol-based metabolic cycle for estrogens in rat liver microsomes. *Drug Metab. Dispos.*, **31**, 701–704.
15. Dumas, J., Hancur-Bucci, C., Naylor, M., Sites, C., and Newhouse, P. (2008) Estradiol interacts with the cholinergic system to affect verbal memory in postmenopausal women: evidence for the critical period hypothesis. *Horm. Behav.*, **53**, 159–169.
16. Grodstein, F., Manson, J.E., and Stampfer, M.J. (2006) Hormone therapy and coronary heart disease: the role of time since menopause and age at hormone initiation. *J. Women's Health (Larchmt)*, **15**, 35–44.
17. Harman, S.M. (2004) What do hormones have to do with aging? What does aging have to do with hormones. *Ann. N. Y. Acad. Sci.*, **1019**, 299–308.
18. Hodis, H.N., Mack, W.J., and Lobo, R.A. (2003) Randomized controlled trial evidence that estrogen replacement therapy reduces the progression of subclinical atherosclerosis in healthy postmenopausal women without preexisting cardiovascular disease. *Circulation*, **108**, e5; author reply e5.
19. Sontag, E., Luangpirom, A., Hladik, C., Mudrak, I., Ogris, E., Speciale, S., and White, C.L. III (2004) Altered expression levels of the protein phosphatase 2A ABalphaC enzyme are associated with Alzheimer disease pathology. *J. Neuropathol. Exp. Neurol.*, **63**, 287–301.
20. Coker, L.H., Hogan, P.E., Bryan, N.R., Kuller, L.H., Margolis, K.L., Bettermann, K., Wallace, R.B., Lao, Z., Freeman, R., Stefanick, M.L., and Shumaker, S.A. (2009) Postmenopausal hormone therapy and subclinical cerebrovascular disease: the WHIMS-MRI study. *Neurology*, **72**, 125–134.
21. Dubey, R.K., Imthurn, B., Barton, M., and Jackson, E.K. (2005) Vascular consequences of menopause and hormone therapy: importance of timing of treatment and type of estrogen. *Cardiovasc. Res.*, **66**, 295–306.
22. Maki, P.M. (2006) Potential importance of early initiation of hormone therapy for cognitive benefit. *Menopause*, **13**, 6–7.
23. Manson, J.E., Allison, M.A., Rossouw, J.E., Carr, J.J., Langer, R.D., Hsia, J., Kuller, L.H., Cochrane, B.B., Hunt, J.R., Ludlam, S.E., Pettinger, M.B., Gass, M., Margolis, K.L., Nathan, L., Ockene, J.K., Prentice, R.L., Robbins, J., and Stefanick, M.L. (2007) Estrogen therapy and coronary-artery calcification. *N. Engl. J. Med.*, **356**, 2591–2602.
24. Resnick, S.M., Espeland, M.A., Jaramillo, S.A., Hirsch, C., Stefanick, M.L., Murray, A.M., Ockene, J., and Davatzikos, C. (2009) Postmenopausal hormone therapy and regional brain volumes: the WHIMS-MRI study. *Neurology*, **72**, 135–142.
25. Salpeter, S.R., Walsh, J.M., Greyber, E., and Salpeter, E.E. (2006) Brief report: coronary heart disease events associated with hormone therapy in younger and older women. A meta-analysis. *J. Gen. Intern. Med.*, **21**, 363–366.
26. Zandi, P.P., Carlson, M.C., Plassman, B.L., Welsh-Bohmer, K.A., Mayer, L.S., Steffens, D.C., and Breitner, J.C. (2002) Hormone replacement therapy and incidence of Alzheimer disease in older women: the Cache County Study. *J. Am. Med. Assoc.*, **288**, 2123–2129.
27. Brinton, R.D. (2004) Requirements of a brain selective estrogen: advances and remaining challenges for developing a NeuroSERM. *J. Alzheimers Dis.*, **6**, S27–S35.

28. Shelly, W., Draper, M.W., Krishnan, V., Wong, M., and Jaffe, R.B. (2008) Selective estrogen receptor modulators: an update on recent clinical findings. *Obstet. Gynecol. Surv.*, **63**, 163–181.
29. Perez, E., Cai, Z.Y., Covey, D.F., and Simpkins, J.W. (2005) Neuroprotective effects of estratriene analogs: structure-activity relationships and molecular optimization. *Drug Dev. Res.*, **66**, 78–92.
30. Behl, C., Skutella, T., Lezoualc'h, F., Post, A., Widmann, M., Newton, C.J., and Holsboer, F. (1997) Neuroprotection against oxidative stress by estrogens: structure-activity relationship. *Mol. Pharmacol.*, **51**, 535–541.
31. Moosmann, B. and Behl, C. (1999) The antioxidant neuroprotective effects of estrogens and phenolic compounds are independent from their estrogenic properties. *Proc. Natl. Acad. Sci. U.S.A.*, **96**, 8867–8872.
32. Cegelski, L., Rice, C.V., O'Connor, R.D., Caruano, A.L., Tochtrop, G.P., Cai, Z.Y., Covey, D.F., and Schaefer, J. (2005) Mapping the locations of estradiol and potent neuroprotective analogues in phospholipid bilayers by REDOR NMR 2006. *Drug Dev. Res.*, **66**, 93–102.
33. Mooradian, A.D. (1993) Antioxidant properties of steroids. *J. Steroid Biochem. Mol. Biol.*, **45**, 509–511.
34. Romer, W., Oettel, M., Droescher, P., and Schwarz, S. (1997) Novel "scavestrogens" and their radical scavenging effects, iron-chelating, and total antioxidative activities: delta 8,9-dehydro derivatives of 17 alpha-estradiol and 17 beta-estradiol. *Steroids*, **62**, 304–310.
35. Romer, W., Oettel, M., Menzenbach, B., Droescher, P., and Schwarz, S. (1997) Novel estrogens and their radical scavenging effects, iron-chelating, and total antioxidative activities: 17 alpha-substituted analogs of delta 9(11)-dehydro-17 beta-estradiol. *Steroids*, **62**, 688–694.
36. Miller, C.P., Jirkovsky, I., Hayhurst, D.A., and Adelman, S.J. (1996) In vitro antioxidant effects of estrogens with a hindered 3-OH function on the copper-induced oxidation of low density lipoprotein. *Steroids*, **61**, 305–308.
37. Ruiz-Larrea, B., Leal, A., Martin, C., Martinez, R., and Lacort, M. (1995) Effects of estrogens on the redox chemistry of iron: a possible mechanism of the antioxidant action of estrogens. *Steroids*, **60**, 780–783.
38. Ruiz-Larrea, M.B., Martin, C., Martinez, R., Navarro, R., Lacort, M., and Miller, N.J. (2000) Antioxidant activities of estrogens against aqueous and lipophilic radicals; differences between phenol and catechol estrogens. *Chem. Phys. Lipids*, **105**, 179–188.
39. Ryan, T.P., Steenwyk, R.C., Pearson, P.G., and Petry, T.W. (1993) Inhibition of in vitro lipid peroxidation by 21-aminosteroids. Evidence for differential mechanisms. *Biochem. Pharmacol.*, **46**, 877–884.
40. Dicko, A., Morissette, M., Ben Ameur, S., Pezolet, M., and Di Paolo, T. (1999) Effect of estradiol and tamoxifen on brain membranes: investigation by infrared and fluorescence spectroscopy. *Brain Res. Bull.*, **49**, 401–405.
41. Liang, Y., Belford, S., Tang, F., Prokai, L., Simpkins, J.W., and Hughes, J.A. (2001) Membrane fluidity effects of estratrienes. *Brain Res. Bull.*, **54**, 661–668.
42. Hughes, R.A., Harris, T., Altmann, E., McAllister, D., Vlahos, R., Robertson, A., Cushman, M., Wang, Z., and Stewart, A.G. (2002) 2-Methoxyestradiol and analogs as novel antiproliferative agents: analysis of three-dimensional quantitative structure-activity relationships for DNA synthesis inhibition and estrogen receptor binding. *Mol. Pharmacol.*, **61**, 1053–1069.
43. Liu, R., Yang, S.H., Perez, E., Yi, K.D., Wu, S.S., Eberst, K., Prokai, L., Prokai-Tatrai, K., Cai, Z.Y., Covey, D.F., Day, A.L., and Simpkins, J.W. (2002) Neuroprotective effects of a novel non-receptor-binding estrogen analogue: in vitro and in vivo analysis. *Stroke*, **33**, 2485–2491.
44. Simpkins, J.W., Rajakumar, G., Zhang, Y.Q., Simpkins, C.E., Greenwald, D., Yu, C.J., Bodor, N., and Day, A.L. (1997) Estrogens may reduce mortality and ischemic damage caused by middle cerebral artery occlusion in the female rat. *J. Neurosurg.*, **87**, 724–730.

45. Watters, J.J., Campbell, J.S., Cunningham, M.J., Krebs, E.G., and Dorsa, D.M. (1997) Rapid membrane effects of steroids in neuroblastoma cells: effects of estrogen on mitogen activated protein kinase signalling cascade and c-fos immediate early gene transcription. *Endocrinology*, **138**, 4030–4033.
46. Xia, S., Cai, Z.Y., Thio, L.L., Kim-Han, J.S., Dugan, L.L., Covey, D.F., and Rothman, S.M. (2002) The estrogen receptor is not essential for all estrogen neuroprotection: new evidence from a new analog. *Neurobiol. Dis.*, **9**, 282–293.
47. Yang, S.H., He, Z., Wu, S.S., He, Y.J., Cutright, J., Millard, W.J., Day, A.L., and Simpkins, J.W. (2001) 17-beta estradiol can reduce secondary ischemic damage and mortality of subarachnoid hemorrhage. *J. Cereb. Blood Flow Metab.*, **21**, 174–181.
48. Perez, E., Liu, R., Yang, S.H., Cai, Z.Y., Covey, D.F., and Simpkins, J.W. (2005) Neuroprotective effects of an estratriene analog are estrogen receptor independent in vitro and in vivo. *Brain Res.*, **1038**, 216–222.
49. Migliaccio, A., Di Domenico, M., Castoria, G., de Falco, A., Bontempo, P., Nola, E., and Auricchio, F. (1996) Tyrosine kinase/p21ras/MAP-kinase pathway activation by estradiol-receptor complex in MCF-7 cells. *EMBO J.*, **15**, 1292–1300.
50. Kelly, M.J., Lagrange, A.H., Wagner, E.J., and Ronnekleiv, O.K. (1999) Rapid effects of estrogen to modulate G protein-coupled receptors via activation of protein kinase A and protein kinase C pathways. *Steroids*, **64**, 64–75.
51. Singer, C.A., Figueroa-Masot, X.A., Batchelor, R.H., and Dorsa, D.M. (1999) The mitogen-activated protein kinase pathway mediates estrogen neuroprotection after glutamate toxicity in primary cortical neurons. *J. Neurosci.*, **19**, 2455–2463.
52. Singh, M., Setalo, G. Jr., Guan, X., Warren, M., and Toran-Allerand, C.D. (1999) Estrogen-induced activation of mitogen-activated protein kinase in cerebral cortical explants: convergence of estrogen and neurotrophin signaling pathways. *J. Neurosci.*, **19**, 1179–1188.
53. Toran-Allerand, C.D., Singh, M., and Setalo, G. Jr. (1999) Novel mechanisms of estrogen action in the brain: new players in an old story. *Front. Neuroendocrinol.*, **20**, 97–121.
54. Zhang, L., Rubinow, D.R., Xaing, G., Li, B.S., Chang, Y.H., Maric, D., Barker, J.L., and Ma, W. (2001) Estrogen protects against beta-amyloid-induced neurotoxicity in rat hippocampal neurons by activation of Akt. *Neuroreport*, **12**, 1919–1923.
55. Lee, S.J. and McEwen, B.S. (2001) Neurotrophic and neuroprotective actions of estrogens and their therapeutic implications. *Annu. Rev. Pharmacol. Toxicol.*, **41**, 569–591.
56. Yi, K.D., Cai, Z.Y., Covey, D.F., and Simpkins, J.W. (2008) Estrogen receptor-independent neuroprotection via protein phosphatase preservation and attenuation of persistent extracellular signal-regulated kinase 1/2 activation. *J. Pharmacol. Exp. Ther.*, **324**, 1188–1195.
57. Zhu, X., Lee, H.G., Raina, A.K., Perry, G., and Smith, M.A. (2002) The role of mitogen-activated protein kinase pathways in Alzheimer's disease. *Neurosignals*, **11**, 270–281.
58. Ferrer, I., Friguls, B., Dalfo, E., and Planas, A.M. (2003) Early modifications in the expression of mitogen-activated protein kinase (MAPK/ERK), stress-activated kinases SAPK/JNK and p38, and their phosphorylated substrates following focal cerebral ischemia. *Acta Neuropathol.*, **105**, 425–437.
59. Zhu, J.H., Kulich, S.M., Oury, T.D., and Chu, C.T. (2002) Cytoplasmic aggregates of phosphorylated extracellular signal-regulated protein kinases in Lewy body diseases. *Am. J. Pathol.*, **161**, 2087–2098.
60. Slevin, M., Krupinski, J., Slowik, A., Rubio, F., Szczudlik, A., and Gaffney, J. (2000) Activation of MAP kinase (ERK-1/ERK-2), tyrosine kinase and VEGF in the human brain following acute ischaemic stroke. *Neuroreport*, **11**, 2759–2764.

61. Sarkar, S.N., Huang, R.Q., Logan, S.M., Yi, K.D., Dillon, G.H., and Simpkins, J.W. (2008) Estrogens directly potentiate neuronal L-type Ca^{2+} channels. *Proc. Natl. Acad. Sci. U.S.A.*, **105**, 15148–15153.
62. Yi, K.D., Covey, D.F., and Simpkins, J.W. (2009) Mechanism of okadaic acid-induced neuronal death and the effect of estrogens. *J. Neurochem.*, **108**, 732–740.
63. Yi, K.D., Chung, J., Pang, P., and Simpkins, J.W. (2005) Role of protein phosphatases in estrogen-mediated neuroprotection. *J. Neurosci.*, **25**, 7191–7198.
64. Yi, K.D. and Simpkins, J.W. (2008) Protein phosphatase 1, protein phosphatase 2A, and calcineurin play a role in estrogen-mediated neuroprotection. *Endocrinology*, **149**, 5235–5243.
65. Perez, E., Cai, Z.Y., Covey, D.F., and Simpkins, J.W. (2006) Neuroprotective effects of estratriene analogs: structure-activity relationships and molecular optimization. *Drug Dev. Res.*, **68**, 1–15.

3
Progestins and Neuroprotection: Why the Choice of Progestin Matters

Meharvan Singh

3.1
Introduction

Gonadal steroid hormones such as estradiol and progesterone have effects well beyond the strict confines of reproductive function and are known to exert their effects on a variety of target tissues including the bone, the heart, and the brain. With respect to the brain, there is a considerable body of literature that supports the protective effects of estradiol and progesterone. However, results from the Women's Health Initiative-Memory Study (WHIMS) highlighted the fact that important caveats must exist in translating the preclinical data to humans. These are based on the concept that the type of hormone used may influence the outcome. Most preclinical studies that supported the neuroprotective potential of progestins evaluated the effects of progesterone, rather than those of the synthetic progestin, medroxy progesterone acetate (MPA) used in the WHIMS. Here we review the mechanism of progesterone's protective effects and contrast them with that of MPA, and in so doing, offer a biological basis for why progesterone, but not MPA, is neuroprotective.

It is estimated that by 2010, the population of women between the ages of 45 and 64 will reach approximately 42 million (U.S. Census Bureau. Projected population of the United States, by Age and Sex: 2000–2050. *www.census.gov/ipc/www/usinterimproj/* Internet release date: March 18, 2004). Among the health-related changes and decisions these women will need to consider includes whether or not to consider the use of hormone therapy (HT) for not just the management of menopausal symptoms, but potentially, to help maintain a healthy brain. And though numerous basic science, epidemiological, and some clinical studies have supported the potential benefit of HT in reducing the incidence of age-associated brain dysfunction (including reducing the risk for Alzheimer's disease), recent results from the WHIMS failed to reveal beneficial effects in reducing the risk of Alzheimer's disease or "all-cause" dementia. As a consequence, these reports left the field unsettled as to the future of HT. Since the publication of these studies, it has become apparent that there were important

---

Hormones in Neurodegeneration, Neuroprotection, and Neurogenesis.
Edited by Achille G. Gravanis and Synthia H. Mellon
Copyright © 2011 WILEY-VCH Verlag GmbH & Co. KGaA, Weinheim
ISBN: 978-3-527-32627-3

caveats to the data that need to be considered. Among these include consideration of the type of hormone used.

Here, by outlining differences in the neurobiology of two major progestins, the "natural" progestin (progesterone) and MPA (the most commonly used progestin in HT regimens), we suggest that MPA is not an effective choice of progestin for the brain. And, while progesterone and MPA may be equally effective at reducing the uterotrophic effects of unopposed estrogen treatment, their effects on the brain are far from identical.

3.2
The Biology of Progesterone

Progesterone, the natural progestin, is a major gonadal hormone that is synthesized primarily by the ovary in the female, and the testes and adrenal cortex in the male. While the level of progesterone is generally higher in the female, it is worth noting that levels of progesterone during the female follicular phase of the menstrual cycle are similar to those seen in males [1], and thus, may be equally important in males. The "classical" mechanism by which progesterone elicits its effects is via the progesterone receptor (PR), which like the estrogen receptor (ER), has classically been described as a nuclear transcription factor, acting through specific progesterone response elements (PREs) within the promoter region of target genes to regulate transcription. Two major isoforms of the classical PR exist, PR-B, and its N-terminally truncated form, PR-A (for review, see [2]). The latter has been shown to exert negative control of not only PR-B-mediated transcription, but that mediated by the ER and glucocorticoid receptor as well [3]. This negative regulation of ER function by a PR may underlie, at least in part, the mechanism by which progestins functionally antagonize the effects of estrogen. For example, progesterone can inhibit estrogen's ability to increase serum levels of 1,25-dihydroxy vitamin D [4], whose consequence may be to antagonize estrogen's beneficial effects on the bone. Relevant to HT, the functional antagonism exerted by progestins on estrogen's actions also underlie the rationale for combined estrogen and progestin therapy in women with an intact uterus, as the addition of a progestin reduces the risk of uterine cancer associated with unopposed estrogen therapy [5]. However, the relationship between PR and ER may not always be antagonistic. For example, Migliaccio *et al.* demonstrated not only a physical interaction between the PR and the ER, but also that this association was necessary for progesterone to elicit the activation of a signal transduction pathway and mitogen-activated protein kinase (MAPK) pathway in mammary tumor cells [6].

In addition to the regulation of gene transcription, progesterone can also elicit its effects via nongenomic mechanisms such as through the activation of signal transduction pathways. Among these second messenger/signal transduction systems now known to be activated by progesterone include cAMP/PKA (protein kinase A) [7], MAPK (ERK1/2) [6, 8], and the (phosphatidylinositol 3-kinase) PI-3K/Akt pathway [8]. Activation of such signaling pathways may not only be relevant to how progesterone regulates cellular function related to reproductive function, but may also play an important role in its neuroprotective effects.

3.3
Membrane-Associated Progesterone Receptors

While the classical, intracellular/nuclear PR certainly plays an important role in mediating the effects of progesterone, evidence also exists for alternative mechanisms of action, including that which involves integral membrane progesterone receptors (mPRs). If nothing else, progesterone's high degree of lipophilicity (having a log P value, or octanol/water partition coefficient, of approximately 4), may be consistent with the idea that progesterone interacts with a plasma membrane–associated receptor.

Specific and displaceable binding sites for steroid hormones have been known for some time [9, 10]. In the brain, for example, the use of a bovine serum albumin (BSA)-conjugated, iodinated progesterone (the BSA conjugation is believed to render the molecule membrane impermeable) bound specifically to several regions of brain including the cerebral cortex, brainstem, cerebellum, striatum, and hypothalamus [9]. These putative progesterone/progestin receptors have now been more carefully characterized/cloned. Among these are a single-transmembrane spanning receptor termed *25-Dx*, and three mPR, which interestingly, have a predicted seven transmembrane spanning domain.

25-Dx was first described in rat liver as a gene regulated by the environmental toxin, 2,3,7,8-tetrachlorodibenzo-*p*-dioxin [11], whose protein product was subsequently described as a functional progesterone binding protein [12]. This novel PR is expressed in the brain [12–14] and is involved in numerous aspects of cell function ranging from neuronal development [15], steroidogenesis [16], regulation of reproductive behavior [12], regulation of CSF production, and osmoregulation [17]. More recently, three isoforms of a membrane-associated PR (termed *mPR* α, β, and γ) were recently cloned and found to be expressed in humans, mice, and teleosts; the latter being the first species in which this membrane PR was identified [18, 19]. Interestingly, unlike 25-Dx, which, based on hydrophobicity plots, is believed to have only a single-transmembrane domain, the mPR has seven transmembrane spanning domains, and exhibits characteristics that would suggest that the mPR is a G-protein-coupled receptor [18].

As it relates to progesterone-induced neuroprotection, it is plausible that the membrane-associated receptors and the "classical," intracellular receptors play complementary roles in promoting neuroprotection. For example, our laboratory has determined that the ability of progesterone to increase brain-derived neurotrophic factor (BDNF) requires the classical PR [20]. In contrast, a putative ligand of membrane-associated PRs, P4-BSA, that does not bind to the classical PR, fails to increase BDNF levels, but yet is effective in increasing the phosphorylation of ERK (extracellular signal-related kinase) [20], another proposed mediator of progesterone's neuroprotective effects [21]. As such, the ability of a progestin to have maximal neuroprotective efficacy may depend on the array of PRs that it is capable of binding/activating.

Progesterone can also elicit its effects through its metabolites. For example, the 5α-reduced metabolite of progesterone, allopregnanolone, binds to discrete sites within the hydrophobic domain of the $GABA_A$ receptor complex, which results in the potentiation of GABA-induced chloride conductance (see [22] for review). This, in turn, may regulate cellular excitability and thus, excitotoxicity. Thus, progesterone's ability to interact with specific sites within the membrane (either membrane binding sites (receptors) or with the $GABA_A$ receptor), as well as with specific cytosolic signal transducers, may help explain some of the rapid effects of progesterone, which in addition to its classical genomic mechanisms, may be important for regulating cell viability.

And finally, progesterone has also been found to interact with sigma 1 (σ_1) receptor [11, 23]. Given the purported role of the sigma 1 receptor in neuroprotection (for review, see [24]), this mechanism may also be relevant to progesterone's protective actions.

3.4
Progesterone-Induced Protection

Progesterone has been reported to exert protective effects in a variety of experimental models that mimic certain pathogenic aspects of brain dysfunction seen with advanced age- or age-related neurodegenerative diseases such as Alzheimer's disease. For example, progesterone has been shown to significantly attenuate oxidative injury resulting from glutamate [21, 25–27] and glucose deprivation–induced toxicity [28], and also protects against $FeSO_4$ and amyloid β-peptide–induced toxicity in primary hippocampal cultures [28].

Progesterone is also an effective neuroprotectant in animal models of stroke. For example, Jiang et al. illustrated that the administration of progesterone before middle cerebral artery occlusion (MCAO) resulted in a marked reduction in cerebral infarction and reduced impairments that resulted from the occlusion [29]. Interestingly, postischemic administration of progesterone was also found to be protective [30, 31], and resulted in improvements in various functional measures, including the rotarod test, and adhesive-backed somatosensory and neurological scores [32]. The ability of progesterone to protect even after the insult (albeit within a relatively narrow window) may suggest that both rapid/immediate and long-term mechanisms of progesterone action are involved in the protective effects of progesterone. Progesterone has also been shown to reduce the amount of cell death following an acute episode of global ischemia [33], and is thought to be related to the ability of progesterone to reduce lipid peroxidation, the generation of isoprostanes [34], and the expression of proinflammatory genes [35].

Another model in which progesterone has been shown to exert protective effects is in the traumatic brain injury (TBI) model. The administration of progesterone reduces cerebral edema for up to 24 h after injury. In a rodent model of medial frontal cortex impact injury, progesterone reduced complement factor C3, glial fibrillary acidic protein (GFAP), and nuclear factor kappa beta (NFκB) [35], all of which can be interpreted as protective mechanisms. Progesterone also decreased the levels of lipid peroxidation in male rats when administered after TBI [36].

Interestingly, the severity of impairment following TBI appears to vary with the gender. Females appeared to have less spatial learning impairments when compared to their male counterparts. And though the lesion size was similar, females exhibited less ventricular dilation indicating lower edema and water retention [37]. In fact, direct assessments of edema revealed that progesterone treatment significantly attenuated the level of edema seen in injured animals in contrast to nonprogesterone treated animals that had undergone experimental TBI [38].

The protective effects of progesterone are also evident in other regions of the central nervous system in addition to the hippocampus and cerebral cortex. For example, progesterone has also been shown to have a beneficial effect on spinal cord contusion injuries as supported by the work of Thomas *et al.*, who found that there was a marked reduction in the size of the lesion and a prevention of secondary neuronal loss with progesterone treatment [39]. Further support for progesterone's protective actions in the spinal cord comes from the observation that progesterone has been shown to promote morphological and functional recovery in the Wobbler mouse, an animal model of spinal cord degeneration [40, 41]. Progesterone can also induce re-myelination as supported by the increased expression of myelin proteins in the damaged sciatic nerves of both young adult rats and in 22–24 month old males [42]. Thus, progesterone may be of potential therapeutic benefit in diseases where demyelination is an important component of its pathogenesis.

While the studies described above were all derived from animal models and cell/tissue culture models, it is worth mentioning that a relatively recently completed phase II, randomized, double-blind, placebo-controlled clinical trial assessing the efficacy of progesterone treatment for acute TBI yielded promising results. The data suggested that progesterone treatment can improve functional recovery, at least when administered to those who experienced moderate, but not severe, TBI [43].

3.5
Mechanisms Underlying Progesterone's Protective Effects

Numerous mechanisms of action likely underlie the protective effects of progesterone. The classical genomic mechanism of progesterone action, for example, may be involved in the regulation of neurotrophin expression [21], which in turn could promote cell survival. Alternatively, progesterone may act through novel receptor systems, such as the membrane PR, the sigma receptor, to activate certain signal transduction pathways, which in turn, triggers cellular events that are relevant and important for neuroprotection. Additionally, major metabolites of progesterone, such as allopregnanolone, have been reported to participate in the neuroprotective effects of progesterone [44].

With respect to "nongenomic" or cell signaling mechanisms underlying progesterone's protective effects, progesterone has been shown to elicit rapid effects on specific signaling pathways including the cAMP/PKA [7], MAPK (ERK1/2) [6, 8, 25], and the PI-3K/Akt pathway [8], all of which have been implicated in mediating neuroprotective effects. Progesterone-induced neuroprotection has not only been correlated with activation of the MAPK and Akt signaling pathways [25, 27] but has also been shown to depend on the activation of the MAPK pathway [21]. Activation of these signaling pathways, in turn, may also lead to increased expression of antiapoptotic proteins such as Bcl-2, as has been described by Nilsen and Brinton [25].

Another mechanism through which progesterone can exert protective effects is through its metabolites, which in turn, can interact with membrane-associated receptors coupled to ion-channels, such as the $GABA_A$ receptor system (see [22] for review). Such metabolites include allopregnanolone (or $3\alpha,5\alpha$-tetrahydroprogesterone), which bind to discrete sites within the hydrophobic domain of the $GABA_A$ receptor complex, and result in the potentiation of GABA-induced chloride conductance. Indeed, allopregnanolone has been suggested to play a role in mediating the protective effects of progesterone [45–50]. In addition to the effects of allopregnanolone on the $GABA_A$ receptor, as outlined above, allopregnanolone may also elicit its protective effects through its actions on the mitochondria [51]. For example, allopregnanolone was reported to inhibit currents associated with the opening of the mitochondrial permeability transition pore (mtPTP) [49], and as such, may help reduce the potential apoptotic consequences of mtPTP opening (such as cytochrome c release) during insult or injury.

Alternatively, the parent compound, progesterone, may also have nonallosteric influences on the $GABA_A$ receptor. Progesterone may influence the $GABA_A$ receptor via the activation of a signal transduction pathway, which in turn, influences GABA-gated currents through phosphorylation of discrete sites within certain subunits of the $GABA_A$ receptor [52, 53]. Since the regulation of the $GABA_A$ receptor has been shown to modulate cell survival, particularly in models of excitotoxicity, the regulation of the $GABA_A$ receptor by progesterone may be relevant to the protective effect of progesterone seen against kainate-induced seizure activity and subsequent cell death [54].

3.6
Medroxyprogesterone Acetate

The major form of progestin used in HT is MPA, a synthetic progestin derived from 17α-hydroxyprogesterone, which is often used in conjunction with estrogens to reduce the risk of certain cancers (e.g., cervical cancer) resulting from unopposed estrogen therapy [5, 55]. It is certainly clear that there are pharmacological and pharmacokinetic differences between MPA and progesterone. For example, orally administered MPA does not undergo any first pass effects [56], unlike progesterone. Furthermore, MPA has little binding affinity toward sex hormone binding globulin [56]. In addition to differences in bioavailability and half-life, MPA also displays many nonprogestagenic effects [56], including the ability of MPA to bind to the androgen receptor (AR) where it acts as a partial agonist [57] with a binding affinity (K_d) of approximately 2.1 nM [58]. Progesterone, in contrast, does not bind to the AR [56]. MPA can also bind to, and activate, glucocorticoid receptors [56, 59] with an effective concentration (EC_{50}), that is, nearly 300-fold lower than that for progesterone [59]. However, at this time, it is still unclear if any of the pharmacological/pharmacokinetic differences described above can explain differences in the biological actions of these two progestins, particularly as it relates to neuroprotection.

What is clear is that MPA appears to be ineffective as a neuroprotectant. For example, Nilsen and Brinton described that MPA was ineffective at protecting primary hippocampal neurons against glutamate toxicity, whereas progesterone was neuroprotective [25]. Mechanistically, the protective effects of progesterone appeared to be mediated, in part, by attenuating the glutamate-induced increase in intracellular Ca^{2+} levels. In contrast, MPA failed to alter the glutamate-induced influx of Ca^{2+}. Of significance was that MPA not only failed to elicit protective effects but also blocked the beneficial effect of estradiol. In sharp contrast, progesterone did not inhibit the effect of estradiol [26].

As stated earlier, progesterone's protective effects, in at least two neuronal models, were dependent on activation of the ERK/MAPK pathway. And while both progesterone and MPA can elicit ERK phosphorylation, only progesterone treatment resulted in nuclear translocation of ERK [27], the consequence of which is likely to regulate key genes, whose protein products may enable more long-term/sustainable protection. In fact, progesterone, but not MPA, increased the expression of the antiapoptotic Bcl-2 protein. And as observed in the model of glutamate-induced Ca^{2+} influx, MPA not only failed to increase the expression of Bcl-2, but actually inhibited that elicited by estradiol [26].

Another potential explanation of the disparity in the neuroprotective potential of progesterone and MPA is based on differences in their respective capacity to regulate BDNF. Our studies showed that while progesterone increases both the mRNA and protein levels of BDNF in the cerebral cortex, MPA treatment resulted in a substantial inhibition [20]. Combined with the observation that progesterone's protective effects may be dependent on neurotrophin signaling [21], this inhibition of BDNF expression by MPA may underscore the fact that MPA may not just be without effect, but may actually have adverse consequences on brain function.

The disparity between the effects of P4 and MPA has also been observed *in vivo*. For example, a study using rhesus monkeys illustrated that the combined treatment with estradiol and progesterone protects against coronary vasospasm, whereas treatment with estradiol and MPA did not [60]. And once again, in contrast to the antagonistic effects of MPA on estrogen's effects, progesterone enhanced the protective effects of estrogen against exercise-induced myocardial ischemia in postmenopausal women, whereas MPA did not [61]. Moreover, in a model of stroke (reversible focal stroke using the intraluminal filament model followed by 22 h of reperfusion), MPA diminished the protective effects of conjugated equine estrogens (CEEs) and MPA diminished estrogen's ability to reduce stroke damage. The functional antagonistic effects of MPA were also noted in the cholinergic system of monkeys, where MPA administered in conjunction with CEE reduced choline acetyl transferase (ChAT) in such cognition-relevant areas of the brain as the medial septum [62]. Similar consequences of MPA were seen in the cardiovascular system of cynomolgus monkeys. Adams *et al.* demonstrated that monkeys treated with CEE showed a 72% reduction in coronary artery atherosclerosis whereas, there were no benefits observed in CEE plus MPA group [63]. Interestingly, with regard to the TBI model, MPA required a larger dose than P4 to accomplish a comparable reduction in cerebral edema. However, regardless of the dose of MPA, MPA did not favor a better behavioral recovery than progesterone (reviewed in [64]).

While most studies have suggested that progesterone does not interfere with the beneficial effects of estrogens, some studies have noted that progesterone can antagonize the protective effects of estrogen. For example, Aguirre and Baudry showed that progesterone inhibited the neuroprotective effects of estradiol by antagonizing the effects of estradiol on BDNF in rat hippocampal cultures [65]. Further, Murphy and Segal demonstrated that progesterone antagonizes the effect of E2 on hippocampal spine density [66]. Further analysis of such issues that may influence the neuroprotective efficacy of progesterone when administered in conjunction with estrogen, which include timing of progesterone treatment relative to estrogen, dose of progesterone, and potential regional differences in the effects of combined estrogen/progesterone treatment (cerebral cortex vs hippocampus), to name a few, is needed to better understand the biological basis for such discrepancies.

Nevertheless, MPA appears to have the capacity to prevent estrogen's beneficial effects in the CNS consistently. As a consequence, these data require us to consider the possibility that some of the negative consequences of HT observed in the WHI trials may have been a result of the choice of progestin used in the HT regimen.

Collectively, the information presented here supports the conclusion that progesterone is protective and that such protection can be afforded through multiple mechanisms. While both the synthetic progestins (i.e., MPA) and the natural hormone, progesterone, can elicit similar effects in peripheral targets of these progestins (i.e., both can inhibit the uterotrophic effects of estrogen), it is clear that these hormones are not equivalent when considering their neuroprotective efficacies/potential. Such differences may be important in considering the results of the WHIMS studies which used MPA rather than progesterone, and, further, could provide critical insight into the development of the most effective therapeutic formulations for the treatment of menopause and various diseases whose incidence increases during the postmenopausal period.

Acknowledgments

Supported by NIH grants AG022550, AG023330, AG026672, AG027956, an Investigator-Initiated Research Grant (IIRG) from the Alzheimer's Association, and a generous gift from the Garvey Texas Foundation.

References

1. Strauss, J. and Barbieri, R. (2004) *Yen and Jaffe's Reproductive Endocrinology*, 5th edn, Saunders.
2. Conneely, O.M. and Lydon, J.P. (2000) Progesterone receptors in reproduction: functional impact of the A and B isoforms. *Steroids*, **65**, 571–577.
3. Vegeto, E., Shahbaz, M.M., Wen, D.X., Goldman, M.E., O'Malley, B.W., and McDonnell, D.P. (1993) Human progesterone receptor A form is a cell- and promoter-specific repressor of human progesterone receptor B function. *Mol. Endocrinol.*, **7**, 1244–1255.
4. Bikle, D.D., Halloran, B.P., Harris, S.T., and Portale, A.A. (1992) Progestin antagonism of estrogen stimulated 1,25-dihydroxyvitamin D levels. *J. Clin. Endocrinol. Metab.*, **75**, 519–523.
5. Hirvonen, E. (1996) Progestins. *Maturitas*, **23** (Suppl), S13–S18.
6. Migliaccio, A., Piccolo, D., Castoria, G., Di Domenico, M., Bilancio, A., Lombardi, M., Gong, W. *et al.* (1998) Activation of the Src/p21ras/Erk pathway by progesterone receptor via cross-talk with estrogen receptor. *EMBO J.*, **17**, 2008–2018.
7. Collado, M.L., Rodriguez-Manzo, G., and Cruz, M.L. (1985) Effect of progesterone upon adenylate cyclase activity and cAMP levels on brain areas. *Pharmacol. Biochem. Behav.*, **23**, 501–504.
8. Singh, M. (2001) Ovarian hormones elicit phosphorylation of Akt and extracellular-signal regulated kinase in explants of the cerebral cortex. *Endocrine*, **14**, 407–415.
9. Ke, F.C. and Ramirez, V.D. (1990) Binding of progesterone to nerve cell membranes of rat brain using progesterone conjugated to 125I-bovine serum albumin as a ligand. *J. Neurochem.*, **54**, 467–472.
10. Towle, A.C. and Sze, P.Y. (1983) Steroid binding to synaptic plasma membrane: differential binding of glucocorticoids and gonadal steroids. *J. Steroid. Biochem.*, **18**, 135–143.
11. Selmin, O., Lucier, G.W., Clark, G.C., Tritscher, A.M., Vanden Heuvel, J.P., Gastel, J.A., Walker, N.J. *et al.* (1996) Isolation and characterization of a novel gene induced by 2,3,7,8-tetrachlorodibenzo-p-dioxin in rat liver. *Carcinogenesis*, **17**, 2609–2615.
12. Krebs, C.J., Jarvis, E.D., Chan, J., Lydon, J.P., Ogawa, S., and Pfaff, D.W. (2000) A membrane-associated progesterone-binding protein, 25-Dx, is regulated by progesterone in brain regions involved in female reproductive behaviors. *Proc. Natl. Acad. Sci. U.S.A.*, **97**, 12816–12821.
13. Meyer, C., Schmid, R., Scriba, P.C., and Wehling, M. (1996) Purification and partial sequencing of high-affinity progesterone-binding site(s) from porcine liver membranes. *Eur. J. Biochem.*, **239**, 726–731.
14. Falkenstein, E., Schmieding, K., Lange, A., Meyer, C., Gerdes, D., Welsch, U., and Wehling, M. (1998) Localization of a putative progesterone membrane binding protein in porcine hepatocytes. *Cell. Mol. Biol. (Noisy-le-grand)*, **44**, 571–578.
15. Sakamoto, H., Ukena, K., Takemori, H., Okamoto, M., Kawata, M., and Tsutsui, K. (2004) Expression and localization of 25-Dx, a membrane-associated putative progesterone-binding protein, in the developing Purkinje cell. *Neuroscience*, **126**, 325–334.
16. Min, L., Takemori, H., Nonaka, Y., Katoh, Y., Doi, J., Horike, N., Osamu, H. *et al.* (2004) Characterization of the adrenal-specific antigen IZA (inner zone

17. Meffre, D., Delespierre, B., Gouezou, M., Leclerc, P., Vinson, G.P., Schumacher, M., Stein, D.G. et al. (2005) The membrane-associated progesterone-binding protein 25-Dx is expressed in brain regions involved in water homeostasis and is up-regulated after traumatic brain injury. *J. Neurochem.*, **93**, 1314–1326.
18. Zhu, Y., Bond, J., and Thomas, P. (2003) Identification, classification, and partial characterization of genes in humans and other vertebrates homologous to a fish membrane progestin receptor. *Proc. Natl. Acad. Sci. U.S.A.*, **100**, 2237–2242.
19. Zhu, Y., Rice, C.D., Pang, Y., Pace, M., and Thomas, P. (2003) Cloning, expression, and characterization of a membrane progestin receptor and evidence it is an intermediary in meiotic maturation of fish oocytes. *Proc. Natl. Acad. Sci. U.S.A.*, **100**, 2231–2236.
20. Jodhka, P.K., Kaur, P., Underwood, W., Lydon, J.P., and Singh, M. (2009) The differences in neuroprotective efficacy of progesterone and medroxyprogesterone acetate correlate with their effects on brain-derived neurotrophic factor expression. *Endocrinology*, **150**, 3162–3168.
21. Kaur, P., Jodhka, P.K., Underwood, W.A., Bowles, C.A., de Fiebre, N.C., de Fiebre, C.M., and Singh, M. (2007) Progesterone increases brain-derived neuroptrophic factor expression and protects against glutamate toxicity in a mitogen-activated protein kinase- and phosphoinositide-3 kinase-dependent manner in cerebral cortical explants. *J. Neurosci. Res.*, **85**, 2441–2449.
22. Deutsch, S.I., Mastropaolo, J., and Hitri, A. (1992) GABA-active steroids: endogenous modulators of GABA-gated chloride ion conductance. *Clin. Neuropharmacol.*, **15**, 352–364.
23. Seth, P., Fei, Y.J., Li, H.W., Huang, W., Leibach, F.H., and Ganapathy, V. (1998) Cloning and functional characterization of a sigma receptor from rat brain. *J. Neurochem.*, **70**, 922–931.
24. Maurice, T., Gregoire, C., and Espallergues, J. (2006) Neuro(active)steroids actions at the neuromodulatory sigma1 (sigma1) receptor: biochemical and physiological evidences, consequences in neuroprotection. *Pharmacol. Biochem. Behav.*, **84**, 581–597.
25. Nilsen, J. and Brinton, R.D. (2002) Impact of progestins on estrogen-induced neuroprotection: synergy by progesterone and 19-norprogesterone and antagonism by medroxyprogesterone acetate. *Endocrinology*, **143**, 205–212.
26. Nilsen, J. and Brinton, R.D. (2002) Impact of progestins on estradiol potentiation of the glutamate calcium response. *Neuroreport*, **13**, 825–830.
27. Nilsen, J. and Brinton, R.D. (2003) Divergent impact of progesterone and medroxyprogesterone acetate (Provera) on nuclear mitogen-activated protein kinase signaling. *Proc. Natl. Acad. Sci. U.S.A.*, **100**, 10506–10511.
28. Goodman, Y., Bruce, A.J., Cheng, B., and Mattson, M.P. (1996) Estrogens attenuate and corticosterone exacerbates excitotoxicity, oxidative injury, and amyloid beta-peptide toxicity in hippocampal neurons. *J. Neurochem.*, **66**, 1836–1844.
29. Jiang, N., Chopp, M., Stein, D., and Feit, H. (1996) Progesterone is neuroprotective after transient middle cerebral artery occlusion in male rats. *Brain Res.*, **735**, 101–107.
30. Kumon, Y., Kim, S.C., Tompkins, P., Stevens, A., Sakaki, S., and Loftus, C.M. (2000) Neuroprotective effect of postischemic administration of progesterone in spontaneously hypertensive rats with focal cerebral ischemia. *J. Neurosurg.*, **92**, 848–852.
31. Morali, G., Letechipia-Vallejo, G., Lopez-Loeza, E., Montes, P., Hernandez-Morales, L., and Cervantes, M. (2005) Post-ischemic administration of progesterone in rats exerts neuroprotective effects on the hippocampus. *Neurosci. Lett.*, **382**, 286–290.
32. Chen, J., Chopp, M., and Li, Y. (1999) Neuroprotective effects of progesterone after transient middle cerebral artery occlusion in rat. *J. Neurol. Sci.*, **171**, 24–30.
33. Cervantes, M., Gonzalez-Vidal, M.D., Ruelas, R., Escobar, A., and Morali, G.

(2002) Neuroprotective effects of progesterone on damage elicited by acute global cerebral ischemia in neurons of the caudate nucleus. *Arch. Med. Res.*, **33**, 6–14.

34. Roof, R.L., Hoffman, S.W., and Stein, D.G. (1997) Progesterone protects against lipid peroxidation following traumatic brain injury in rats. *Mol. Chem. Neuropathol.*, **31**, 1–11.

35. Pettus, E.H., Wright, D.W., Stein, D.G., and Hoffman, S.W. (2005) Progesterone treatment inhibits the inflammatory agents that accompany traumatic brain injury. *Brain Res.*, **1049**, 112–119.

36. Roof, R.L. and Hall, E.D. (2000) Gender differences in acute CNS trauma and stroke: neuroprotective effects of estrogen and progesterone. *J. Neurotrauma*, **17**, 367–388.

37. Attella, M.J., Nattinville, A., and Stein, D.G. (1987) Hormonal state affects recovery from frontal cortex lesions in adult female rats. *Behav. Neural. Biol.*, **48**, 352–367.

38. Roof, R.L., Duvdevani, R., Heyburn, J.W., and Stein, D.G. (1996) Progesterone rapidly decreases brain edema: treatment delayed up to 24 hours is still effective. *Exp. Neurol.*, **138**, 246–251.

39. Thomas, A.J., Nockels, R.P., Pan, H.Q., Shaffrey, C.I., and Chopp, M. (1999) Progesterone is neuroprotective after acute experimental spinal cord trauma in rats. *Spine*, **24**, 2134–2138.

40. Gonzalez Deniselle, M.C., Lopez Costa, J.J., Gonzalez, S.L., Labombarda, F., Garay, L., Guennoun, R., Schumacher, M. et al. (2002) Basis of progesterone protection in spinal cord neurodegeneration. *J. Steroid. Biochem. Mol. Biol.*, **83**, 199–209.

41. Gonzalez Deniselle, M.C., Lopez-Costa, J.J., Saavedra, J.P., Pietranera, L., Gonzalez, S.L., Garay, L., Guennoun, R. et al. (2002) Progesterone neuroprotection in the Wobbler mouse, a genetic model of spinal cord motor neuron disease. *Neurobiol. Dis.*, **11**, 457–468.

42. Ibanez, C., Shields, S.A., El-Etr, M., Leonelli, E., Magnaghi, V., Li, W.W., Sim, F.J. et al. (2003) Steroids and the reversal of age-associated changes in myelination and remyelination. *Prog. Neurobiol.*, **71**, 49–56.

43. Wright, D.W., Kellermann, A.L., Hertzberg, V.S., Clark, P.L., Frankel, M., Goldstein, F.C., Salomone, J.P. et al. (2007) ProTECT: a randomized clinical trial of progesterone for acute traumatic brain injury. *Ann. Emerg. Med.*, **49**, 391–402, 402e1–402e2.

44. Ciriza, I., Azcoitia, I., and Garcia-Segura, L.M. (2004) Reduced progesterone metabolites protect rat hippocampal neurones from kainic acid excitotoxicity in vivo. *J. Neuroendocrinol.*, **16**, 58–63.

45. Ardeshiri, A., Kelley, M.H., Korner, I.P., Hurn, P.D., and Herson, P.S. (2006) Mechanism of progesterone neuroprotection of rat cerebellar Purkinje cells following oxygen-glucose deprivation. *Eur. J. Neurosci.*, **24**, 2567–2574.

46. Djebaili, M., Hoffman, S.W., and Stein, D.G. (2004) Allopregnanolone and progesterone decrease cell death and cognitive deficits after a contusion of the rat pre-frontal cortex. *Neuroscience*, **123**, 349–359.

47. He, J., Evans, C.O., Hoffman, S.W., Oyesiku, N.M., and Stein, D.G. (2004) Progesterone and allopregnanolone reduce inflammatory cytokines after traumatic brain injury. *Exp. Neurol.*, **189**, 404–412.

48. He, J., Hoffman, S.W., and Stein, D.G. (2004) Allopregnanolone, a progesterone metabolite, enhances behavioral recovery and decreases neuronal loss after traumatic brain injury. *Restor. Neurol. Neurosci.*, **22**, 19–31.

49. Sayeed, I., Parvez, S., Wali, B., Siemen, D., and Stein, D.G. (2009) Direct inhibition of the mitochondrial permeability transition pore: a possible mechanism for better neuroprotective effects of allopregnanolone over progesterone. *Brain Res.*, **1263**, 165–173.

50. Vitarbo, E.A., Chatzipanteli, K., Kinoshita, K., Truettner, J.S., Alonso, O.F., and Dietrich, W.D. (2004) Tumor necrosis factor alpha expression and protein levels after fluid percussion injury in rats: the effect of injury severity and brain temperature. *Neurosurgery*, **55**, 416–424; discussion 424–425.

51. Robertson, C.L., Puskar, A., Hoffman, G.E., Murphy, A.Z., Saraswati, M., and Fiskum, G. (2006) Physiologic progesterone reduces mitochondrial dysfunction and hippocampal cell loss after traumatic brain injury in female rats. *Exp. Neurol.*, **197**, 235–243.

52. Bell-Horner, C.L., Dohi, A., Nguyen, Q., Dillon, G.H., and Singh, M. (2006) ERK/MAPK pathway regulates GABAA receptors. *J. Neurobiol.*, **66**, 1467–1474.

53. Vasan, R., Vali, M., Bell-Horner, C., Kaur, P., Dillon, G.H., and Singh, M. (2003) Regulation of the GABA-A receptor by the MAPK pathway and progesterone. 33rd Annual Society for Neuroscience Meeting, New Orleans, LA, p. 472.12.

54. Hoffman, G.E., Moore, N., Fiskum, G., and Murphy, A.Z. (2003) Ovarian steroid modulation of seizure severity and hippocampal cell death after kainic acid treatment. *Exp. Neurol.*, **182**, 124–134.

55. Gambrell, R.D. Jr. (1986) The role of hormones in the etiology and prevention of endometrial cancer. *Clin. Obstet. Gynaecol.*, **13**, 695–723.

56. Schindler, A.E., Campagnoli, C., Druckmann, R., Huber, J., Pasqualini, J.R., Schweppe, K.W., and Thijssen, J.H. (2003) Classification and pharmacology of progestins. *Maturitas*, **46** (Suppl 1), S7–S16.

57. Winneker, R.C., Bitran, D., and Zhang, Z. (2003) The preclinical biology of a new potent and selective progestin: trimegestone. *Steroids*, **68**, 915–920.

58. Hackenberg, R., Hofmann, J., Wolff, G., Holzel, F., and Schulz, K.D. (1990) Down-regulation of androgen receptor by progestins and interference with estrogenic or androgenic stimulation of mammary carcinoma cell growth. *J. Cancer Res. Clin. Oncol.*, **116**, 492–498.

59. Koubovec, D., Ronacher, K., Stubsrud, E., Louw, A., and Hapgood, J.P. (2005) Synthetic progestins used in HRT have different glucocorticoid agonist properties. *Mol. Cell. Endocrinol.*, **242**, 23–32.

60. Miyagawa, K., Vidgoff, J., and Hermsmeyer, K. (1997) Ca^{2+} release mechanism of primate drug-induced coronary vasospasm. *Am. J. Physiol.*, **272**, H2645–H2654.

61. Rosano, G.M., Webb, C.M., Chierchia, S., Morgani, G.L., Gabraele, M., Sarrel, P.M., de Ziegler, D. *et al.* (2000) Natural progesterone, but not medroxyprogesterone acetate, enhances the beneficial effect of estrogen on exercise-induced myocardial ischemia in postmenopausal women. *J. Am. Coll. Cardiol.*, **36**, 2154–2159.

62. Gibbs, R.B., Nelson, D., Anthony, M.S., and Clarkson, T.B. (2002) Effects of long-term hormone replacement and of tibolone on choline acetyltransferase and acetylcholinesterase activities in the brains of ovariectomized, cynomologus monkeys. *Neuroscience*, **113**, 907–914.

63. Adams, M.R., Register, T.C., Golden, D.L., Wagner, J.D., and Williams, J.K. (1997) Medroxyprogesterone acetate antagonizes inhibitory effects of conjugated equine estrogens on coronary artery atherosclerosis. *Arterioscler. Thromb. Vasc. Biol.*, **17**, 217–221.

64. Stein, D.G. (2005) The case for progesterone. *Ann. N. Y. Acad. Sci.*, **1052**, 152–169.

65. Aguirre, C.C. and Baudry, M. (2009) Progesterone reverses 17 beta-estradiol-mediated neuroprotection and BDNF induction in cultured hippocampal slices. *Eur. J. Neurosci.*, **29**, 447–454.

66. Murphy, D.D. and Segal, M. (2000) Progesterone prevents estradiol-induced dendritic spine formation in cultured hippocampal neurons. *Neuroendocrinology*, **72**, 133–143.

4
Endogenous and Synthetic Neurosteroids in the Treatment of Niemann–Pick Type C Disease

Synthia H. Mellon, Wenhui Gong, and Marcus D. Schonemann

4.1
Introduction

Neurosteroids are steroids synthesized *de novo* in the brain, or are steroids derived from circulation, which are converted into neuroactive compounds in the brain. In the 1980s, Etienne Baulieu coined the term *neurosteroid* both to distinguish this class of steroids from glucocorticoids, mineralocorticoids, and sex steroids [1, 2], whose essential functions had been well understood for decades, as well as to identify these steroids as endogenously produced in the brain. While neurosteroids were being identified in animal brains, other laboratories were identifying functions for neurosteroids, distinct from their classic and well-appreciated functions as ligands for nuclear receptors [3, 4]. Since the initial identification of neurosteroids in rodent brains, the identification of the steroids in the brains of many species has demonstrated a remarkable similarity in temporal and regional localization. Functions associated with these neuroactive compounds have also been identified (reviewed in [1, 5–26]), and new functions are still being discovered. Despite this increasing list of functions of neurosteroids in both *in vitro* and *in vivo* systems, it is still unknown if these compounds are essential for life.

The synthesis of steroids and neurosteroids from cholesterol occurs through enzymatic steps that are not unique to gonads, adrenals, placenta, or brain [1, 27–31]. The presence or absence of particular steroidogenic enzymes and cofactors dictates which steroid(s) a steroidogenic organ or cell type will produce.

One of the first functions demonstrated for allopregnanolone was its role in augmenting GABA-evoked effects at $GABA_A$ receptors [21]. Owing to specific binding on $GABA_A$ receptors [32] and $GABA_A$ subunits [33] and enhancing neuronal inhibition, allopregnanolone exhibits anxiolytic, anticonvulsant, and anesthetic actions (reviewed in [15, 23, 34, 35]). The neurosteroid pregnenolone, a precursor of allopregnanolone, enhances memory when given intrathecally [36–40]. In rodent models of alcohol intoxication, one mechanism through which alcohol elicits its effects is through increased synthesis of allopregnanolone in the brain [17, 41–47]. Finally, the neurosteroid allopregnanolone has been implicated in a severe form of premenstrual disorder, called *premenstrual dysphoric disorder* [10, 16, 48–56].

Most recently, changes in GABA$_A$ receptor subunit expression and sensitivity to the neurosteroid allopregnanolone have been implicated in changes in behavioral responses seen at puberty [57].

In addition to their behavioral effects, neurosteroids have also been implicated in affecting neuronal function and differentiation [24]. These include neuroprotection against ischemia and stroke [58–64], recovery of motor function after spinal cord injury [65–70], regulation of myelination [13, 19, 71–76], proliferation of neuronal stem cells [77, 78], neurogenesis in the hippocampus [77–82], and induction of analgesia [83, 84]. Many of these actions of neurosteroids are discussed in detail in other chapters of this book.

We have taken several different approaches to understanding the developmental roles of neurosteroids *in vivo*. Several transgenic mouse lines have been created, in which genes encoding neurosteroidogenic enzymes have been ablated. These mouse lines include ablation of the P450scc [85], P450c17 [86], 5α-reductase type I, and 5α-reductase type II [87–89]. Among these knockout mice, the P450c17 knockout mice are unexpectedly embryonic lethal, and die at embryonic day 7, before gastrulation. P450scc mice lack glucocorticoid production and need replacement at birth. Female mice lacking 5α-reductase type I and type II exhibit parturition and fecundity defects similar to those of animals without 5α-reductase type 1; male mice are phenotypically relatively normal, and the data from the knockout mice indicate that testosterone appears to be the only androgen required for differentiation of the male urogenital tract in mice and the synthesis of dihydrotestosterone serves largely as a signal amplification mechanism. Thus, global ablation of these four steroidogenic enzyme genes produced no neural phenotypes and provided no insights into the neural function of their neurosteroid products. Alternatively, these transgenic knockout mice may suggest that the neurosteroids do not play obligate and unique roles in the nervous system.

To identify regions and cells of the nervous system that express the neurosteroidogenic enzymes, promoter–reporter constructs can be prepared using the promoters for the genes encoding steroidogenic enzymes. Studies using a P450scc-cre reporter [90] showed that the P450scc promoter is expressed in the cortex, hippocampus, thalamus, and hypothalamus (dorsomedial ventromedial hypothalamus and arcuate nucleus). We prepared P450c17-Green fluorescent protein (GFP) transgenic mice, using a 1.5 kb promoter region of the rat *P450c17* gene. In the brain, we found GFP expression in axonal tracts projecting rostrally from the midbrain and consolidating in the subcortical plate of the embryonic cortex, similar to the expression of endogenous P450c17 that we identified [91]. We also identified neuronal projections traversing along the dorsal–ventral axis connecting the spinal cord and the brainstem, extending basally toward the medulla and in condensed ganglia of the dorsal root. Like the expression of endogenous P45017, much of the expression of the GFP reporter was identified in fibers. Recently, a transgenic mouse harboring a cyp17-iCRE construct was reported [92]; however, neural expression of Cre recombinase was not reported. These transgenic mice, and others that rely on promoters of steroidogenic genes to target expression of reporter genes, may be useful in identifying and defining neurosteroidogenic neurons and glia throughout development, and under different physiologic conditions.

Another strategy for understanding neurosteroid function *in vivo* is to identify existing mouse lines that may have altered neurosteroid production. We took this approach, and identified a mouse line for a childhood neurodegenerative disease, Niemann–Pick type C (NP-C). We believe that this mouse provides us with a useful model for understanding the consequences of altered neurosteroidogenesis [93].

4.2 Niemann–Pick Type C Disease as a Model of Disrupted Neurosteroidogenesis

NP-C disease is a fatal autosomal recessive, childhood-onset, neurodegenerative disorder. This lysosomal lipid storage disorder is characterized by a defect in intracellular cholesterol trafficking, resulting in lysosomal accumulation of unesterified cholesterol and other materials that traffic through the same intracellular pathway (reviewed in [94]). The accumulation of cholesterol causes hepatomegaly with foamy macrophage infiltration, and chronic neurologic deterioration associated with accumulation of sphingomyelin and other glycolipids in neuronal tissues, leading to seizures, supranuclear ophthalmoplegia, and progressive loss of motor and intellectual function in the second decade of life [95, 96]. NP-C has been linked to two genetic loci, *NPC1* (major locus) and *NPC2* [94, 97–100]. The human *NPC1* gene encodes a protein of 1278 amino acids [101], containing 13 transmembrane domains and 3 large luminal loops [102], which share homology with other proteins that regulate cholesterol homeostasis, including 3-hydroxy-3-methylglutaryl-coenzyme A (HMG CoA) reductase, sterol regulatory element binding protein cleavage-activating protein (SCAP), with Patched, the receptor for sonic hedgehog [103], and the resistance-nodulation division (RND) family of prokaryotic permeases, [104]. More recent studies have demonstrated a cholesterol and oxysterol amino terminal binding domain in the first luminal loop of NPC1 protein [102, 105, 106]. More than 95% cases of NP-C are caused by mutations in *NPC1* [101, 107]. NPC2, first identified as human epididymal protein 1 (HE1), is a widely expressed 151-amino acid lysosomal glycoprotein that binds cholesterol (but not oxysterols) [105]. About 5% of NP-C is caused by mutations in *NPC2*. NP-C patients from both complementation groups have similar clinical and biochemical phenotypes, originally suggesting that NPC1 and NPC2 may interact or function sequentially in a common metabolic pathway. Recent studies have shown mechanistically and structurally how cholesterol is transferred from NPC2 to NPC1 [106].

Much of the work on the NPC1 protein, neuronal histology, and cholesterol utilization has come from the mouse model of NP-C [108], a strain of BALB/c mice with a retroposon insertion in *NPC1* [103, 108]. These mice have defects in cholesterol metabolism morphologically and biochemically similar to human NP-C, and show most of the same neurological phenotypes as human beings with NP-C, although the neurological demise is much more rapid in the mouse than in human beings. Nevertheless, both the murine model and patients with NP-C show similar widespread histopathological abnormalities in the central and peripheral nervous systems, including cerebellar degeneration [108, 109], Purkinje cell degeneration, irregular dendritic trees, decreased numbers of dendritic spines [110], and progressive dysmyelination of the CNS [111–114], and reduction in myelin seen by diffusion tensor imaging [115] suggesting progressively defective utilization of cholesterol. The mechanisms of neuronal dysfunction and degeneration are not fully understood. Cholesterol content does not appear to be elevated in cortical neurons, even though these cells exhibit neuronal storage abnormalities [116, 117], and the rate of sterol synthesis and loss is lower in NP-C mice [114]. Human NP-C brains have cortical neurons with distended cytoplasm, ballooned neurons [96, 118, 119], and neurofibrillary tangles [120]. Cholesterol and sphingomyelin are decreased in white matter due to demyelination [114, 119]. In addition to accumulating cholesterol, cells from NP-C mice also accumulate gangliosides and glycosphingolipids [121].

4.3
Steroidogenesis and Neurosteroidogenesis in NP-C

In addition to abnormal cholesterol trafficking in NP-C neurons, NP-C mice also show abnormalities in testicular steroidogenesis [122] and ovarian steroidogenesis [123], although plasma concentrations of progesterone, corticosterone, or testosterone were not reduced [124]. Since we believe that neurosteroids are necessary for neuronal and glial function, we hypothesized that alterations in sequestration of intracellular cholesterol would result in altered neurosteroidogenesis, which we hypothesize would subsequently alter neuronal and glial function. Adult NP-C mouse brains showed significantly reduced concentrations of pregnenolone, dehydroepiandrosterone (DHEA), and allopregnanolone in comparison to brains of age-matched wild-type mice [93].

We determined whether neurons and glia that express the steroidogenic enzymes required for allopregnanolone production are also diminished in brains of NP-C mice. Using immunohistochemistry, we found that adult NP-C brains had significantly diminished expression of P450scc, 3-beta hydroxysteroid dehydrogenase (3β-HSD), 5α-reductase, and 3α-HSD. This reduction in expression of these neurosteroidogenic enzymes was seen in the cortex and in the cerebellum. In the cerebellum, Purkinje neurons that express these enzymes are lost in the adult NP-C mouse. In the cortex, it is unknown which particular neurons or glia express these enzymes. Thus, it is unknown if those cells are also lost in NP-C mouse brains, or if there is a reduction in expression of neurosteroidogenic enzymes.

Analysis of neurosteroidogenic enzyme activity throughout the life of the NP-C mouse showed that neurosteroidogenesis (allopregnanolone production) is normal at least at the end of gestation. However, at birth, we found that NP-C mouse brains had a significant reduction in 3α-HSD activity, and hence could not convert dihydroprogesterone to allopregnanolone. While not explicitly tested as substrates in these studies, it is also likely that NP-C mouse brains cannot convert corticosterone to tetrahydrodeoxycorticosterone, or testosterone to androstenediol, two other neuroactive steroids. These conversions use the same enzymes as those used to convert progesterone to allopregnanolone (i.e., 5α-reductase and 3α-HSD).

The diminution in enzyme activity was seen in the cortex, midbrain, and hindbrain, indicating that there was not region-specific reduction in enzymatic activity. Several weeks later, we also demonstrated a significant reduction in 5α-reductase activity (conversion of progesterone to dihydroprogesterone) in all brain regions. This reduction in 5α-reductase and 3α-HSD activities preceded, by several weeks, the onset of behavioral symptoms of ataxia, tremor, and weight loss.

4.4
Treatment of NP-C Mice with Allopregnanolone

If the loss of allopregnanolone production contributed to the neuropathology of NP-C, we reasoned that appropriately timed treatment of NP-C mice with allopregnanolone should reduce the symptoms and pathology seen in untreated NP-C mice. We tried several approaches to treatment with allopregnanolone – providing the neurosteroid in drinking water, as a timed-release pellet, and as an injection, and each treatment had some efficacy [93]. Efficacy was assessed by survival of mice, by time of onset of neurological symptoms, as well as by weekly assessment of locomotor function and coordination. All these markers of effective treatment changed in parallel with effective treatment; that is, effective treatment delayed weight loss and onset of ataxia and tremor, prolonged locomotor function, and increased survival.

We reasoned that since there was a reduction in neurosteroidogenic enzyme activity in the early neonatal period, the lack of allopregnanolone at that time might be crucial for appropriate brain development. Hence, we treated mice during the first two weeks of life [93]. Our results indicated that a single injection of allopregnanolone beginning at weaning (postnatal day 23) or earlier (from postnatal day 7 to postnatal day 23) was effective, and that efficacy depended upon the day at which treatment was given. Treatment at postnatal day 7 was the most effective time to treat NP-C mice, and resulted in a doubling of lifespan, a five week delay in loss of locomotor function, and a three to four week delay in onset of symptoms. Other groups have also reproduced these results, demonstrating similar effects of allopregnanolone treatment on their colonies of NP-C mice [125–128]. Additional studies treating mice at postnatal day 0 or day 3 indicated that treatment at these times was less effective than treatment at day 7 (unpublished results), again suggesting a time-dependency to treatment.

Allopregnanolone treatment significantly increased survival and locomotor function in NP-C mice. We assessed whether there were changes in neuronal survival in the mice, as allopregnanolone had been shown to increase hippocampal neurogenesis [77]. Untreated NP-C mice have substantial loss of cerebellar Purkinje neurons at the end of life (~60 days). Analysis of brains of NP-C mice treated with allopregnanolone at postnatal day 7 indicated that those mice had substantial survival of cerebellar Purkinje neurons, which was seen in all lobes of the cerebellum [93, 126]. *In vitro*, allopregnanolone, but not its 3β-stereoisomer, also increased Purkinje cell survival [93], through a mechanism involving $GABA_A$ receptors, since the $GABA_A$ channel blocker, bicuculline, inhibited this effect. It is not known if this effect is due to increased proliferation or reduced apoptosis, but others have shown that allopregnanolone increases cerebellar neurogenesis *in vitro* by increasing calcium influx through voltage-gated calcium channels and $GABA_A$ receptor activation [79], similar to that seen for effects on hippocampal neurogenesis.

4.5
Mechanism of Allopregnanolone Action: $GABA_A$ Receptor

One mechanism by which allopregnanolone functions is by augmentation of the $GABA_A$ receptor channel opening, through alteration of the kinetics of entry to and exit from desensitized states of the receptor [129, 130]. We used another GABAergic neurosteroid, ganaxolone, to treat NP-C mice. Ganaxolone is a C3-β-methyl derivative of allopregnanolone, and was developed as an orally effective neurosteroid, presumably because of the inhibition of C3-hydroxyl oxidation due to the presence of the methyl group [131–134]. Pharmacokinetic and *in vivo* studies have shown that ganaxolone has greater efficacy than allopregnanolone at $GABA_A$ receptors [131–141]. We treated NP-C mice at postnatal day 7, using the same dose of ganaxolone as we used for allopregnanolone (25 mg kg^{-1}), and assessed onset of symptoms, locomotor function and coordination, and survival of mice. Data for survival are shown in Figure 4.1a. Ganaxolone-treated mice had a significant increase in longevity (average of 98 ± 23 days vs 71 ± 11 days in untreated NP-C mice or 80 ± 15 days in β-cyclodextrin-treated mice). However, this increased longevity was not as great as that seen with allopregnanolone treatment (allopregnanolone, 114 ± 16 days vs ganaxolone 98 ± 23 days, Figure 4.1a). In addition to increasing longevity, ganaxolone treatment resulted in a significant delay in tremor, ataxia, and weight loss in treated NP-C mice. In parallel with increased survival, ganaxolone treatment significantly delayed loss of locomotor function, which was much more gradual in ganaxolone-treated mice than it was in untreated mice. However, in comparison to allopregnanolone-treated mice, ganaxolone-treated mice had an earlier loss of locomotor function and motor coordination. These results indicate that ganaxolone, a synthetic neurosteroid, which is being developed clinically for treatment of pediatric seizure disorders, is an effective treatment in NP-C mice. However, the differences we found in the effects of ganaxolone and allopregnanolone treatment suggest that the mechanism(s) by which allopregnanolone elicits its effects is not completely identical to that of ganaxolone. Since ganaxolone is thought to elicit its effects solely through $GABA_A$-mediated mechanisms, our data suggest that in addition to affecting $GABA_A$ receptors, allopregnanolone likely has other mechanisms of action *in vivo* that are not mimicked in entirety by ganaxolone. Treatment

4.5 Mechanism of Allopregnanolone Action: GABA$_A$ Receptor

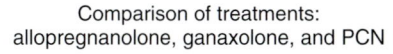

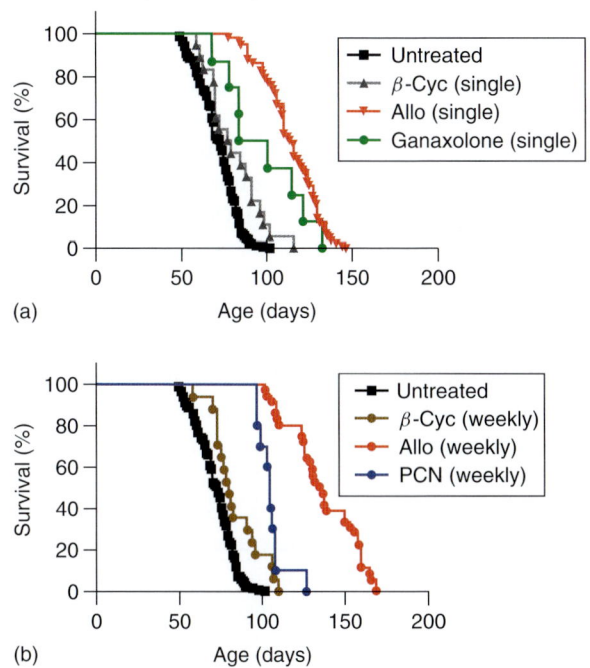

Figure 4.1 Effect of allopregnanolone and synthetic neurosteroids on the survival of NP-C mice. (a) Allopregnanolone (Allo) or ganaxolone (25 mg kg^{-1} in 20% hydroxypropyl β-cyclodextrin), β-cyclodextrin, or nothing (untreated) was administered subcutaneously in a single injection at postnatal day 7. (b) Allopregnanolone (Allo) or pregnenolone-16α-carbonitrile (PCN, 25 mg kg^{-1} in 20% β-cyclodextrin), β-cyclodextrin, or nothing (untreated) was administered subcutaneously once weekly, beginning at postnatal day 7. Data are means ± SD $N = 168$ for untreated, $n = 18$ for β-cyclodextrin-treated (singly) or $n = 17$ for β-cyclodextrin-treated once weekly, $n = 58$ for allopregnanolone-treated mice; $n = 8$ for ganaxolone-treated mice, $n = 10$ for PCN-treated mice.

of NP-C mice with other ligands of the GABA$_A$ receptor, such as benzodiazepines, has not been tested.

Another reason that ganaxolone may not be as effective as allopregnanolone may be due to ganaxolone's inability to be metabolized to other neuroactive compounds, as is possible for allopregnanolone. Allopregnanolone can be converted to dihydroprogesterone (5α-pregnan-3,20-dione) by the enzyme 3α-hydroxysteroid dehydrogenase. This is a reversible enzymatic reaction, although the reduction (production of allopregnanolone) is favored enzymatically. Dihydroprogesterone, unlike allopregnanolone, is active at nuclear progesterone receptors. Because of the 3β-methyl group on ganaxolone, it cannot be converted to a similar compound. Hence, some of the additional benefits of allopregnanolone may be due to its metabolism to other neuroactive compounds.

4.6
Mechanism of Allopregnanolone Action: Pregnane-X Receptor

Allopregnanolone may also elicit its effects through pregnane-X receptors (PXRs) [142–145]. This activation requires micromolar concentrations of allopregnanolone in *in vitro* assays. (By contrast, allopregnanolone activates $GABA_A$ receptors at nanomolar concentrations [21].) PXR is found mainly in the liver, where its target genes include cytochrome P450s. Allopregnanolone treatment of NP-C mice increased expression of cyp3A13 mRNA in the brain, a PXR target gene, which occurred within 24 h of treatment, and persisted for 28 days. Thus, induction of PXR may be an additional mechanism through which allopregnanolone elicits its effects, and results in beneficial treatment of NP-C.

We determined if another PXR ligand, pregnenolone-16α-carbonitrile (PCN), could also affect NP-C mouse survival and neurologic function. We treated NP-C mice with a dose of PCN identical to that used for allopregnanolone or ganaxolone (25 mg kg^{-1} in 20% β-cyclodextrin), beginning at postnatal day 7, and likewise evaluated the mice for onset of neurological symptoms and weight gain/loss. PCN treatment resulted in increased survival (average survival = 105 ± 9 days vs 71 ± 11 days untreated) (Figure 4.1b). PCN-treated mice also have delayed onset of tremor, ataxia, and weight loss, and delayed loss of locomotor function. However, when compared with the results from allopregnanolone treatment, PCN treatment was not as effective (average survival: PCN, 105 ± 9 days vs allopregnanolone, 137 ± 20 days). Differences may be due to pharmacologic effects such as suboptimal dosing of PCN or bioavailability of PCN in the brain. Alternatively, the differences in efficacy of PCN versus allopregnanolone may be due to different biological actions of PCN versus allopregnanolone, and may suggest that the mechanisms by which allopregnanolone elicits its effects are not limited to activation of PXR for which PCN is an optimal ligand.

4.7
Mechanism of Allopregnanolone Action: Reduction of Cellular Oxidative Stress

Analysis of fibroblasts from patients with NP-C suggested that there were significant increases in reactive oxygen species (ROS) and in lipid peroxidation when compared with fibroblasts from control subjects. In addition, NP-C fibroblasts were more susceptible to apoptosis, in a process mediated via activation of NF-κB. Allopregnanolone treatment of fibroblasts from NP-C patients reduced accumulation of ROS, and reduced apoptosis. These effects on reducing ROS were rapid, occurring within 30 min, and were not due to allopregnanolone acting as a free-radical scavenger. As fibroblasts lack $GABA_A$ receptors, these effects were not mediated through that receptor. The reduction in apoptosis may be mediated through increased cytoplasmic IκB and decreased nuclear NF-κB. It is not known how this occurs, but suggests additional targets for allopregnanolone action.

4.8
Conclusions – Mechanisms of Allopregnanolone Action in Treatment of NP-C and Other Neurodegenerative Diseases

The results from these studies have demonstrated that a lack of allopregnanolone synthesis in the early neonatal period may contribute to the neuropathology seen in NP-C mice. The treatment studies suggest that allopregnanolone may function in the early postnatal period in the brain of mice, and that allopregnanolone's actions in NP-C mice may be time specific. These actions may be related to specific cellular development, proliferation, and migration. The action of allopregnanolone on reducing cellular accumulation of gangliosides may likewise contribute to the beneficial effects of allopregnanolone treatment. The mechanism(s) through which allopregnanolone functions in NP-C is unknown. Our studies, using the synthetic $GABA_A$ receptor ligand ganaxolone, suggest that this receptor may certainly play a role in the beneficial actions of allopregnanolone. Indeed, we have shown that allopregnanolone can mediate beneficial actions in cultured Purkinje neurons that are mediated through $GABA_A$ receptors. Effective treatment of NP-C mice with PCN and activation of PXR-regulated genes in the brains of allopregnanolone-treated mice also suggest that the PXR may mediate some of the effects of allopregnanolone.

Studies from many laboratories have provided evidence for additional cellular mechanisms of allopregnanolone action. These effects are depicted in a cartoon in Figure 4.2. First and foremost, the receptors mediating the intracellular effects of allopregnanolone are not completely known. Our studies in NP-C mice, and studies from others using additional rodent models, have demonstrated effects of allopregnanolone that can be attributed to $GABA_A$ and PXR receptors. Allopregnanolone also affects voltage-gated calcium channels, coupled to $GABA_A$ receptors. These responses evoke intracellular effects on cell cycle genes, thereby resulting in increased cellular proliferation [146, 147]. Allopregnanolone may also mediate its proliferative effects through the membrane progesterone receptor, progesterone receptor membrane component 1 (PGRMC1) [81].

Other studies by our group demonstrated that allopregnanolone has a broad effect on reducing cellular oxidative stress and consequent apoptosis [152], which may be some of the mechanisms by which allopregnanolone is effective in NP-C. The receptor(s) initiating these effects on reducing cellular ROS and reducing activation of NF-κB remain to be determined.

Others have also demonstrated effects of allopregnanolone on phosphorylation of Akt (protein kinase B, PKB) in P19 neurons, which may be mediated through either stimulation of $GABA_A$ receptors or inhibition of NMDA receptors [150, 151]. Allopregnanolone treatment of P19 neurons reduced apoptosis by maintaining activation of Akt kinase, and interfering with the mitochondrial apoptotic pathway, preserving cytochrome c in the mitochondria and Bax in the cytoplasm. Similar effects of allopregnanolone on reducing caspase 3 activation and increasing cytoplasmic Bax were also found in another model of brain injury [59]. These effects of allopregnanolone may be through inhibition of the mitochondrial permeability

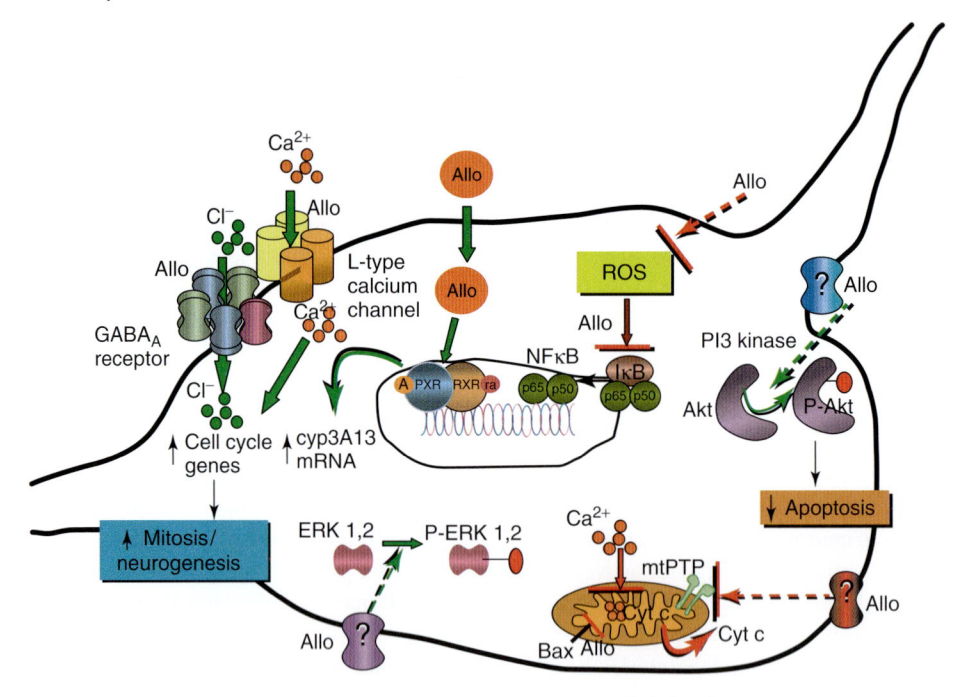

Figure 4.2 Potential mechanisms of allopregnanolone action. Allopregnanolone increases duration of GABA$_A$ receptor channel opening [3], augments calcium entry into the cell through L-type calcium channels [77, 147], and induces increases in cell cycle proteins, and increased neurogenesis [77, 147]. Allopregnanolone also elicits effects through pregnane-X receptors, to increase expression of cytochrome P450s in the liver [145] and brain [126]. Allopregnanolone reduces intracellular reactive oxygen species (ROS) through mechanisms involving reducing NF-κB signaling [148] and augmenting mitochondrial function, by reducing calcium entry into mitochondria; inhibiting the mitochondrial permeability transition pore [148, 149], inhibiting caspase 3 activation, and Bax entry into mitochondria; inhibiting cytochrome c release; and reducing apoptotic DNA fragmentation [59]. Allopregnanolone may elicit some effects through the Akt [150, 151] or ERK pathways (our unpublished data).

transition pore currents, resulting in inhibition of mitochondrial swelling due to Ca^{2+} entry, and prevention of mitochondrial cytochrome c release [149]. Since these effects are seen in uninjured cultured neurons and in models of neurotrauma, these data further suggest that these beneficial effects of allopregnanolone may be broad based.

Understanding the mechanisms through which allopregnanolone functions in normal brain development would greatly assist our understanding of its reparative potential in neurodegenerative disorders. Our data and that of others [82] clearly demonstrate that allopregnanolone has additional effects in neuroprotection in animal models of human neurodegenerative diseases. Elucidating these functions will unlock its full potential in the treatment of a wide variety of brain disorders.

Acknowledgments

This work was supported by grants **NIH NS 049462, NSF 0090995, March of Dimes FY06-340/FY079-554**, grants from The Lysosomal Storage Disease Research Consortium and The National MPS Society to SHM, and a fellowship from **NIH Training Grant DK07161** to MDS. We thank Dr. Kelvin Gee, University of California, Irvine, for the ganaxolone.

References

1. Compagnone, N.A. and Mellon, S.H. (2000) Neurosteroids: Biosynthesis and function of these novel neuromodulators. *Front. Neuroendocrinol.*, **21**, 1–58.
2. Baulieu, E. E., Robel, P., and Schumacher, M. (eds) (1999) *Neurosteroids: A New Regulatory Function in the Nervous System*, Human Press, Inc., Totowa, NJ.
3. Majewska, M.D., Harrison, N.L., and Schwartz, R.D. (1986) Steroid hormone metabolites are barbiturate-like modulators of the GABA receptor. *Science*, **232**, 1004–1007.
4. Harrison, N.L. and Simmonds, M.A. (1984) Modulation of GABA receptor complex by a steroid anesthetic. *Brain Res.*, **323**, 284–293.
5. Mensah-Nyagan, A.G., Beaujean, D., Luu-The, V., Pelletier, G., and Vaudry, H. (2001) Anatomical and biochemical evidence for the synthesis of unconjugated and sulfated neurosteroids in amphibians. *Brain Res. Brain Res. Rev.*, **37**, 13–24.
6. Morrow, A.L., VanDoren, M.J., Fleming, R., and Penland, S. (2001) Ethanol and neurosteroid interactions in the brain. *Int. Rev. Neurobiol.*, **46**, 349–377.
7. Vallee, M., Mayo, W., Koob, G.F., and Le Moal, M. (2001) Neurosteroids in learning and memory processes. *Int. Rev. Neurobiol.*, **46**, 273–320.
8. Reddy, D.S. (2002) The clinical potentials of endogenous neurosteroids. *Drugs Today (Barc)*, **38**, 465–485.
9. Rogawski, M.A. and Reddy, D.S. (2002) Neurosteroids and infantile spasms: the deoxycorticosterone hypothesis. *Int. Rev. Neurobiol.*, **49**, 199–219.
10. Backstrom, T., Andersson, A., Andree, L., Birzniece, V., Bixo, M., Bjorn, I., Haage, D., Isaksson, M., Johansson, I.M., Lindblad, C., Lundgren, P., Nyberg, S., Odmark, I.S., Stromberg, J., Sundstrom-Poromaa, I., Turkmen, S., Wahlstrom, G., Wang, M., Wihlback, A.C., Zhu, D., and Zingmark, E. (2003) Pathogenesis in menstrual cycle-linked CNS disorders. *Ann. N. Y. Acad. Sci.*, **1007**, 42–53.
11. Guarneri, P., Cascio, C., Russo, D., D'Agostino, S., Drago, G., Galizzi, G., De Leo, G., Piccoli, F., Guarneri, M., and Guarneri, R. (2003) Neurosteroids in the retina: neurodegenerative and neuroprotective agents in retinal degeneration. *Ann. N. Y. Acad. Sci.*, **1007**, 117–128.
12. Lambert, J.J., Belelli, D., Peden, D.R., Vardy, A.W., and Peters, J.A. (2003) Neurosteroid modulation of GABA(A) receptors. *Prog. Neurobiol.*, **71**, 67–80.
13. Schumacher, M., Weill-Engerer, S., Liere, P., Robert, F., Franklin, R.J., Garcia-Segura, L.M., Lambert, J.J., Mayo, W., Melcangi, R.C., Parducz, A., Suter, U., Carelli, C., Baulieu, E.E., and Akwa, Y. (2003) Steroid hormones and neurosteroids in normal and pathological aging of the nervous system. *Prog. Neurobiol.*, **71**, 3–29.
14. Stoffel-Wagner, B. (2003) Neurosteroid biosynthesis in the human brain and its clinical implications. *Ann. N. Y. Acad. Sci.*, **1007**, 64–78.
15. Barbaccia, M.L. (2004) Neurosteroidogenesis: relevance to neurosteroid actions in brain and modulation by

16. Bernardi, F., Pluchino, N., Begliuomini, S., Lenzi, E., Palumbo, M., Luisi, M., and Genazzani, A.R. (2004) Disadaptive disorders in women: allopregnanolone, a sensitive steroid. *Gynecol Endocrinol.*, **19**, 344–353.
17. Finn, D.A., Ford, M.M., Wiren, K.M., Roselli, C.E., and Crabbe, J.C. (2004) The role of pregnane neurosteroids in ethanol withdrawal: behavioral genetic approaches. *Pharmacol. Ther.*, **101**, 91–112.
18. Reddy, D.S. (2004) Role of neurosteroids in catamenial epilepsy. *Epilepsy Res.*, **62**, 99–118.
19. Schumacher, M., Guennoun, R., Robert, F., Carelli, C., Gago, N., Ghoumari, A., Gonzalez Deniselle, M.C., Gonzalez, S.L., Ibanez, C., Labombarda, F., Coirini, H., Baulieu, E.E., and De Nicola, A.F. (2004) Local synthesis and dual actions of progesterone in the nervous system: neuroprotection and myelination. *Growth Horm. IGF Res.*, **14** (Suppl A), S18–S33.
20. Tsutsui, K., Sakamoto, H., Shikimi, H., and Ukena, K. (2004) Organizing actions of neurosteroids in the Purkinje neuron. *Neurosci. Res.*, **49**, 273–279.
21. Belelli, D. and Lambert, J.J. (2005) Neurosteroids: endogenous regulators of the GABA(A) receptor. *Nat. Rev. Neurosci.*, **6**, 565–575.
22. Uzunova, V., Sampson, L., and Uzunov, D.P. (2006) Relevance of endogenous 3α-reduced neurosteroids to depression and antidepressant action. *Psychopharmacology (Berl)*, **186**, 351–361.
23. Belelli, D., Herd, M.B., Mitchell, E.A., Peden, D.R., Vardy, A.W., Gentet, L., and Lambert, J.J. (2006) Neuroactive steroids and inhibitory neurotransmission: mechanisms of action and physiological relevance. *Neuroscience*, **138**, 821–829.
24. Brinton, R.D. and Wang, J.M. (2006) Therapeutic potential of neurogenesis for prevention and recovery from Alzheimer's disease: allopregnanolone as a proof of concept neurogenic agent. *Curr. Alzheimer. Res.*, **3**, 185–190.
25. Rupprecht, R. and Holsboer, F. (1999) Neuropsychopharmacological properties of neuroactive steroids. *Steroids*, **64**, 83–91.
26. Tsutsui, K. and Mellon, S.H. (2006) Neurosteorids in the brain neuron: biosynthesis, action and medicinal impact on neurodegenerative disease. *Cent. Nerv. Syst. Agents Med. Chem.*, **6**, 73–82.
27. Miller, W. and Tyrell, J. (1994) in *Endocrinology & Metabolism*, 3rd edn (eds P. Felig, J. Baxter, and L. Frohman), McGraw-Hill, New York, pp. 555–711.
28. Miller, W.L. (1988) Molecular biology of steroid hormone synthesis. *Endocr. Rev.*, **9**, 295–318.
29. Mellon, S.H. and Deschepper, C.F. (1993) Neurosteroid biosynthesis: genes for adrenal steroidogenic enzymes are expressed in the brain. *Brain Res.*, **629**, 283–292.
30. Mellon, S.H. and Vaudry, H. (2001) Biosynthesis of neurosteroids and regulation of their synthesis. *Int. Rev. Neurobiol.*, **46**, 33–78.
31. Do Rego, J.L., Seong, J.Y., Burel, D., Leprince, J., Luu-The, V., Tsutsui, K., Tonon, M.C., Pelletier, G., and Vaudry, H. (2009) Neurosteroid biosynthesis: enzymatic pathways and neuroendocrine regulation by neurotransmitters and neuropeptides. *Front Neuroendocrinol.*, **30**, 259–301.
32. Hosie, A.M., Wilkins, M.E., da Silva, H.M., and Smart, T.G. (2006) Endogenous neurosteroids regulate GABAA receptors through two discrete transmembrane sites. *Nature*, **444**, 486–489.
33. Smith, S.S., Shen, H., Gong, Q.H., and Zhou, X. (2007) Neurosteroid regulation of GABA(A) receptors: Focus on the alpha4 and delta subunits. *Pharmacol. Ther.*, **116**, 58–76.
34. Reddy, D.S. (2003) Pharmacology of endogenous neuroactive steroids. *Crit. Rev. Neurobiol.*, **15**, 197–234.
35. Rupprecht, R. (2003) Neuroactive steroids: mechanisms of action and

neuropsychopharmacological properties. *Psychoneuroendocrinology*, **28**, 139–168.

36. Robel, P., Young, J., Corpechot, C., Mayo, W., Perche, F., Haug, M., Simon, H., and Baulieu, E.E. (1995) Biosynthesis and assay of neurosteroids in rats and mice: functional correlates. *J. Steroid. Biochem. Mol. Biol.*, **53**, 355–360.

37. Flood, J.F., Morley, J.E., and Roberts, E. (1995) Pregnenolone sulfate enhances post-training memory processes when injected in very low doses into limbic system structures: the amygdala is by far the most sensitive. *Proc. Natl. Acad. Sci. U.S.A.*, **92**, 10806–10810.

38. Mathis, C., Paul, S.M., and Crawley, J.N. (1994) The neurosteroid pregnenolone sulfate blocks NMDA antagonist-induced deficits in a passive avoidance memory task. *Psychopharmacology (Berl.)*, **116**, 201–206.

39. Mayo, W., Dellu, F., Robel, P., Cherkaoui, J., Le Moal, M., Baulieu, E.E., and Simon, H. (1993) Infusion of neurosteroids into the nucleus basalis magnocellularis affects cognitive processes in the rat. *Brain Res.*, **607**, 324–328.

40. Flood, J.F., Morley, J.E., and Roberts, E. (1992) Memory-enhancing effects in male mice of pregnenolone and steroids metabolically derived from it. *Proc. Natl. Acad. Sci. U.S.A.*, **89**, 1567–1571.

41. Grobin, A.C., VanDoren, M.J., Porrino, L.J., and Morrow, A.L. (2005) Cortical 3α-hydroxy-5α-pregnan-20-one levels after acute administration of Delta 9-tetrahydrocannabinol, cocaine and morphine. *Psychopharmacology (Berl.)*, **179**, 544–550.

42. Follesa, P., Biggio, F., Caria, S., Gorini, G., and Biggio, G. (2004) Modulation of GABA(A) receptor gene expression by allopregnanolone and ethanol. *Eur. J. Pharmacol.*, **500**, 413–425.

43. Caldeira, J.C., Wu, Y., Mameli, M., Purdy, R.H., Li, P.K., Akwa, Y., Savage, D.D., Engen, J.R., and Valenzuela, C.F. (2004) Fetal alcohol exposure alters neurosteroid levels in the developing rat brain. *J. Neurochem.*, **90**, 1530–1539.

44. Finn, D.A., Roberts, A.J., Long, S., Tanchuck, M., and Phillips, T.J. (2003) Neurosteroid consumption has anxiolytic effects in mice. *Pharmacol. Biochem. Behav.*, **76**, 451–462.

45. VanDoren, M.J., Matthews, D.B., Janis, G.C., Grobin, A.C., Devaud, L.L., and Morrow, A.L. (2000) Neuroactive steroid 3α-hydroxy-5α-pregnan-20-one modulates electrophysiological and behavioral actions of ethanol. *J. Neurosci.*, **20**, 1982–1989.

46. Janis, G.C., Devaud, L.L., Mitsuyama, H., and Morrow, A.L. (1998) Effects of chronic ethanol consumption and withdrawal on the neuroactive steroid 3α-hydroxy-5α-pregnan-20-one in male and female rats. *Alcohol Clin. Exp. Res.*, **22**, 2055–2061.

47. Brot, M.D., Akwa, Y., Purdy, R.H., Koob, G.F., and Britton, K.T. (1997) The anxiolytic-like effects of the neurosteroid allopregnanolone: interactions with GABA(A) receptors. *Eur. J. Pharmacol.*, **325**, 1–7.

48. Smith, S.S., Gong, Q.H., Hsu, F.-C., Markowitz, R.S., ffrench-Mullen, J.M.H., and Li, X. (1998) GABA$_A$ receptor α4 subunit suppression prevents withdrawal properties of an endogenous steroid. *Nature*, **392**, 926–930.

49. Schmidt, P.J., Purdy, R.H., Moore, P.H. Jr., Paul, S.M., and Rubinow, D.R. (1994) Circulating levels of anxiolytic steroids in the luteal phase in women with premenstrual syndrome and in control subjects. *J. Clin. Endocrinol. Metab.*, **79**, 1256–1260.

50. Bixo, M., Andersson, A., Winblad, B., Purdy, R.H., and Backstrom, T. (1997) Progesterone, 5α-pregnane-3,20-dione and 3α-hydroxy-5α-pregnane-20-one in specific regions of the human female brain in different endocrine states. *Brain Res.*, **764**, 173–178.

51. Monteleone, P., Luisi, S., Tonetti, A., Bernardi, F., Genazzani, A.D., Luisi, M., Petraglia, F., and Genazzani, A.R. (2000) Allopregnanolone concentrations and premenstrual syndrome. *Eur. J. Endocrinol.*, **142**, 269–273.

52. Friedman, L., Gibbs, T.T., and Farb, D.H. (1993) Gamma-aminobutyric acid A receptor regulation: chronic treatment with pregnanolone uncouples allosteric interactions between steroid and benzodiazepine recognition sites. *Mol. Pharmacol.*, **44**, 191–197.
53. Bicikova, M., Dibbelt, L., Hill, M., Hampl, R., and Starka, L. (1998) Allopregnanolone in women with premenstrual syndrome. *Horm. Metab. Res.*, **30**, 227–230.
54. Rapkin, A.J., Morgan, M., Goldman, L., Brann, D.W., Simone, D., and Mahesh, V.B. (1997) Progesterone metabolite allopregnanolone in women with premenstrual syndrome. *Obstet. Gynecol.*, **90**, 709–714.
55. Girdler, S.S., Straneva, P.A., Light, K.C., Pedersen, C.A., and Morrow, A.L. (2001) Allopregnanolone levels and reactivity to mental stress in premenstrual dysphoric disorder. *Biol. Psychiatry.*, **49**, 788–797.
56. Epperson, C.N., Haga, K., Mason, G.F., Sellers, E., Gueorguieva, R., Zhang, W., Weiss, E., Rothman, D.L., and Krystal, J.H. (2002) Cortical gamma-aminobutyric acid levels across the menstrual cycle in healthy women and those with premenstrual dysphoric disorder: a proton magnetic resonance spectroscopy study. *Arch. Gen. Psychiatry.*, **59**, 851–858.
57. Shen, H., Gong, Q.H., Aoki, C., Yuan, M., Ruderman, Y., Dattilo, M., Williams, K., and Smith, S.S. (2007) Reversal of neurosteroid effects at alpha4beta2delta GABA(A) receptors triggers anxiety at puberty. *Nat. Neurosci.*, **10**, 469–477.
58. Lapchak, P.A. (2004) The neuroactive steroid 3-α-ol-5-beta-pregnan-20-one hemisuccinate, a selective NMDA receptor antagonist improves behavioral performance following spinal cord ischemia. *Brain Res.*, **997**, 152–158.
59. Djebaili, M., Guo, Q., Pettus, E.H., Hoffman, S.W., and Stein, D.G. (2005) The neurosteroids progesterone and allopregnanolone reduce cell death, gliosis, and functional deficits after traumatic brain injury in rats. *J. Neurotrauma.*, **22**, 106–118.
60. Cutler, S.M., Pettus, E.H., Hoffman, S.W., and Stein, D.G. (2005) Tapered progesterone withdrawal enhances behavioral and molecular recovery after traumatic brain injury. *Exp. Neurol.*, **195**, 423–429.
61. Hoffman, S.W., Virmani, S., Simkins, R.M., and Stein, D.G. (2003) The delayed administration of dehydroepiandrosterone sulfate improves recovery of function after traumatic brain injury in rats. *J. Neurotrauma.*, **20**, 859–870.
62. Meffre, D., Pianos, A., Liere, P., Eychenne, B., Cambourg, A., Schumacher, M., Stein, D.G., and Guennoun, R. (2007) Steroid profiling in brain and plasma of male and pseudopregnant female rats after traumatic brain injury: analysis by gas chromatography/mass spectrometry. *Endocrinology*, **148**, 2505–2517.
63. Shear, D.A., Galani, R., Hoffman, S.W., and Stein, D.G. (2002) Progesterone protects against necrotic damage and behavioral abnormalities caused by traumatic brain injury. *Exp. Neurol.*, **178**, 59–67.
64. VanLandingham, J.W., Cutler, S.M., Virmani, S., Hoffman, S.W., Covey, D.F., Krishnan, K., Hammes, S.R., Jamnongjit, M., and Stein, D.G. (2006) The enantiomer of progesterone acts as a molecular neuroprotectant after traumatic brain injury. *Neuropharmacology*, **51**, 1078–1085.
65. Pomata, P.E., Colman-Lerner, A.A., Baranao, J.L., and Fiszman, M.L. (2000) In vivo evidences of early neurosteroid synthesis in the developing rat central nervous system and placenta. *Brain Res. Dev. Brain. Res.*, **120**, 83–86.
66. Patte-Mensah, C., Penning, T.M., and Mensah-Nyagan, A.G. (2004) Anatomical and cellular localization of neuroactive 5α/3α-reduced steroid-synthesizing enzymes in the spinal cord. *J. Comp. Neurol.*, **477**, 286–299.
67. di Michele, F., Lekieffre, D., Pasini, A., Bernardi, G., Benavides, J., and Romeo, E. (2000) Increased neurosteroids synthesis after brain and spinal

cord injury in rats. *Neurosci. Lett.*, **284**, 65–68.
68. Labombarda, F., Pianos, A., Liere, P., Eychenne, B., Gonzalez, S., Cambourg, A., De Nicola, A.F., Schumacher, M., and Guennoun, R. (2006) Injury elicited increase in spinal cord neurosteroid content analyzed by gas chromatography mass spectrometry. *Endocrinology*, **147**, 1847–1859.
69. Fiore, C., Inman, D.M., Hirose, S., Noble, L.J., Igarashi, T., and Compagnone, N.A. (2004) Treatment with the neurosteroid dehydroepiandrosterone promotes recovery of motor behavior after moderate contusive spinal cord injury in the mouse. *J. Neurosci. Res.*, **75**, 391–400.
70. Compagnone, N.A. (2008) Treatments for spinal cord injury: is there hope in neurosteroids? *J. Steroid. Biochem. Mol. Biol.*, **109**, 307–313.
71. Gago, N., Akwa, Y., Sananes, N., Guennoun, R., Baulieu, E.E., El-Etr, M., and Schumacher, M. (2001) Progesterone and the oligodendroglial lineage: stage-dependent biosynthesis and metabolism. *Glia*, **36**, 295–308.
72. Schumacher, M., Akwa, Y., Guennoun, R., Robert, F., Labombarda, F., Desarnaud, F., Robel, P., De Nicola, A.F., and Baulieu, E.E. (2000) Steroid synthesis and metabolism in the nervous system: trophic and protective effects. *J. Neurocytol.*, **29**, 307–326.
73. Le Goascogne, C., Eychenne, B., Tonon, M.C., Lachapelle, F., Baumann, N., and Robel, P. (2000) Neurosteroid progesterone is up-regulated in the brain of jimpy and shiverer mice. *Glia*, **29**, 14–24.
74. Ghoumari, A.M., Ibanez, C., El-Etr, M., Leclerc, P., Eychenne, B., O'Malley, B.W., Baulieu, E.E., and Schumacher, M. (2003) Progesterone and its metabolites increase myelin basic protein expression in organotypic slice cultures of rat cerebellum. *J. Neurochem.*, **86**, 848–859.
75. Ghoumari, A.M., Baulieu, E.E., and Schumacher, M. (2005) Progesterone increases oligodendroglial cell proliferation in rat cerebellar slice cultures. *Neuroscience*, **135**, 47–58.
76. Chavez-Delgado, M.E., Gomez-Pinedo, U., Feria-Velasco, A., Huerta-Viera, M., Castaneda, S.C., Toral, F.A., Parducz, A., Anda, S.L., Mora-Galindo, J., and Garcia-Estrada, J. (2005) Ultrastructural analysis of guided nerve regeneration using progesterone- and pregnenolone-loaded chitosan prostheses. *J. Biomed. Mater. Res. B: Appl. Biomater.*, **74**, 589–600.
77. Wang, J.M., Johnston, P.B., Ball, B.G., and Brinton, R.D. (2005) The neurosteroid allopregnanolone promotes proliferation of rodent and human neural progenitor cells and regulates cell-cycle gene and protein expression. *J. Neurosci.*, **25**, 4706–4718.
78. Suzuki, M., Wright, L.S., Marwah, P., Lardy, H.A., and Svendsen, C.N. (2004) Mitotic and neurogenic effects of dehydroepiandrosterone (DHEA) on human neural stem cell cultures derived from the fetal cortex. *Proc. Natl. Acad. Sci. U.S.A.*, **101**, 3202–3207.
79. Keller, E.A., Zamparini, A., Borodinsky, L.N., Gravielle, M.C., and Fiszman, M.L. (2004) Role of allopregnanolone on cerebellar granule cells neurogenesis. *Brain Res. Dev. Brain Res.*, **153**, 13–17.
80. Wang, J.M., Liu, L., Irwin, R.W., Chen, S., and Brinton, R.D. (2008) Regenerative potential of allopregnanolone. *Brain Res. Rev.*, **57**, 398–409.
81. Liu, L., Wang, J., Zhao, L., Nilsen, J., McClure, K., Wong, K., and Brinton, R.D. (2009) Progesterone increases rat neural progenitor cell cycle gene expression and proliferation via extracellularly regulated kinase and progesterone receptor membrane components 1 and 2. *Endocrinology*, **150**, 3186–3196.
82. Wang, J.M., Singh, C., Liu, L., Irwin, R.W., Chen, S., Chung, E.J., Thompson, R.F., and Brinton, R.D. (2010) Allopregnanolone reverses neurogenic and cognitive deficits in mouse model of Alzheimer's disease. *Proc. Natl. Acad. Sci. U.S.A.*, **107**, 6498–6503.
83. Todorovic, S.M., Pathirathna, S., Brimelow, B.C., Jagodic, M.M., Ko, S.H., Jiang, X., Nilsson, K.R.,

Zorumski, C.F., Covey, D.F., and Jevtovic-Todorovic, V. (2004) 5beta-reduced neuroactive steroids are novel voltage-dependent blockers of T-type Ca^{2+} channels in rat sensory neurons in vitro and potent peripheral analgesics in vivo. *Mol. Pharmacol.*, **66**, 1223–1235.

84. Pathirathna, S., Brimelow, B.C., Jagodic, M.M., Krishnan, K., Jiang, X., Zorumski, C.F., Mennerick, S., Covey, D.F., Todorovic, S.M., and Jevtovic-Todorovic, V. (2005) New evidence that both T-type calcium channels and GABAA channels are responsible for the potent peripheral analgesic effects of 5α-reduced neuroactive steroids. *Pain*, **114**, 429–443.

85. Hu, M.C., Hsu, N.C., El Hadj, N.B., Pai, C.I., Chu, H.P., Wang, C.K., and Chung, B.C. (2002) Steroid deficiency syndromes in mice with targeted disruption of Cyp11a1. *Mol. Endocrinol.*, **16**, 1943–1950.

86. Bair, S.R. and Mellon, S.H. (2004) Deletion of the mouse P450c17 gene causes early embryonic lethality. *Mol. Cell. Biol.*, **24**, 5383–5390.

87. Mahendroo, M., Wilson, J.D., Richardson, J.A., and Auchus, R.J. (2004) Steroid 5α-reductase 1 promotes 5α-androstane-3 alpha,17 beta-diol synthesis in immature mouse testes by two pathways. *Mol. Cell Endocrinol.*, **222**, 113–120.

88. Mahendroo, M.S., Cala, K.M., Hess, D.L., and Russell, D.W. (2001) Unexpected virilization in male mice lacking steroid 5α-reductase enzymes. *Endocrinology*, **142**, 4652–4662.

89. Mahendroo, M.S., Cala, K.M., Landrum, D.P., and Russell, D.W. (1997) Fetal death in mice lacking 5α-reductase type 1 caused by estrogen excess. *Mol. Endocrinol.*, **11**, 917–927.

90. Wu, H.S., Lin, H.T., Wang, C.K., Chiang, Y.F., Chu, H.P., and Hu, M.C. (2007) Human CYP11A1 promoter drives Cre recombinase expression in the brain in addition to adrenals and gonads. *Genesis*, **45**, 59–65.

91. Compagnone, N.A., Bulfone, A., Rubenstein, J.L.R., and Mellon, S.H. (1995) Steroidogenic enzyme P450c17 is expressed in the embryonic central nervous system. *Endocrinology*, **136**, 5212–5223.

92. Bridges, P.J., Koo, Y., Kang, D.W., Hudgins-Spivey, S., Lan, Z.J., Xu, X., DeMayo, F., Cooney, A., and Ko, C. (2008) Generation of Cyp17iCre transgenic mice and their application to conditionally delete estrogen receptor alpha (Esr1) from the ovary and testis. *Genesis*, **46**, 499–505.

93. Griffin, L.D., Gong, W., Verot, L., and Mellon, S.H. (2004) Niemann-Pick type C disease involves disrupted neurosteroidogenesis and responds to allopregnanolone. *Nat. Med.*, **10**, 704–711.

94. Patterson, M.C., Vanier, M.T., Suzuki, K., Morris, J.A., Carstea, E., Neufeld, E.B., Blanchette-Mackie, J.E., and Pentchev, P.G. (2001) in *The Metabolic and Molecular Bases of Inherited Disease*, 8th edn (eds C.R. Scriver, A.L. Beaudet, W.S. Sly, and D. Valle), McGraw-Hill, New York, pp. 3611–3633.

95. Fink, J.K., Filling-Katz, M.R., Sokol, J., Cogan, D.G., Pikus, A., Sonies, B., Soong, B., Pentchev, P.G., Comly, M.E., Brady, R.O. et al. (1989) Clinical spectrum of Niemann-Pick disease type C. *Neurology*, **39**, 1040–1049.

96. Norman, R.M., Forrester, R.M., and Tingey, A.H. (1967) The juvenile form of Niemann-Pick disease. *Arch. Dis. Child.*, **42**, 91–96.

97. Pentchev, P.G., Vanier, M.T., Suzuki, K., and Patterson, M.C. (1995) in *The Metabolic and Molecular Basis of Inherited Disease*, 7th edn (eds C.R. Scriver, A.L. Beaudet, W.S. Sly, and D.D Valle), McGraw-Hill, New York, pp. 2625–2639.

98. Vanier, M.T., Duthel, S., Rodriguez-Lafrasse, C., Pentchev, P., and Carstea, E.D. (1996) Genetic heterogeneity in Niemann-Pick C disease: a study using somatic cell hybridization and linkage analysis. *Am. J. Hum. Genet.*, **58**, 118–125.

99. Naureckiene, S., Sleat, D.E., Lackland, H., Fensom, A., Vanier, M.T., Wattiaux, R., Jadot, M., and Lobel, P.

(2000) Identification of HE1 as the second gene of Niemann-Pick C disease. *Science*, **290**, 2298–2301.
100. Millat, G., Chikh, K., Naureckiene, S., Sleat, D.E., Fensom, A.H., Higaki, K., Elleder, M., Lobel, P., and Vanier, M.T. (2001) Niemann-Pick disease type C: spectrum of HE1 mutations and genotype/phenotype correlations in the NPC2 group. *Am. J. Hum. Genet.*, **69**, 1013–1021.
101. Carstea, E.D., Morris, J.A., Coleman, K.G., Loftus, S.K., Zhang, D., Cummings, C., Gu, J., Rosenfeld, M.A., Pavan, W.J., Krizman, D.B., Nagle, J., Polymeropoulos, M.H., Sturley, S.L., Ioannou, Y.A., Higgins, M.E., Comly, M., Cooney, A., Brown, A., Kaneski, C.R., Blanchette-Mackie, E.J., Dwyer, N.K., Neufeld, E.B., Chang, T.Y., Liscum, L., Tagle, D.A. et al. (1997) Niemann-Pick C1 disease gene: homology to mediators of cholesterol homeostasis. *Science*, **277**, 228–231.
102. Infante, R.E., Radhakrishnan, A., Abi-Mosleh, L., Kinch, L.N., Wang, M.L., Grishin, N.V., Goldstein, J.L., and Brown, M.S. (2008) Purified NPC1 protein: II. Localization of sterol binding to a 240-amino acid soluble luminal loop. *J. Biol. Chem.*, **283**, 1064–1075.
103. Loftus, S.K., Morris, J.A., Carstea, E.D., Gu, J.Z., Cummings, C., Brown, A., Ellison, J., Ohno, K., Rosenfeld, M.A., Tagle, D.A., Pentchev, P.G., and Pavan, W.J. (1997) Murine model of Niemann-Pick C disease: mutation in a cholesterol homeostasis gene. *Science*, **277**, 232–235.
104. Davies, J.P., Chen, F.W., and Ioannou, Y.A. (2000) Transmembrane molecular pump activity of Niemann-Pick C1 protein. *Science*, **290**, 2295–2298.
105. Infante, R.E., Abi-Mosleh, L., Radhakrishnan, A., Dale, J.D., Brown, M.S., and Goldstein, J.L. (2008) Purified NPC1 protein. I. Binding of cholesterol and oxysterols to a 1278-amino acid membrane protein. *J. Biol. Chem.*, **283**, 1052–1063.
106. Kwon, H.J., Abi-Mosleh, L., Wang, M.L., Deisenhofer, J., Goldstein, J.L., Brown, M.S., and Infante, R.E. (2009) Structure of N-terminal domain of NPC1 reveals distinct subdomains for binding and transfer of cholesterol. *Cell*, **137**, 1213–1224.
107. Bauer, P., Knoblich, R., Bauer, C., Finckh, U., Hufen, A., Kropp, J., Braun, S., Kustermann-Kuhn, B., Schmidt, D., Harzer, K., and Rolfs, A. (2002) NPC1: complete genomic sequence, mutation analysis, and characterization of haplotypes. *Hum. Mutat.*, **19**, 30–38.
108. Morris, M.D., Bhuvaneswaran, C., Shio, H., and Fowler, S. (1982) Lysosome lipid storage disorder in NCTR-BALB/c mice. I. Description of the disease and genetics. *Am. J. Pathol.*, **108**, 140–149.
109. Gilbert, E.F., Callahan, J., Viseskul, C., and Opitz, J.M. (1981) Niemann-Pick disease type C. Pathological, histochemical, ultrastructural and biochemical studies. *Eur. J. Pediatr.*, **136**, 263–274.
110. Higashi, Y., Murayama, S., Pentchev, P.G., and Suzuki, K. (1993) Cerebellar degeneration in the Niemann-Pick type C mouse. *Acta Neuropathol. (Berl.)*, **85**, 175–184.
111. Weintraub, H., Abramovici, A., Sandbank, U., Booth, A.D., Pentchev, P.G., and Sela, B. (1987) Dysmyelination in NCTR-Balb/C mouse mutant with a lysosomal storage disorder. Morphological survey. *Acta Neuropathol. (Berl.)*, **74**, 374–381.
112. Weintraub, H., Abramovici, A., Sandbank, U., Pentchev, P.G., Brady, R.O., Sekine, M., Suzuki, A., and Sela, B. (1985) Neurological mutation characterized by dysmyelination in NCTR-Balb/C mouse with lysosomal lipid storage disease. *J. Neurochem.*, **45**, 665–672.
113. Higashi, Y., Murayama, S., Pentchev, P.G., and Suzuki, K. (1995) Peripheral nerve pathology in Niemann-Pick type C mouse. *Acta Neuropathol. (Berl.)*, **90**, 158–163.
114. Xie, C., Burns, D.K., Turley, S.D., and Dietschy, J.M. (2000) Cholesterol is sequestered in the brains of mice with Niemann-Pick type C disease but turnover is increased. *J. Neuropathol. Exp. Neurol.*, **59**, 1106–1117.

115. Lope-Piedrafita, S., Totenhagen, J.W., Hicks, C.M., Erickson, R.P., and Trouard, T.P. (2008) MRI detects therapeutic effects in weaning Niemann-Pick type C mice. *J. Neurosci. Res.*, **86**, 2802–2807.

116. Spence, M.W. and Callahan, J.W. (1989) in *The Metabolic Basis of Inherited Disease* (eds C.R. Scriver, A.L. Beaudet, V.S. Sly, and D. Valle), McGraw-Hill, New York, pp. 1655–1676.

117. Vanier, M.T., Pentchev, P., Rodriguez-Lafrasse, C., and Rousson, R. (1991) Niemann-Pick disease type C: an update. *J. Inherit. Metab. Dis.*, **14**, 580–595.

118. Anzil, A.P., Blinzinger, K., Mehraein, P., and Dozic, S. (1973) Niemann-Pick disease type C: case report with ultrastructural findings. *Neuropadiatrie*, **4**, 207–225.

119. Braak, H., Braak, E., and Goebel, H.H. (1983) Isocortical pathology in type C Niemann-Pick disease. A combined Golgi-pigmentoarchitectonic study. *J. Neuropathol. Exp. Neurol.*, **42**, 671–687.

120. Suzuki, K., Parker, C.C., Pentchev, P.G., Katz, D., Ghetti, B., D'Agostino, A.N., and Carstea, E.D. (1995) Neurofibrillary tangles in Niemann-Pick disease type C. *Acta Neuropathol. (Berl.)*, **89**, 227–238.

121. Zervas, M., Dobrenis, K., and Walkley, S.U. (2001) Neurons in Niemann-Pick disease type C accumulate gangliosides as well as unesterified cholesterol and undergo dendritic and axonal alterations. *J. Neuropathol. Exp. Neurol.*, **60**, 49–64.

122. Roff, C.F., Strauss, J.F.R., Goldin, E., Jaffe, H., Patterson, M.C., Agritellis, G.C., Hibbs, A.M., Garfield, M., Brady, R.O., and Pentchev, P.G. (1993) The murine Niemann-Pick type C lesion affects testosterone production. *Endocrinology*, **133**, 2913–2923.

123. Gevry, N.Y., Lopes, F.L., Ledoux, S., and Murphy, B.D. (2004) Aberrant intracellular cholesterol transport disrupts pituitary and ovarian function. *Mol. Endocrinol.*, **18**, 1778–1786.

124. Xie, C., Richardson, J.A., Turley, S.D., and Dietschy, J.M. (2006) Cholesterol substrate pools and steroid hormone levels are normal in the face of mutational inactivation of NPC1 protein. *J. Lipid. Res.*, **47**, 953–963.

125. Ahmad, I., Lope-Piedrafita, S., Bi, X., Hicks, C., Yao, Y., Yu, C., Chaitkin, E., Howison, C.M., Weberg, L., Trouard, T.P., and Erickson, R.P. (2005) Allopregnanolone treatment, both as a single injection or repetitively, delays demyelination and enhances survival of Niemann-Pick C mice. *J. Neurosci. Res.*, **82**, 811–821.

126. Langmade, S.J., Gale, S.E., Frolov, A., Mohri, I., Suzuki, K., Mellon, S.H., Walkley, S.U., Covey, D.F., Schaffer, J.E., and Ory, D.S. (2006) Pregnane X receptor (PXR) activation: A mechanism for neuroprotection in a mouse model of Niemann-Pick C disease. *Proc. Natl. Acad. Sci. U.S.A.*, **103**, 13807–13812.

127. Liao, G., Cheung, S., Galeano, J., Ji, A.X., Qin, Q., and Bi, X. (2009) Allopregnanolone treatment delays cholesterol accumulation and reduces autophagic/lysosomal dysfunction and inflammation in Npc1−/−mouse brain. *Brain Res.*, **1270**, 140–151.

128. Davidson, C.D., Ali, N.F., Micsenyi, M.C., Stephney, G., Renault, S., Dobrenis, K., Ory, D.S., Vanier, M.T., and Walkley, S.U. (2009) Chronic cyclodextrin treatment of murine Niemann-Pick C disease ameliorates neuronal cholesterol and glycosphingolipid storage and disease progression. *PLoS ONE*, **4**, e6951.

129. Zhu, W.J. and Vicini, S. (1997) Neurosteroid prolongs GABAA channel deactivation by altering kinetics of desensitized states. *J. Neurosci.*, **17**, 4022–4031.

130. Zhu, W.J., Wang, J.F., Krueger, K.E., and Vicini, S. (1996) Delta subunit inhibits neurosteroid modulation of GABAA receptors. *J. Neurosci.*, **16**, 6648–6656.

131. Gee, K.W. (1996) Epalons as anticonvulsants: actions mediated by the GABAA receptor complex. *Proc. West Pharmacol. Soc.*, **39**, 55.

132. Beekman, M., Ungard, J.T., Gasior, M., Carter, R.B., Dijkstra, D., Goldberg, S.R., and Witkin, J.M. (1998) Reversal of behavioral effects of pentylenetetrazol by the neuroactive steroid ganaxolone. *J. Pharmacol. Exp. Ther.*, **284**, 868–877.

133. Carter, R.B., Wood, P.L., Wieland, S., Hawkinson, J.E., Belelli, D., Lambert, J.J., White, H.S., Wolf, H.H., Mirsadeghi, S., Tahir, S.H., Bolger, M.B., Lan, N.C., and Gee, K.W. (1997) Characterization of the anticonvulsant properties of ganaxolone (CCD 1042; 3α-hydroxy-3beta-methyl-5α-pregnan-20-one), a selective, high-affinity, steroid modulator of the gamma-aminobutyric acid(A) receptor. *J. Pharmacol. Exp. Ther.*, **280**, 1284–1295.

134. Gasior, M., Ungard, J.T., Beekman, M., Carter, R.B., and Witkin, J.M. (2000) Acute and chronic effects of the synthetic neuroactive steroid, ganaxolone, against the convulsive and lethal effects of pentylenetetrazol in seizure-kindled mice: comparison with diazepam and valproate. *Neuropharmacology*, **39**, 1184–1196.

135. Monaghan, E.P., Navalta, L.A., Shum, L., Ashbrook, D.W., and Lee, D.A. (1997) Initial human experience with ganaxolone, a neuroactive steroid with antiepileptic activity. *Epilepsia*, **38**, 1026–1031.

136. Ungard, J.T., Beekman, M., Gasior, M., Carter, R.B., Dijkstra, D., and Witkin, J.M. (2000) Modification of behavioral effects of drugs in mice by neuroactive steroids. *Psychopharmacology (Berl.)*, **148**, 336–343.

137. Reddy, D.S. and Rogawski, M.A. (2000) Chronic treatment with the neuroactive steroid ganaxolone in the rat induces anticonvulsant tolerance to diazepam but not to itself. *J. Pharmacol. Exp. Ther.*, **295**, 1241–1248.

138. Kerrigan, J.F., Shields, W.D., Nelson, T.Y., Bluestone, D.L., Dodson, W.E., Bourgeois, B.F., Pellock, J.M., Morton, L.D., and Monaghan, E.P. (2000) Ganaxolone for treating intractable infantile spasms: a multicenter, open-label, add-on trial. *Epilepsy Res.*, **42**, 133–139.

139. Laxer, K., Blum, D., Abou-Khalil, B.W., Morrell, M.J., Lee, D.A., Data, J.L., and Monaghan, E.P. (2000) Assessment of ganaxolone's anticonvulsant activity using a randomized, double-blind, presurgical trial design. Ganaxolone Presurgical Study Group. *Epilepsia*, **41**, 1187–1194.

140. Reddy, D.S. and Rogawski, M.A. (2000) Enhanced anticonvulsant activity of ganaxolone after neurosteroid withdrawal in a rat model of catamenial epilepsy. *J. Pharmacol. Exp. Ther.*, **294**, 909–915.

141. Robichaud, M. and Debonnel, G. (2005) Allopregnanolone and ganaxolone increase the firing activity of dorsal raphe nucleus serotonergic neurons in female rats. *Int. J. Neuropsychopharmacol.*, **9**, 1–10.

142. Kliewer, S.A., Goodwin, B., and Willson, T.M. (2002) The nuclear pregnane X receptor: a key regulator of xenobiotic metabolism. *Endocr. Rev.*, **23**, 687–702.

143. Watkins, R.E., Wisely, G.B., Moore, L.B., Collins, J.L., Lambert, M.H., Williams, S.P., Willson, T.M., Kliewer, S.A., and Redinbo, M.R. (2001) The human nuclear xenobiotic receptor PXR: structural determinants of directed promiscuity. *Science*, **292**, 2329–2333.

144. Moore, L.B., Parks, D.J., Jones, S.A., Bledsoe, R.K., Consler, T.G., Stimmel, J.B., Goodwin, B., Liddle, C., Blanchard, S.G., Willson, T.M., Collins, J.L., and Kliewer, S.A. (2000) Orphan nuclear receptors constitutive androstane receptor and pregnane X receptor share xenobiotic and steroid ligands. *J. Biol. Chem.*, **275**, 15122–15127.

145. Lamba, V., Yasuda, K., Lamba, J.K., Assem, M., Davila, J., Strom, S., and Schuetz, E.G. (2004) PXR (NR1I2): splice variants in human tissues, including brain, and identification of neurosteroids and nicotine as PXR activators. *Toxicol. Appl. Pharmacol.*, **199**, 251–265.

146. Wang, C.Y., Li, C.W., Chen, J.D., and Welsh, W.J. (2006) Structural

147. Wang, J.M. and Brinton, R.D. (2008) Allopregnanolone-induced rise in intracellular calcium in embryonic hippocampal neurons parallels their proliferative potential. *BMC Neurosci.*, **9** (Suppl 2), S11.
148. Zampieri, S., Mellon, S.H., Butters, T.D., Nevyjel, M., Covey, D.F., Bembi, B., and Dardis, A. (2009) Oxidative stress in NPC1 deficient cells: protective effect of allopregnanolone. *J. Cell Mol. Med.*, **13**, 3786–3796.
149. Sayeed, I., Parvez, S., Wali, B., Siemen, D., and Stein, D.G. (2009) Direct inhibition of the mitochondrial permeability transition pore: a possible model reveals key interactions in the assembly of the pregnane X receptor/corepressor complex. *Mol. Pharmacol.*, **69**, 1513–1517.
mechanism for better neuroprotective effects of allopregnanolone over progesterone. *Brain Res.*, **1263**, 165–173.
150. Xilouri, M. and Papazafiri, P. (2008) Induction of Akt by endogenous neurosteroids and calcium sequestration in P19 derived neurons. *Neurotox Res.*, **13**, 209–219.
151. Xilouri, M., Avlonitis, N., Calogeropoulou, T., and Papazafiri, P. (2007) Neuroprotective effects of steroid analogues on P19-N neurons. *Neurochem. Int.*, **50**, 660–670.
152. Zampieri, S., Mellon, S.H., Butters, T.D., Nevyjel, M., Covey, D.F., Bembi, B., and Dardis, A. (2010) Oxidative stress in NPC1 deficient cells: protective effect of allopregnanolone. *J. Cell. Mol. Med.*, **13**, 3786–3796.

Part II
Glucocorticoids, Dehydroepiandrosterone, Neuroprotection and Neuropathy

5
Glucocorticoids, Developmental "Programming," and the Risk of Affective Dysfunction

Bayanne Olabi and Jonathan Seckl

> *"The child is father to the man."*
>
> *How can he be? The words are wild.*
>
> *Suck any sense from that who can:*
>
> *"The child is father to the man."*
>
> Gerald Manley Hopkins (1918)

Hopkins' musings on Wordsworth's enigmatic line is now reflected in some surprising biology. Conventional wisdom suggests that the common psychiatric and neurodegenerative disorders of adult life are caused by the genes individuals inherit from their parents and the adult environment in which they live. However, recent data indicate a major influence of early life events on subsequent disease risk many years after the environmental challenge has passed. Such "developmental programming" reflects the action of a factor(s) during critical sensitive windows of development to alter cell and tissue structure and function. While many such effects have only transient impacts, some influences persist throughout the lifespan and perhaps even into a subsequent generation (transgenerational effects), thus mimicking classical genetic "inheritance." In fact this biology is common, perhaps ubiquitous, in the animal kingdom, presumably reflecting drivers operational throughout evolution. Programming impacts on most and probably all body cells and organs. The brain and its neuroendocrine and behavioral outputs is no exception. Indeed, since the "purpose" of programming may reflect attempts to anticipate and adapt to the likely environment pertaining through the lifespan to maximize "fitness," key adaptive systems such as the brain are particularly susceptible. Here we review the relevance of developmental programming to the brain and, in particular, to neuroendocrine stress responses and affective behavior. The data suggest that the child (fetus) is indeed "father" (poetically) to the man's risk of stress-related disorders.

Hormones in Neurodegeneration, Neuroprotection, and Neurogenesis.
Edited by Achille G. Gravanis and Synthia H. Mellon
Copyright © 2011 WILEY-VCH Verlag GmbH & Co. KGaA, Weinheim
ISBN: 978-3-527-32627-3

5.1
Introduction to Programming

More than half a century ago, Levine first described that early life influences can permanently alter the function of the hypothalamic-pituitary-adrenocortical (HPA) axis [1], which regulates the major neuroendocrine response to stress. The vulnerability of the HPA axis to permanent organizational effects during development has since been repeatedly shown [2–7]. Given the close integration of HPA function and behavior, particularly behavioral responses to stress [8], it is unsurprising that affective and cognitive processes are also subject to developmental programming [3, 4, 6].

For many years, these observations remained within the province of behavioral neuroendocrinology and psychology. However, the pioneering epidemiological studies of David Barker and since then, others [9–11] have revealed that early life events, as marked by low birth weight and other early life anthropometric parameters suggestive of a challenging intrauterine environment, associate with a markedly elevated risk of subsequent glucose intolerance, hypertension, dyslipidemia (collectively known as the *metabolic syndrome*), ischemic heart disease, as well as HPA axis hyperactivity, anxiety-related and affective disorders in adult life [10, 12–14].

The concept of early life physiological "programming" has been advanced to explain the associations between perinatal environmental events, altered fetal/postnatal growth and development and later pathophysiology [15–17]. Programming reflects the action of an environmental "factor" during a sensitive or vulnerable developmental period or "window" to exert effects on the structure and function of tissues that persist throughout life. Different cells and tissues are sensitive at different developmental stages, so the effects of environmental challenges will have distinct effects depending not only on the challenge involved but also upon its timing. Recent evidence indicates that the timing of the stressor determines the ensuing neuroendocrine and psychopathological phenotype, as prenatal stress at different stages of gestation has been shown to differentially affect the expression of key regulators of HPA axis activity [18], a pattern that is also emerging in human studies. The occurrence of associations between early life environmental manipulations and later physiology and disease risk in in-bred rodent models are well documented and imply causation via nongenetic processes. However, the underlying molecular mechanisms seem to be evolutionarily conserved implying that genetic processes have maintained the ability to show developmental plasticity in the face of early life environment.

As the body's most complex and vulnerable organ, the brain is particularly susceptible to such early life influences. Many markers of an adverse early life environment (low birth weight, prematurity, exposure to toxins or pharmacological agents, postnatal illness, or trauma) have been associated with persisting effects on brain structure and function. However, these findings reflect the "disease diathesis" approach. This neglects the underlying biological subtleties of processes conserved through evolution. This consideration implies that permanent changes induced by transient effects in early life confer adaptive Darwinian advantages. The biological "purpose" of behavioral programming is perhaps easiest to conceptualize in animal populations. If a pregnant animal is exposed to a hostile environment that requires increased vigilance, it is plausibly beneficial to its offspring's survival for a "stress signal" to be transmitted to the fetus. This then acts to program the development of the fetal HPA axis and associated behaviors, leading to an enhanced ability to survive after birth. Plausibly, humans have inherited from vertebrate evolution a sophisticated mechanism to adapt our offspring to the environment into which they are to be born. Thus, the outcome of a compromised pregnancy will be the modification of endocrine, behavioral, cardiovascular, and metabolic regulation. Providing this improves survival and reproductive fitness, it is likely that the underlying processes that mediate such effects will be maintained. Of course, the prediction of the subsequent environment, especially in long-lived species such as our own, is likely to be inexact. It has been suggested that there is mismatch between the expected and actual ambient environment that underpins later disease [19]. However, many of the associations of low birth weight are with degenerative disorders in the later part of life, after the reproductive years have peaked, and when deleterious effects are less amenable to selectional influences.

In contrast to Neel's thrifty genotype hypothesis [20] to address the persistence of high levels of diabetes and obesity in human populations (genes which led to rapid growth and calorie storage could promote survival in populations exposed to regular feast and famine cycles), Hales and Barker proposed the thrifty phenotype [21], which suggested that diabetes and obesity reflect a mismatch between fetal metabolic adaptations to environmental predictions of a life exposed to famine and a modern reality of calorie excess. In this vein, the behavioral adaptations could be termed the *shifty phenotype*; the fetus responds to maternal signals predicting a hazardous environment with anxiety and distractable patterns of behavior, which are "inappropriate" to the rigors of contemporary education systems and their quiescent behavioral norms.

5.2
Programming

5.2.1
Epidemiology

Numerous human epidemiological studies have shown an association between lower, but still normal birth weight and the subsequent development of the common cardiometabolic disorders of adult life, notably hypertension, type 2 diabetes, and cardiovascular disease deaths [15, 22–32]. The associations with adult disorders are continuous through the normal range of birth weight [15, 22, 25, 26], do not reflect prematurity or twinning [33], occur in all races studied, and persist after adjustment for confounders such as maternal smoking, alcohol intake, and social class [34, 35]. Importantly, birth weight is an important predictor of adult morbidity [22, 25, 26, 36] and mortality [37–40]. Indeed, the importance of intrauterine growth is highlighted by the observations that postnatal catch-up growth amplifies the risk of adult cardiovascular disease [23, 28, 29, 41–44]. Although this area has been popularized in relationship to prenatal events impacting birth weight, development proceeds from preconception to puberty and "programming" has been demonstrated or suggested at all stages of this continuum [45–47].

5.2.2
Birth Weight and Neuropsychiatric Disorders

Birth weight is associated with affective disorders in adults and children; effects apparently independent of maternal mental state and perhaps more marked in female offspring [14, 48–50]. Similarly, birth weight has been linked with schizophrenia [51, 52] and indeed birth weight appears to compound genetic influences, at least of the catechol-O-methyltransferase gene association with antisocial behavior and attention-deficit/hyperactivity disorder [53]. Slower fetal growth associates with toddler temperament traits, indicating prenatal programming of psychological development [54], and at least one report suggests a link between birth weight and subsequent vulnerability to posttraumatic stress disorder (PTSD) [55]. Such repeated, though by no means ubiquitous, reports of links between as relatively crude a measure of an adverse prenatal environment as low birth weight and later behavioral and psychiatric disorders merit some exploration.

Two major mechanistic hypotheses have been proposed to explain fetal programming; materno-fetal undernutrition and overexposure of the fetus to glucocorticoids/stress [15–17]. Here, the focus is on glucocorticoids and their CNS targets. However, it should be recognized that maternal starvation and stress are difficult to separate, indeed starvation induces glucocorticoid hypersecretion and maternal stress and glucocorticoid therapy reduce food intake. The offspring of mothers subject to starvation in the Dutch hunger winter of 1944–1945 and the Chinese famine of 1959–1961 have substantially increased rates of major psychiatric disorders [56–58].

5.3
Glucocorticoids and Fetal Development

Glucocorticoids have potent effects upon fetal tissue development, accelerating maturation of the lung [59] and exerting a wide spectrum of organizational effects for normal brain development [60]. Glucocorticoids are important for normal maturation in most regions of the developing CNS [4, 61, 62], including the hippocampus [63] and the cerebral cortex [64], initiating terminal maturation, synapse formation [65], remodeling axons and dendrites, and determining cell survival [66]. These short-term developmental trajectories may indeed reflect the

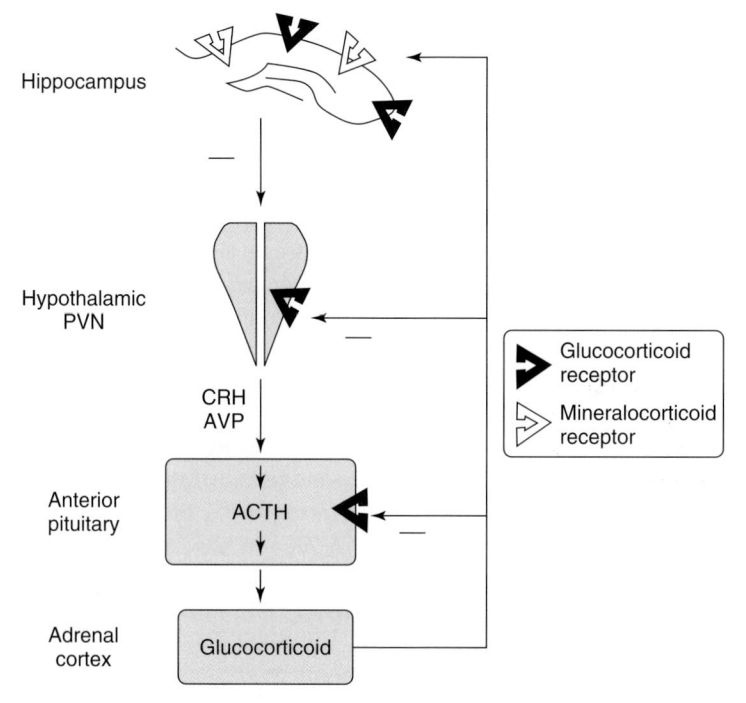

Figure 5.1 The hypothalamic-pituitary-adrenal (HPA) axis. Parvocellular neurons in the PVN produce corticotrophin-releasing hormone (CRH) and vasopressin (AVP), which in turn stimulate adrenocorticotrophin (ACTH) synthesis and release from the anterior pituitary corticotroph cells. ACTH then initiates the production and release of cortisol from the adrenal cortex. Glucocorticoids not only act at multiple loci within the body to maintain homeostasis, but also act in the brain to modify behavior and learning. Owing to the damaging effects of extended glucocorticoid exposure the HPA axis is tightly regulated. Glucocorticoids feedback via glucocorticoid and mineralocorticoid receptors in the limbic system, and glucocorticoid receptors in the PVN and anterior pituitary, to prevent further HPA activity. (Adapted from [60].)

routes by which fetal overexposure to glucocorticoids may have persisting effects on brain structure and function [4].

Glucocorticoids are synthesized and secreted from the adrenal cortex following activation of the HPA axis, the key neuroendocrine response to stress. The anatomical mediators of this response comprise hypothalamic paraventricular nucleus (PVN), the anterior pituitary gland, and the adrenal cortex. Higher control of both HPA axis stimulation and its subsequent inhibitory feedback are afforded by a number of cortico-limbic and brain stem structures (Figure 5.1).

Peripherally, glucocorticoids have permissive effects on various metabolic and physiologic processes, which are genomically mediated predominantly by intracellular glucocorticoid receptors (GRs). Upon binding to glucocorticoid, GR undergoes a conformational change (activation) and translocates to the nucleus, where it homodimerizes with another ligand-bound GR. This dimer binds to palindromic DNA sequences, the glucocorticoid response element (GRE), to modulate target gene expression, often in association with steroid receptor coactivators or corepressors [67, 68]. *In vitro* and in some tissues *in vivo* (e.g., heart, hippocampus), glucocorticoids also bind to mineralocorticoid receptors (MRs), which have 10-fold higher affinity for physiological glucocorticoids than GR. MRs can homodimerize to bind to identical or similar response elements in target genes and may also form heterodimers with the activated GR [69, 70]. Furthermore, an emerging strand of literature suggests that MRs and perhaps GRs are also expressed on cell membranes, notably in the CNS, where they mediate rapid nongenomic effects on electrophysiological and other functions, effects lost in mice lacking MR in the brain [71].

In the adult brain, GRs are ubiquitously expressed, with high levels in the hippocampus, cortex, cerebellum, hypothalamic nuclei including the PVN and the pituitary [8, 72, 73]. In the adult forebrain, MR expression is highly circumscribed, with high expression confined to the hippocampus and septum. MR levels are very low in other regions, at least under basal circumstances [74, 75]. Most fetal tissues have GR expression from early embryonal stages [76, 77], whereas expression of MR has a more limited tissue distribution in development and is only present at later gestational stages, at least in rodents [78]. Additionally, GRs are expressed in the placenta [79] where they mediate metabolic, anti-inflammatory, and parturitional effects.

In the early human embryo, specific expression of GR mRNA has been identified in the metanephros, gut, muscle, spinal cord and dorsal root ganglia, periderm, sex chords of the testis, and adrenal by 8–10 weeks of life [80, 81]. Expression is also high in the human lung by 12 weeks of gestation. Clearly, systems to transduce glucocorticoid actions upon the genome exist from early developmental stages, with complex cell-specific patterns of expression and presumable sensitivity to the steroid ligands [77]. This is also reflected in the complexity of developmental profiles of corticosteroid receptors during critical periods of fetal brain growth [78, 82–84]. Unfortunately, however, the detailed ontogeny of GRs and MRs in the human fetus, later in gestation, or at any stage of brain development are unknown.

5.4
Glucocorticoids: the Endocrine Programming Factor

Glucocorticoids have become a primary candidate for fetal programming during prenatal stress. Glucocorticoid treatment during pregnancy reduces birth weight in animal models, in nonhuman primates [85–90], and humans [88, 91]. Birth weight reduction is most notable when glucocorticoids are administered in the later stages of pregnancy [86]. In human pregnancy, endogenous fetal cortisol levels are increased in intrauterine growth retardation and in preeclampsia, implicating endogenous glucocorticoids in fetal growth retardation [92, 93].

5.4.1
Placental 11β-HSD2: a Barrier to Maternal Glucocorticoids

Under normal circumstances, access of maternal endogenous glucocorticoid to the fetus is relatively low. Stress will lead to a large number of cardiovascular and endocrine changes in the mother, including increases in plasma adrenocorticotrophin (ACTH), β-endorphin, glucocorticoid, and catecholamine concentrations. To protect the fetus, the placenta forms a structural and biochemical barrier to many of these maternal factors and glucocorticoids are no exception.

11β-Hydroxysteroid dehydrogenase type 2 (11β-HSD2), an NAD-dependent, 11β-dehydrogenase, catalyzes the rapid inactivation of physiological glucocorticoids (cortisol, corticosterone) to inert 11-keto forms (cortisone, 11-dehydrocorticosterone) [94]. The biology of 11β-HSD2 was first understood in the distal nephron where the enzyme "protects" otherwise nonselective MRs from glucocorticoids, allowing the nonsubstrate aldosterone exclusive access to these sites. 11β-HSD2 is also highly expressed in the placenta. It is located at the interface between maternal and fetal circulations, in the syncytiotrophoblast in humans [78, 95] and the labyrinthine zone in rodents [96, 97]. Placental 11β-HSD2 forms a potent barrier to maternal glucocorticoids [98–100]. This barrier is apparently incomplete since 10–20% of maternal cortisol crosses intact to the fetus (Figure 5.2). Indeed, in rodents the peak of the circadian rhythm of plasma corticosterone penetrates the 11β-HSD2 barrier

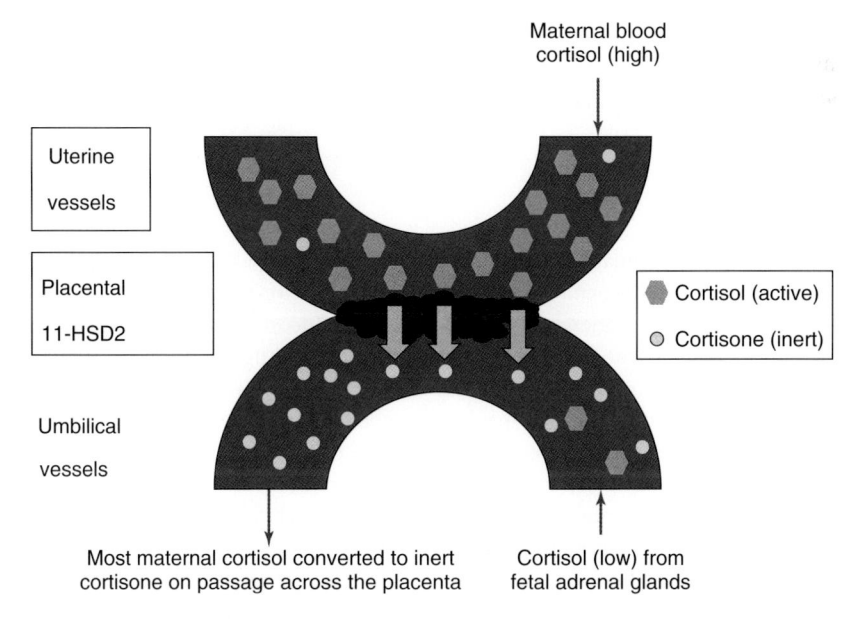

Figure 5.2 Placental 11β-hydroxysteroid dehydrogenase type 2. Glucocorticoids restrain fetal growth and alter the trajectory of fetal tissue maturation. Concentrations of the active glucocorticoid cortisol are high in maternal blood during pregnancy. This placenta cannot stop lipophilic steroids crossing to the fetus, but uses placental 11β-HSD2 to rapidly inactivate cortisol to inert cortisone, thus minimizing fetal exposure.

to an appreciable extent [101], and in guinea pigs a decrease in placental 11β-HSD2 at term is associated with increased transfer of maternal cortisol to the fetus [102]. This is presumably necessary for normal prebirth developmental processes. The synthetic glucocorticoids dexamethasone and betamethasone are poor substrates for 11β-HSD2 and therefore pass the placenta [78, 103]. In contrast, 11β-HSD2 inactivates prednisolone to inert prednisone, attenuating its impact upon the fetus *in vivo* [104].

The activity of placental 11β-HSD2 near term shows considerable interindividual variation in humans and rats [105, 106]. A relative deficiency of 11β-HSD2, with consequent reduced placental inactivation of maternal corticosteroids, is hypothesized to lead to overexposure of the fetus to glucocorticoids, retardation of fetal growth and program responses leading to later disease [16]. In support of this idea, lower placental 11β-HSD2 activity in rats associates with the smallest fetuses [106]. Similar associations have been reported in humans [105, 107–109], although not all studies have concurred [110, 111]. Additionally, markers of fetal exposure to glucocorticoids such as cord blood levels of osteocalcin (a glucocorticoid-sensitive osteoblast product that does not cross the placenta), also correlate with placental 11β-HSD2 activity [112].

Humans homozygous (or compound heterozygous) for deleterious mutations of the 11β-HSD2 gene have very low birth weight [113], averaging 1.2 kg less than their mostly heterozygous sibs. Although an initial report suggested that 11β-HSD2 null mice have normal fetal weight in late gestation [114], this appears to have reflected the "genetic noise" of the crossed (129×MF1) strain background of the original 11β-HSD2 null mouse. In congenic mice with the C57Bl/6 strain background, 11β-HSD2 nullizygosity lowers birth weight [115].

5.4.2
Glucocorticoid Programming

All these models of prenatal glucocorticoid exposure also have persisting peripheral effects through the lifespan. Thus, prenatal dexamethasone, stress or 11β-HSD2 inhibition programs higher adult blood pressure, glucose, and insulin levels in rats, sheep, and other model species [106, 116–125] including nonhuman primates [126]. Moreover, in humans, reduced placental 11β-HSD2 activity, as determined by a higher cortisol to cortisone ratio in venous umbilical cord blood, is associated with higher blood pressure at three years [127]. Intriguingly, this biology is not restricted to placental mammals since glucocorticoids can program neuroendocrine stress axis activity in nonmammalian species such as the frog [128] and bird [129].

An elegant series of experiments has shown that the glucocorticoid effects can be mediated directly on the fetus and its placenta rather than indirectly via alterations in maternal food intake or other aspects of her biology. A "heterozygote" cross of 11β-HSD2$^{+/-}$ mice allows a single mother to bear wild-type, heterozygous, and null offspring in the same pregnancy; note that placental 11β-HSD2 is in the fetal component of the placenta. Birth weight follows the feto-placental genotype, as does offspring behavior, with 11β-HSD2 null offspring having the lowest birth weight and exhibiting the most "anxious" adult phenotype compared with wild-type littermates [115]. Importantly, heterozygotes show birth weights intermediate between their wild-type and 11β-HSD2$^{-/-}$ homozygous littermates suggesting that variation and not only complete absence of feto-placental 11β-HSD2 determines fetal growth and plausibly programming. Indeed, because maternal glucocorticoid levels are much higher (2–10-fold higher) than those of the fetus, subtle changes in placental 11β-HSD2 activity may have profound effects on fetal glucocorticoid exposure [98, 99].

5.4.3
Transgenerational Effects

Programming may not be restricted to the immediate offspring of a challenged pregnancy but may also affect a subsequent generation, the so-called "transgenerational effects." Thus, exposure of pregnant rats to dexamethasone for five days at the end of gestation reduces birth weight and produces persisting cardiometabolic and neuroendocrine changes in the offspring (F1). Mating such programmed F1 offspring produces similar reductions in birth weight and subsequent pathophysiology in their offspring (F2) [130]. While this might be a "uterine effect," after all, the F2 offspring are developing in a uterus within a mildly hypertensive, hyperglycemic, insulin resistant, and hypercorticosteronemic mother, the effect does not recur with further crosses (F3), though the F2 phenotype is at least as marked as F1. Moreover, while the same low birth weight, programmed phenotype is seen with dexamethasone F1 females crossed with control F1 males, it also follows the paternal line such that dexamethasone-exposed males (F1) mated with control females give rise to low birth weight and hyperglycemic offspring. This implies an epigenetic (chromatin) or other nucleic acid-associated mechanisms, since there is little else delivered by the F1 sperm to the next generation.

5.4.4
The Placenta

Of course, 11β-HSD2 may also regulate glucocorticoid effects directly on the placenta, which highly expresses GRs (and some MRs). Indeed, the placentas of 11β-HSD2$^{-/-}$ offspring of 11β-HSD2$^{+/-}$ (heterozygous crosses) have reduced weight, reduced fetal capillary surface area, plausibly due to lower expression of VEGF (inhibited by glucocorticoids), and reduced glucose and amino acid transport to the fetus compared with placentas of wild-type littermates [131].

5.4.5
A Common Mechanism?

Maternal undernutrition/malnutrition or maternal illness in pregnancy also causes low birth weight and offspring cardiometabolic and neuroendocrine changes

[132–134]. Intriguingly, glucocorticoid exposure may represent a common underlying mechanism. Dietary protein restriction during rat pregnancy selectively attenuates 11β-HSD2, but apparently not other placental enzymes [135–137]. In the maternal protein restriction model, offspring hypertension can be prevented by treating the pregnant dam with glucocorticoid synthesis inhibitors, and can be recreated by concurrent administration of corticosterone, at least in the female offspring [138].

Placental 11β-HSD2 is similarly reduced by several key maternal environmental factors. Although the precise control molecular basis for its control have yet to be fully delineated, *in vitro* "placental" cell 11β-HSD2 is regulated by glucocorticoids [139], catecholamines acting via adrenergic receptors [140], components of tobacco smoke [141], and various inflammatory signals [142, 143]. *In vivo* placental 11β-HSD2 activity is down-regulated by stress, illness, inflammation, infection, and hypoxia [144–146]. The implication of stress-induced down-regulation of placental 11β-HSD2 is reduced inactivation of active glucocorticoids to their inert 11-keto forms, consequently allowing overexposure of the fetus to glucocorticoids, thus retarding fetal growth. Placental 11β-HSD2 may thus represent one of only a few common mechanisms "signaling" between maternal and fetal environments.

5.5
Fetal Tissue Glucocorticoid Sensitivity

The placenta is not the only level of control of exposure of developing fetal tissues to glucocorticoids. Intracellular receptor density has complex cell- and temporal-specific patterns of expression throughout fetal development, which presumably reflects sensitivity to the steroid ligands [77]. However, whether or not fetal GRs and MRs are occupied by endogenous glucocorticoids until late gestation is uncertain, as there is plentiful 11β-HSD2 in many fetal tissues including the CNS at midgestation [78, 83, 147], which presumably "protects" vulnerable developing cells from premature glucocorticoid action. 11β-HSD2 expression is dramatically switched-off at the end of midgestation in the rat and mouse brain, coinciding with the terminal stage of neurogenesis [78, 83]. Similarly, in the human fetal brain 11β-HSD2 appears to be silenced between gestational weeks 19 and 26 [78, 148]. 11β-HSD2 knockout mice are sensitive to cerebellar effects of exogenous corticosterone postnatally, while wild-type animals are not [149]. Thus, there appears to be an exquisitely timed system of protection and then exposure of developing brain regions to circulating glucocorticoids.

Clearly, penetration of glucocorticoids into the fetal brain represents an important determinant of acute and persistent effects on neural development and structural and functional integrity. The blood–brain barrier (BBB) limits access of synthetic glucocorticoids into the adult rodent brain. Multidrug resistance (Mdr)1A (also called P-glycoprotein, Pgp) expression in the choroid plexus attenuates the penetration of dexamethasone into the brain [150] by actively pumping the steroid across membranes and out of cells [151] (Figure 5.3). Mdr1a is also expressed in the

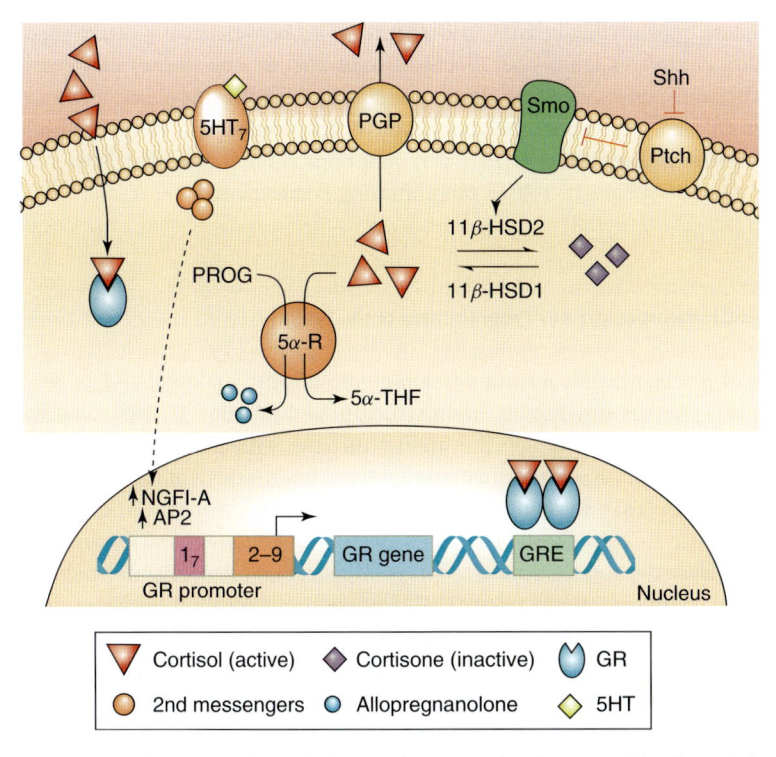

Figure 5.3 Glucocorticoid metabolism and signaling in neurons. Glucocorticoids are lipophilic and cross the plasma membrane to bind with intracellular glucocorticoid receptors (GR) and mineralocorticoid receptors (not shown). For their effects on gene expression, receptors dimerize and interact at glucocorticoid response elements (GRE). Serotonergic (5HT) signaling involves binding of 5HT to the G-protein-coupled receptor, $5HT_7$, which is positively coupled to cAMP generation. This then stimulates nuclear transcription factors, most notably NGFI-A and AP-2, to bind to the GR promoter, and regulate GR gene transcription. Methylation status at exon I_7 of the GR promoter is important in regulating transcription factor binding, and thus gene expression. Glucocorticoid signaling is regulated at several levels, including through P-glycoprotein (PGP), which attenuates the penetration of glucocorticoids by actively pumping the steroid across membranes and out of cells. Also, intracellular steroid metabolism by 11β-hydroxysteroid dehydrogenase (11β-HSD) types 1 and 2 determines the ratio between active (cortisol) and inactive (cortisone) glucocorticoid in the cell. Sonic hedgehog (Shh) signaling is a potent inducer of 11β-HSD2, but not 11β-HSD1, acting through Patched (Ptch) and Smoothened (Smo) receptors. 5α-Reductase (5α-R) represents the rate-limiting step in the conversion of cortisol to 5α-tetrahydrocortisol (5α-THF) and of progesterone (PROG) to the anxiolytic, neuroprotective neurosteroid, allopregnanolone.

fetal brain [152]. However, the injection of pregnant guinea pigs with the Mdr1a substrate steroids dexamethasone or betamethasone results in a dose-dependent reduction in hypothalamic corticotrophin-releasing hormone (CRH) mRNA [153], confirming that appreciable amounts of synthetic glucocorticoid pass into the fetal brain. Therefore, protection of the brain during this developmental stage is incomplete, presumably because the BBB is yet to become firmly established. Preliminary data also suggest that glucocorticoids and various stimuli that evoke cellular stress responses can alter Pgp expression at the BBB [154–156], although the effect on fetal Pgp expression has yet to be determined. Furthermore, high Mdr expression in the placenta [157] may represent another potential barrier to maternal glucocorticoids. Its role in programming is unexplored.

5.6
Stress and Glucocorticoids: Key Programmers of the Brain

Elevation of glucocorticoids during development following maternal and/or fetal stressors can permanently modify brain structure and function [4, 158], altering developmental trajectories of specific brain structures with persistent effects [3, 159]. Prenatal glucocorticoid administration retards brain weight at birth in sheep [160] and rodents [161], delaying maturation of neurons, myelination, glia, and vasculature [162, 163]. In rhesus monkeys, treatment with antenatal dexamethasone causes dose-dependent degeneration of hippocampal neurons and reduces hippocampal volume, which persists [164]. Fetuses receiving multiple lower-dose injections showed more severe damage than those receiving a single large injection. Decreased synaptic density in the hippocampus has also been demonstrated in juvenile rat offspring born to prenatally stressed mothers [165]. Prenatal stress in rats reduces neurogenesis in the dentate gyrus and impairs hippocampus-related spatial tasks, and blocks the increase of learning-induced neurogenesis in adult offsprings [166, 167]. Human and animal studies have demonstrated that altered hippocampal structure may be associated with a number of consequences for memory and behavior [168–170].

5.6.1
Programming the HPA Axis

The HPA axis, and its key limbic regulator, the hippocampus [171], are particularly sensitive to glucocorticoids and their perinatal programming actions [159, 172–174]. Prenatal stress may impact upon the development of neurons of the hypothalamic PVN in the fetal rat, including the CRH-containing neurons, with implications for HPA axis programming [175]. Alterations in HPA activity throughout life will impact adult health due to altered tissue exposure to endogenous glucocorticoid. Prenatal glucocorticoid exposure permanently increases basal and stress-stimulated plasma corticosterone levels in adult rats [121, 159] and primates [126, 176]. Adrenal hypertrophy in prenatally stressed offspring may contribute to their tendency for greater stress reactivity [177, 178], and the density of GRs and MRs is permanently reduced in the hippocampus; changes that are anticipated to attenuate HPA axis feedback sensitivity. Maternal undernutrition in rats [135] and sheep [179] also affects adult HPA axis function, suggesting that HPA programming may be a common outcome of prenatal environmental challenge, perhaps acting in part via alterations in placental 11β-HSD2 activity, which is selectively down-regulated by maternal low-protein/high-carbohydrate diets [135, 136].

In such prenatal environmental programming models, chronic plasma glucocorticoid excess in the offspring has been associated with affective and cognitive impairments, as well as hypertension, hyperglycemia, increased risk of diabetes, hypercholesterolemia, and immunosuppression [180, 181].

5.6.2
Sex-Specific Effects

HPA axis programming also illustrates an important variable; it often differs between male and female offspring of the same litter. Sex-specific programming of the HPA axis has been reported for prenatal stress in rats [182, 183]. In male guinea pigs, short-term prenatal exposure to dexamethasone significantly elevates subsequent basal plasma cortisol levels, whereas similarly exposed females have reduced HPA responses to stress. Conversely, exposure to longer courses of prenatal glucocorticoids exhibit reduced plasma cortisol levels in adulthood, while similarly exposed females have higher plasma cortisol levels as adults in the follicular and early luteal phases of their oestrus cycles. In terms of mechanisms of such sexual dimorphism, there may be greater glucocorticoid transfer across the placenta of female compared to male fetuses [184], perhaps due to sex differences in placental 11β-HSD2, the presence and interaction of gonadal steroids [185–188], and basal sex differences in epigenetic machinery, discussed below [189].

5.6.3
Programming Behavior

Overexposure to glucocorticoids *in utero* leads to alterations in adult behavior. Late gestational dexamethasone in rats apparently impairs coping in aversive situations later in life [2]. Prenatal glucocorticoid exposure also affects the developing dopaminergic system [190, 191] with implications for understanding of the developmental contributions to schizo-affective, attention-deficit hyperactivity and extrapyramidal disorders. Stressful events in the second trimester of human pregnancy associate with an increased incidence of offspring schizophrenia [192]. Prenatal exposure to glucocorticoids may exert more widespread neuronal effects since they also increase the susceptibility of the cochlea to acoustic noise trauma in adulthood [193].

Rodents prenatally stressed give birth to offspring with elevated anxiety behavior. These animals show increased emotionality and timidity in elevated plus, open field, and other tests of "anxiety-related" behaviors [182, 194–198]. Other prenatal stress paradigms have supported this, with offspring showing reduced grooming and rearing [199–201]. Behavioral changes in adults exposed prenatally to glucocorticoids may be associated with altered functioning of the amygdala, a structure key to the expression of fear and anxiety. Intra-amygdala administration of CRH is anxiogenic [202]. Prenatal glucocorticoid exposure increases adult CRH levels specifically in the central nucleus of the amygdala [159, 199]. Prenatal stress similarly programs increased "anxiety-related" behaviors with elevated CRH in the amygdala [203]. Moreover, corticosteroids facilitate CRH mRNA expression in this nucleus [204] and increase GR and/or MR in the amygdalae [159, 199]. The amygdala stimulates the HPA axis via a CRH signal [205]. Thus, an elevated corticosteroid signal in the amygdala due to hypercorticosteronemia in the adult offspring of dexamethasone-treated dams may produce the increased CRH levels in adulthood. A direct relationship between brain corticosteroid receptor levels and anxiety-like behavior is supported by the phenotype of transgenic mice with selective loss of GR gene expression in the forebrain (including amygdala and hippocampus but not hypothalamus), which show markedly reduced "anxiety" [206].

5.7
CNS Programming Mechanisms

Indications of the molecular mechanisms by which early life environmental factors program lifespan physiology come from the studies of the processes underpinning early postnatal environmental programming of the HPA axis [1, 62, 207, 208]. In these models, short daily handling of rat pups [208] or merely more attentive maternal care to the pups [209] during the first two weeks of life permanently increases GR density in the hippocampus and prefrontal cortex, but not in other brain regions. This increase in receptor density potentiates the HPA axis sensitivity to glucocorticoid negative feedback and results in lower plasma glucocorticoid levels throughout life, a state compatible with a tighter HPA recovery from environmental stress [210, 211]. It is important to note that this adaptation may be perceived as advantageous or disadvantageous depending on the actual postnatal challenges experienced. Neonatal glucocorticoid exposure may have similar effects [212].

Such postnatal events act via ascending serotonergic (5HT) pathways from the midbrain raphe nuclei to the hippocampus [213]. Activation of 5HT induces GR gene expression in fetal hippocampal neurons *in vitro* [214] and in hippocampal neurons *in vivo* [215]. The induction of 5HT requires thyroid hormones, which are elevated by appropriate maternal cues. At the postsynaptic hippocampal neuron, early postnatal events mediate their effects likely via the $5HT_7$ receptor subtype, which is regulated by glucocorticoids [216] and positively coupled to cAMP generation [217]. The next step appears to involve stimulation of cAMP-associated and other transcription factors, most notably NGFI-A and AP-2 [217]. NGFI-A and AP-2 may bind to the GR gene promoter [218] (Figure 5.3). This pathway might also be involved in some *prenatal* programming paradigms affecting the HPA axis since last trimester dexamethasone exposure increases 5HT transporter expression in the rat brain [219, 220], an effect predicted to reduce 5HT availability in the hippocampus and elsewhere. This may well induce a fall of GRs and MRs, the converse of postnatal higher level maternal care.

Recent studies have demonstrated that the detrimental effects of excess glucocorticoid exposure on neural development involves the Sonic hedgehog (Shh) signaling pathway, a crucial regulator of neural stem/progenitor cells [221, 222]. Glucocorticoids suppress Shh-induced proliferation of cerebellar progenitor cells in postnatal mice [223]. Shh signaling is also protective against glucocorticoid-induced neonatal cerebellar injury by inducing 11β-HSD2 [224] (Figure 5.3), which is expressed in the developing CNS, including in cerebellar granule neuron precursors [225]. 11β-HSD2 expression is maintained later in the cerebellum than in other CNS regions and its function is necessary for normal cerebellar development [115], inactivating corticosterone and prednisolone, but not dexamethasone. This may provide the basis of further studies in the therapeutic potential of neuroprotective Shh agonists to prevent glucocorticoid-induced prenatal brain injury, the benefits of which have been observed in preliminary rodent studies [226]. Additionally, this provides the rationale for the therapeutic use of 11β-HSD2-sensitive glucocorticoids in humans, such as hydrocortisone (cortisol).

5.7.1
The GR Gene: a Common Programming Target?

Transgenic mice with a reduction of 30–50% in tissue levels of GRs have striking neuroendocrine, metabolic, and immunological abnormalities [227] and, while complete knockout is lethal [76], tissue-specific knockout also has profound effects upon the manipulated cells and organs [206, 228]. The level of expression of GRs is thus critical for cell function. A series of studies have shown that early life events permanently influence the levels of expression of GRs in target tissues. Thus, antenatal glucocorticoid exposure in the rat permanently elevates the GR mRNA in adult offspring periportal hepatic cells, visceral adipose tissue, and amygdala, but reduces the GR in hippocampal neurons [86, 199, 229, 230].

The GR promoter is complex, with multiple tissue-specific alternate untranslated first exons in rats [231], mice [76], and humans [232], most within a transcriptionally active CpG island. All the GR mRNA species give rise to the same receptor protein encoded by exons 2–9. The alternate untranslated first exons are spliced onto the common translated sequence beginning at exon 2. In the rat, two of the alternate exons are present in all tissues that have been studied; however, others are tissue-specific [231]. This permits considerable complexity of tissue-specific variation in the control of GR expression, without allowing any tissue to become GR deplete.

Tissue-specific exon 1 usage is regulated by perinatal environment manipulations. Indeed, neonatal handling, which increases aspects of maternal behavior to her pups, notably licking and grooming, permanently programs increased expression of only one of the six alternate first exons (exon 1_7) utilized in the hippocampus [231]. Exon 1_7 contains sites that are appropriate to bind the very third messenger/intermediate early gene transcription factors (AP-2, NGF1-A) themselves induced by the neonatal manipulation [217, 233].

5.7.2
Epigenetics

The next question is how discrete perinatal events can permanently alter gene expression. Evidence has recently emerged for selective methylation/demethylation of specific CpG dinucleotides of the *GR* gene, notably around the putative NGFI-A site of exon 1_7. These sites are subject to differential and permanent demethylation just after birth, in association with the level of maternal care. Thus, transcription from this brain-enriched promoter is reduced with lower-density maternal care [233]. Such changes affect NGFI-A binding to the promoter, with methylation, as expected, inhibiting NGFI-A binding [234]. NGFI-A binding to its exon 1_7 promoter sequence is required for epigenetic reprogramming. Indeed, knockdown of NGFI-A in primary hippocampal cell culture prevents serotonin-induced demethylation of exon 1_7 and decreases its expression. Furthermore, GRs under some circumstances can mediate differential demethylation of target gene promoters, which persist after steroid withdrawal, at least in liver-derived cells [235].

Such epigenetic effects are unlikely to be confined to the hippocampus or the GR. Thus, GRs in liver-derived cells mediate differential demethylation of target gene promoters, effects which persist after steroid withdrawal [235]. During development, such target promoter demethylation occurs before birth and may fine-tune the promoter to "remember" regulatory events occurring during development. Indeed, maternal malnutrition reduces methylation of the *GR* gene in postnatal liver [236], perhaps underpinning the increased GR expression seen in adults following prenatal glucocorticoids [86, 237].

An analogous effect of maternal care has been reported for the estrogen receptor alpha (ERα) gene. One of its alternate first exons/promoters, ER$α_{1b}$, shows methylation differences. These associate with variations in the binding and effects of the transcription factor Stat5b upon expression of the receptor in the medial preoptic area and thus maternal behaviors [238]. Similarly, methylation of the *PPARα* gene in liver is reduced following prenatal calorie restriction [235] though functional links are as yet unproven. In humans the biology may be conserved since low birth weight plus lower levels of maternal care associate with reduced adult hippocampal volume, a marker of vulnerability to affective and other neuropsychiatric disorders, albeit only in females [239].

5.8
Glucocorticoid Programming in Humans

5.8.1
Clinical Use of Prenatal Glucocorticoid Therapy

The effect of fetal overexposure to glucocorticoids is highly clinically relevant because pregnant women at risk of preterm delivery are routinely treated with synthetic glucocorticoids. This regimen has been demonstrated to substantially reduce respiratory distress syndrome in newborns at 24–34 weeks gestation [240]. Betamethasone and dexamethasone are most commonly administered as they have little MR activity and the greatest efficacy in improving postnatal respiratory outcomes. Prenatal synthetic glucocorticoid treatment may also reduce gross neuroanatomical deficits such as intraventricular hemorrhage, periventricular leucomalacia, and cerebral palsy [240–242]. Although there have been no reports of adverse effects of a single course of antenatal glucocorticoid treatment in children, it has been well-recognized that multiple courses can have adverse neurological outcomes [243], associating with decreased weight, length, and head circumference at birth, without improving neonatal mortality or the respiratory and neuroanatomical deficits [244]. This common obstetric practice [245, 246] seems deleterious for postnatal health. In human pregnancy, glucocorticoids are also used long term in the antenatal treatment of fetuses at risk of congenital adrenal hyperplasia (CAH) due to 21-hydroxylase deficiency.

5.8.2
Consequences of Human Fetal Glucocorticoid Overexposure

High-dose glucocorticoid treatment during pregnancy reduces birth weight [88, 91, 247], although normal birth weight has been reported in infants at risk of CAH whose mothers received relatively low-dose dexamethasone *in utero* from the first trimester [248, 249]. Short-term antenatal glucocorticoid administration has been linked with higher blood pressure in adolescence [250] and mild insulin resistance in adults [251], but perhaps not in the context of early gestational treatment of pregnancies at risk of CAH [252]. A number of studies aimed at establishing the long-term neurological and developmental effects of antenatal glucocorticoid exposure have been complicated by the fact that most of the children studied were born before term and were therefore already at risk of delayed neurological development. It is also difficult to separate the confounding effects of prematurity from the use of antenatal glucocorticoids, often for lung maturation. In a group of six year old children, antenatal glucocorticoid exposure was associated with subtle effects on neurological function, including reduced visual closure and visual memory [253]. Children exposed to dexamethasone in early pregnancy because they were at risk of CAH, and who were born at term, showed increased emotionality, unsociability, avoidance, and behavioral problems [254]. These effects were seen in genetically unaffected glucocorticoid-exposed offspring. Furthermore, a recent study has shown that multiple doses of antenatal glucocorticoids given to women at risk of preterm delivery were associated with reduced head circumference in the offspring [88], which may indicate compromised cognitive function [255], and a small study utilizing MRI analysis revealed a reduction in cortical folding and surface area in infants exposed to repeated synthetic glucocorticoids [256].

There are also effects on behavior; three or more courses of glucocorticoids associate with an increased risk of externalizing behavior problems, distractibility, and inattention [88]. Also, increased hyperkinetic, aggressive behavior was observed in children aged three and six years after three or more courses of prenatal glucocorticoids [257]. A recent controlled trial of postnatal dexamethasone in premature babies showed associations with lower subsequent IQ and decrements of other cortical functions [258].

However, these findings are not consistently reported, and follow-up studies of children treated from the first trimester with dexamethasone in suspected cases of CAH revealed no adverse effects of prolonged glucocorticoid exposure on cognitive and developmental outcomes at 12 years of age [259, 260]. Similarly, in a randomized trial comparing single and repeat courses of antenatal corticosteroids, two to three year follow-up of children who had been exposed to repeat as compared with single courses of antenatal corticosteroids did not differ significantly in physical or neurocognitive measures [261]. A long-term follow-up study revealed no effect of a single course of prenatal betamethasone on intellectual capacity, sex-specific cognitive function, and psychoneuroticism at 20 years of age [262]. Overall, the jury is still out, with short-term late gestational glucocorticoids probably safe and certainly efficacious in preventing respiratory distress, but longer-term antenatal glucocorticoid treatment more likely to have persisting adverse effects for uncertain additional benefits.

As in other mammals, the human HPA axis appears to be programmed by the early life environment. Higher plasma and urinary glucocorticoid levels are found in children and adults who were of lower birth weight [263, 264]. This appears to occur in disparate populations [265] and may precede overt adult disease [13], at least in a socially disadvantaged South African population. Additionally, adult HPA responses to ACTH stimulation are exaggerated in those of low birth weight [13, 266], reflecting the stress axis biology elucidated in animal models. The HPA axis activation is associated with higher blood pressure, insulin resistance, glucose intolerance, and hyperlipidemia [266]. Finally, the human *GR* gene promoter has multiple alternate untranslated first exons [232] analogous to the rat and mouse promoter structure. Crucially, in humans, childhood abuse associates with lower levels of hippocampal GR mRNA, fewer NGFI-A-responsive GR exon 1F transcripts, and increased methylation of this region [267], findings closely analogous to those in the rat.

In sum, the machinery in humans and rats seems similar, and at least one programming mechanism is analogous. Further, chronically high glucocorticoid exposure has programming effects in humans, whereas low or brief exposure probably does not. The next question is whether this links to brain disorders.

5.8.3
Programming and Posttraumatic Stress Disorder (PTSD)

It has been estimated that 10–40% of individuals who have been exposed to extreme trauma, such as the Holocaust, combat, rape, abuse, or a road traffic accident, develop PTSD [268]. Reduced urinary, plasma, and salivary cortisol levels in PTSD have been reported in several studies of trauma survivors [269], perhaps reflecting the increased tissue sensitivity to glucocorticoids in PTSD [270] and hence enhanced feedback. As such, these individuals show higher pituitary sensitivity to the corticotrophin-suppressing effects of dexamethasone [269]. Recently, lower levels of salivary cortisol were reported in mothers who developed PTSD after being present at or near to the World Trade center atrocity on 11 September 2001 in New York, than in mothers who did not develop PTSD. Crucially, the one year old offspring of mothers with PTSD also had lower salivary cortisol levels [271]. The changes were most apparent in babies born to mothers who were in the last three months of their pregnancies when the trauma occurred. Intriguingly there was no effect on birth weight, though the incidence of intrauterine growth retardation was increased [272] and head circumference reduced [273]. Singleton-bearing nonhuman primates exposed to low doses of dexamethasone in the second half of pregnancy also show persisting hypercortisolemia (as seen in the analogous rodent models), but no alteration in birth weight [126]. The direction of change in glucocorticoids may differ between perinatal challenges, their gestational timing, and the species involved (some data imply that hypocortisolemia with PTSD associates with increased tissue sensitivity to glucocorticoids so that the target actions may be increased), but the principle of particular vulnerability of the HPA axis to developmental "programming" seems consistent. The findings implicate *in utero* effects as major contributors to a possible biological risk factor (hypocortisolemia) for PTSD.

Intriguingly, the children of Holocaust survivors with PTSD have significantly lower 24-h mean urinary cortisol excretion than the children of Holocaust survivors without PTSD [269] suggesting similar intergenerational transmission as seen in glucocorticoid-exposed rats [130]. Recent data on Holocaust survivors suggest a link between the age of trauma exposure and the persisting phenotype, with trauma at younger ages linking with changes in cortisol metabolism but lesser PTSD and metabolic syndrome–related pathologies in later life [274]. Such effects may transmit into subsequent generations, since healthy adult children of Holocaust survivors with PTSD (and therefore lower plasma cortisol levels) themselves have lower cortisol levels though no PTSD. This appears to be confined to the children of Holocaust-exposed mothers with PTSD [275]. The implication is that the marker of altered HPA axis functioning, itself perhaps a vulnerability factor, is transmitted into a subsequent generation, suggesting epigenetic "inheritance" across at least one generation, as seen in rodent models. The Holocaust survivors also showed a reduction in renal 11β-HSD2 activity [274]. This change may reflect underlying processes to maintain glucocorticoid action in key target tissues, such as the kidney, to increase sodium retention, and maximize survival. Whether or not there are parallel effects on placental 11β-HSD2, which is largely driven by the fetal genotype though modulated by the maternal environment, remains an intriguingly open question in the light of possible transgenerational effects.

5.8.4
Programming other Glucocorticoid Metabolizing Enzymes

It is well understood that intracellular steroid metabolism plays an important role in tissue sensitivity to the action of glucocorticoids. While we have concentrated on 11β-HSD2 here, metabolism by A-ring reductases represents an important glucocorticoid metabolic pathway, and interestingly, the process of 5α-reduction may also serve as a potential target in developmental programming. 5α-Reductase exists as two isoforms, both of which are expressed in the developing and adult brain [276–279], as well as the liver and male reproductive organs [280]. The type 2 isozyme has been associated with sexually dimorphic functions of the male, whereas the type 1 form has been proposed as a constitutive enzyme that essentially plays a catabolic and neuroprotective role [281]. The enzyme is responsible for the conversion of free cortisol to 5α-tetrahydrocortisol, and represents the rate-limiting step in allopregnanolone production (Figure 5.3), a key anxiolytic neurosteroid, which allosterically activates $GABA_A$ receptors in the CNS. Allopregnanolone and related neurosteroids exert paracrine effects important for neuronal survival and neurogenesis [282], enhance neurogenesis [283], contribute to synapse stabilization [284], and inhibit toxin-induced cell death [285]. Interestingly, 5α-reductase is sensitive to permanent "programming" by prenatal stress. For example, hepatic 5α-reductase is irreversibly "programmed" in rodents following prenatal stress [286] and in the brain; daily immobilization stress of pregnant rats lowers 5α-reductase (isoform 1) activity in the male offspring [287]. The consequence would be a fall in allopregnanolone production, potentially attenuating its neuroprotective and anxiolytic roles.

Offspring of Holocaust survivors with PTSD showed significant lowering of 5α-reductase [274], the immediate consequence of which would be a decrease in cortisol breakdown, potentially amplifying glucocorticoid effects in the liver and other metabolic tissues, without altering circulating cortisol. A fall in 5α-reductase activity could also prolong cortisol action in the CNS, with potential detrimental catabolic consequences on neuronal integrity [288]. Thus, this may represent a mechanism by which prenatal stress programs the kinetics of glucocorticoid metabolism in the brain, predisposing the individual to the development of affective disorders in later life, through the aforementioned neurodegenerative effects and/or via reduced anxiolytic neurosteroid synthesis in adult life. These intriguing possibilities require study.

5.9
Future Perspectives and Therapeutic Opportunities

Perinatal overexposure of glucocorticoids contributes to the development of cardiometabolic, neuroendocrine, and behavioral disorders. The brain is particularly susceptible to these programming effects, and recent studies aiming to attenuate neuronal damage that resulted from excess GR activation in a rat model [289] have exploited various genetic strategies to attenuate hippocampal lesion size, rendering excess glucocorticoids protective rather than destructive. A reversal of programmed up-regulation of GRs in the liver has also been observed following metformin administration [290], a widely used antidiabetic agent, thus providing therapeutic promise. Moreover, in earlier life, leptin administration to mother or her offspring appears to reverse some of the effects of early life challenge, perhaps because this adipose tissue hormone sends an "all's well" message to the CNS otherwise lacking in starved/stressed offspring [291–293].

This also affords the possibility of personalizing therapy, aiming to dissect the genetic and epigenetic, early life and adult factors, thus targeting treatment to the individual's repertoire of causation. For example, metformin may be appropriate for "programmed" metabolic syndrome, whereas 5a-reductase inducers/neurosteroid analogs may perhaps be more suitable for developmentally programmed affective disorders, and leptin administration may a more suitable approach toward managing the consequences of malnourished/undernourished pregnancies.

Further, in humans, as in animal models, reinstating appropriate HPA signaling seems to be a promising treatment approach. On the basis of the rationale that endogenous cortisol provides an inadequate signal to contain the stress reactions of PTSD patients, some pilot studies have shown that cortisol administration ameliorates PTSD symptoms [294, 295].

Epigenetic programming is becoming an increasingly important aspect of this biology. Epigenetic modifications are "metastable" and can be modulated in adult life. Pharmacological manipulations including histone deacetylase inhibition and L-methionine (methyl donor) administration in adulthood [296, 297] have recently been shown to reverse the effect of lower hippocampal GR expression induced by lower-density maternal care. This epigenetic reprogramming reduces anxiety behavior. However, many genes and their promoters are methylated and histones acetylated, and "beneficial" effects on one system may be counterbalanced by less advantageous effects on others.

Of course, future treatment strategies that also implement preventative measures may benefit from neuroendocrine testing as a "vulnerability indicator" for the development of affective disorders. It has recently been shown that HPA axis hyperactivity can serve as a suicide predictor in adolescents [298] and elderly mood disorder in patients [299], highlighting the importance of the dexamethasone suppression test (DST) nonsuppression to identify individuals with increased vulnerability to the adversities and stressors of life. With ongoing studies of the molecular, genetic, neruoendocrine, and clinical aspects of "programming" biology, we can be optimistic of understanding of the causation of stress-related diseases, and also of new targets for treatment.

5.10
Overview

Prenatal exposure to glucocorticoids may "program" a range of tissue-specific pathophysiologies. The fetus may be exposed to exogenous glucocorticoids, active steroids of maternal origin, or its own adrenal products. The outcomes in a host of species and models are remarkably consistent with cardiometabolic and CNS effects predominating. Once traumatic life events, in combination with genetic disposition, have engrained long-lasting changes in MR and GR signaling, a vulnerable phenotype emerges. Work on candidate mechanisms, GR gene programming, has illuminated a potential fundamental mechanism to underlie this rapidly emerging biology. Such fine-tuning of fetal physiology by the environment is conserved and therefore apparently important. Studies are unraveling the underlying processes, a prerequisite to rational treatments for the consequences of an adverse perinatal environment.

References

1. Levine, S. (1957) Maternal and environmental influences on the adrenocortical response to stress in weaning rats. *Science*, **156**, 258–260.
2. Welberg, L.A. and Seckl, J.R. (2001) Prenatal stress, glucocorticoids and the programming of the brain. *J. Neuroendocrinol.*, **13**, 113–128.
3. Weinstock, M. (2001) Alterations induced by gestational stress in brain morphology and behaviour in offspring. *Prog. Neurobiol.*, **65**, 427–451.
4. Matthews, S.G. (2000) Antenatal glucocorticoids and programming of the developing CNS. *Paediatr. Res.*, **47**, 291–300.
5. Sloboda, D.M., Moss, T.J., Gurrin, L.C., Newnham, J.P., and Challis, J.R. (2002) The effect of prenatal betamethasone administration on postnatal administration on postnatal ovine hypothalamic-pituitary-adrenal function. *J. Endocrinol.*, **172**, 71–81.
6. Meaney, M.J. (2001) Maternal care, gene expression, and the transmission of individual differences in stress reactivity across generations. *Annu. Rev. Neurosci.*, **24**, 1161–1192.
7. Schneider, M.L., Moore, C.F., Kraemer, G.W., Roberts, A.D., and DeJesus, O.T. (2002) The impact of prenatal stress, foetal alcohol exposure, or both on development: perspectives from a primate model. *Psychoneuroendocrinology*, **27**, 285–298.
8. De Kloet, E.R., Vreugdenhill, E., Oitzl, M.S., and Joels, M. (1998) Brain corticosteroid receptor balance in health and disease. *Endocr. Rev.*, **19**, 269–301.
9. Barker, D.J., Osmond, C., Golding, J., Kuh, D., and Wadsworth, M.E. (1989) Growth in utero, blood pressure in childhood and adult life, and mortality from cardiovascular disease. *Br. Med. J.*, **298**, 564–567.
10. Barker, D.J. (2002) Foetal programming of coronary heart disease. *Trends. Endocrinol. Metab.*, **13**, 364–368.
11. Godfrey, K.M. and Barker, D.J. (2000) Foetal nutrition and adult disease. *Am. J. Clin. Nutr.*, **71** (Suppl 5), 1344S–1352S.
12. Ravelli, A.C., Vander der Meulen, J.H., Michels, R.P., Osmond, C., Barker, D.J., Hales, C.N. et al. (1998) Glucose tolerance in adults after prenatal exposure to famine. *Lancet*, **351**, 173–137.
13. Levitt, N.S., Lambert, E.V., Woods, D., Hales, C.N., Andrew, R., and Seckl, J.R. (2001) Impaired glucose tolerance and elevated blood pressure in low birth weight, non obese, young South African adults: early programming of cortisol axis. *J. Clin. Endocrinol. Metab.*, **85**, 4611–4618.
14. Thompson, C., Syddall, H., Rodin, I., Osmond, C., and Barker, D.J. (2001) Birth weight and the risk of depressive disorder in late life. *Br. J. Psychiatry*, **179**, 450–455.
15. Barker, D.J.P., Hales, C.N., Fall, C.H.D., Osmond, C., Phipps, K., and Clarke, P.M.S. (1993) Type 2 (non-insulin dependent) diabetes mellitus, hypertension and hyperlipidaemia (syndrome X): relation to reduced fetal growth. *Diabetologia*, **36**, 62–67.
16. Edwards, C.R.W., Benediktsson, R., Lindsay, R., and Seckl, J.R. (1993) Dysfunction of the placental glucocorticoid barrier: a link between the foetal environment and adult hypertension? *Lancet*, **341**, 355–357.
17. Seckl, J.R. (1998) Physiologic programming of the fetus. *Clin. Perinatol.*, **25**, 939–964.

18. Kapoor, A., Leen, J., and Matthews, S.G. (2008) Molecular regulation of the hypothalamic-pituitary-adrenal axis in adult male guinea pigs after prenatal stress at different stages of gestation. *J. Physiol.*, **586** (Pt 17), 4317–4326.
19. Godfrey, K.M., Lillycrop, K.A., Burdge, G.C., Gluckman, P.D., and Hanson, M.A. (2007) Epigenetic mechanisms and the mismatch concept of the developmental origins of health and disease. *Pediatr. Res.*, **61** (5, Pt 2), 5R–10R.
20. Neel, J.V. (1962) Diabetes mellitus: a "thrifty" genotype rendered detrimental by "progress"? *Am. J. Hum. Genet.*, **14**, 353–362.
21. Hales, C.N. and Barker, D.J. (2001) The thrifty phenotype hypothesis. *Br. Med. Bull.*, **60**, 5–20.
22. Barker, D.J.P. (1990) Fetal and infant origins of adult disease. *Br. Med. J.*, **301**, 1111.
23. Barker, D.J.P., Gluckman, P.D., Godfrey, K.M., Harding, J.E., Owens, J.A., and Robinson, J.S. (1993) Fetal nutrition and cardiovascular disease in adult life. *Lancet*, **341**, 938–941.
24. Lithell, H.O., McKeigue, P.M., Berglund, L., Mohsen, R., Lithell, U.B., and Leon, D.A. (1996) Relation of size at birth to non-insulin dependent diabetes and insulin concentrations in men aged 50–60 years. *Br. Med. J.*, **312**, 406–410.
25. Curhan, G.C., Chertown, M., Willett, W.C., Spiegelman, D., Colditz, G.A., Manson, J.E., Speizer, F.E., and Stampfer, M.J. (1996) Birth-weight and adult hypertension and obesity in women. *Circulation*, **94**, 1310–1315.
26. Curhan, G.C., Willett, W.C., Rimm, E.B., Spiegelman, D., Ascherio, A.L., and Stampfer, M.J. (1996) Birth weight and adult hypertension, diabetes mellitus, and obesity in US Men. *Circulation*, **94**, 3246–3250.
27. Fall, C.H.D., Osmond, C., Barker, D.J.P., Clark, P.M.S., Hales, C.N., Stirling, Y., and Meade, T.W. (1995) Fetal and infant growth and cardiovascular risk factors in women. *Br. Med. J.*, **310**, 428–432.
28. Forsen, T., Eriksson, J.G., Tuomilehto, J., Teramo, K., Osmond, C., and Barker, D.J.P. (1997) Mother's weight in pregnancy and coronary heart disease in a cohort of Finnish men: Follow up study. *Br. Med. J.*, **315**, 837–840.
29. Leon, D.A., Koupilova, I., Lithell, H.O., Berglund, L., Mohsen, R., Vagero, D., Lithell, U.B., and McKeigue, P.M. (1996) Failure to realise growth potential in utero and adult obesity in relation to blood pressure in 50 year old Swedish men. *Br. Med. J.*, **312**, 401–406.
30. Moore, V.M., Miller, A.G., Boulton, T.J.C., Cockington, R.A., Craig, I.H., Magarey, A.M., and Robinson, J.S. (1996) Placental weight, birth measurements, and blood pressure at age 8 years. *Arch. Dis. Child.*, **74**, 538–541.
31. RichEdwards, J.W., Stampfer, M.J., Manson, J.E., Rosner, B., Hankinson, S.E., Colditz, G.A., Willett, W.C., and Hennekens, C.H. (1997) Birth weight and risk of cardiovascular disease in a cohort of women followed up since 1976. *Br. Med. J.*, **315**, 396–400.
32. Yajnik, C.S., Fall, C.H.D., Vaidya, U., Pandit, A.N., Bavdekar, A., Bhat, D.S., Osmond, C., Hales, C.N., and Barker, D.J.P. (1995) Fetal growth and glucose and insulin metabolism in four-year-old Indian children. *Diabet. Med.*, **12**, 330–336.
33. Baird, J., Osmond, C., MacGregor, A., Snieder, H., Hales, C., and Phillips, D. (2001) Testing the fetal origins hypothesis in twins: the Birmingham twin study. *Diabetologia*, **44**, 33–39.
34. Phillips, D.I., Barker, D.J., Hales, C.N., Hirst, S., and Osmond, C. (1994) Thinness at birth and insulin resistance in adult life. *Diabetologia*, **37**, 150–154.
35. Whincup, P.H., Kaye, S.J., Owen, C.G., Huxley, R., Cook, D.G., Anazawa, S., Barrett-Connor, E. et al. (2008) Birth weight and risk of type 2 diabetes: a systematic review. Birth weight and risk of type 2 diabetes: a systematic review. *J. Am. Med. Assoc.*, **300**, 2886–2897.
36. Irving, R.J., Belton, N.R., Elton, R.A., and Walker, B.R. (2000) Adult cardiovascular risk factors in premature babies. *Lancet*, **355**, 2135–2136.

37. McCormick, M.C. (1985) The contribution of low birth weight to infant mortality and childhood morbidity. N. Engl. J. Med., 312, 82–90.
38. Horbar, J.D., Badger, G.J., Carpenter, J.H., Fanaroff, A.A., Kilpatrick, S., LaCorte, M., Phibbs, R. et al. (2002) Trends in mortality and morbidity for very low birth weight infants, 1991–1999. Pediatrics, 110, 143–151.
39. Bernstein, I.M., Horbar, J.D., Badger, G.J., Ohlsson, A., and Golan, A. (2000) Morbidity and mortality among very-low-birth-weight neonates with intrauterine growth restriction. The Vermont Oxford Network. Am. J. Obstet. Gynecol., 182, 198–206.
40. Fanaroff, A.A., Stoll, B.J., Wright, L.L., Carlo, W.A., Ehrenkranz, R.A., Stark, A.R., and Bauer, C.R. (2007) Trends in neonatal morbidity and mortality for very low birthweight infants. Am. J. Obstet. Gynecol., 196, 147e1–147e8.
41. Osmond, C., Barker, D.J.P., Winter, P.D., Fall, C.H.D., and Simmonds, S.J. (1993) Early growth and death from cardiovascular disease in women. Br. Med. J., 307, 1524–1527.
42. Levine, R.S., Hennekens, C.H., and Jesse, M.J. (1994) Blood pressure in prospective population based cohort of newborn and infant twins. Br. Med. J., 308, 298–302.
43. Bavdekar, A., Yajnik, C.S., Fall, C.H., Bapat, S., Pandit, A.N., Deshpande, V., Bhave, S., Kellingray, S.D., and Joglekar, C. (1999) Insulin resistance syndrome in 8-year-old Indian children: small at birth, big at 8 years, or both? Diabetes, 48, 2422–2429.
44. Law, C.M., Shiell, A.W., Newsome, C.A., Syddall, H.E., Shinebourne, E.A., Fayers, P.M., Martyn, C.N., and de Swiet, M. (2002) Fetal, infant, and childhood growth and adult blood pressure: a longitudinal study from birth to 22 years of age. Circulation, 105, 1088–1092.
45. Watkins, A.J., Wilkins, A., Cunningham, C., Perry, V.H., Seet, M.J., Osmond, C., Eckert, J.J. et al. (2008) Low protein diet fed exclusively during mouse oocyte maturation leads to behavioural and cardiovascular abnormalities in offspring. J. Physiol., 586, 2231–2244.
46. Watkins, A.J., Papenbrock, T., and Fleming, T.P. (2008) The preimplantation embryo: handle with care. Semin. Reprod. Med., 26, 175–185.
47. Watkins, A.J., Platt, D., Papenbrock, T., Wilkins, A., Eckert, J.J., Kwong, W.Y., Osmond, C. et al. (2007) Mouse embryo culture induces changes in postnatal phenotype including raised systolic blood pressure. Proc. Natl. Acad. Sci. U.S.A., 104, 5449–5454.
48. Costello, E.J., Worthman, C., Erkanli, A., and Angold, A. (2007) Prediction from low birth weight to female adolescent depression: a test of competing hypotheses. Arch. Gen. Psychiatry, 64, 338–344.
49. Alati, R., Lawlor, D.A., Al Mamun, A., Williams, G.M., Najman, J.M., O'Callaghan, M., and Bor, W. (2007) Is there a fetal origin of depression? Evidence from the mater university study of pregnancy and its outcomes. Am. J. Epidemiol., 165, 575–582.
50. Wiles, N.J., Peters, T.J., Leon, D.A., and Lewis, G. (2005) Birth weight and psychological distress at age 45–51 years – Results from the Aberdeen children of the 1950s cohort study. Br. J. Psychiatry, 187, 21–28.
51. Cannon, M., Jones, P.B., and Murray, R.M. (2002) Obstetric complications and schizophrenia: historical and meta-analytic review. Am. J. Psychiatry, 159, 1080–1092.
52. Jones, P.B., Rantakallio, P., Hartikainen, A.L., Isohanni, M., and Sipila, P. (1998) Schizophrenia as a long-term outcome of pregnancy, delivery, and perinatal complications: a 28-year follow-up of the 1966 North Finland general population birth cohort. Am. J. Psychiatry., 155, 355–364.
53. Thapar, A., Langley, K., Fowler, T., Rice, F., Turic, D., Whittinger, N., Aggleton, J. et al. (2005) Catechol O-methyltransferase gene variant and birth weight predict early-onset antisocial behavior in children with attention-deficit/hyperactivity disorder. Arch. Gen. Psychiatry, 62, 1275–1278.

54. Pesonen, A.K., Räikkönen, K., Lano, A., Peltoniemi, O., Hallman, M., and Kari, M.A. (2009) Antenatal betamethasone and fetal growth in prematurely born children: implications for temperament traits at the age of 2 years. *Pediatrics*, **123**, e31–e37.
55. Famularo, R. and Fenton, T. (1994) Early developmental history and pediatric posttraumatic-stress-disorder. *Arch. Pediatr. Adolesc. Med.*, **148**, 1032–1038.
56. Susser, E., Neugebauer, R., Hoek, H.W., Brown, A.S., Lin, S., Labovitz, D., and Gorman, J.M. (1996) Schizophrenia after prenatal famine. *Arch. Gen. Psychiatry*, **53**, 25–31.
57. St Clair, D., Xu, M.Q., Wang, P., Yu, Y.Q., Fang, Y.R., Zhang, F., Zheng, X.Y. et al. (2005) Rates of adult schizophrenia following prenatal exposure to the Chinese famine of 1959–1961. *J. Am. Med. Assoc.*, **294**, 557–562.
58. Susser, E.S. and Lin, S.P. (1992) Schizophrenia after prenatal exposure to the Dutch hunger winter of 1944–1945. *Arch. Gen. Psychiatry*, **49**, 983–988.
59. Ward, R.M. (1994) Pharmacologic enhancement of fetal lung maturation. *Clin. Perinatol.*, **21**, 523–542.
60. Owen, D., Andrews, M.H., and Matthews, S.G. (2005) Maternal adversity, glucocorticoids and programming of neuroendocrine function and behaviour. *Neurosci. Biobehav. Rev.*, **29**, 209–226.
61. Korte, S.M. (2001) Corticosteroids in relation to fear, anxiety and psychopathology. *Neurosci. Biobehav. Rev.*, **25**, 117–142.
62. Meaney, M.J., Diorio, J., Francis, D., Widdowson, J., LaPlante, P., Caldji, C., Sharma, S. et al. (1996) Early environmental regulation of forebrain glucocorticoid receptor gene expression: Implications for adrenocortical responses to stress. *Dev. Neurosci.*, **18**, 49–72.
63. Rua, C., Trejo, J.L., Machin, C., and Arahuetes, R.M. (1995) Effects of maternal adrenalectomy and glucocorticoid administration on the development of rat hippocampus. *J. Himforschung*, **36**, 473–483.
64. Trejo, J.L., Machin, C., Arachuetes, R.M., and Rua, C. (1995) Influence of maternal adrenalectomy and glucocorticoid administration on the development of the rat cerebral cortex. *Anat. Embryol.*, **192**, 89–99.
65. Antonow-Schlorke, I., Schwab, M., Li, C., and Nathanielsz, P.W. (2003) Glucocorticoid exposure at the dose used clinically alters cytoskeletal proteins and presynaptic terminals in the fetal baboon brain. *J. Physiol.*, **547**, 117–123.
66. Meyer, J.S. (1983) Early adrenalectomy stimulates subsequent growth and development of the rat brain. *Exp. Neurol.*, **82**, 432–446.
67. Webster, J.C. and Cidlowski, J.A. (1999) Mechanisms of glucocorticoid receptor mediated repression of gene expression. *Trends Endocrinol. Metab.*, **10**, 396–402.
68. Jenkins, B.D., Pullen, C.B., and Darimont, B.D. (2001) Novel glucocorticoid receptor coactivator effector mechanisms. *Trends Endocrinol. Metab.*, **12**, 122–126.
69. Trapp, T. and Holsboer, F. (1996) Heterodimerisation between mineralocorticoid and glucocorticoid receptors increases the functional diversity of corticosteroid action. *Trends Pharmacol. Sci.*, **17**, 145–149.
70. Savory, J.G., Prefontaine, G.G., Lamprecht, C., Liao, M., Walther, R.F., Lefebvre, Y.A., and Haché, R.J. (2001) Glucocorticoid receptor homodimers and glucocorticoid-mineralococrticoid receptor heterodimers form in the cytoplasm through alternative dimerisation interaces. *Mol. Cell Biol.*, **21**, 781–793.
71. Karst, H., Berger, S., Turiault, M., Tronche, F., Schütz, G., and Joëls, M. (2005) Mineralocorticoid receptors are indispensable for nongenomic modulation of hippocampal glutamate transmission by corticosterone. *Proc. Natl. Acad. Sci. U.S.A.*, **102**, 19204–19207.
72. Matthews, S.G. (1998) Dynamic changes in glucocorticoid and mineralocorticoid receptor mRNA in the

developing guinea pig brain. *Dev. Brain Res.*, **107**, 123–132.

73. Reul, J.M. and De Kloet, E.R. (1985) Two receptor systems for corticosterone in rat brain: microdistribution and differential occupation. *Endocrinology*, **117**, 2505–2511.

74. Macleod, M.R., Johansson, I.M., Söderström, I., Lai, M., Gidö, G., Wieloch, T., Seckl, J.R., and Olsson, T. (2003) Mineralocorticoid receptor expression and increased survival following neuronal injury. *Eur. J. Neurosci.*, **17**, 1549–1555.

75. Lai, M., Horsburgh, K., Bae, S.E., Carter, R.N., Stenvers, D.J., Fowler, J.H., Yau, J.L. et al. (2007) Forebrain mineralocorticoid receptor overexpression enhances memory, reduces anxiety and attenuates neuronal loss in cerebral ischaemia. *Eur. J. Neurosci.*, **25**, 1832–1842.

76. Cole, T.J. (1995) Cloning of the mouse 11 beta-hydroxysteroid dehydrogenase type 2 gene: tissue specific expression and localization in distal convoluted tubules and collecting ducts of the kidney. *Endocrinology*, **136**, 4693–4696.

77. Speirs, H., Seckl, J.R., and Brown, R. (2004) Ontogeny of glucocorticoid receptor and 11 beta-hydroxysteroid dehydrogenase type 1 gene expression identifies potential critical periods of glucocorticoid susceptibility during development. *J. Endocrinol.*, **181**, 105–116.

78. Brown, R.W., Diaz, R., Robson, A.C., Kotelevtsev, Y., Mullins, J.J., Kaufman, M.H., and Seckl, J.R. (1996) The ontogeny of 11β-hydroxysteroid dehydrogenase type 2 and mineralocorticoid receptor gene expression reveal intricate control of glucocorticoid action in development. *Endocrinology*, **137**, 794–797.

79. Sun, K., Yang, K., and Challis, J.R. (1997) Differential expression of 11 beta-hydroxysteroid dehydrogenase types 1 and 2 in human placenta and fetal membranes. *J. Clin. Endocrinol. Metab.*, **82**, 300–305.

80. Costa, A., Rocci, M.P., Arisio, R., Benedetto, C., Fabris, C., Bertino, E., Botta, G. et al. (1996) Glucocorticoid receptors immunoreactivity in tissue of human embryos. *J. Endocrinol. Invest.*, **19**, 92–98.

81. Condon, J., Gosden, C., Gardener, D., Nickson, P., Hewison, M., Howie, A.J., and Stewart, P.M. (1998) Expression of type 2 11 beta-hydroxysteroid dehydrogenase and corticosteroid hormone receptors in early human life. *J. Clin. Endocrinol. Metab.*, **83**, 4490–4407.

82. Fuxe, K., Wikstrom, A.-C., Okret, S., Agnati, L.F., Harfstrand, A., Yu, Z.-Y., Granholm, L. et al. (1985) Mapping of glucocorticoid receptor immunoreactive neurons in the rat tel- and diencephalon using a monoclonal antibody against rat liver glucocorticoid receptor. *Endocrinology*, **117**, 1803–1812.

83. Diaz, R., Brown, R.W., and Seckl, J.R. (1998) Distinct ontogeny of glucocorticoid and mineralocorticoid receptor and 11 β-hydroxysteroid dehydrogenase types I and II mRNAs in the foetal rat brain suggest a complex role of glucocorticoid actions. *J. Neurosci.*, **18**, 2570–2580.

84. Kitraki, E., Kittas, C., and Stylianopoulou, F. (1997) Glucocorticoid receptor gene expression during rat embryogenesis: an in situ hybridization study. *Differentiation*, **62**, 21–31.

85. Reinisch, J.M., Simon, N.G., Karwo, W.G., and Gandelman, R. (1978) Prenatal exposure to prednisone in humans and animals retards intra-uterine growth. *Science*, **202**, 436–438.

86. Nyirenda, M.J., Lindsay, R.S., Kenyon, C.J., Burchell, A., and Seckl, J.R. (1998) Glucocorticoid exposure in late gestation permanently programmes rat hepatic phosphoenolpyruvate carboxykinase and glucocorticoid receptor expression and causes glucose intolerance in adult offspring. *J. Clin. Invest.*, **101**, 2174–2181.

87. Ikegami, M., Jobe, A.H., Newnham, J., Polk, D.H., Willet, K.E., and Sly, P. (1997) Repetitive prenatal glucocorticoids improve lung function and decrease growth in preterm lambs. *Am. J. Respir. Crit. Care Med.*, **156**, 178–184.

88. French, N.P., Hagan, R., Evans, S.F., Godfrey, M., and Newnham, J.P. (1999) Repeated antenatal corticosteroids: size at birth and subsequent development. *Am. J. Obstet. Gynecol.*, **180**, 114–121.
89. Newnham, J.P., Evans, S.F., Godfrey, M., Huang, W., Ikegami, M., and Jobe, A. (1999) Maternal, but not fetal, administration of corticosteroids restricts fetal growth. *J. Matern. Fetal Med.*, **8**, 81–87.
90. Newnham, J.P. and Moss, T.J. (2001) Antenatal glucocorticoids and growth: single versus multiple doses in animal and human studies. *Semin. Neonatol.*, **6**, 285–292.
91. Bloom, S.L., Sheffield, J.S., McIntire, D.D., and Leveno, K.J. (2001) Antenatal dexamethasone and decreased birth weight. *Obstet. Gynecol.*, **97**, 485–490.
92. Goland, R.S., Jozak, S., Warren, W.B., Conwell, I.M., Stark, R.I., and Tropper, P.J. (1993) Elevated levels of umbilical cord plasma corticotropin-releasing hormone in growth-retarded fetuses. *J. Clin. Endocrinol. Metab.*, **77**, 1174–1179.
93. Goland, R.S., Tropper, P.J., Warren, W.B., Stark, R.I., Jozak, S.M., and Conwell, I.M. (1995) Concentrations of corticotropin-releasing hormone in the umbilical-cord blood of pregnancies complicated by preeclampsia. *Reprod. Fertil. Dev.*, **7**, 1227–1230.
94. White, P.C., Mune, T., and Agarwal, A.K. (1997) 11 beta-Hydroxysteroid dehydrogenase and the syndrome of apparent mineralocorticoid excess. *Endocr. Rev.*, **18**, 135–156.
95. Brown, R.W., Kotolevtsev, Y., Leckie, C., Lindsay, R.S., Lyons, V., Murad, P., Mullins, J.J. et al. (1996) Isolation and cloning of human placental 11β-hydroxysteroid dehydrogenase-2 cDNA. *Biochem. J.*, **313**, 1007–1017.
96. Waddell, B., Benediktsson, R., Brown, R., and Seckl, J.R. (1998) Tissue-specific mRNA expression of 11B-hydroxysteroid dehydrogenase types 1 and 2 and the glucocorticoid receptor within rat placenta suggest exquisite local control of glucocorticoid action. *Endocrinology*, **139**, 1517–1523.
97. Waddell, B.J. and Atkinson, H.C. (1994) Production rate, metabolic clearance rate and uterine extraction of corticosterone during rat pregnancy. *J. Endocrinol.*, **143**, 183–190.
98. Lopez-Bernal, A., Flint, A.P.F., Anderson, A.B.M., and Turnbull, A.C. (1980) 11β-hydroxysteroid dehydrogenase activity (E.C.1.1.1.146) in human placenta and decidua. *J. Steroid Biochem.*, **13**, 1081–1087.
99. Lopez Bernal, A. and Craft, I.L. (1981) Corticosteroid metabolism in vitro by human placenta, foetal membranes and decidua in early and late gestation. *Placenta*, **2**, 279–285.
100. Benediktsson, R., Calder, A.A., Edwards, C.R.W., and Seckl, J.R. (1997) Placental 11β-hydroxysteroid dehydrogenase type 2 is the placental barrier to maternal glucocorticoids: ex vivo studies. *Clin. Endocrinol.*, **46**, 161–166.
101. Venihaki, M.A., Carrigan, P., Dikkes, P., and Majzoub, J. (2000) Circadian rise in maternal glucocorticoid prevents pulmonary dysplasia in fetal mice with adrenal insufficiency. *Proc. Natl. Acad. Sci. U.S.A.*, **97**, 7336–7341.
102. Sampath-Kumar, R., Matthews, S.G., and Yang, K. (1998) 11beta-hydroxysteroid dehydrogenase type 2 is the predominant isoenzyme in the guinea pig placenta: decreases in messenger ribonucleic acid and activity at term. *Biol. Reprod.*, **59**, 1378–1384.
103. Albiston, A.L., Obeyesekere, V.R., Smith, R.E., and Krozowski, Z.S. (1994) Cloning and tissue distribution of the human 11β-hydroxysteroid dehydrogenase type 2 enzyme. *Mol. Cell. Endocrinol.*, **105**, R11–R17.
104. Murphy, V.E., Fittock, R.J., Zarzycki, P.K., Delahunty, M.M., Smith, R., and Clifton, V.L. (2007) Metabolism of synthetic steroids by the human placenta. *Placenta*, **28**, 39–46.
105. Stewart, P.M., Rogerson, F.M., and Mason, J.I. (1995) Type 2 11β-hydroxysteroid dehydrogenase messenger RNA and activity in human placenta and foetal membranes: its relationship to birth weight and putative role in foetal steroidogenesis. *J. Clin. Endocrinol. Metab.*, **86**, 4979–4983.

106. Benediktsson, R., Lindsay, R., Noble, J., Seckl, J.R., and Edwards, C.R. (1993) Glucocorticoid exposure in utero: a new model for adult hypertension. Lancet, 341, 339–341.
107. McTernan, C.L., Draper, N., Nicholson, H., Chalder, S.M., Driver, P., Hewison, M., Kilby, M.D., and Stewart, P.M. (2001) Reduced placental 11 beta-hydroxysteroid dehydrogenase type 2 mRNA levels in human pregnancies complicated by intrauterine growth restriction: an analysis of possible mechanisms. J. Clin. Endocrinol. Metab., 86, 4979–4983.
108. Shams, M., Kilby, M.D., Somerset, D.A., Howie, A.J., Gupta, A., Wood, P.J., Afnan, M., and Stewart, P.M. (1998) 11beta hydroxysteroid dehydrogenase type 2 in human pregnancy and reduced expression in intrauterine growth retardation. Hum. Reprod., 13, 799–804.
109. Murphy, V.E., Zakar, T., Smith, R., Giles, W.B., Gibson, P.G., and Clifton, V.L. (2002) Reduced 11beta-hydroxysteroid dehydrogenase type 2 activity is associated with decreased birth weight centile in pregnancies complicated by asthma. J. Clin. Endocrinol. Metab., 87, 1660–1668.
110. Rogerson, F.M., Kayes, K., and White, P.C. (1996) No correlation in human placenta between activity or mRNA for the K (type 2) isozyme of 11β-hydroxysteroid dehydrogenase and fetal or placental weight. Tenth Int. Congr. Endocrinol. Abstr., P1-231, 193.
111. Rogerson, F.M., Kayes, K.M., and White, P.C. (1997) Variation in placental type 2 11beta-hydroxysteroid dehydrogenase activity is not related to birth weight or placental weight. Mol. Cell. Endocrinol., 128, 103–109.
112. Benediktsson, R., Brennand, J., Tibi, L., Calder, A.A., Seckl, J.R., and Edwards, C.R.W. (1995) Fetal osteocalcin levels are related to placental 11β-hydroxysteroid dehydrogenase activity. Clin. Endocrinol. (Oxf.), 42, 551–555.
113. Dave-Sharma, S., Wilson, R.C., Harbison, M.D., Newfield, R., Azar, M., Krozowski, Z.S., Funder, J.W. et al. (1998) Extensive personal experience – examination of genotype and phenotype relationships in 14 patients with apparent mineralocorticoid excess. J. Clin. Endocrinol. Metab., 83, 2244–2254.
114. Kotelevtsev, Y., Brown, R.W., Fleming, S., Kenyon, C.J., Edwards, C.R.W., Seckl, J.R., and Mullins, J.J. (1999) Hypertension in mice lacking 11β-hydroxysteroid dehydrogenase type 2. J. Clin. Invest., 103, 683–689.
115. Holmes, M.C., Abrahamsen, C.T., French, K.L., Paterson, J.M., Mullins, J.J., and Seckl, J.R. (2006) The mother or the fetus? 11 beta-hydroxysteroid dehydrogenase type 2 null mice provide evidence for direct fetal programming of behavior by endogenous glucocorticoids. J. Neurosci., 26, 3840–3844.
116. Dodic, M., Hantzis, V., Duncan, J., Rees, S., Koukoulas, I., Johnson, K., Wintour, E.M., and Moritz, K. (2002) Programming effects of short prenatal exposure to cortisol. FASEB J., 16, 1017–1026.
117. Dodic, M., May, C.N., Wintour, E.M., and Coghlan, J.P. (1998) An early prenatal exposure to excess glucocorticoid leads to hypertensive offspring in sheep. Clin. Sci., 94, 149–155.
118. Dodic, M., Moritz, K., Koukoulas, I., and Wintour, E.M. (2002) Programmed hypertension: kidney, brain or both? Trends Endocrinol. Metab., 13, 403–408.
119. Dodic, M., Peers, A., Coghlan, J., May, C., Lumbers, E., Yu, Z., and Wintour, E. (1999) Altered cardiovascular haemodynamics and baroreceptor-heart rate reflex in adult sheep after prenatal exposure to dexamethasone. Clin. Sci., 97, 103–109.
120. Jensen, E.C., Gallaher, B.W., Breier, B.H., and Harding, J.E. (2002) The effect of a chronic maternal cortisol infusion on the late-gestation fetal sheep. J. Endocrinol., 174, 27–36.
121. Levitt, N., Lindsay, R.S., Holmes, M.C., and Seckl, J.R. (1996) Dexamethasone in the last week of pregnancy attenuates hippocampal glucocorticoid receptor gene expression and elevates blood pressure in the adult offspring

in the rat. *Neuroendocrinology*, **64**, 412–418.
122. Sugden, M.C., Langdown, M.L., Munns, M.J., and Holness, M.J. (2001) Maternal glucocorticoid treatment modulates placental leptin and leptin receptor expression and materno-fetal leptin physiology during late pregnancy, and elicits hypertension associated with hyperleptinaemia in the early-growth-retarded adult offspring. *Eur. J. Endocrinol.*, **145**, 529–539.
123. Gatford, K.L., Wintour, E.M., de Blasio, M.J., Owens, J.A., and Dodic, M. (2000) Differential timing for programming of glucose homeostasis, sensitivity to insulin and blood pressure by in utero exposure to dexamethasone in sheep. *Clin. Sci.*, **98**, 553–560.
124. Lindsay, R.S., Lindsay, R.M., Edwards, C.R.W., and Seckl, J.R. (1996) Inhibition of 11β-hydroxysteroid dehydrogenase in pregnant rats and the programming of blood pressure in the offspring. *Hypertension*, **27**, 1200–1204.
125. Lindsay, R.S., Lindsay, R.M., Waddell, B., and Seckl, J.R. (1996) Programming of glucose tolerance in the rat: role of placental 11β-hydroxysteroid dehydrogenase. *Diabetologia*, **39**, 1299–1305.
126. de Vries, A., Holmes, M., Heijnis, A., Seier, J., van Heerden, J., Louw, J., Wolfe-Coote, S. *et al.* (2007) Prenatal dexamethasone exposure induces changes in offspring cardio – metabolic and hypothalamic– pituitary-adrenal axis function without alteration of birth weight in a non-human primate, the African vervet, Chlorocebus aethiops. *J. Clin. Invest.*, **117**, 1058–1067.
127. Huh, S.Y., Andrew, R., Rich-Edwards, J.W., Kleinman, K.P., Seckl, J.R., and Gillman, M.W. (2008) Association between umbilical cord glucocorticoids and blood pressure at age 3 years. *BMC Med.*, **6**, 25.
128. Hu, F., Crespi, E.J., and Denver, R.J. (2008) Programming neuroendocrine stress axis activity by exposure to glucocorticoids during postembryonic development of the frog, Xenopus laevis. *Endocrinology*, **149**, 5470–5481.
129. Spencer, K.A., Evans, N.P., and Monaghan, P. (2009) Postnatal stress in birds: a novel model of glucocorticoid programming of the hypothalamic-pituitary-adrenal axis. *Endocrinology*, **150**, 1931–1934.
130. Drake, A.J., Walker, B.R., and Seckl, J.R. (2005) Intergenerational consequences of fetal programming by in utero exposure to glucocorticoids in rats. *Am. J. Physiol. Regul. Integr. Comp. Physiol.*, **288**, R34–R38.
131. Wyrwoll, C.S., Seckl, J.R., and Holmes, M.C. (2009) Altered placental function of 11beta-hydroxysteroid dehydrogenase 2 knockout mice. *Endocrinology*, **150**, 1287–1293.
132. Le Clair, C., Abbi, T., Sandhu, H., and Tappia, P.S. (2009) Impact of maternal undernutrition on diabetes and cardiovascular disease risk in adult offspring. *Can. J. Physiol. Pharmacol.*, **87**, 161–179.
133. Jimenez-Chillaron, J.C., Isganaitis, E., Charalambous, M., Gesta, S., Pentinat-Pelegrin, T., Faucette, R.R., Otis, J.P. *et al.* (2009) Intergenerational transmission of glucose intolerance and obesity by in utero undernutrition in mice. *Diabetes*, **58**, 460–468.
134. Victora, C.G., Adair, L., Fall, C., Hallal, P.C., Martorell, R., Richter, L., and Sachdev, H.S. (2008) Maternal and child undernutrition study group. Maternal and child undernutrition: consequences for adult health and human capital. *Lancet*, **371**, 340–357.
135. Langley-Evans, S.C., Philips, G., Benediktsson, R., Gardner, D., Edwards, C.R.W., Jackson, A.A., and Seckl, J.R. (1996) Maternal dietary protein restriction, placental glucocorticoid metabolism and the programming of hypertension. *Placenta*, **17**, 169–172.
136. Bertram, C., Trowern, A.R., Copin, N., Jackson, A.A., and Whorwood, C.B. (2001) The maternal diet during pregnancy programs altered expression of the glucocorticoid receptor and type 2 11beta-hydroxysteroid dehydrogenase: potential molecular mechanisms underlying the programming of hypertension in utero. *Endocrinology*, **142**, 2841–2853.

137. Lesage, J., Blondeau, B., Grino, M., Breant, B., and Dupouy, J.P. (2001) Maternal undernutrition during late gestation induces fetal overexposure to glucocorticoids and intrauterine growth retardation, and disturbs the hypothalamo-pituitary adrenal axis in the newborn rat. *Endocrinology*, **142**, 1692–1702.
138. Langley-Evans, S.C. (1997) Hypertension induced by foetal exposure to a maternal low-protein diet, in the rat, is prevented by pharmacological blockade of maternal glucocorticoid synthesis. *J. Hypertens.*, **15**, 537–544.
139. van Beek, J.P., Guan, H., Julan, L., and Yang, K. (2004) Glucocorticoids stimulate the expression of 11β-hydroxysteroid dehydrogenase type 2 in cultured human placental trophoblast cells. *J. Clin. Endocrinol. Metab.*, **89**, 5614–5621.
140. Sarkar, S., Tsai, S.W., Nguyen, T.T., Plevyak, M., Padbury, J.F., and Rubin, L.P. (2001) Inhibition of placental 11 beta-hydroxysteroid dehydrogenase type 2 by catecholamines via alpha-adrenergic signalling. *Am. J. Physiol. Regul. Integr. Comp. Physiol.*, **281**, R1966–R1974.
141. Yang, K., Julan, L., Rubio, F., Sharma, A., and Guan, H. (2006) Cadmium reduces 11 beta-hydroxysteroid dehydrogenase type 2 activity and expression in human placental trophoblast cells. *Am. J. Physiol. Endocrinol. Metab.*, **290**, E135–E142.
142. Julan, L., Guan, H., van Beek, J.P., and Yang, K. (2005) Peroxisome proliferator-activated receptor (delta) suppresses 11 (beta)-hydroxysteroid dehydrogenase type two gene expression in human placental trophoblast cells. *Endocrinology*, **146**, 1482–1490.
143. Hardy, D.B., Pereria, L.E., and Yang, K. (1999) Prostaglandins and leukotriene B4 are potent inhibitors of 11beta-hydroxysteroid dehydrogenase type 2 activity in human choriocarcinoma JEG-3 cells. *Biol. Reprod.*, **61**, 40–45.
144. Hardy, D.B. and Yang, K. (2002) The expression of 11 beta-hydroxysteroid dehydrogenase type 2 is induced during trophoblast differentiation: effects of hypoxia. *J. Clin. Endocrinol. Metab.*, **87**, 3696–3701.
145. Welberg, L.A.M., Thrivikraman, K.V., and Plotsky, P.M. (2005) Chronic maternal stress inhibits the capacity to up-regulate placental 11 β-hydroxysteroid dehydrogenase type 2 activity. *J. Endocrinol.*, **186**, R7–R12.
146. Mairesse, J., Lesage, J., Breton, C., Bréant, B., Hahn, T., Darnaudéry, M., Dickson, S.L. et al. (2007) Maternal stress alters endocrine function of the feto-placental unit in rats. *Am. J. Physiol. Endocrinol. Metab.*, **292**, E1526–E1533.
147. Robson, A.C., Leckie, C., Seckl, J.R., and Holmes, M.C. (1998) Expression of 11β-hydroxysteroid dehydrogenase type 2 in the postnatal and adult rat brain. *Mol. Brain. Res.*, **61**, 1–10.
148. Stewart, P.M., Murry, B.A., and Mason, J.I. (1994) Type 2 11β-hydroxysteroid dehydrogenase in human fetal tissues. *J. Clin. Endocrinol. Metab.*, **78**, 1529–1532.
149. Holmes, M., Sangra, M., French, K., Whittle, I., Paterson, J., Mullins, J., and Seckl, J.R. (2005) 11beta-Hydroxysteroid dehydrogenase type 2 protects the neonatal cerebellum from deleterious effects of glucocorticoids. *Neuroscience*, **137**, 865–873.
150. Meijer, O.C., de Lange, E.C., Breimer, D.D., de Boer, A.G., Workel, J.O., and de Kloet, E.R. (1998) Penetration of dexamethasone into brain glucocorticoid targets is enhanced in mdr1A P-glycoprotein knockout mice. *Endocrinology*, **139**, 1789–1793.
151. Schinkel, A.H. (1997) The physiological function of drug-transporting P-glycoproteins. *Semin. Cancer Biol.*, **8**, 161–170.
152. Erdeljan, P. and Matthews, S.G. (1999) Development of MDR1a expression in the guinea pig brain: implications for foetal glucocorticoid exposure. *Soc. Neurosci.*, 583–610 (Abstract).
153. McCabe, L., Marash, D., Li, A., and Matthews, S.G. (2001) Repeated antenatal glucocorticoid treatment decreases hypothalamic corticotrophin releasing

hormone m RNA but not corticosteroid receptor m RNA expression in the foetal guinea pig brain. *J. Neuroendocrinol.*, **13**, 425–431.
154. Seree, E., Villard, P.H., Hever, A., Guigal, N., Puyoou, F., Charvet, B., Point-Scomma, H. et al. (1998) Modulation of MDR1 and CYP3A expression by dexamethasone: evidence for an inverse regulation in adrenals. *Biochem. Biophys. Res. Commun.*, **252**, 392–395.
155. Johnstone, R.W., Ruefli, A.A., and Smyth, M.J. (2000) Multiple physiological functions for multidrug transporter P-glycoprotein? *Trends Biochem. Sci.*, **25**, 1–6.
156. Sukhai, M. and Piquette, M. (2000) Regulation of the multidrug resistance genes by stress signals. *J. Pharm. Sci.*, **3**, 268–280.
157. Tanabe, M., Ieiri, I., Nagata, N., Inoue, K., Ito, S., Kanamori, Y., Takahashi, M. et al. (2001) Expression of P-glycoprotein in human placenta: relation to genetic polymorphism of the multidrug resistance (MDR)-1 gene. *J. Pharmacol. Exp. Ther.*, **297**, 1137–1143.
158. Fuxe, K., Diaz, R., Cintra, A., Bhatnagar, M., Tinner, B., Gustafsson, J.A. et al. (1996) On the role of glucocorticoid receptors in brain plasticity. *Cell. Mol. Neurobiol.*, **16**, 239–258.
159. Welberg, L.A.M., Seckl, J.R., and Holmes, M.C. (2001) Prenatal glucocorticoid programming of brain corticosteroid receptors and corticotrophin-releasing hormone: possible implications for behaviour. *Neuroscience*, **104**, 71–79.
160. Huang, W.L., Beazley, L.D., Quinlivan, J.A., Evans, S.F., Newnham, J.P., and Dunlop, S.A. (1999) Effect of corticosteroids on brain growth in fetal sheep. *Obstet. Gynecol.*, **94**, 213–218.
161. Duksal, F., Kilic, I., Tufan, A.C., and Akdogan, I. (2009) Effects of different corticosteroids on the brain weight and hippocampal neuronal loss in rats. *Brain Res.*, **1250**, 75–80.
162. Huang, W.L., Harper, C.G., Evans, S.F., Newnham, J.P., and Dunlop, S.A. (2001) Repeated prenatal corticosteroid administration delays astrocyte and capillary tight junction maturation in fetal sheep. *Int. J. Dev. Neurosci.*, **19**, 487–493.
163. Huang, W.L., Harper, C.G., Evans, S.F., Newnham, J.P., and Dunlop, S.A. (2001) Repeated prenatal corticosteroid administration delays myelination of the corpus callosum in fetal sheep. *Int. J. Dev. Neurosci.*, **19**, 415–425.
164. Uno, H., Lohmiller, L., Thieme, C., Kemnitz, J.W., Engle, M.J., Roecker, E.B., and Farrell, P.M. (1990) Brain damage induced by prenatal exposure to dexamethasone in fetal rhesus macaques. I. hippocampus. *Dev. Brain Res.*, **53**, 157–167.
165. Hayashi, A., Nagaoka, M., Yamada, K., Ichitani, Y., Miake, Y., and Okado, N. (1998) Maternal stress induces synaptic loss and developmental disabilities of offspring. *Int. J. Dev. Neurosci.*, **16**, 209–216.
166. Lemaire, V., Koehl, M., Le Moal, M., and Abrous, D.N. (2000) Prenatal stress produces learning deficits associated with an inhibition of neurogenesis in the hippocampus. *Proc. Natl. Acad. Sci. U.S.A.*, **97**, 11032–11037.
167. Koo, J.W., Park, C.H., Choi, S.H., Kim, N.J., Kim, H.S., Choe, J.C. et al. (2003) The postnatal environment can counteract prenatal effects on cognitive ability, cell proliferation, and synaptic protein expression. *FASEB J.*, **17**, 1556–1558.
168. Sheline, Y.I., Wang, P.W., Gado, M.H., Csernansky, J.G., and Vannier, M.W. (1996) Hippocampal atrophy in recurrent major depression. *Proc. Natl. Acad. Sci. U.S.A.*, **93**, 3908–3913.
169. Stein, M.B., Koverola, C., Hanna, C., Torchia, M.G., and McClarty, B. (1997) Hippocampal volume in women victimized by childhood sexual abuse. *Psychol. Med.*, **27**, 951–959.
170. Bremner, J.D., Randall, P., Scott, T.M., Bronen, R.A., Seibyl, J.P., Southwick, S.M., Delaney, R.C. et al. (1995) MRI-based measurement of hippocampal volume in patients with combat-related posttraumatic stress disorder. *Am. J. Psychiatry*, **152**, 973–981.

171. Jacobson, L. and Sapolsky, R. (1991) The role of the hippocampus in feedback regulation of the hypothalamic-pituitary-adrenal axis. *Endocr. Rev.*, **12**, 118–134.

172. Gould, E., Woolley, C.S., Cameron, H.A., Daniels, D.C., and McEwen, B.S. (1991) Adrenal steroids regulate postnatal development in the rat dentate gyrus: II. Effects of glucocorticoids on cell birth. *J. Comp. Neurol.*, **313**, 486–493.

173. Gould, E., Woolley, C.S., and McEwen, B.S. (1991) Adrenal steroids regulate postnatal development in the rat dentate gyrus: I. Effects of glucocorticoids on cell death. *J. Comp. Neurol.*, **313**, 479–485.

174. Bohn, M.C. (1980) Granule cell genesis in the hippocampus of rats treated neonatally with hydrocortisone. *Neuroscience*, **5**, 2003–2012.

175. Fujioka, T., Sakata, Y., Yamaguchi, K., Shibasaki, T., Kato, H., and Nakamura, S. (1999) The effects of prenatal stress on the development of hypothalamic paraventricular neurones in foetal rats. *Neuroscience*, **92**, 1079–1088.

176. Uno, H., Eisele, S., Sakai, A., Shelton, S., Baker, E., de Jesus, O., and Holden, J. (1994) Neurotoxicity of glucocorticoids in the primate brain. *Horm. Behav.*, **28**, 336–348.

177. Ward, H.E., Johnson, E.A., Salm, A.K., and Birkle, D.L. (2000) Effects of prenatal stress on defensive withdrawal behaviour and corticotrophin releasing factor systems in rat brain. *Physiol. Behav.*, **70**, 359–366.

178. Llorente, E., Brito, M.L., Machado, P., and Gonzalez, M.C. (2002) Effect of prenatal stress on the hormonal response to acute and chronic stress on the hormonal response to acute and chronic stress and on immune parameters in the offspring. *J. Physiol. Biochem.*, **58**, 143–149.

179. Hawkins, P., Steyn, C., McGarrigle, H.H., Calder, N.A., Saito, T., Stratford, L.L., Noakes, D.E., and Hanson, M.A. (2000) Cardiovascular and hypothalamic-pituitary-adrenal axis development in late gestation fetal sheep and young lambs following modest maternal nutrient restriction in early gestation. *Reprod. Fertil. Dev.*, **12**, 443–456.

180. Sapolsky, R.M., Romero, L.M., and Munck, A.U. (2000) How do glucocorticoids influence stress responses? Integrating permissive, suppressive, and preparative actions. *Endocr. Rev.*, **21**, 55–89.

181. Lupien, S.J. and Lepage, M. (2001) Stress, memory, and the hippocampus: can't live with it, can't live without it. *Behav. Brain. Res.*, **127**, 137–158.

182. Weinstock, M., Matlina, E., Maor, G.I., Rosen, H., and McEwen, B.S. (1992) Prenatal stress selectively alters the reactivity of the hypothalamic-pituitary adrenal system in the female rat. *Brain Res.*, **595**, 195–200.

183. McCormick, C.M., Smythe, J.W., Sharma, S., and Meaney, M.J. (1995) Sex-specific effects of prenatal stress on hypothalamic-pituitary-adrenal responses to stress and brain glucocorticoid receptor density in adult rats. *Dev. Brain Res.*, **84**, 55–61.

184. Montano, M.M., Wang, M.-H., and vom Saal, F.S. (1993) Sex differences in plasma corticosterone in mouse foetuses are mediated by differential placental transport from the mother and eliminated by maternal adrenalectomy or stress. *J. Reprod. Fertil.*, **99**, 283–290.

185. Handa, R.J., Nunley, K.M., Lorens, S.A., Louie, J.P., McGivern, R.F., and Bollnow, M.R. (1994) Androgen regulation of adrenocorticotropin and corticosterone secretion in the male rat following novelty and foot shock stressors. *Physiol. Behav.*, **55**, 117–124.

186. Viau, V. and Meaney, M.J. (1996) The inhibitory effect of testosterone on hypothalamic-pituitary-adrenal responses to stress is mediated by the medial preoptic area. *J. Neurosci.*, **16**, 1866–1876.

187. Viau, V. and Sawchenko, P.E. (2002) Hypophysiotropic neurones of the paraventricular nucleus respond in spatially, temporally, and phenotypically differentiated manners to acute vs. repeated restraint stress: rapid

publication. *J. Comp. Neurol.*, **445**, 293–307.
188. Viau, V. (2002) Functional crosstalk between the hypothalamic-pituitary-gonadal and – adrenal axes. *J. Neuroendocrinol.*, **14**, 506–513.
189. Mueller, B.R. and Bale, T.L. (2008) Sex-specific programming of offspring emotionality after stress early in pregnancy. *J. Neurosci.*, **28**, 9055–9065.
190. Diaz, R., Fuxe, K., and Ogren, S.O. (1997) Prenatal corticosterone treatment induces long-term changes in spontaneous and apomorphine-mediated motor activity in male and female rats. *Neuroscience*, **81**, 129–140.
191. Diaz, R., Ogren, S.O., Blum, M., and Fuxe, K. (1995) Prenatal corticosterone increases spontaneous and d-amphetamine induced locomotor activity and brain dopamine metabolism in prepubertal male and female rats. *Neuroscience*, **66**, 467–473.
192. Koenig, J.I., Kirkpatrick, B., and Lee, P. (2002) Glucocorticoid hormones and early brain development in schizophrenia. *Neuropsychopharmacology*, **27**, 309–318.
193. Canlon, B., Erichsen, S., Nemlander, E., Chen, M., Hossain, A., Celsi, G., and Ceccatelli, S. (2003) Alterations in the intrauterine environment by glucocorticoids modifies the development programme of the auditory system. *Eur. J. Neurosci.*, **17**, 2035–2041.
194. Vallee, M., Mayo, W., Dellu, F., Le Moal, M., Simon, H., and Maccari, S. (1997) Prenatal stress induced high anxiety and postnatal handling induced low anxiety in adult offspring: correlation with stress-induced corticosterone secretion. *J. Neurosci.*, **17**, 2626–2636.
195. Fride, E. and Weinstock, M. (1988) Prenatal stress increases anxiety related behaviour and alters cerebral lateralisation of dopamine activity. *Life Sci.*, **42**, 1059–1065.
196. Poltyrev, T., Keshet, G.I., Kay, G., and Weinstock, M. (1996) Role of experimental conditions in determining differences in exploratory behaviour of prenatally stressed rats. *Dev. Psychobiol.*, **29**, 453–462.
197. Wakshlak, A. and Weinstock, M. (1990) Neonatal handling reverses behavioural abnormalities induced by rats by prenatal stress. *Physiol. Behav.*, **48**, 289–292.
198. Suchecki, D. and Palermo, N.J. (1991) Prenatal stress and emotional response of adult offspring. *Physiol. Behav.*, **48**, 289–292.
199. Welberg, L.A.M., Seckl, J.R., and Holmes, M.C. (2000) Inhibition of 11β-hydroxysteroid dehydrogenase, the feto-placental barrier to maternal glucocorticoids, permanently programs amygdala glucocorticoid receptor mRNA expression and anxiety-like behavior in the offspring. *Eur. J. Neurosci.*, **12**, 1047–1054.
200. Lehmann, J., Stohr, T., and Feldon, J. (2000) Long-term effects of prenatal stress experiences and postnatal maternal separation on emotionality and attentional processes. *Behav. Brain Res.*, **107**, 133–144.
201. Fujioka, T., Fujioka, A., Tan, N., Chowdhury, G.M., Mouri, H., Sakata, Y., and Nakamura, S. (2001) Mild prenatal stress enhances learning performance in the non-adopted rat offspring. *Neuroscience*, **103**, 301–307.
202. Dunn, A. and Berridge, C. (1990) Physiological and behavioral-responses to corticotropin-releasing factor administration – is crf a mediator of anxiety or stress responses. *Brain Res. Rev.*, **15**, 71–100.
203. Cratty, M.S., Ward, H.E., Johnson, E.A., Azzaro, A.J., and Birkle, D.L. (1995) Prenatal stress increases corticotropin-releasing factor (Crf) content and release in rat amygdala minces. *Brain Res.*, **675**, 297–302.
204. Hsu, D., Chen, F., Takahashi, L., and Kalin, N. (1998) Rapid stress-induced elevations in corticotropin-releasing hormone mRNA in rat central amygdala nucleus and hypothalamic paraventricular nucleus: an in situ hybridization analysis. *Brain Res.*, **788**, 305–310.
205. Feldman, S. and Weidenfeld, J. (1998) The excitatory effects of the amygdala

on hypothalamo-pituitary- adrenocortical responses are mediated by hypothalamic norepinephrine, serotonin, and CRF-41. *Brain Res. Bull.*, **45**, 389–393.
206. Tronche, F., Kellendonk, C., Kretz, O., Gass, P., Anlag, K., Orban, P.C., Bock, R. *et al.* (1999) Disruption of the glucocorticoid receptor gene in the nervous system results in reduced anxiety. *Nat. Genet.*, **23**, 99–103.
207. Levine, S. (1962) Plasma-free corticosteroid response to electric shock in rats stimulated in infancy. *Science*, **135**, 795–796.
208. Meaney, M.J., Aitken, D.H., van Berkel, C., Bhatnagar, S., and Sapolsky, R.M. (1988) Effect of neonatal handling on age-related impairments associated with the hippocampus. *Science*, **239**, 766–768.
209. Liu, D., Diorio, J., Tannenbaum, B., Caldji, C., Francis, D., Freedman, A., Sharma, S. *et al.* (1997) Maternal care, hippocampal glucocorticoid receptors, and hypothalamic-pituitary-adrenal responses to stress. *Science*, **277**, 1659–1662.
210. Meaney, M.J., Aitken, D.H., Sharma, S., and Viau, V. (1992) Basal ACTH, corticosterone and corticosterone-binding globulin levels over the diurnal cycle, and hippocampal corticosteroid receptors in young and aged, handled and non-handled rats. *Neuroendocrinology*, **55**, 204–213.
211. Meaney, M.J., Aitken, D.H., Viau, V., Sharma, S., and Sarrieau, A. (1989) Neonatal handling alters adrenocortical negative feedback sensitivity and hippocampal type II glucocorticoid receptor binding in the rat. *Neuroendocrinology*, **50**, 597–604.
212. Catalani, A., Marinelli, M., Scaccianoce, S., Nicolai, R., Muscolo, L.A.A., Porcu, A., Koranyi, L. *et al.* (1993) Progeny of mothers drinking corticosterone during lactation has lower stress-induced corticosterone secretion and better cognitive performance. *Brain Res.*, **624**, 209–215.
213. Smythe, J.W., Rowe, W.B., and Meaney, M.J. (1994) Neonatal handling alters serotonin (5-HT) turnover and 5-HT2 receptor binding in selected brain regions: relationship to the handling effect on glucocorticoid receptor expression. *Brain Res. Dev. Brain Res.*, **80**, 183–189.
214. Mitchell, J.B., Rowe, W., Boksa, P., and Meaney, M.J. (1990) Serotonin regulates type II corticosteroid receptor binding in hippocampal cell cultures. *J. Neurosci.*, **10**, 1745–1752.
215. Yau, J.L.W., Noble, J., and Seckl, J.R. (1997) Site-specific regulation of corticosteroid and serotonin receptor subtype gene expression in the rat hippocampus following methylenedioxymethamphetamine: role of corticosterone and serotonin. *Neuroscience*, **78**, 111–121.
216. Yau, J.L.W., Noble, J., Widdowson, J., and Seckl, J.R. (1997) Impact of adrenalectomy on 5-HT6 and 5-HT7 receptor gene expression in the rat hippocampus. *Mol. Brain Res.*, **45**, 182–186.
217. Meaney, M.J., Diorio, J., Francis, D., Weaver, S., Yau, J.L.W., Chapman, K.E., and Seckl, J.R. (2000) Postnatal handling increases the expression of cAMP-inducible transcription factors in the rat hippocampus: the effects of thyroid hormones and serotonin. *J. Neurosci.*, **20**, 3926–3935.
218. Encio, I.J. and Detera-Wadleigh, S.D. (1991) The genomic structure of the human glucocorticoid receptor. *J. Biol. Chem.*, **266**, 7182–7188.
219. Slotkin, T.A., Barnes, G.A., McCook, E.C., and Seidler, F.J. (1996) Programming of brainstem serotonin transporter development by prenatal glucocorticoids. *Brain Res. Dev. Brain Res.*, **93**, 155–161.
220. Fumagalli, F., Jones, S.R., Caron, M.G., Seidler, F.J., and Slotkin, T.A. (1996) Expression of mRNA coding for the serotonin transporter in aged versus young rat brain: differential effects of glucocorticoids. *Brain Res.*, **719**, 225–228.
221. Oliver, T.G., Grasfeder, L.L., Carroll, A.L., Kaiser, C., Gillingham, C.L., Lin, S.M., Wickramasinghe, R. *et al.* (2003) Transcriptional profiling of the Sonic hedgehog response: a critical role for

N-myc in proliferation of neuronal precursors. *Proc. Natl. Acad. Sci. U.S.A.*, **100**, 7331–7336.
222. Machold, R., Hayashi, S., Rutlin, M., Muzumdar, M.D., Nery, S., Corbin, J.G., Gritli-Linde, A. et al. (2003) Sonic hedgehog is required for progenitor cell maintenance in telencephalic stem cell niches. *Neuron*, **39**, 937–950.
223. Gulino, A., de Smaele, E., and Ferretti, E. (2009) Glucocorticoids and neonatal brain injury: the hedgehog connection. *J. Clin. Invest.*, **119**, 243–246.
224. Heine, V.M. and Rowitch, D.H. (2009) Hedgehog signaling has a protective effect in glucocorticoid-induced mouse neonatal brain injury through an 11betaHSD2-dependent mechanism. *J. Clin. Invest.*, **119**, 267–277.
225. Holmes, M.C. and Seckl, J.R. (2006) The role of 11beta-hydroxysteroid dehydrogenases in the brain. *Mol. Cell. Endocrinol*, **248**, 9–14.
226. Roper, R.J., Baxter, L.L., Saran, N.G., Klinedinst, D.K., Beachy, P.A., and Reeves, R.H. (2006) Defective cerebellar response to mitogenic Hedgehog signaling in Down syndrome mice. *Proc. Natl. Acad. Sci. U.S.A.*, **103**, 1452–1456.
227. Pepin, M.C., Pothier, F., and Barden, N. (1992) Impaired glucocorticoid receptor function in transgenic mice expressing antisense RNA. *Nature*, **355**, 725–728.
228. Bhattacharyya, S., Brown, D., Brewer, J., Vogt, S., and Muglia, L. (2007) Macrophage glucocorticoid receptors regulate Toll-like receptor-4-mediated inflammatory responses by selective inhibition of p38 MAP kinase. *Blood*, **109**, 4313–4319.
229. Cleasby, M.E., Kelly, P.A.T., Walker, B.R., and Seckl, J.R. (2003) Programming of rat muscle and fat metabolism by in utero overexposure to glucocorticoids. *Endocrinology*, **144**, 999–1007.
230. Noorlander, C.W., de Graan, P.N., Middeldorp, J., van Beers, J.J., and Visser, G.H. (2006) Ontogeny of hippocampal corticosteroid receptors: effects of antenatal glucocorticoids in human and mouse. *J. Comp. Neurol.*, **499**, 924–932.
231. McCormick, J., Lyons, V., Jacobson, M., Diorio, J., Nyirenda, M., Weaver, S., Yau, J.L.W. et al. (2000) 5′-heterogeneity of glucocorticoid receptor mRNA is tissue-specific; differential regulation of variant promoters by early life events. *Mol. Endocrinol.*, **14**, 506–517.
232. Turner, J.D. and Muller, C.P. (2005) Structure of the glucocorticoid receptor (NR3C1) gene 5′ untranslated region: identification, and tissue distribution of multiple new human exon 1. *J. Mol. Endocrinol.*, **35**, 283–292.
233. Weaver, I., Cervoni, N., Champagne, F., D'Alessio, A., Sharma, S., Seckl, J., Dymov, S. et al. (2004) Epigenetic programming by maternal behavior. *Nat. Neurosci.*, **7**, 847–854.
234. Weaver, I.C.G., D'Alessio, A.C., Brown, S.E., Hellstrom, I.C., Dymov, S., Sharma, S., Szyf, M., and Meaney, M.J. (2007) The transcription factor nerve growth factor-inducible protein A mediates epigenetic programming: altering epigenetic marks by immediate-early genes. *J. Neurosci.*, **27**, 1756–1768.
235. Thomassin, H., Flavin, M., Espinas, M., and Grange, T. (2001) Glucocorticoid-induced DNA demethylation and gene memory during development. *EMBO J.*, **20**, 1974–1983.
236. Lillycrop, K.A., Phillips, E.S., Jackson, A.A., Hanson, M.A., and Burdge, G.C. (2005) Dietary protein restriction of pregnant rats induces and folic acid supplementation prevents epigenetic modification of hepatic gene expression in the offspring. *J. Nutr.*, **135**, 1382–1386.
237. Nyirenda, M., Dean, S., Lyons, V., Chapman, K., and Seckl, J.R. (2006) Prenatal programming of hepatocyte nuclear factor 4 alpha in the rat: a key mechanism in the 'fetal origins of hyperglycemia'? *Diabetologia*, **49**, 1412–1420.
238. Champagne, F.A., Weaver, I.C.G., Diorio, J., Dymov, S., Szyf, M., and Meaney, M.J. (2006) Maternal care associated with methylation of the estrogen receptor-alpha 1b promoter and estrogen receptor-alpha expression in the medial preoptic area of

female offspring. *Endocrinology*, **147**, 2909–2915.
239. Buss, C., Lord, C., Wadiwalla, M., Hellhammer, D.H., Lupien, S.J., Meaney, M.J., and Pruessner, J.C. (2007) Maternal care modulates the relationship between prenatal risk and hippocampal volume in women but not in men. *J. Neurosci.*, **27**, 2592–2595.
240. Aghajafari, F., Murphy, K., Willan, A., Ohlsson, A., Amankwah, K., Matthews, S., and Hannah, M. (2001) Multiple courses of antenatal corticosteroids: a systematic review and meta-analysis. *Am. J. Obstat. Gynecol.*, **185**, 1073–1080.
241. NIH Consensus Development Conference (1995) Effect of corticosteroids for foetal maturation and perinatal outcomes. *Am. J. Obstet. Gynecol.*, **173**, 253–344.
242. NIH Consensus Development Conference Statement (2001) Antenatal corticosteroids revisited: repeat courses. *Obstet. Gynecol.*, **98**, 144–150.
243. Andrews, M.H. and Matthews, S.G. (2003) Antenatal glucocorticoids: is there a cause for concern? *Foetal Matern. Med. Rev.*, **14**, 329–354.
244. Murphy, K.E., Hannah, M.E., Willan, A.R., Hewson, S.A., Ohlsson, A., Kelly, E.N., Matthews, S.G. et al. (2008) Multiple courses of antenatal corticosteroids for preterm birth (MACS): a randomised controlled trial. *Lancet*, **372**, 2143–2151.
245. Quinlivan, J.A., Evans, S.F., Dunlop, S.A., Beazley, L.D., and Newnham, J.P. (1998) Use of corticosteroids by Australian obstetricians – a survey of clinical practice. *Aust. N.Z. J. Obstet. Gynecol.*, **38**, 1–7.
246. Brocklehurst, O., Gates, S., McKenzie-McHard, K., Alfirevic, Z., and Chamberlain, G. (1999) Are we prescribing multiple courses of antenatal corticosteroids? A survey of practice in the UK. *Br. J. Obstet. Gynaecol.*, **106**, 977–979.
247. Seckl, J.R. and Meaney, M.J. (2004) Glucocorticoid programming. *Ann. N. Y. Acad. Sci.*, **1032**, 63–84.
248. Forest, M.G., David, M., and Morel, Y. (1993) Prenatal diagnosis and treatment of 21-hydroxylase deficiency. *J. Steroid Biochem. Mol. Biol.*, **45**, 75–82.
249. Mercado, A.B., Wilson, R.C., Cheng, K.C., Wei, J.Q., and New, M.I. (1995) Prenatal treatment and diagnosis of congenital adrenal hyperlasia owing to 21-hydroxylase deficiency. *J. Clin. Endocrinol. Metab.*, **80**, 2014–2020.
250. Doyle, L.W., Ford, G.W., Davis, N.M., and Callanan, C. (2000) Antenatal corticosteroid therapy and blood pressure at 14 years of age in preterm children. *Clin. Sci.*, **98**, 137–142.
251. Entringer, S., Wüst, S., Kumsta, R., Layes, I.M., Nelson, E.L., Hellhammer, D.H., and Wadhwa, P.D. (2008) Prenatal psychosocial stress exposure is associated with insulin resistance in young adults. *Am. J. Obstet. Gynecol.*, **199** (5), 498.e1–498.e7.
252. Finken, M.J., Keijzer-Veen, M.G., Dekker, F.W., Frölich, M., Walther, F.J., Romijn, J.A., van der Heijden, B.J. et al. (2008) Antenatal glucocorticoid treatment is not associated with long-term metabolic risks in individuals born before 32 weeks of gestation. *Arch. Dis. Child. Fetal Neonatal Ed.*, **93**, F442–F447.
253. MacArthur, B.A., Howie, R.N., Dezoete, J.A., and Elkins, J. (1981) Cognitive and psychosocial development of 4-year-old children whose mothers were treated antenatally with betamethasone. *Pediatrics*, **68**, 638–643.
254. Trautman, P.D., Meyer-Bahlburg, H.F.L., Postelnek, J., and New, M.I. (1995) Effects of early prenatal dexamethasone on the cognitive and behavioral development of young children: results of a pilot study. *Psychoneuroendocrinology*, **20**, 439–449.
255. Hasbargen, U., Reber, D., Versmold, H., and Schulze, A. (2001) Growth and development of children to 4 years of age after repeated antenatal steroid administration. *Eur. J. Pediatr.*, **160**, 552–555.
256. Modi, N., Lewis, H., Al Naqeeb, N., Ajati-Obe, M., Dore, C.J., and

Rutherford, M. (2001) The effects of repeated antenatal glucocorticoid therapy on the developing brain. *Pediatr. Res.*, **50**, 581–585.

257. French, N.P., Hagan, R., Evans, S.F., Mullan, A., and Newnam, J.P. (2004) Repeated antenatal corticosteroids: effects on cerebral palsy and childhood behaviour. *Am. J. Obstet. Gynecol.*, **190**, 588–595.

258. Yeh, T., Lin, Y., Lin, H., Huang, C., Hsieh, W., Lin, C., and Tsai, C. (2004) Outcomes at school age after postnatal dexamethasone therapy for lung disease of prematurity. *New Engl. J. Med.*, **350**, 1304–1313.

259. Meyer-Bahlburg, H.F., Dolezal, C., Baker, S.W., Carlson, A.D., Obeid, J.S., and New, M.I. (2004) Cognitive and motor development of children with and without congenital adrenal hyperplasia after early-prenatal dexamethasone. *J. Clin. Endocrinol. Metab.*, **89**, 610–614.

260. Lajic, S., Nordenstrom, A., and Hirvikoski, T. (2008) Long-term outcome of prenatal treatment of congenital adrenal hyperplasia. *Endocr. Dev.*, **13**, 82–98.

261. Wapner, R.J., Sorokin, Y., Mele, L., Johnson, F., Dudley, D.J., Spong, C.Y., Peaceman, A.M. et al. (2007) Long-term outcomes after repeat doses of antenatal corticosteroids. *New Engl. J. Med.*, **357**, 1190–1198.

262. Dessens, A.B., Haas, H.S., and Koppe, J.G. (2000) Twenty-year follow up of antenatal corticosteroid administration. *Pediatrics*, **105**, E77.

263. Clark, P.M., Hindmarsh, P.C., Shiell, A.W., Law, C.M., Honour, J.W., and Barker, D.J.P. (1996) Size at birth and adrenocortical function in childhood. *Clin. Endocrinol. (Oxf.)*, **45**, 721–726.

264. Phillips, D.I., Fall, C.H.D., Whorwood, C.B., Seckl, J.R., Wood, P.J., Barker, D.J.P., and Walker, B.R. (1998) Elevated plasma cortisol concentrations: an explanation for the relationship between low birthweight and adult cardiovascular risk factors. *J. Clin. Endocrinol. Metab.*, **83**, 757–760.

265. Phillips, D.I.W., Walker, B.R., Reynolds, R.M., Flanagan, D.E.H., Wood, P.J., Osmond, C., Barker, D.J.P., and Whorwood, C.B. (2000) Low birth weight predicts elevated plasma cortisol concentrations in adults from 3 populations. *Hypertension*, **35**, 1301–1306.

266. Reynolds, R.M., Walker, B.R., Syddall, H.E., Andrew, R., Wood, P.J., Whorwood, C.B., and Phillips, D.I.W. (2001) Altered control of cortisol secretion in adult men with low birth weight and cardiovascular risk factors. *J. Clin. Endocrinol. Metab.*, **86**, 245–250.

267. McGowan, P.O., Sasaki, A., D'Alessio, A.C., Dymov, S., Labonté, B., Szyf, M., Turecki, G., and Meaney, M.J. (2009) Epigenetic regulation of the glucocorticoid receptor in human brain associates with childhood abuse. *Nat. Neurosci.*, **12**, 342–348.

268. Davidson, J.R., Stein, D.J., Shalev, A.Y., and Yehuda, R. (2004) Posttraumatic stress disorder: acquisition, recognition, course and treatment. *J. Neuropsychiatry Clin. Neurosci.*, **16**, 135–147.

269. Yehuda, R. (2002) Current concepts: post-traumatic stress disorder. *New Engl. J. Med*, **346**, 108–114.

270. Yehuda, R., Golier, J.A., Yang, R.K., and Tischler, L. (2004) Enhanced sensitivity to glucocorticoids in peripheral mononuclear leukocytes in posttraumatic stress disorder. *Biol. Psychiatry*, **55**, 1110–1116.

271. Yehuda, R., Engel, S.M., Brand, S.R., Seckl, J.R., Marcus, S.M., and Berkowitz, G.S. (2005) Transgenerational effects of posttraumatic stress disorder in babies of mothers exposed to the world trade center attacks during pregnancy. *J. Clin. Endocrinol. Metab.*, **90**, 4115–4118.

272. Berkowitz, G.S., Wolff, M.S., Janevic, T.M., Holzman, I.R., Yehuda, R., and Landrigan, P.J. (2003) The World Trade Center disaster and intrauterine growth restriction. *J. Am. Med. Assoc.*, **290**, 595–596.

273. Engel, S.M., Berkowitz, G.S., Wolff, M.S., and Yehuda, R. (2005) Psychological trauma associated with the World Trade Center attacks and its effect on

pregnancy outcome. *Paediatr. Perinat. Epidemiol.*, **19**, 334–341.
274. Yehuda, R., Bierer, L.M., Andrew, R., Schmeidler, J., and Seckl, J.R. (2009) Enduring effects of severe developmental adversity, including nutritional deprivation, on cortisol metabolism in aging Holocaust survivors. *J. Psychiatr. Res.*, **49**, 877–883.
275. Yehuda, R., Teicher, M., Seckl, J., Grossman, R., Morris, A., and Bierer, L. (2007) Parental PTSD is a 'vulnerability' factor for low cortisol trait in offspring of Holocaust survivors. *Arch. Gen. Psychiatry*, **64**, 1040–1048.
276. Tsuruo, Y., Miyamoto, T., Yokoi, H., Kitagawa, K., Futaki, S., and Ishimura, K. (1996) Immunohistochemical presence of 5 alpha-reductase rat type 1-containing cells in the rat brain. *Brain Res.*, **722**, 207–211.
277. Poletti, A., Negri-Cesi, P., Melcangi, R.C., Colciago, A., Martini, L., and Celotti, F. (1997) Expression of androgen-activating enzymes in cultured cells of developing rat brain. *J. Neurochem.*, **68**, 1298–1303.
278. Vallarino, M., Mathieu, M., do-Rego, J.L., Bruzzone, F., Chartrel, N., Luu-The, V., Pelletier, G., and Vaudry, H. (2005) Ontogeny of 3 beta-hydroxysteroid dehydrogenase and 5 alpha-reductase in the frog brain. *Ann. N. Y. Acad. Sci.*, **1040**, 490–493.
279. Torres, J.M. and Ortega, E. (2003) Differential regulation of steroid 5α-reductase isozymes expression by androgens in the adult rat brain. *FASEB J.*, **17**, 148–1433.
280. Russell, D.W. and Wilson, J.D. (1994) Steroid 5α-reductase:two genes/two enzymes. *Annu. Rev. Biochem.*, **63**, 25–61.
281. Sanchez, P., Torres, J.M., del Moral, R.G., and Ortega, E. (2006) Effects of testosterone on brain mRNA levels of steroid 5α-reductase isozymes in early postnatal life of rat. *Neurochem. Int.*, **49**, 626–630.
282. Charalampopoulos, I., Remboutsika, E., Margioris, A.N., and Gravanis, A. (2008) Neurosteroids as modulators of neurogenesis and neuronal survival. *Trends Endocrinol. Metab.*, **19**, 300–307.
283. Brinton, R.D. and Wang, J.M. (2006) Therapeutic potential of neurogenesis for prevention and recovery from Alzheimer's disease: allopregnanolone as a proof of concept neurogenic agent. *Curr. Alzheimer Res.*, **3**, 185–190.
284. Brinton, R.D. (1994) The neurosteroid 3 alpha-hydroxy-5 alpha-pregnan-20-one induces cytoarchitectural regression in cultured foetal hippocampal neurones. *J. Neurosci.*, **14**, 2763–2774.
285. Charalampopoulos, I., Tsatsanis, C., Dermitzaki, E., Alexaki, V.I., Castanas, E., Margioris, A.N., and Gravanis, A. (2004) Dehydroepiandrosterone and allopregnanolone protect sympathoadrenal medulla cells against apoptosis via antiapoptopic Bcl-2 proteins. *Proc. Natl. Acad. Sci. U.S.A.*, **101**, 8209–8214.
286. Gustafsson, J.A. and Stenberg, A. (1974) Irreversible androgenic programming at birth of microsomal and soluble rat liver enzymes active on 4-androstane-3, 17-dione a 5α-and rostane-3α, 17β-diol. *J. Biol. Chem.*, **249**, 711–718.
287. Ordyan, N.E. and Pivina, S.G. (2005) Effects of prenatal stress on the activity of an enzyme involved in neurosteroid synthesis during the "critical period" of sexual differentiation of the brain in male rats. *Neurosci. Behav. Physiol.*, **35**, 931–935.
288. Datson, N.A., Morsink, M.C., Meijer, O.C., and de Kloet, E.R. (2008) Central corticosteroid actions: search for gene targets. *Eur. J. Pharmacol.*, **583**, 272–289.
289. Kaufer, D., Ogle, W.O., Pincus, Z.S., Clark, K.L. Nicholas, A.C., Dinkel, K.M., Dumas, T.C. et al. (2004) Restructuring the neuronal stress response with anti-glucocorticoid gene delivery. *Nat. Neurosci.*, **7**, 947–953.
290. Cleasby, M.E., Livingstone, D.E., Nyirenda, M.J., Seckl, J.R., and Walker, B.R. (2003) Is programming of glucocorticoid receptor expression by prenatal dexamethasone in the rat secondary to metabolic derangement in adulthood? *Eur. J. Endocrinol.*, **148**, 129–138.
291. Stocker, C.J., Wargent, E., O'Dowd, J., Cornick, C., Speakman, J.R., Arch, J.R.,

and Cawthorne, M.A. (2007) Prevention of diet-induced obesity and impaired glucose tolerance in rats following administration of leptin to their mothers. *Am. J. Physiol. Regul. Integr. Comp. Physiol.*, **292**, R1810–R1818.

292. Vickers, M.H., Gluckman, P.D., Coveny, A.H., Hofman, P.L., Cutfield, W.S., Gertler, A., Breier, B.H., and Harris, M. (2005) Neonatal leptin treatment reverses developmental programming. *Endocrinology*, **146**, 4211–4216.

293. Stocker, C., O'Dowd, J., Morton, N.M., Wargent, E., Sennitt, M.V., Hislop, D., Glund, S. *et al.* (2004) Modulation of susceptibility to weight gain and insulin resistance in low birthweight rats by treatment of their mothers with leptin during pregnancy and lactation. *Int. J. Obes. Relat. Metab. Disord.*, **28**, 129–136.

294. Aerni, A., Traber, R., Hock, C., Roozendaal, B., Schelling, G., Papassotiropoulos, A., Nitsch, R.M. *et al.* (2004) Low-dose cortisol for symptoms of posttraumatic stress disorder. *Am. J. Psychiatry*, **161**, 1488–1490.

295. Schelling, G., Kilger, E., Roozendaal, B., de Quervain, D.J., Briegel, J., Dagge, A., Rothenhäusler, H.B. *et al.* (2004) Stress doses of hydrocortisone, traumatic memories, and symptoms of posttraumatic stress disorder in patients after cardiac surgery. *Biol. Psychiatry*, **55**, 627–633.

296. Weaver, I.C. (2009) Epigenetic effects of glucocorticoids. *Semin. Fetal Neonatal Med.*, **14**, 143–150.

297. Weaver, I.C., Meaney, M.J., and Szyf, M. (2006) Maternal care effects on the hippocampal transcriptome and anxiety-mediated behaviors in the offspring that are reversible in adulthood. *Proc. Natl. Acad. Sci. U.S.A.*, **103**, 3480–3485.

298. Jokinen, J. and Nordström, P. (2009) HPA axis hyperactivity and attempted suicide in young adult mood disorder inpatients. *J. Affect. Disord.*, **116**, 117–120.

299. Jokinen, J. and Nordström, P. (2008) HPA axis hyperactivity as suicide predictor in elderly mood disorder inpatients. *Psychoneuroendocrinology*, **33**, 1387–1393.

6
Regulation of Structural Plasticity and Neurogenesis during Stress and Diabetes; Protective Effects of Glucocorticoid Receptor Antagonists

Paul J. Lucassen, Carlos P. Fitzsimons, Erno Vreugdenhil, Pu Hu, Charlotte Oomen, Yanina Revsin, Marian Joëls, and Edo Ronald de Kloet

In this chapter, we will review changes in structural plasticity of the adult hippocampus during stress and exposure to glucocorticoids (GCs). We further discuss the protective and normalizing role of glucocorticoid receptor (GR) antagonist treatment under these conditions and its implications for disorders such as depression and diabetes mellitus.

6.1
The Stress Response

Stress represents an old, yet essential alarm system for an organism. Whenever a discrepancy occurs between an organism's expectations and the reality it encounters, stress systems are activated; particularly when it involves a threat to, or disturbance of its homeostasis, well being, or health. Loss of control, or unpredictability when faced with predator threat in animals, or psychosocial demands in humans, can all produce stress signals. The same holds true for perturbations of a more physical or biological nature, such as energy crises, injury, or inflammation. Upon exposure to a stressor, various sensory and cognitive signals converge to activate a stress response that triggers several adaptive processes in the body and brain which aim to promote restoration of homeostasis.

In mammals, Selye [1] noted that the effect of stressors develops in a stereotypic manner. The first phase largely involves activation of the sympathoadrenal system through the rapid release of epinephrine and norepinephrine from the adrenal medulla; these hormones elevate basal metabolic rate and increase blood flow to vital organs like the heart and muscles. At a later stage, the limbic hypothalamo-pituitary-adrenal (HPA) system is activated as well, a classic neuroendocrine circuit in which limbic and hypothalamic brain structures integrate emotional, cognitive, neuroendocrine, and autonomic inputs, that together determine the magnitude and duration of the organism's behavioral, neural, and hormonal response to a stressor.

Hormones in Neurodegeneration, Neuroprotection, and Neurogenesis.
Edited by Achille G. Gravanis and Synthia H. Mellon
Copyright © 2011 WILEY-VCH Verlag GmbH & Co. KGaA, Weinheim
ISBN: 978-3-527-32627-3

6.2
HPA Axis and Glucocorticoids

Stress-induced activation of the HPA axis starts with the production of corticotropin-releasing hormone (CRH) in parvocellular neurons of the hypothalamic paraventricular nucleus (PVN), which induces the release of adrenocorticotrophic hormone (ACTH) from the anterior pituitary gland; this in turn causes release of GCs from the adrenal cortex into the general circulation. This stress-induced secretion is superimposed on the hourly pulsatile secretory bursts of steroid release under basal conditions. This ultradian rhythm in the HPA axis is important for maintaining tissue responsiveness. The amplitude of this rhythm is low at the end of the active period and increases again toward the next activity period. The rhythm becomes disordered in old age so that the coordinating and synchronizing action of GCs becomes compromised [2].

6.3
Glucocorticoid Actions

Upon their release, GCs (cortisol in primates, corticosterone in most rodents) cause a wide range of actions and have an important effect on many organs and organ systems in the body, including the brain. In the periphery, GCs mobilize energy by raising glucose levels. They further affect carbohydrate and lipid metabolism and can have a catabolic effect on muscle and bone tissues and affect cognition, for example. GCs also exert a "permissive" influence that enhances sympathoadrenal activity, in contrast to the generally slow genomic action of the latter. If stress becomes chronic, an imbalance may occur and GCs can overshoot and have deleterious effects by inducing muscle wasting, gastrointestinal ulceration, hyperglycemia (diabetes mellitus), and atrophy of the immune system.

6.4
Feedback Regulation

After the discovery that the hippocampal formation holds large numbers of GC receptors [3], a large body of evidence has been gathered, demonstrating that stress via elevated GC levels can affect both, hippocampal structure and function [4–6]. Cerebral impacts of GCs are considered to predominantly involve slow genomic actions following activation of mineralocorticoids (MRs) and GRs. These receptors act as transcriptional regulators of responsive genes. The MR has a high affinity for aldosterone and corticosterone. It is predominantly expressed in the hippocampus, lateral septum, and amygdala. In contrast, the GR has a 10-fold lower affinity for corticosterone and is ubiquitously distributed, with enrichments in the hippocampus, PVN, and pituitary; that is, the main feedback sites through which GCs regulate their own release [7]. Also, other brain regions such as the amygdala and prefrontal cortex, modulate HPA feedback and (re)activity.

Owing to these differences in affinity, the degree of MR and GR occupation depends on circulating GC levels that fluctuate over the course of the day. When at rest, circulating GC levels are low and activate mainly the MR while occupying only a small fraction of the GRs. Only after stress, or at the circadian peak prior to the onset of the activity period, do GRs become activated. Under these circumstances, changes in the degree of MR and GR activation influence gene expression in the hippocampus and may result in persistent changes in the electrophysiological properties of the hippocampal network. The variable MR/GR ratio is particularly relevant for neurons that express both receptor types; that is, CA1 pyramidal neurons and dentate gyrus (DG) granular neurons. In the CA3 region, very few GRs are present.

Recently, fast actions of the GC have been discovered, which are putatively mediated by membrane receptors. These fast actions are thought to contribute to the organization of the stress response [8]. These membrane receptors appear to be variants of the classical nuclear receptors MR and GR. In limbic regions, the membrane MR was found to boost excitatory neurotransmission accompanying the initial stress-induced rise of corticosterone. Electrophysiological, endocrine, and behavioral data suggests that fast-acting corticosterone organizes the initial stage of the stress response through the MR [9]. Interestingly, there is also evidence for a fast-acting GR that requires endocannabinoids [10].

The hippocampus is not only very sensitive to circulating GC levels, it is also important in emotional processing and key aspects of learning and memory, where adult neurogenesis (AN) has also been implicated [6, 7]. Short-term exposure to stressors induces behavioral adaptation and is considered harmless. Although prolonged GC (over)exposure is often thought to be associated with deleterious alterations in hippocampal excitability, long-term potentiation (LTP), and hippocampus-related memory performance, many positive effects of stress have been described as well, that depend on the type of stressor and its convergence in space and time [5, 8, 11].

Prolonged exposure to stress may induce alterations in HPA feedback that can overexpose the brain and body to aberrant GC levels. Even though feedback is largely mediated through the GR, chronic stress may also alter the function of the MR that is implicated in tonic inhibitory control of the HPA axis and modulates AN [5, 7].

6.5
Stress and Depression

Exposure to severe, repetitive, uncontrollable stressors may facilitate the development of psychopathologies. Major depressive disorder is one among these illnesses known to result from an interaction between environmental stressors and genetic/developmental predispositions [12]. Typical observations in depressed patients suggest a hyperactive HPA axis: reduced GR function as tested in the dexamethasone (DEX) suppression test is commonly found, as well as elevated cortisol levels particularly during the trough of the circadian rhythm, increased adrenal size, reduced hippocampal volume, and various other aberrations at different levels of the neuroendocrine system [6, 7, 13].

Many brain structures mediate the different symptoms of depression, and *in vivo* imaging studies on patients with emotional disorders have repeatedly indicated that structures other than the hippocampus are also involved, such as the prefrontal cortex, subgenual cingulate cortex, and amygdala [14]. Altered hippocampal function is likely to influence the activity of other brain structures, in particular, the prefrontal cortex and the amygdala which are key areas in emotional regulation and can, in turn, be influenced by these structures as well. Since the hippocampus indirectly provides negative feedback control of the HPA axis through a poorly understood trans-synaptic network [15], altered hippocampus function may contribute to HPA axis dysregulation, which is common in almost 50% of depressed patients [16].

While depressive and related affective disorders are considered to have a neurochemical basis, recent studies suggest that impairments of structural plasticity, particularly when induced at an early age, contribute to the pathophysiology as well [6, 17–19]. In this review, we will focus on the hippocampal formation, because of the volume changes in depression [20] and the occurrence of AN in this structure.

6.6
Stress-Induced Viability Changes in the Hippocampus: Effect on Function, Volume, Cell Number, and Apoptosis

Loss of hippocampal volume is well documented in stress-related disorders such as depression, and in patients treated with synthetic GCs, or suffering from Cushing's syndrome [20–23]. When stress exposure is prolonged, reductions in neuropil and hippocampal volume have also been reported in animal models. Magnetic resonance imaging (MRI) and morphometric studies, for example, show that chronic psychosocial stressors result in a mild reduction in hippocampal volume, that is around 10% in tree shrews [20]. Effects of stress on hippocampal volume and hippocampal cell number are, however, relatively mild and subregion-specific and occur shortly after stressor onset, but prior to cognitive disturbances [6].

The traditional explanation for this stress-induced hippocampal volume decrease was that elevated GCs in rodents have neurotoxic effects on the hippocampus. Neuronal death in the CA3 and CA1 subregions, in particular, was emphasized [23, 24]. More recent studies however, used state-of-the-art methodology and failed to find evidence for stress-induced massive loss of the principal hippocampal cells [6, 20, 25]. The fact that major neuronal loss cannot explain the hippocampal volume changes observed after stress is consistent with observations that many of the stress-induced structural changes are transient and, for example, spontaneously disappear when animals are subjected to a recovery period [26], or when elevated corticosteroid levels are normalized again [6, 20, 22]. As major cell loss in the CA and DG neuronal layers is not responsible for the hippocampal volume changes observed after stress, these must therefore be derived from other factors. Candidate cellular mechanisms are the somatodendritic components, neurogenesis, and glial changes but factors such as shifts in fluid balance cannot be excluded either [18, 20].

6.7
Effects of Stress on Dendritic Atrophy, Spine, and Synaptic Changes

Structural substrates for the stress-induced functional alterations are not necessarily the same and may involve axonal changes, synapse loss, alterations in postsynaptic densities, and dendritic reorganization. The most thoroughly documented stress-induced structural change is the dendritic reorganization that occurs parallel to the loss of spines and synapses and together with alterations in postsynaptic densities, suggesting general changes in neuronal connectivity. Chronic stress or experimentally increased corticosterone concentrations induce retraction of the apical dendrites of the CA3 and, to a lesser extent, of CA1 pyramidal cells and dentate granule cells [6, 27, 28]. Alterations in CA3 synapses and in the morphology of their mossy fiber terminals have also been described [27–29]. These changes in neuronal morphology are likely to contribute to various cognitive deficits occurring as a result of chronic stress exposure. Another possible functional outcome of dendritic retraction may be a disturbance of HPA axis regulation, leading to up-regulated GC release [29]. Recent studies show that such dendritic alterations in hippocampal neurons can already occur within a very short period of time, particularly in relatively immature cells [30]. This synaptic remodeling may also extend to cortical areas, where changes in cell adhesion, molecule expression, attention, spatial memory, and fear conditioning often occur in parallel. The latter dendritic changes appear to occur relatively early and to last for long periods of time [31].

6.8
Adult Hippocampal Neurogenesis

AN refers to the production of new neurons in an adult brain and is a prominent example of the adult neuroplasticity that occurs in most vertebrate species studied today, including humans. In young adult rodents, thousands of new granule neurons are generated every day, though significant differences exist even within different mouse strains [32]. The process of AN is dynamically regulated by various environmental factors and rapidly declines with age [6, 33]. Neurogenesis also occurs in the subventricular zone (SVZ) of the ventricle wall in many mammals and has been reported in the human brain as well [34, 35].

Several independent groups have observed low levels of neurogenesis in other brain structures also, such as the amygdala, striatum, and neocortex but negative results exist as well [36, 37]. Part of the difficulty in studying AN in these regions as such could reside in the fact that new cortical neurons, for example, probably belong to small subclasses of interneurons dispersed in large neocortical volumes [37].

In contrast to its abundance during embryonic development, neurogenesis in the adult is much less frequent but follows a similar, complex multistep process starting with the proliferation of progenitor cells, followed by their morphological and physiological maturation (often referred to as the *"survival"* process). This ends with a fully functional neuron that is integrated into the preexisting hippocampal network [38]. The existence and number of true multipotent neural stem cells residing in the adult DG is still a disputed issue (Figure 6.1). Experimental data

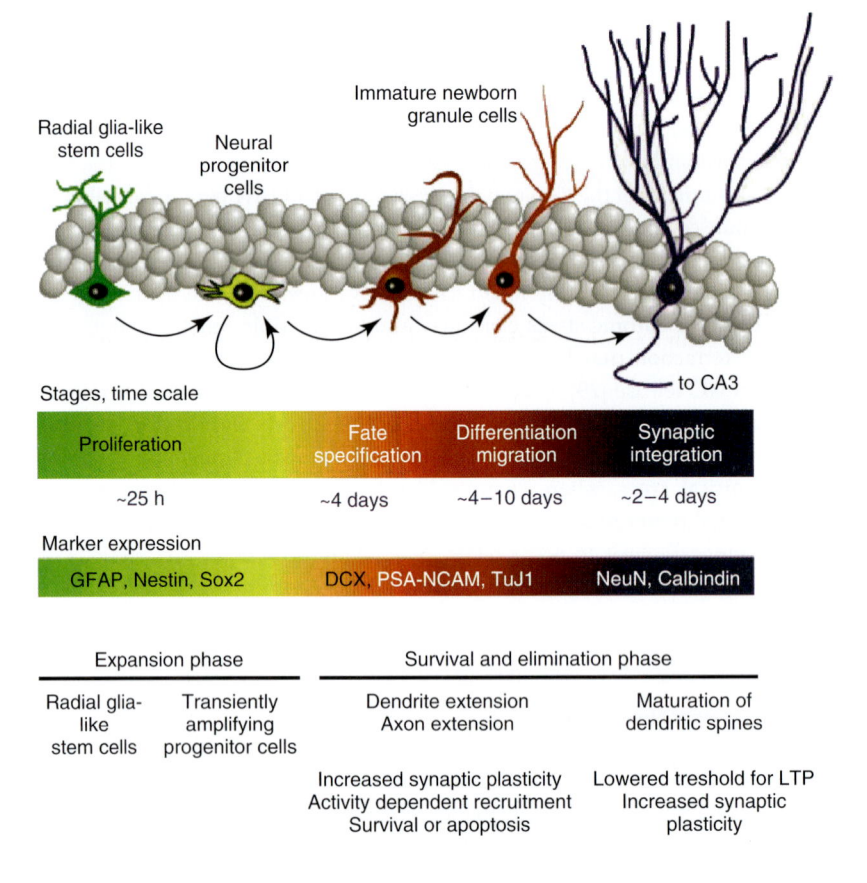

Figure 6.1 Schematic representation of the different stages of neuronal differentiation of newborn cells in the adult dentate gyrus of the hippocampus. Abbreviations: GFAP, glial fibrillary acidic protein; Sox2, sex determining region Y (SRY)-box 2; DCX, doublecortin; PSA-NCAM, polysialylated form of the neural cell adhesion molecule; TuJ1, β-tubulin III; NeuN, neuronal nuclear antigen. (Reproduced from Lucassen et al., 2010.)

report that a heterogeneous population of precursor cells is located and proliferates in the subgranular zone, a narrow layer located between the dentate granule cell layer and hilus. These precursor cells show a characteristic phenotype of radial astrocytes. Daughter cells of these progenitors proliferate at high frequency, often observed as bromodeoxyuridine (BrdU)-positive cell clusters and have been named amplifying neural progenitors (Figure 6.1). As the newborn neurons mature, they extend axons and dendrites, followed by the formation of spines and functional synapses [6, 38].

New neurons display characteristic functional properties such as lower threshold for induction of LTP and robust LTP [39]. Recent data indicate that the subsequent survival of the newly generated neurons is regulated by their input-dependent activity [40]. There is significant overproduction of newborn cells and many of them are rapidly eliminated by apoptotic cell death. A significant turnover of granule cells thus occurs in the DG of young rodents. In monkeys, this turnover rate is significantly lower while no quantitative data are as yet available for humans [41].

AN in the DG is regulated by a large variety of hormonal and environmental factors. AN is, for example, potently stimulated by learning, voluntary exercise, and enriched environmental housing [6]; interestingly, under these circumstances, enhanced neurogenesis is associated with elevated GC levels, whereas in general the reverse is true.

The exact functional role of the newborn granular neurons remains to be determined but numerous reports suggest that AN is involved in learning and memory, especially in the acquisition of spatial learning, in pattern separation, and in anxiety, as reviewed in detail elsewhere [19, 42–44].

6.9
Effect of Stress on Adult Hippocampal Neurogenesis

Stress is one of the most potent inhibitors of AN, as shown in several different species and using various stress paradigms; psychosocial and physical stressors all inhibit one or more phases of the neurogenesis process [6, 26, 28, 45]. Both acute and chronic stress exposure have a potent suppressive effect on proliferation, while continued stress exposure appears to interfere with all stages of neuronal renewal and inhibits both, proliferation and survival, possibly also in depression [18].

Stress-induced reductions in proliferation could result, for example, from apoptosis of progenitor cells or from cell cycle arrest. After acute stress, a reduction in proliferation was paralleled by increased numbers of apoptotic cells, yet no distinction was made between apoptosis of newborn or mature cells. Following chronic stress, both proliferation and apoptosis were reduced, parallel to increases in the cell cycle inhibitor p27Kip1, indicating that more cells had entered cell cycle arrest and that the granule cell turnover had thus slowed down [26, 46].

The exact underlying cellular mechanisms mediating the inhibitory effect of stress are unknown. The adrenal GC hormones have been pointed out as key players in this process and both GR as well as NMDA receptors have been identified on early progenitor cells [47, 48]. At the same time, several examples of a persistent and lasting inhibition of AN exist after an initial stressor, despite a later normalization of GC levels. These findings suggest that while GCs may be involved in the initial suppression of cell proliferation, particularly in early life when neurogenesis is abundant, they are not always necessary for the maintenance of this effect. A large number of other factors may also mediate the stress-induced inhibition of AN. The stress-induced increase in glutamate release via NMDA receptor activation is another leading candidate in this process [47].

Stress is also known to affect the levels of various neurotransmitters that have been implicated in the regulation of AN: GABA, serotonin, noradrenaline, and dopamine, to name a few examples. Other neurotransmitter systems, such as the cannabinoids, opioids, nitric oxide, and various neuropeptides may contribute as well (see [6, 38] for reviews). Furthermore, stress reduces the expression of several growth and neurotrophic factors, such as brain-derived neurotrophic factor (BDNF), insulin-like growth factor-1 (IGF-1), nerve growth factor (NGF), epidermal growth factor (EGF), and vascular endothelial growth factor (VEGF), that all can influence neurogenesis [49]. Gonadal steroids can also be involved [50]. The proximity of the precursors to blood vessels further suggests a strong interaction with the vasculature and it is this population that is particularly sensitive to stress [51]. Also, astrocytes are important in this respect as they support the survival of developing neurons, possess GRs, and are significantly affected by some [52] but not all types of stress [53].

Many of the symptoms of stress or HPA hyperactivity can typically be reversed with antidepressant (AD) treatment in both human and animal models. The observation that different classes of ADs with distinct mechanisms of action can block the behavioral effects of stress and restore normal levels of adult hippocampal neurogenesis [6, 54] supports the possibility that stimulating neurogenesis is a common pathway through which ADs exert their behavioral and therapeutic effects, although exceptions have been reported [18, 54–57].

6.10
Normalization of the Effects of Stress on the Hippocampus by Means of GR Blockade

As outlined above, excessive GC levels and chronic GR activation have been implicated in the pathogenesis of depression because of their role in increased arousal and psychotic disorganization [58, 59]. Thus, a current experimental treatment strategy is to identify drugs blocking (the effects of) the stress system. Indeed, modulators of the HPA axis often approved for different clinical applications can transiently block GC synthesis in the adrenals or block the access of GCs to their receptors in the brain and are associated with beneficial clinical effects in the treatment of psychosis while antidepressant effects have also been reported. These include cortisol synthesis inhibitors such as ketoconazole and metyrapone, corticotropin-releasing factor (CRF) antagonists, and GR antagonists such as mifepristone.

Interestingly, short-term treatment (four days) with mifepristone has been successfully applied to treat/ameliorate symptoms of psychotic depression in clinical trials [60]. It was found that mifepristone not only reduced symptoms in a subset of severely psychotic depressed patients, it was also especially helpful in treating psychosis secondary to Cushing's disease [61]. Given the fact that psychotic depressed patients tend to be the most resistant to the effects of traditional ADs, these findings seem promising. Thus, patients who are unresponsive to ADs alone and only partially responsive to an antidepressant–antipsychotic combination may benefit from treatment with GR antagonists. However, only high doses of mifepristone were effective, doses that are occasionally associated with adverse effects, although not uniformly across patient populations. Even higher doses are associated with skin rash, endometrial hyperplasia, and hot flashes in women. Other inhibitors of HPA axis function are associated with more severe side effects. One concern is mifepristone's clinical efficacy, as highlighted by studies comparing treatment with mifepristone to placebo or other antidepressive and antipsychotic treatments (for a review see [62]).

In a series of recent experimental studies [45, 63–66], mifepristone was also effective in rats that were exposed to 21 days of chronic multiple unpredictable stressors, a paradigm that altered many structural and functional parameters in the hippocampus (Figure 6.2) [5, 8]. Interestingly, treatment with a high dose of mifepristone during the final 4 days of this 21 day paradigm rapidly normalized the stress-induced reduction in adult generated cell numbers and neurogenesis in the hippocampal DG [45, 66]. In similar studies with the same design, the increased amplitude of high-voltage activated Ca-currents in CA1 neurons [63, 64] as well as the impaired induction of LTP in the CA1 region that occurred after 21 days of chronic stress [65] were all normalized already after 4 days of high dose mifepristone application. Interestingly, application of mifepristone alone, that is in the absence of concomitant stress exposure, was in all cases ineffective, indicating that mifepristone selectively interferes with pathways activated by chronic stress only, and, for example, does not appear to stimulate compensatory processes [5, 8]. Accordingly, it is not surprising that "clamping" corticosterone levels throughout life by means of adrenalectomy followed by a fixed suppletion, does not lower neurogenesis [67]. This implies that other mediators of the stress axis may also affect neurogenesis. This is indeed the case; both, a CRH-R1 and a vasopressin-1b antagonist were found to reverse lasting chronic stress-induced suppression of neurogenesis when given three weeks after the start of a seven weeks' mild stress paradigm [68].

A wide variety of different types of ADs including 5-HT reuptake modulators and substance P antagonists have been successfully applied to reverse effects of chronic stress on brain cells such as dendritic retraction in the CA3 region. Almost all ADs reverse suppressions of neurogenesis after chronic stress and promote neurogenesis in naive, nonstressed animals. In general, beneficial effects of ADs are seen only after several weeks of treatment and these effects are thought to be crucial for their clinical effectiveness. As such, they most likely do not

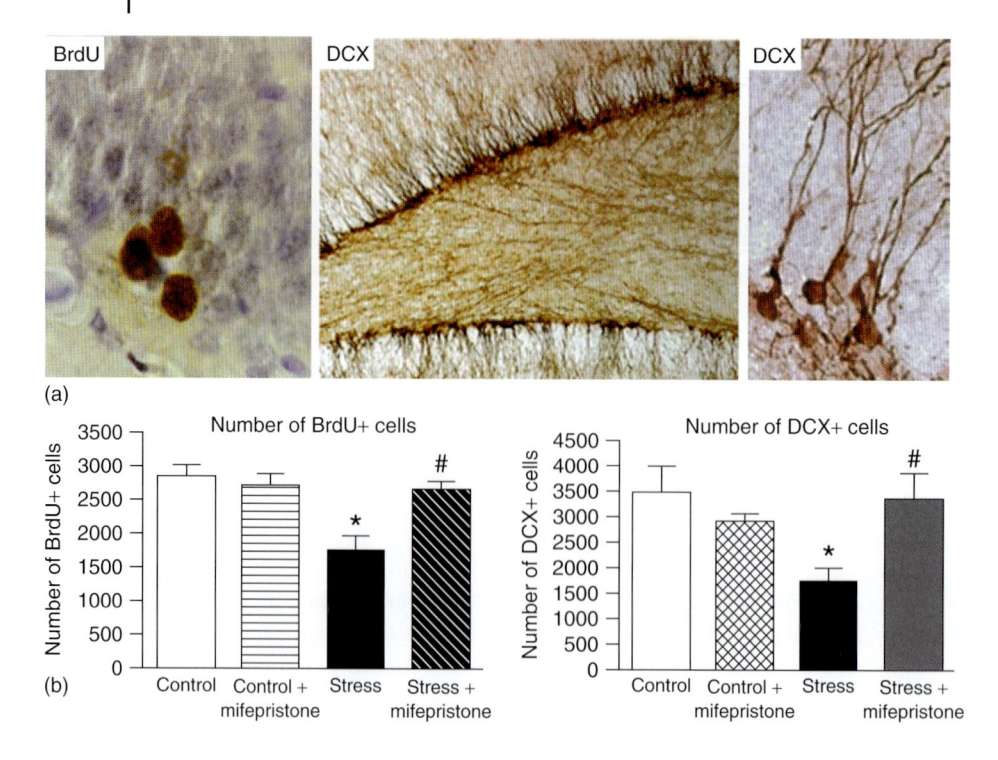

Figure 6.2 (a) Show examples of bromodeoxyuridine (BrdU+) and doublecortin (DCX+) immunostained cells in the DG. DCX-positive somata are located in the subgranular zone (SGZ) with extensions (arrowheads) passing through the granular cell layer. (b) Display BrdU- and DCX-positive cell numbers in rats subjected to 21 days of chronic unpredictable stress. The significant reduction in both BrdU- (21 day old cells) and DCX-positive cell numbers after chronic stress or corticosterone treatment is normalized by four days of high dose treatment with the GR antagonist mifepristone, whereas application of the drug alone to control animals has no effect (see Mayer et al. [66] and Oomen et al. [45] for details). (Reproduced from Lucassen et al. 2010.)

directly interfere with stress-activated pathways but rather exert counteractive or compensatory effects [6, 7, 38, 49, 54].

Compounds that affect local excitability can also prevent chronic stress-induced changes. This has been mainly studied for CA3 dendritic morphology where application of a competitive NMDA receptor antagonist during chronic restraint stress prevented CA3 dendritic retraction. Modulation of this same pathway by agmatine also affected neurogenesis. Benzodiazepine application was effective in preventing stress-induced dendritic atrophy, as was daily treatment of tree shrews or rats with phenytoin both regarding dendritic morphology in CA3 and LTP in CA1. Finally, a variety of nonspecific interferences have been described, ranging from transcranial magnetic stimulation, electroconvulsive seizures to learning experiences. In general, these treatments by themselves evoke an effect opposite to that of chronic stress, for example, on neurogenesis or spinogenesis. Also, a recovery period after chronic stress tends to reverse effects on neurogenesis or dendritic remodeling in the CA3 area and prefrontal cortex, though not in the BLA [5, 6, 8, 26, 28].

In conclusion, many treatments can reverse effects of chronic stress on neurogenesis, cell morphology, or cell function. Yet, nearly always, these treatments need to be maintained for weeks and seem to act in an indirect manner, usually by exerting effects opposite to those of chronic stress.

6.11
Normalization of Hippocampal Alterations during Diabetes Mellitus Using the GR Antagonist Mifepristone

In recent studies, hippocampal morphology and function was studied in a mouse model for type 1 diabetes (T1D) generated by treating mice with streptozotocin, a compound that destroys the β cells in the pancreas and hence, circulating insulin levels drop. The lack of insulin causes hyperglycemia and cellular starvation, a condition known as *metabolic stress*. One of the concomitant hormonal changes to cope with this metabolic stressor is a profound activation of the adrenal GC secretion [69]. Interestingly, this increased secretion of the adrenals develops through hyperreponsiveness of the adrenals to circulating ACTH rather than elevated levels of ACTH *per se* [70, 71]. As a result of hypercorticism, cells including the neurons in the central nervous system are further deprived of glucose, causing a sustained elevation of corticosterone in these mice. Then the question arose whether this endocrine adaptation in T1D leads to a more fragile state of the brain in which GC excess may enhance the potential for damage and attenuate a protective mechanism, thus facilitating cognitive impairment.

Indeed, uncontrolled diabetes is known to produce signs of hippocampal pathology, which proceeds together with changes in other brain structures such as hypothalamus and cerebral cortex. The hippocampus of the diabetic mice exhibited increased neuronal activation, signs of oxidative stress, and astrogliosis [72–74]. Cognitive deficits were also observed in the hippocampus-dependent novel object-placement recognition task (Figure 6.3) [75], in which mild hippocampal alterations can be tested under conditions of novelty exposure. Nondiabetic control mice preferred the exploration of the object placed in a novel location while the diabetic mice did not, indicating impaired spatial object-placement memory. This mild disturbance is typically observed at early stages of diabetes, while during a more prolonged disease state more severe behavioral deficits were established [76, 77].

While seeking to clarify the role of hypercorticism in the hippocampus of streptozotocin (STZ) – diabetic mice, Revsin discovered that the GR antagonist mifepristone administered for four consecutive days (from day 6 to 10 after onset of diabetes in the STZ model) partly prevented and/or reversed hippocampal

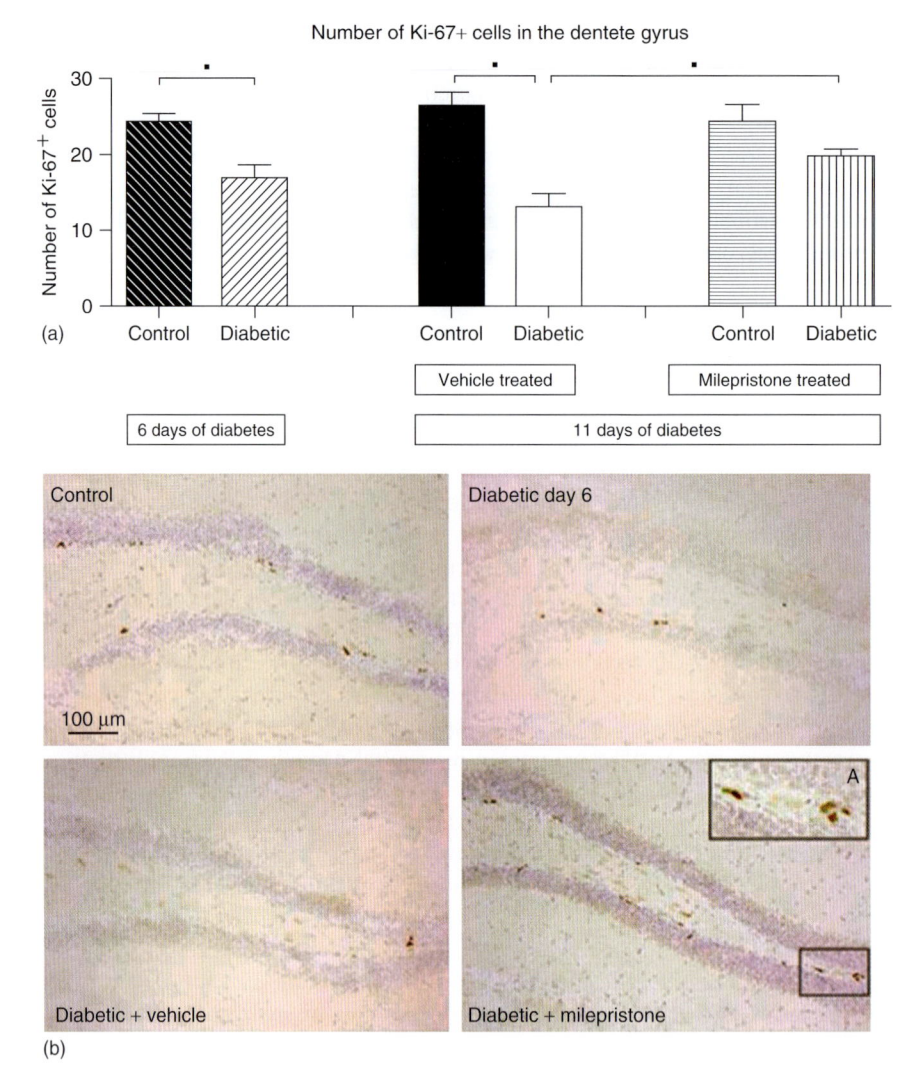

Figure 6.3 Cell proliferation in the subgranular zone (SGZ) of the dentate gyrus of diabetic animals treated with mifepristone. (a) Quantitative analysis of the immunocytochemistry for Ki-67 in the SGZ. Values expressed mean ± SEM, $n = 6–8$, $*p < 0.05$; (b) microphotographs of the different experimental groups. Insert I: 4 times magnification of Ki-67+ cells. Control refers to a control + vehicle-treated mouse.

functional deficits and morphological abnormalities [75]. The antagonist prevented astrogliosis and excessive neuronal activation and reversed the suppressed markers for neurogenesis in the hippocampus induced by the diabetic state and associated high corticosterone [66, 75]. Cognitive deficits observed at day 11 of diabetes were also ameliorated resulting in a performance even better than the untreated controls [75]. These findings are supported by another recent study demonstrating that signs of hippocampal deficits did not develop in the diabetic mouse as long as the circulating corticosterone levels were kept low, rather than the excessively high concentrations seen during diabetes. This was achieved by adrenalectomizing the diabetic animals and replacing with corticosterone to achieve physiological levels of the circulating steroid [78].

Hence, remodeling of the structure of the hippocampus is accelerated and its vulnerability to cognitive dysfunctions is enhanced during diabetes in a process that can be blocked by antiglucocorticoid therapy. Previously, hyperglycemia and insulin deficiency were thought to underlie the hippocampal deficits (see [76, 77] for reviews), but these possibilities can be excluded in view of the experiments of Revsin *et al.* [75, 78], although synergy between these conditions and the deleterious effect of hypercortisolemia cannot be ruled out. A rationale behind this observation is that the blockade of the GR would allow a more prominent function of neuroprotective MR-mediated actions [9]. Hence, this would predict that during GR blockade in the face of high circulating GCs, the maintenance of hippocampal integrity is a necessary condition for hippocampal-dependent behavioral performance.

6.12
Concluding Remarks

Chronic stress and stress hormone exposure alters many aspects of brain structure and function including long-term potentiation, neurogenesis, and dendritic complexity. Similar to its rapid effects in clinical studies with psychotic depressed patients, many of the chronic stress- or chronic corticosterone-induced changes in hippocampal structure and function in rodent models could all be normalized by brief treatment with the GR antagonist mifepristone. The same treatment also prevented the occurrence of hippocampal pathology induced by a diabetic state.

Acknowledgments

The support by the Royal Netherlands Academy of Arts and Sciences (KNAW)(to ERdK) is gratefully acknowledged. PJL is supported by ISAO, KNAW, The Volkswagen Stiftung Germany, Corcept Inc, the European Union (NEURAD consortium) and the Nederlandse HersenStichting. We thank Dr. B. Czéh (MPI Munich) for assistance with the figures.

Disclosure

ERdK is a member of the scientific advisory board of Corcept Therapeutics Inc and owns stock.

References

1. Selye, H. (1952) The general-adaptation-syndrome and the diseases of adaptation. *Can. Nurse*, **48**(1), 14–16.
2. Lightman, S.L., Wiles, C.C., Atkinson, H.C. et al. (2008) The significance of glucocorticoid pulsatility. *Eur. J. Pharmacol.*, **583**, 255–262.
3. De Kloet, R., Wallach, G., and McEwen, B.S. (1975) Differences in corticosterone and dexamethasone binding to rat brain and pituitary. *Endocrinology*, **96**, 598–609.
4. McEwen, B.S. (2006) Protective and damaging effects of stress mediators: central role of the brain. *Dialogues Clin. Neurosci.*, **8**, 367–381.
5. Joëls, M., Karst, H., Krugers, H.J., and Lucassen, P.J. (2007) Chronic stress: implications for neuronal morphology, function and neurogenesis. *Front. Neuroendocrinol.*, **28**(2-3), 72–96.
6. Lucassen, P.J., Meerlo, P., Naylor, A.S., van Dam, A.M., Dayer, A.G., Fuchs, E., Oomen, C.A., and Czéh, B. (2010) Regulation of adult neurogenesis by stress, sleep disruption, exercise and inflammation: implications for depression and antidepressant action. *Eur. Neuropsychopharmacol.*, **20**(1), 1–17.
7. de Kloet, E.R., Joëls, M., and Holsboer, F. (2005) Stress and the brain: from adaptation to disease. *Nat. Rev. Neurosci.*, **6**(6), 463–475.
8. Joëls, M., Krugers, H.J., Lucassen, P.J., and Karst, H. (2009) Corticosteroid effects on cellular physiology of limbic cells. *Brain Res.*, **1293**, 91–100.
9. Joëls, M., Karst, H., DeRijk, R., and de Kloet, E.R. (2008) The coming out of the brain mineralocorticoid receptor. *Trends Neurosci.*, **31**, 1–7.
10. Tasker, J.G., Di, S., and Malcher-Lopes, R. (2006) Minireview: rapid glucocorticoid signaling via membrane-associated receptors. *Endocrinology*, **147**(12), 5549–5556.
11. Joëls, M., Pu, Z., Wiegert, O., Oitzl, M.S., and Krugers, H.J. (2006) Learning under stress: how does it work? *Trends Cogn. Sci.*, **10**(4), 152–158.
12. Kendler, K.S., Karkowsk, L.M., and Prescott, C.A. (1999) Causal relationship between stressful life events and the onset of major depression. *Am. J. Psychiatry*, **156**, 837–841.
13. Campbell, S., Marriott, M., Nahmias, C., and MacQueen, G.M. (2004) Lower hippocampal volume in patients suffering from depression: a meta-analysis. *Am. J. Psychiatry*, **161**, 598–607.
14. Ressler, K.J. and Mayberg, H.S. (2007) Targeting abnormal neural circuits in mood and anxiety disorders: from the laboratory to the clinic. *Nat. Neurosci.*, **10**, 1116–1124.
15. Ulrich-Lai, Y.M. and Herman, J.P. (2009) Neural regulation of endocrine and autonomic stress responses. *Nat. Rev. Neurosci.*, **10**(6), 397–409.
16. Swaab, D.F., Bao, A.M., and Lucassen, P.J. (2005) The stress system in the human brain in depression and neurodegeneration. *Ageing Res. Rev.*, **4**, 141–194.
17. Castrén, E. (2005) Is mood chemistry? *Nat. Rev. Neurosci.*, **6**, 241–246.
18. Lucassen, P.J., Stumpel, M., Wang, Q., and Aronica, E. (2010) Decreased numbers of progenitor cells but no response to antidepressant drugs in the hippocampus of elderly depressed patients. *Neuropharmacology*, **58**, 940–949.
19. Oomen, C.A., Soeters, H., Audureau, N., Vermunt, L., Van Hasselt, F.N., Manders, E.M.M., Joëls, M., Lucassen, P.J., and Krugers, H. (2010) Severe early life stress hampers spatial learning and neurogenesis, but improves hippocampal synaptic plasticity and emotional learning under high-stress conditions in adulthood. *J. Neurosci.*, **30**, 6635–6645.
20. Czéh, B. and Lucassen, P.J. (2007) What causes the hippocampal volume decrease in depression? Are neurogenesis, glial changes and apoptosis implicated? *Eur. Arch. Psychiatry Clin. Neurosci.*, **257**, 250–260.
21. Gianaros, P.J., Jennings, J.R., Sheu, L.K., Greer, P.J., Kuller, L.H., and Matthews, K.A. (2007) Prospective reports of chronic life stress predict decreased

grey matter volume in the hippocampus. *Neuroimage*, **35**, 795–803.
22. Starkman, M.N., Gebarski, S.S., Berent, S., and Schteingart, D.E. (1992) Hippocampal formation volume, memory dysfunction, and cortisol levels in patients with Cushing's syndrome. *Biol. Psychiatry*, **32**, 756–765.
23. Sapolsky, R.M. (2000) Glucocorticoids and hippocampal atrophy in neuropsychiatric disorders. *Arch. Gen. Psychiatry*, **57**, 925–935.
24. Sapolsky, R.M., Uno, H., Rebert, C.S., Finch, C.E., and Neurosci, J. (1990) Hippocampal damage associated with prolonged glucocorticoid exposure in primates. *J. Neurosci.*, **10**, 2897–2902.
25. Lucassen, P.J., Müller, M.B., Holsboer, F., Bauer, J., Holtrop, A., Wouda, J., Hoogendijk, W.J., Pathol, J. et al. (2001) Hippocampal apoptosis in major depression is a minor event and absent from subareas at risk for glucocorticoid overexposure. *Am. J. Pathol.*, **158**, 453–468.
26. Heine, V.M., Maslam, S., Zareno, J., Joëls, M., Lucassen, P.J., and Neurosci, J. (2004c) Suppressed proliferation and apoptotic changes in the rat dentate gyrus after acute and chronic stress are reversible. *Eur. J. Neurosci.*, **19**, 131–144.
27. Stewart, M.G., Davies, H.A., Sandi, C., Kraev, I.V., Rogachevsky, V.V., Peddie, C.J., Rodriguez, J.J. et al. (2005) Stress suppresses and learning induces plasticity in CA3 of rat hippocampus: a three-dimensional ultrastructural study of thorny excrescences and their postsynaptic densities. *Neuroscience*, **131**(1), 43–54.
28. Fuchs, E., Flugge, G., and Czeh, B. (2006) Remodeling of neuronal networks by stress. *Front. Biosci.*, **11**, 2746–2758.
29. Conrad, C.D. (2006) What is the functional significance of chronic stress-induced CA3 dendritic retraction within the hippocampus? *Behav. Cogn. Neurosci. Rev.*, **5**, 41–60.
30. Alfarez, D.N., De Simoni, A., Velzing, E.H., Bracey, E., Joëls, M., Edwards, F.A., and Krugers, H.J. (2009) Corticosterone reduces dendritic complexity in developing hippocampal CA1 neurons. *Hippocampus*, **19**(9), 828–836.
31. Holmes, A. and Wellman, C.L. (2009) Stress-induced prefrontal reorganization and executive dysfunction in rodents. *Neurosci. Biobehav. Rev.*, **33**(6), 773–783.
32. Kempermann, G. and Gage, F.H. (2002) Genetic determinants of adult hippocampal neurogenesis correlate with acquisition, but not probe trial performance, in the water maze task. *Eur. J. Neurosci.*, **16**, 129–136.
33. Heine, V.M., Maslam, S., Joëls, M., and Lucassen, P.J. (2004a) Prominent decline of newborn cell proliferation, differentiation, and apoptosis in the aging dentate gyrus, in absence of an age-related hypothalamus-pituitary-adrenal axis activation. *Neurobiol. Aging*, **25**, 361–375.
34. Eriksson, P.S., Perfilieva, E., Bjork-Eriksson, T., Alborn, A.M., Nordborg, C., Peterson, D.A., and Gage, F.H. (1998) Neurogenesis in the adult human hippocampus. *Nat. Med.*, **4**, 1313–1317.
35. Curtis, M.A., Kam, M., Nannmark, U., Anderson, M.F., Axell, M.Z., Wikkelso, C., Holtas, S. et al. (2007) Human neuroblasts migrate to the olfactory bulb via a lateral ventricular extension. *Science*, **315**, 1243–1249.
36. Gould, E. (2007) How widespread is adult neurogenesis in mammals? *Nat. Rev. Neurosci.*, **8**, 481–488.
37. Cameron, H.A. and Dayer, A.G. (2008) New interneurons in the adult neocortex: small, sparse, but significant? *Biol. Psychiatry*, **63**, 650–655.
38. Balu, D.T. and Lucki, I. (2009) Adult hippocampal neurogenesis: regulation, functional implications, and contribution to disease pathology. *Neurosci. Biobehav. Rev.*, **33**(3), 232–252.
39. Schmidt-Hieber, C., Jonas, P., and Bischofberger, J. (2004) Enhanced synaptic plasticity in newly generated granule cells of the adult hippocampus. *Nature*, **429**, 184–187.
40. Tashiro, A., Sandler, V.M., Toni, N., Zhao, C., and Gage, F.H. (2006) NMDA-receptor mediated, cell-specific integration of new neurons in adult dentate gyrus. *Nature*, **442**, 929–933.

41. Kornack, D.R. and Rakic, P. (1999) Continuation of neurogenesis in the hippocampus of the adult macaque monkey. *Proc. Natl. Acad. Sci. U.S.A.*, **96**, 5768–5773.
42. Dupret, D., Revest, J.M., Koehl, M., Ichas, F., De Giorgi, F., Costet, P., Abrous, D.N. et al. (2008) Spatial relational memory requires hippocampal adult neurogenesis. *PLoS ONE*, **3**(4), e1959.
43. Revest, J.M., Dupret, D., Koehl, M., Funk-Reiter, C., Grosjean, N., Piazza, P.V., and Abrous, D.N. (2009) Adult hippocampal neurogenesis is involved in anxiety-related behaviors. *Mol. Psychiatry*, **14**(10), 959–956.
44. Clelland, C.D., Choi, M., Romberg, C., Clemenson, G.D. Jr., Fragniere, A., Tyers, P., Jessberger, S. et al. (2009) A functional role for adult hippocampal neurogenesis in spatial pattern separation. *Science*, **325**(5937), 210–213.
45. Oomen, C.A., Mayer, J.L., de Kloet, E.R., Joëls, M., and Lucassen, P.J. (2007) Brief treatment with the glucocorticoid receptor antagonist mifepristone normalizes the reduction in neurogenesis after chronic stress. *Eur. J. Neurosci.*, **26**, 3395–3401.
46. Heine, V.M., Maslam, S., Joëls, M., and Lucassen, P.J. (2004b) Increased P27KIP1 protein expression in the dentate gyrus of chronically stressed rats indicates G1 arrest involvement. *Neuroscience*, **129**, 593–601.
47. Nacher, J. and McEwen, B.S. (2006) The role of N-methyl-D-asparate receptors in neurogenesis. *Hippocampus*, **16**, 267–270.
48. Garcia, A., Steiner, B., Kronenberg, G., Bick-Sander, A., and Kempermann, G. (2004) Age-dependent expression of glucocorticoid- and mineralocorticoid receptors on neural precursor cell populations in the adult murine hippocampus. *Aging Cell*, **3**(6), 363–371.
49. Schmidt, H.D. and Duman, R.S. (2007) The role of neurotrophic factors in adult hippocampal neurogenesis, antidepressant treatments and animal models of depressive-like behavior. *Behav. Pharmacol.*, **18**, 391–418.
50. Galea, L.A. (2008) Gonadal hormone modulation of neurogenesis in the dentate gyrus of adult male and female rodents. *Brain Res. Rev.*, **57**, 332–341.
51. Heine, V.M., Zareno, J., Maslam, S., Joëls, M., and Lucassen, P.J. (2005) Chronic stress in the adult dentate gyrus reduces cell proliferation near the vasculature and VEGF and Flk-1 protein expression. *Eur. J. Neurosci.*, **21**, 1304–1314.
52. Czéh, B., Simon, M., Schmelting, B., Hiemke, C., and Fuchs, E. (2006) Astroglial plasticity in the hippocampus is affected by chronic psychosocial stress and concomitant fluoxetine treatment. *Neuropsychopharmacology*, **31**, 1616–1626.
53. Oomen, C.A., Girardi, C.E., Cahyadi, R., Verbeek, E.C., Krugers, H., Joëls, M., and Lucassen, P.J. (2009) Opposite effects of early maternal deprivation on neurogenesis in male versus female rats. *PLoS ONE*, **4**(1), e3675.
54. Sahay, A. and Hen, R. (2007) Adult hippocampal neurogenesis in depression. *Nat. Neurosci.*, **10**, 1110–1115.
55. Holick, K.A., Lee, D.C., Hen, R., and Dulawa, S.C. (2008) Behavioral effects of chronic fluoxetine in BALB/cJ mice do not require adult hippocampal neurogenesis or the serotonin 1A receptor. *Neuropsychopharmacology*, **33**(2), 406–417.
56. David, D.J., Samuels, B.A., Rainer, Q., Wang, J.W., Marsteller, D., Mendez, I., Drew, M., Craig, D.A., Guiard, B.P., Guilloux, J.P., Artymyshyn, R.P., Gardier, A.M., Gerald, C., Antonijevic, I.A., Leonardo, E.D., and Hen, R. (2009) Neurogenesis-dependent and -independent effects of fluoxetine in an animal model of anxiety/depression. *Neuron*, **62**, 479–493.
57. Marlatt, M.W., Lucassen, P.J., and Van Praag, H. (2010) A comparison of the neurogenic effects of fluoxetine, duloxetine and running in mice. *Brain Res.*, **1341**, 93–99.
58. Sachar, E.J., Puig-Antich, J., Ryan, N.D., Asnis, G.M., Rabinovich, H., Davies, M., and Halpern, F.S. (1985) Three tests of cortisol secretion in adult endogenous

depressives. *Acta Psychiatr. Scand.*, **71**(1), 1–8.
59. Holsboer, F. (2000) The corticosteroid receptor hypothesis of depression. *Neuropsychopharmacology*, **23**(5), 477–501.
60. Flores, B.H., Kenna, H., Keller, J., Solvason, H.B., and Schatzberg, A.F. (2006) Clinical and biological effects of mifepristone treatment for psychotic depression. *Neuropsychopharmacology*, **31**(3), 628–636.
61. Chu, J.W., Matthias, D.F., Belanoff, J., Schatzberg, A., Hoffman, A.R., and Feldman, D. (2001) Successful long-term treatment of refractory Cushing's disease with high-dose mifepristone (RU486). *J. Clin. Endocrinol. Metab.*, **86**(8), 3568–3573.
62. Fitzsimons, C.P., van Hooijdonk, L.W., Morrow, J.A., Peeters, B.W., Hamilton, N., Craighead, M., and Vreugdenhil, E. (2009) Antiglucocorticoids, neurogenesis and depression. *Mini Rev. Med. Chem.*, **9**(2), 249–264.
63. Karst, H. and Joëls, M. (2007) Brief RU 38486 treatment normalizes the effects of chronic stress on calcium currents in rat hippocampal CA1 neurons. *Neuropsychopharmacology*, **32**(8), 1830–1839.
64. Van Gemert, N.G. and Joëls, M. (2006) Effect of chronic stress and mifepristone treatment on voltage-dependent Ca^{2+} currents in rat hippocampal dentate gyrus. *J. Neuroendocrinol.*, **18**(10), 732–741.
65. Krugers, H.J., Goltstein, P.M., van der Linden, S., and Joëls, M. (2006) Blockade of glucocorticoid receptors rapidly restores hippocampal CA1 synaptic plasticity after exposure to chronic stress. *Eur. J. Neurosci.*, **23**(11), 3051–3055.
66. Mayer, J.L., Klumpers, L., Maslam, S., de Kloet, E.R., Joëls, M., and Lucassen, P.J. (2006) Brief treatment with the glucocorticoid receptor antagonist mifepristone normalises the corticosterone-induced reduction of adult hippocampal neurogenesis. *J. Neuroendocrinol.*, **18**, 629–631.
67. Montaron, M.F., Drapeau, E., Dupret, D., Kitchener, P., Aurousseau, C., Le Moal, M., Piazza, P.V., and Abrous, D.N. (2006) Lifelong corticosterone level determines age-related decline in neurogenesis and memory. *Neurobiol. Aging*, **27**(4), 645–654.
68. Alonso, R., Griebel, G., Pavone, G., Stemmelin, J., Le Fur, G., and Soubrié P. (2004) Blockade of CRF(1) or V(1b) receptors reverses stress-induced suppression of neurogenesis in a mouse model of depression. *Mol. Psychiatry*, **9**(3), 224, 278–286.
69. Cameron, O.G., Kronfol, Z., Greden, J.F., and Carroll, B.J. (1984) Hypothalamic-pituitary-adrenocortical activity in patients with diabetes mellitus. *Arch. Gen. Psychiatry*, **41**, 1090–1095.
70. Beauquis, J., Homo-Delarche, F., Revsin, Y., de Nicola, A.F., and Saravia, F. (2008) Brain alterations in auto-immune and pharmacological models of diabetes mellitus: focus on hypothalamic-pituitary-adrenocortical axis disturbances. *Neuroimmunomodulation*, **15**, 61–67.
71. Revsin, Y., van Wijk, D., Saravia, F.E., Oitzl, M.S., De Nicola, A.F., and de Kloet, E.R. (2008) Adrenal hypersensitivity precedes chronic hypercorticism in Streptozotocin-induced diabetes mice. *Endocrinology*, **149**, 3531–3539.
72. Magariños, A.M. and McEwen, B.S. (2000) Experimental diabetes in rats causes hippocampal dendritic and synaptic reorganization and increased glucocorticoid reactivity to stress. *Proc. Natl. Acad. Sci. U.S.A.*, **97**(20), 11056–11061.
73. Saravia, F.E., Revsin, Y., Gonzalez Deniselle, M.C., Gonzalez, S.L., Roig, P., Lima, A., Homo-Delarche, F., and De Nicola, A.F. (2002) Increased astrocyte reactivity in the hippocampus of murine models of type 1 diabetes: the nonobese diabetic (NOD) and streptozotocin-treated mice. *Brain Res.*, **957**, 345–353.
74. Revsin, Y., Saravia, F., Roig, P., Lima, A., de Kloet, E.R., Homo-Delarche, F., and De Nicola, A.F. (2005) Neuronal and astroglial alterations in the hippocampus of a mouse model for type 1 diabetes. *Brain Res.*, **1038**, 22–31.

75. Revsin, Y., Rekers, N.V., Louwe, M.C. et al. (2009) Glucocorticoid receptor blockade normalizes hippocampal alterations and cognitive impairment in streptozotocin-induced type 1 diabetes mice. *Neuropsychopharmacology*, **34**, 747–758.
76. Biessels, G.J., Kappelle, A.C., Bravenboer, B., Erkelens, D.W., and Gispen, W.H. (1994) Cerebral function in diabetes mellitus. *Diabetologia*, **37**, 643–650.
77. Biessels, G.J., Deary, I.J., and Ryan, C.M. (2008) Cognition and diabetes: a lifespan perspective. *Lancet Neurol.*, **7**(2), 184–190.
78. Stranahan, A.M., Arumugam, T.V., Cutler, R.G., Lee, K., Egan, J.M., and Mattson, M.P. (2008) Diabetes impairs hippocampal function through glucocorticoid-mediated effects on new and mature neurons. *Nat. Neurosci.*, **11**, 309–317.

7
Neuroactive Steroids and Peripheral Neuropathy

Roberto C. Melcangi, Silvia Giatti, Marzia Pesaresi, Donatella Caruso, and Marc J. Tetel

Peripheral neuropathy, either inherited or acquired, is a very common disorder for which effective clinical treatments are not yet available. Neuroactive steroids such as progesterone, testosterone, and their reduced metabolites represent a promising therapeutic option for peripheral neuropathy. Peripheral nerves are able to synthesize and metabolize neuroactive steroids and are a target for these molecules since they express classical and nonclassical steroid receptors. Neuroactive steroids modulate the expression of key transcription factors (TFs) for Schwann cell function, regulate Schwann cell proliferation, and promote the expression of myelin proteins. Interestingly, they are also able to counteract biochemical, morphological, and functional alterations of peripheral nerves in different experimental models of neuropathy, including the alterations caused by aging, diabetic neuropathy, and physical injury. Therefore, neuroactive steroids and pharmacological agents that are able to increase their local synthesis and the synthetic ligands for their receptors offer a promising potential for the treatment of different forms of peripheral neuropathy.

7.1
Introduction

Peripheral neuropathy, a process affecting the function of one or more peripheral nerves, is one of the most common disorders with a prevalence of about 2.4% in the general population that rises with aging to 8% [1]. Inherited forms of peripheral neuropathy are a group of disorders collectively referred to as *Charcot–Marie–Tooth* (*CMT*) disease that include demyelinating and axonal variants. Acquired peripheral neuropathy may occur during the aging process, in systemic or metabolic disorders; after physical injury, in infections and autoimmune disorders; after exposure to toxic compounds; and during drug treatment. The major clinical manifestations include negative sensory symptoms (e.g., loss of touch, thermal, and pain sensation) and positive sensory symptoms (e.g., burning,

Hormones in Neurodegeneration, Neuroprotection, and Neurogenesis.
Edited by Achille G. Gravanis and Synthia H. Mellon
Copyright © 2011 WILEY-VCH Verlag GmbH & Co. KGaA, Weinheim
ISBN: 978-3-527-32627-3

tingling, and numbness), muscle waste and weakness, and autonomic impairment. While axonal damage is more frequent than primary demyelination, these pathological features frequently coexist. Evidence of a demyelinating neuropathy increases the possibility of defining the etiology and starting an effective treatment with immunomodulatory and/or immunosuppressant drugs. Conversely, there is no causative treatment for most of the axonal neuropathies. When indicated, patients are limited to following symptomatic treatment for pain and physical therapy.

A possible clue to identify targets for more effective treatment would be the identification of relevant molecular events occurring in the peripheral nerves under physiologic and pathologic conditions. Recent findings indicate that peripheral nerves are able to synthesize and metabolize neuroactive steroids. Neuroactive steroids are steroids acting in the nervous system and produced by the nervous system (neurosteroids) and classical steroidogenic tissues (i.e., gonads and adrenal glands). In addition, peripheral nerves express receptors for neuroactive steroids and thus, are a target for them. Data summarized here indicate that neuroactive steroids are involved in the regulation of different functions of peripheral nerves, including Schwann cell proliferation and their cellular products. Moreover, this chapter highlights findings from a variety of experimental models that reveal that neuroactive steroids exert neuroprotective effects on peripheral neuropathies, and suggest new possible therapeutic strategies.

7.2
Regulation of Neuroactive Steroid Responsiveness in Peripheral Nerves

7.2.1
Synthesis and Metabolism of Neuroactive Steroids

Several observations have demonstrated that Schwann cells express peripheral benzodiazepine receptor (PBR) which has now been renamed as translocator protein-18 kDa (TSPO) [2], its endogenous ligand, octadecaneuropeptide [3, 4], and the steroidogenic acute regulatory protein [5]. These molecules participate in the transport of cholesterol from intracellular stores to the inner mitochondrial membrane, the location of cytochrome P450scc which is the enzyme that converts cholesterol to pregnenolone (PREG). Both cytochrome P450scc and 3β-hydroxysteroid dehydrogenase (the enzyme that converts PREG into progesterone (P)) are present in Schwann cells [6–12]. Moreover, metabolism of native steroids into their 5α- and 3α-hydroxy-5α reduced derivatives via the enzymatic complex formed by the 5α-reductase (5α-R) and the 3α-hydroxysteroid dehydrogenase (3α-HSD) [13, 14] also occurs in peripheral nerves. In particular, P can be converted into dihydroprogesterone (DHP) and subsequently into tetrahydroprogesterone (THP), testosterone (T) into dihydrotestosterone (DHT) and then into 5α-androstane-3α, 17β-diol (3α-diol) [14].

7.2.2
Classical and Nonclassical Steroid Receptors are Expressed in Peripheral Nerves

The classical steroid receptors for P (PR), estrogens, androgens (AR), glucocorticoids, and mineralocorticoids are members of a large superfamily of transcriptional activators [15, 16]. In a variety of species, including rodents [17], monkeys, and humans [18, 19], PR are expressed in two forms; the full-length PR-B and the N-terminally truncated PR-A. These PR isoforms are the product of a single gene located on chromosome 11 at q22–23 that undergoes transcription via alternate promoters and internal translational start sites [19]. PR, as well as other steroid receptors, can function in a classical genomic mechanism by acting as ligand-dependent nuclear TFs and at the membrane or in the cytosol as mediators of growth factor-initiated signaling pathways [20]. Progestins are thought to act via membrane-bound receptors related to classical PR and through interactions with neurotransmitter receptors [20, 21]. Finally, in the nervous system, neuroactive steroids have been proposed to interact with nonclassical steroid receptors, such as γ-amino butyric acid type A and B (GABA$_A$ receptor, GABA$_B$ receptor), and modulate a variety of neurotransmitter receptors including serotonin type 3 (5-HT3), N-methyl-D-aspartate (NMDA), α-amino-3-hydroxy-5-methyl-4-isoxazolepropionic acid (AMPA), kainate receptor, and an atypical intracellular receptor like the sigma 1 [22–28].

PR, as well as other steroid receptors, are expressed in rat sciatic nerve and in Schwann cells [29–33]. In addition, peripheral nerves and Schwann cells express the nonclassical steroid receptors for GABA$_A$ and GABA$_B$ [34, 35] and a variety of neurotransmitter receptors, including NMDA receptor 1 subunit, glutamate receptor 1 (GluR1) AMPA subunit, and GluR 2–7 subunits [36–38]. Thus, neuroactive steroids elicit their biological effects in the PNS via a variety of cell signaling mechanisms.

7.3
Schwann Cell Responses to Neuroactive Steroids

Neuroactive steroids modulate the expression of myelin proteins of the PNS, including myelin protein zero (P0) and the peripheral myelin protein 22 (PMP22). P0 and PMP22 are important for the maintenance of the multilamellar structure of PNS myelin [39]. P0, a member of the immunoglobulin gene superfamily (IgCAM), accounts for more than half of the total peripheral myelin proteins [40], and is predominantly confined to the compact portion of the mature myelin. P0-null mice exhibit pathological alterations affecting both myelin and axonal compartments, similar to those occurring in some dominantly inherited human peripheral neuropathies such as CMT 1B and Déjérine–Sottas syndrome (DSS) [41–43]. PMP22 is also localized in the compact myelin [44] and may interact with P0 [39, 45] to hold adjacent Schwann cell membranes together and stabilize myelin compaction. Mutations of the PMP22 gene have been associated with hereditary peripheral neuropathies in humans (CMT 1A; DSS; hereditary neuropathy with liability to pressure palsies, HNPPs); and with the Trembler and Trembler J alleles in mice [46, 47].

In vivo and *in vitro* studies reveal that P and its metabolites modulate the expression of these two important myelin proteins. Expression of P0 in sciatic nerve of adult male rats and in cultures of rat Schwann cells is increased by treatment with P, DHP, or THP [35, 48, 49]. In support of PR-mediated expression of P0, a PR antagonist (RU486) blocks the stimulatory effects of P or DHP on P0 expression in cultured rat Schwann cells. Furthermore, treatment with RU486 on postnatal (PN) day 1 decreases P0 expression at PN day 20 [50]. Taken together, these data suggest that P and its derivatives are acting through PR in a classic genomic mechanism of action to enhance P0 expression in Schwann cells.

In contrast to P0, PMP22 expression is enhanced only by THP, suggesting that this effect may be mediated by the $GABA_A$ receptor. A $GABA_A$ receptor antagonist like bicuculline completely abolished the stimulatory effect exerted by THP on PMP22 in Schwann cell cultures, while a $GABA_A$ receptor agonist like muscimol had a stimulatory effect on PMP22 that was comparable to that of THP [51]. In further support of the effect of THP on the $GABA_A$ receptor is the finding that isopregnanolone, which does not interact with $GABA_A$ receptor, does not alter PMP22 expression. Moreover, among the T derivatives tested so far, only 3α-diol which binds to the $GABA_A$ receptor increases PMP22 mRNA levels [52, 53]. Taken together, these data indicate that $GABA_A$ receptors are important mediators of the effects of THP on PMP22 expression.

In addition to the effects of progestins, the neuroactive derivatives of T influence P0 and PMP22 expression. In adult male rats, castration decreases the expression of P0 mRNA in the sciatic nerve, while subsequent treatment with DHT or 3α-diol restores P0 mRNA to precastration levels [30, 53]. Castration also decreases PMP22 mRNA levels in the sciatic nerve, but only 3α-diol treatment restores PMP22 levels [53]. In support of these findings in the male sciatic nerve, DHT and 3α-diol increase P0 and PMP22 mRNA levels, respectively, in cultured rat Schwann cells [30, 52]. In confirmation of a role for AR in controlling P0 expression, *in vivo* treatment with an AR antagonist like flutamide decreases P0 synthesis in the rat sciatic nerve [53]. Interestingly, inhibition of AR influences P0 synthesis in adult animals only. It is possible that P derivatives may be necessary for inducing P0 synthesis during the initial phases of the myelination process, while the subsequent intervention of T derivatives assists in the maintenance of this process. Thus, P0 expression is under the control of classical steroid receptors (e.g., PR and AR), while PMP22 expression involves nonclassical steroid receptors ($GABA_A$ receptors) [48], indicating that neuroactive steroids regulate the expression of these important myelin proteins through a variety of mechanisms.

In further support of the concept that P and its derivatives are acting through PR in a classic genomic mechanism of action to regulate P0 expression, nuclear receptor coactivators are involved in this regulation. Nuclear receptor coactivators dramatically increase the transcriptional activity of steroid receptors and other nuclear receptors through a variety of mechanisms, including acetylation, methylation, phosphorylation, and chromatin remodeling [54]. *In vitro* studies using antibodies against nuclear receptor coactivators indicate that recruitment of coactivators is rate-limiting in steroid receptor-mediated gene transcription [54]. Moreover, squelching, or the repression of the transcriptional activity of one steroid receptor by another, is reversed by the addition of coactivators [55]. Steroid receptor coactivator-1(SRC-1), also known as *NcoA-1*, was one of the first nuclear receptor coactivators that was found to interact and function with hormone-bound steroid receptors, including PR [55]. SRC-1 and other coactivators are critical for steroid action in the PNS and CNS and the expression of hormone-dependent behaviors [56]. In a culture of an immortalized line of Schwann cells (MCS80), overexpression of SRC-1 potentiated the DHP-induced increase in P0 expression, while underexpression of SRC-1 eliminated this increase in P0 expression [57]. These findings suggest that SRC-1 functions with activated PR to regulate P0 gene expression and maintain myelination. Moreover, putative P responsive elements have been identified on the P0 gene [30], further supporting the idea that P0 is regulated by a classical steroid genomic mechanism.

The profound effects of neuroactive steroids and their receptors on the expression of P0 and PMP22 suggest that expression of these myelin proteins is sexually dimorphic. Indeed, P or DHP treatment induces a stimulatory effect on P0 mRNA levels in primary Schwann cell cultures from male, but not female, rats [58]. In contrast, treatment with THP increases gene expression of P0 in Schwann cells from female rats, but not in cells from males. Similar to this sexually dimorphic expression of P0, expression of PMP22 is stimulated by P in primary Schwann cell cultures from males, while it is enhanced by THP in cultures from females [58].

In addition to myelin proteins, neuroactive steroids regulate the expression of TFs that are critical in Schwann cell physiology and their myelinating program. In cultured rat Schwann cells, P stimulates the gene expression of Krox-20, Krox-24, Egr-3, and FosB [59, 60]. Moreover, Krox-20 expression is stimulated by treatment with DHP or THP, while Sox-10 expression is enhanced only by DHP [61]. Taken together, these findings suggest that P and its derivative coordinate Schwann cell myelinating programs via different intracellular pathways. Finally, in support of P stimulating myelin synthesis, P accelerates the time of initiation and enhances the rate of myelin synthesis in Schwann cells cocultured with dorsal root ganglia neurons [8].

Neuroactive steroids not only influence the expression of myelin proteins in Schwann cells but they also affect their proliferation. P stimulates the proliferation of Schwann cells *in vitro* [62, 63]. Interestingly, this effect of P is sex-dependent given that P increases Schwann cell proliferation in cultures of segments of rat sciatic nerve from females, but is ineffective in cultures from males [63]. An effect of androgens on Schwann cell proliferation is also evident. Castration decreases the number of terminal Schwann cells unsheathing the synaptic junction between motor nerve endings and muscles, while T replacement counteracts this effect [64].

In addition to regulation of Schwann cell proliferation by neuroactive steroids, this process appears to be modulated by nuclear receptor coactivators. We have demonstrated that in MCS80 cells, overexpression of SRC-1 attenuates Schwann cell proliferation [48]. In contrast, overexpression of another coactivator, steroid receptor RNA activator, potentiates the proliferation of MSC80 cells [48]. These coactivators may be critical in fine-tuning the responsiveness of Schwann cells to neuroactive steroid regulation of proliferation in the PNS.

7.4
Sexually Dimorphic Changes of Neuroactive Steroid Levels Induced by Pathology in Peripheral Nerves

Similarly to what has been observed in CNS [65–68], in peripheral nerves, the levels of neuroactive steroids are also modified by trauma or pathology. In a crush injury model, the levels of PREG and P metabolites (DHP and THP) evaluated by liquid chromatography-tandem mass spectrometry in the sciatic nerve are decreased by trauma [69]. A general decrease of neuroactive steroid levels also occurs in diabetic neuropathy. Indeed, in an experimental model of diabetic neuropathy induced by a single injection of streptozotocin (STZ), the levels of PREG, DHP, THP, and T and its derivatives are reduced in the brachial nerve [70]. Interestingly, we recently found that neuroactive steroid levels in sciatic nerve are sexually dimorphic. PREG, P, and its derivatives (DHP, THP, and isopregnanolone) are higher in the sciatic nerve of females than in males. In contrast, T and DHT levels were higher in males than in females. Moreover, the effect of diabetes is different in the two sexes. The levels of PREG, T, DHT, and 3α-diol are decreased in males but not in females, while P, THP, and isopregnanolone are reduced only in females [71]. Finally, a sex difference has been observed in an experimental model of inherited peripheral neuropathy, CMT1A, which is induced by overexpression of PMP22. In this model, isopregnanolone and 3α-diol levels were decreased in females and males, respectively [72].

7.5
Neuroactive Steroids as Protective Agents in PNS

Neuroactive steroids may act as protective agents in several experimental models of neurodegenerative disorders of the CNS, as well as influencing many parameters of the PNS. Thus, the efficacy of neuroactive steroids has been tested in different experimental models of peripheral neuropathy.

7.5.1
Aging Process

Aging induces important biochemical and morphological changes in peripheral nerves. For instance, aging is associated with a decrease in the synthesis of P0 and PMP22 and large myelinated fibers undergo atrophy, while myelin sheaths increase in thickness and show various irregularities, including myelin ballooning, splitting, infolding, reduplication, and remyelination [73, 74]. Neuroactive steroids, such as P and its derivatives, stimulate the expression of P0 and PMP22 in the sciatic nerve of aged rats. Thus, treatment with P or DHP increases the low levels of P0 present in aged male rats, while THP increases PMP22 levels [35, 49, 73, 75].

P, DHP, or THP have a clear effect on the number and shape of myelinated fibers, as well as on the frequency of myelin abnormalities [73, 74]. In particular, these neuroactive steroids increase the number of myelinated fibers of small caliber (<5 µm). In addition, P or its derivatives act on small myelinated fibers to increase their g-ratio, which is the ratio between the axon and the entire fiber diameter. These findings suggest that the increase in the number of myelinated fibers reflects an increased remyelination of small fibers in aged sciatic nerves. Treatment with P, DHP, or THP also reduces the frequency of axons with myelin abnormalities (myelin in-foldings) and the proportion of fibers with irregular shapes [73, 74]. All these effects seem to be specific to P and its derivatives because neither T, DHT nor 3α-diol are able to influence these morphological parameters [73, 74].

7.5.2
Physical Injury

Peripheral nerves are frequently exposed to physical damage, including crushing, cutting, and entrapment with nerve compression. These injuries can range from mild physical injury involving subtle alterations in myelin structure, to more severe injury that can lead to a complete interruption of the nerve axon and myelin. Moreover, injury of peripheral nerves can result in severe functional impairment and decreased quality of life because of the loss of sensory and motor functions.

Protective and regenerative effects of neuroactive steroids have been well characterized in experimental models of degeneration occurring after physical injury of peripheral nerves. For example, following nerve transection, P or DHP increases P0 mRNA levels in the portion of the sciatic nerve distal to the cut [52]. Moreover, local administration of P or PREG attenuates the decrease in myelin membranes induced by a cryolesion in the mouse sciatic nerve [12]. Furthermore, the best result on guided regeneration of the facial nerve of rabbits is obtained with P when impregnated in a biodegradable prosthesis, such as chitosan [76]. In this experimental model, P induces an increase in the number of Schwann cell nuclei, number of nonmyelinated and myelinated nerve fibers, and the g-ratio of the myelinated nerve fibers. In a more recent study using a nerve crush model, we demonstrated that P or DHP can counteract biochemical alterations induced by injury, such as impairment of myelin proteins and Na^+, K^+-ATPase pump, and stimulate expression of reelin, an extracellular matrix protein [69]. These two neuroactive steroids also counteract nociception impairment, and DHP treatment decreases the up-regulation of myelinated fiber density occurring in crushed nerves. Altogether, these findings suggest that P and DHP should be considered as protective agents following nerve crush injury [69]. Finally, several other neuroactive steroids including T, DHT, dehydroepiandrosterone (DHEA), and estradiol provide similar protective effects in rodent peripheral nerve injury models [77–83].

7.5.3
Diabetic Neuropathy

More than 50% of diabetic patients show different types of peripheral nerve impairment, including mononeuropathy, plexopathy, mononeuritis multiplex, and distal symmetric sensory-motor polyneuropathy. The most common distal polyneuropathy is associated with a spectrum of functional and structural changes in the peripheral nerves, including a slowing in nerve conduction velocity (NCV), axonal degeneration, paranodal demyelination, and loss of myelinated fibers [84, 85]. Decreased Na^+, K^+-ATPase activity in peripheral nerves and reduced intraepidermal nerve fiber density associated with impaired nociceptive threshold are well characterized in humans and in different experimental models of diabetic neuropathy, including a STZ-induced model in rats [86–89]. STZ-induced diabetes produces several morphological alterations in the myelinated fibers of the peripheral nerves, including myelin invaginations in the axoplasm (in-foldings) and evaginations in the cytoplasm (out-foldings) of Schwann cells, alterations in myelin compaction and abnormal separation of myelin lamellae. As in aged rats, the most abundant myelin abnormality observed in STZ-treated rats is the presence of myelin in-foldings in the axoplasm [90].

Previous observations obtained in our laboratory have indicated that neuroactive steroids, such as P, T, and their derivatives, exert important protective effects in STZ rats. We found that P or DHP attenuates the increase in the number of fibers with myelin in-foldings induced by STZ treatment in the sciatic nerve [90], which is similar to the morphological alterations of myelin in the sciatic nerve of aged rats discussed above [73, 74].

Neuroactive steroids are also able to influence biochemical and functional parameters of peripheral nerves. For example, DHEA prevents vascular and neuronal dysfunction in the sciatic nerve of STZ rats [91]. Moreover, chronic steroid treatment with P (or with its derivatives), DHP, and THP counteracted the impairment of NCV and thermal threshold, restoring skin innervation density and P0 and PMP22 mRNA levels, and improving Na^+, K^+-ATPase activity [92]. In addition, T and its derivatives including DHT and 3α-diol can reverse behavioral, neurophysiological, morphological, and biochemical alterations induced by peripheral diabetic neuropathy [93].

7.6
Chemotherapy-Induced Peripheral Neuropathy

The increasing awareness of chemotherapy-induced peripheral neurotoxicity has revealed the need for neuroprotective agents that function in the PNS. While there are currently no effective treatments for peripheral neuropathy induced by antineoplastic drugs, preclinical models have attempted to prevent neurotoxicity with agents such as thiols, neurotrophic factors, and antioxidants [94–96]. For example, docetaxel, which is a semisynthetic taxane widely employed as an antineoplastic agent for the treatment of breast, ovarian, and nonsmall cell lung cancers, exerts specific toxicity on the PNS [94, 95, 97–99]. In experimental models, treatment with P or DHP alleviated docetaxel-induced neuropathy by preventing nerve conduction and thermal threshold changes and degeneration of skin nerves in the footpad [100]. Neuroactive steroids also counteracted the changes in gene expression of several myelin proteins and calcitonin gene-related peptides induced by docetaxel in the sciatic nerve and lumbar spinal cord, respectively. While most nerve abnormalities induced by docetaxel spontaneously recovered after drug withdrawal in our animal model, similar to what occurs in patients, we found that rats treated with DHP or P recovered faster from neuropathy than control animals [100]. These findings indicate that neuroactive steroids exert a protective effect on peripheral nerves at a variety of levels and suggest that neuroactive steroids represent a new therapeutic frontier for patients with chemotherapy-induced neuropathy.

7.7
Concluding Remarks

Over the past decade, a variety of studies have dramatically enhanced our understanding of the protective effects of neuroactive steroids in the PNS. In addition, our knowledge of steroid receptor mechanisms and their coregulators have increased our understanding of how these important neuroactive steroids function in peripheral nerves. As reported here, neuroactive steroids, or synthetic ligands of their receptors, represent potential and important therapeutic tools for peripheral neuropathies, especially given that in many situations there are no effective treatments that can prevent or reverse peripheral nerve damage. An additional intriguing therapeutic strategy is the use of pharmacological agents to directly increase the synthesis of neuroactive steroids in the nervous system. Indeed, recent observations indicate that ligands of TSPO, which increase the synthesis of neuroactive steroids, exert protective effects on peripheral nerves [101, 102]. In summary, these findings indicate the promising potential of these important molecules for the treatment of a variety of peripheral neuropathies and emphasize the need for a sustained research and development effort in the future.

Acknowledgments

The financial supports of PRIN (20074SPYCM_002) and PUR from University of Milan, Italy, to R.C.M. are gratefully acknowledged.

References

1. England, J.D. and Asbury, A.K. (2004) Peripheral neuropathy. *Lancet*, **363**, 2151–2161.

2. Papadopoulos, V., Baraldi, M., Guilarte, T.R., Knudsen, T.B., Lacapere, J.J., Lindemann, P., Norenberg, M.D.,

Nutt, D., Weizman, A., Zhang, M.R., and Gavish, M. (2006) Translocator protein (18k Da): new nomenclature for the peripheral-type benzodiazepine receptor based on its structure and molecular function. *Trends Pharmacol. Sci.*, **27**, 402–409.

3. Lacor, P., Gandolfo, P., Tonon, M.C., Brault, E., Dalibert, I., Schumacher, M., Benavides, J., and Ferzaz, B. (1999) Regulation of the expression of peripheral benzodiazepine receptors and their endogenous ligands during rat sciatic nerve degeneration and regeneration: a role for PBR in neurosteroidogenesis. *Brain Res.*, **815**, 70–80.

4. Lacor, P., Benavides, J., and Ferzaz, B. (1996) Enhanced expression of the peripheral benzodiazepine receptor (PBR) and its endogenous ligand octadecaneuropeptide (ODN) in the regenerating adult rat sciatic nerve. *Neurosci. Lett.*, **220**, 61–65.

5. Benmessahel, Y., Troadec, J.D., Cadepond, F., Guennoun, R., Hales, D.B., Schumacher, M., and Groyer, G. (2004) Downregulation of steroidogenic acute regulatory protein (StAR) gene expression by cyclic AMP in cultured Schwann cells. *Glia*, **45**, 213–228.

6. Guennoun, R., Schumacher, M., Robert, F., Delespierre, B., Gouezou, M., Eychenne, B., Akwa, Y., Robel, P., and Baulieu, E.E. (1997) Neurosteroids: expression of functional 3beta-hydroxysteroid dehydrogenase by rat sensory neurons and Schwann cells. *Eur. J. Neurosci.*, **9**, 2236–2247.

7. Coirini, H., Gouezou, M., Delespierre, B., Liere, P., Pianos, A., Eychenne, B., Schumacher, M., and Guennoun, R. (2003) Characterization and regulation of the 3 beta-hydroxysteroid dehydrogenase isomerase enzyme in the rat sciatic nerve. *J. Neurochem.*, **84**, 119–126.

8. Chan, J.R., Rodriguez-Waitkus, P.M., Ng, B.K., Liang, P., and Glaser, M. (2000) Progesterone synthesized by Schwann cells during myelin formation regulates neuronal gene expression. *Mol. Biol. Cell*, **11**, 2283–2295.

9. Chan, J.R., Phillips, L.J. II, and Glaser, M. (1998) Glucocorticoids and progestins signal the initiation and enhance the rate of myelin formation. *Proc. Natl. Acad. Sci. U.S.A.*, **95**, 10459–10464.

10. Schumacher, M., Guennoun, R., Mercier, G., Desarnaud, F., Lacor, P., Benavides, J., Ferzaz, B., Robert, F., and Baulieu, E.E. (2001) Progesterone synthesis and myelin formation in peripheral nerves. *Brain Res. Brain Res. Rev.*, **37**, 343–359.

11. Rodriguez-Waitkus, P.M., Lafollette, A.J., Ng, B.K., Zhu, T.S., Conrad, H.E., and Glaser, M. (2003) Steroid hormone signaling between Schwann cells and neurons regulates the rate of myelin synthesis. *Ann. N. Y. Acad. Sci.*, **1007**, 340–348.

12. Koenig, H.L., Schumacher, M., Ferzaz, B., Thi, A.N., Ressouches, A., Guennoun, R., Jung-Testas, I., Robel, P., Akwa, Y., and Baulieu, E.E. (1995) Progesterone synthesis and myelin formation by Schwann cells. *Science*, **268**, 1500–1503.

13. Yokoi, H., Tsuruo, Y., and Ishimura, K. (1998) Steroid 5α-reductase type 1 immunolocalized in the rat peripheral nervous system and paraganglia. *Histochem. J.*, **30**, 731–739.

14. Melcangi, R.C., Magnaghi, V., Galbiati, M., and Martini, L. (2001) Formation and effects of neuroactive steroids in the central and peripheral nervous system. *Int. Rev. Neurobiol.*, **46**, 145–176.

15. Tsai, M.J. and O'Malley, B.W. (1994) Molecular mechanisms of action of steroid/thyroid receptor superfamily members. *Annu. Rev. Biochem.*, **63**, 451–486.

16. Mangelsdorf, D.J., Thummel, C., Beato, M., Herrlich, P., Schütz, G., Umesono, K., Blumberg, B., Kastner, P., Mark, M., Chambon, P., and Evans, R.M. (1995) The nuclear receptor superfamily: the second decade. *Cell*, **83**, 835–839.

17. Schott, D.R., Shyamala, G., Schneider, W., and Parry, G. (1991) Molecular cloning, sequence analyses, and

expression of complementary DNA encoding murine progesterone receptor. *Biochemistry*, **30**, 7014–7020.
18. Duffy, D.M., Wells, T.R., Haluska, G.J., and Stouffer, R.L. (1997) The ratio of progesterone receptor isoforms changes in the monkey corpus luteum during the luteal phase of the menstrual cycle. *Biol. Reprod.*, **57**, 693–699.
19. Kastner, P., Krust, A., Turcotte, B., Stropp, U., Tora, L., Gronemeyer, H., and Chambon, P. (1990) Two distinct estrogen-regulated promoters generate transcripts encoding the two functionally different human progesterone receptor forms A and B. *EMBO J.*, **9**, 1603–1614.
20. Tetel, M.J. and Lange, C.A. (2009) in *Hormones, Brain and Behavior* (eds D.W. Pfaff, A.P. Arnold, A.M. Etgen, S.E. Fahrbach, and R.T. Rubin), Academic Press, San Diego, pp. 1439–1465.
21. Mani, S.K., Portillo, W., and Reyna, A. (2009) Steroid hormone action in the brain: cross-talk between signalling pathways. *J. Neuroendocrinol.*, **21**, 243–247.
22. Rupprecht, R., di Michele, F., Hermann, B., Strohle, A., Lancel, M., Romeo, E., and Holsboer, F. (2001) Neuroactive steroids: molecular mechanisms of action and implications for neuropsychopharmacology. *Brain Res. Brain Res. Rev.*, **37**, 59–67.
23. Romieu, P., Martin-Fardon, R., Bowen, W.D., and Maurice, T. (2003) Sigma 1 receptor-related neuroactive steroids modulate cocaine-induced reward. *J. Neurosci.*, **23**, 3572–3576.
24. Maurice, T., Urani, A., Phan, V.L., and Romieu, P. (2001) The interaction between neuroactive steroids and the sigma1 receptor function: behavioral consequences and therapeutic opportunities. *Brain Res. Brain Res. Rev.*, **37**, 116–132.
25. Lambert, J.J., Belelli, D., Peden, D.R., Vardy, A.W., and Peters, J.A. (2003) Neurosteroid modulation of GABAA receptors. *Prog. Neurobiol.*, **71**, 67–80.
26. Falkenstein, E. and Wehling, M. (2000) Nongenomically initiated steroid actions. *Eur. J. Clin. Invest.*, **30** (Suppl 3), 51–54.
27. Al-Dahan, M.I. and Thalmann, R.H. (1996) Progesterone regulates gamma-aminobutyric acid B (GABAB) receptors in the neocortex of female rats. *Brain Res.*, **727**, 40–48.
28. Al-Dahan, M.I., Jalilian Tehrani, M.H., and Thalmann, R.H. (1994) Regulation of gamma-aminobutyric acid B (GABAB) receptors in cerebral cortex during the estrous cycle. *Brain Res.*, **640**, 33–39.
29. Melcangi, R.C., Magnaghi, V., Galbiati, M., and Martini, L. (2001) Glial cells: a target for steroid hormones. *Prog. Brain Res.*, **132**, 31–40.
30. Magnaghi, V., Cavarretta, I., Zucchi, I., Susani, L., Rupprecht, R., Hermann, B., Martini, L., and Melcangi, R.C. (1999) Po gene expression is modulated by androgens in the sciatic nerve of adult male rats. *Brain Res. Mol. Brain Res.*, **70**, 36–44.
31. Jung-Testas, I., Schumacher, M., Robel, P., and Baulieu, E.E. (1996) Demonstration of progesterone receptors in rat Schwann cells. *J. Steroid Biochem. Mol. Biol.*, **58**, 77–82.
32. Islamov, R.R., Hendricks, W.A., Katwa, L.C., McMurray, R.J., Pak, E.S., Spanier, N.S., and Murashov, A.K. (2003) Effect of 17 beta-estradiol on gene expression in lumbar spinal cord following sciatic nerve crush injury in ovariectomized mice. *Brain Res.*, **966**, 65–75.
33. Groyer, G., Eychenne, B., Girard, C., Rajkowski, K., Schumacher, M., and Cadepond, F. (2006) Expression and functional state of the corticosteroid receptors and 11 beta-hydroxysteroid dehydrogenase type 2 in Schwann cells. *Endocrinology*, **147**, 4339–4350.
34. Magnaghi, V., Ballabio, M., Cavarretta, I.T., Froestl, W., Lambert, J.J., Zucchi, I., and Melcangi, R.C. (2004) GABAB receptors in Schwann cells influence proliferation and myelin protein expression. *Eur. J. Neurosci.*, **19**, 2641–2649.
35. Melcangi, R.C., Magnaghi, V., Cavarretta, I., Zucchi, I., Bovolin, P., D'Urso, D., and Martini, L. (1999) Progesterone derivatives are able to influence peripheral myelin protein 22 and P0 gene expression: possible

mechanisms of action. *J. Neurosci. Res.*, **56**, 349–357.
36. Dememes, D., Lleixa, A., and Dechesne, C.J. (1995) Cellular and subcellular localization of AMPA-selective glutamate receptors in the mammalian peripheral vestibular system. *Brain Res.*, **671**, 83–94.
37. Verkhratsky, A. and Steinhauser, C. (2000) Ion channels in glial cells. *Brain Res. Brain Res. Rev.*, **32**, 380–412.
38. Coggeshall, R.E. and Carlton, S.M. (1998) Ultrastructural analysis of NMDA, AMPA, and kainate receptors on unmyelinated and myelinated axons in the periphery. *J. Comp. Neurol.*, **391**, 78–86.
39. D'Urso, D., Brophy, P.J., Staugaitis, S.M., Gillespie, C.S., Frey, A.B., Stempak, J.G., and Colman, D.R. (1990) Protein zero of peripheral nerve myelin: biosynthesis, membrane insertion, and evidence for homotypic interaction. *Neuron*, **4**, 449–460.
40. Ishaque, A., Roomi, M.W., Szymanska, I., Kowalski, S., and Eylar, E.H. (1980) The PO glycoprotein of peripheral nerve myelin. *Can. J. Biochem.*, **58**, 913–921.
41. Zielasek, J., Martini, R. and Toyka, K.V. (1996) Functional abnormalities in P0-deficient mice resemble human hereditary neuropathies linked to P0 gene mutations. *Muscle Nerve*, **19**, 946–952.
42. Martini, R., Zielasek, J., Toyka, K.V., Giese, K.P., and Schachner, M. (1995) Protein zero (P0)-deficient mice show myelin degeneration in peripheral nerves characteristic of inherited human neuropathies. *Nat. Genet.*, **11**, 281–286.
43. Giese, K.P., Martini, R., Lemke, G., Soriano, P., and Schachner, M. (1992) Mouse P0 gene disruption leads to hypomyelination, abnormal expression of recognition molecules, and degeneration of myelin and axons. *Cell*, **71**, 565–576.
44. Pareek, S., Suter, U., Snipes, G.J., Welcher, A.A., Shooter, E.M., and Murphy, R.A. (1993) Detection and processing of peripheral myelin protein PMP22 in cultured Schwann cells. *J. Biol. Chem.*, **268**, 10372–10379.
45. D'Urso, D., Ehrhardt, P., and Muller, H.W. (1999) Peripheral myelin protein 22 and protein zero: a novel association in peripheral nervous system myelin. *J. Neurosci.*, **19**, 3396–3403.
46. Suter, U. and Scherer, S.S. (2003) Disease mechanisms in inherited neuropathies. *Nat. Rev. Neurosci.*, **4**, 714–726.
47. Naef, R. and Suter, U. (1998) Many facets of the peripheral myelin protein PMP22 in myelination and disease. *Microsc. Res. Tech.*, **41**, 359–371.
48. Melcangi, R.C., Cavarretta, I.T., Ballabio, M., Leonelli, E., Schenone, A., Azcoitia, I., Miguel Garcia-Segura, L., and Magnaghi, V. (2005) Peripheral nerves: a target for the action of neuroactive steroids. *Brain Res. Brain Res. Rev.*, **48**, 328–338.
49. Melcangi, R.C., Magnaghi, V., Cavarretta, I., Martini, L., and Piva, F. (1998) Age-induced decrease of glycoprotein Po and myelin basic protein gene expression in the rat sciatic nerve. Repair by steroid derivatives. *Neuroscience*, **85**, 569–578.
50. Melcangi, R.C., Leonelli, E., Magnaghi, V., Gherardi, G., Nobbio, L., and Schenone, A. (2003) Mifepristone (RU 38486) influences expression of glycoprotein Po and morphological parameters at the level of rat sciatic nerve: in vivo observations. *Exp. Neurol.*, **184**, 930–938.
51. Magnaghi, V., Cavarretta, I., Galbiati, M., Martini, L., and Melcangi, R.C. (2001) Neuroactive steroids and peripheral myelin proteins. *Brain Res. Brain Res. Rev.*, **37**, 360–371.
52. Melcangi, R.C., Magnaghi, V., Galbiati, M., Ghelarducci, B., Sebastiani, L., and Martini, L. (2000) The action of steroid hormones on peripheral myelin proteins: a possible new tool for the rebuilding of myelin? *J. Neurocytol.*, **29**, 327–339.
53. Magnaghi, V., Ballabio, M., Gonzalez, L.C., Leonelli, E., Motta, M., and Melcangi, R.C. (2004) The synthesis of glycoprotein Po and peripheral

myelin protein 22 in sciatic nerve of male rats is modulated by testosterone metabolites. *Brain Res. Mol. Brain. Res.*, **126**, 67–73.
54. O'Malley, B.W. (2006) Molecular biology. Little molecules with big goals. *Science*, **313**, 1749–1750.
55. Oñate, S.A., Tsai, S.Y., Tsai, M.J., and O'Malley, B.W. (1995) Sequence and characterization of a coactivator for the steroid hormone receptor superfamily. *Science*, **270**, 1354–1357.
56. Tetel, M.J. (2009) Modulation of steroid action in the central and peripheral nervous systems by nuclear receptor coactivators. *Psychoneuroendocrinology*, **345**, 59–519.
57. Cavarretta, I.T., Martini, L., Motta, M., Smith, C.L., and Melcangi, R.C. (2004) SRC-1 is involved in the control of the gene expression of myelin protein Po. *J. Mol. Neurosci.*, **24**, 217–226.
58. Magnaghi, V., Veiga, S., Ballabio, M., Gonzalez, L.C., Garcia-Segura, L.M., and Melcangi, R.C. (2006) Sex-dimorphic effects of progesterone and its reduced metabolites on gene expression of myelin proteins by rat Schwann cells. *J. Peripher. Nerv. Syst.*, **11**, 111–118.
59. Mercier, G., Turque, N., and Schumacher, M. (2001) Early activation of transcription factor expression in Schwann cells by progesterone. *Brain Res. Mol. Brain Res.*, **97**, 137–148.
60. Guennoun, R., Benmessahel, Y., Delespierre, B., Gouezou, M., Rajkowski, K.M., Baulieu, E.E., and Schumacher, M. (2001) Progesterone stimulates Krox-20 gene expression in Schwann cells. *Brain Res. Mol. Brain Res.*, **90**, 75–82.
61. Magnaghi, V., Ballabio, M., Roglio, I., and Melcangi, R.C. (2007) Progesterone derivatives increase expression of Krox-20 and Sox-10 in rat Schwann cells. *J. Mol. Neurosci.*, **31**, 149–157.
62. Bartolami, S., Auge, C., Travo, C., Venteo, S., Knipper, M., and Sans, A. (2003) Vestibular Schwann cells are a distinct subpopulation of peripheral glia with specific sensitivity to growth factors and extracellular matrix components. *J. Neurobiol.*, **57**, 270–290.
63. Fex Svenningsen, A. and Kanje, M. (1999) Estrogen and progesterone stimulate Schwann cell proliferation in a sex- and age-dependent manner. *J. Neurosci. Res.*, **57**, 124–130.
64. Lubischer, J.L., and Bebinger, D.M. (1999) Regulation of terminal Schwann cell number at the adult neuromuscular junction. *J. Neurosci.*, **19**, RC46.
65. di Michele, F., Lekieffre, D., Pasini, A., Bernardi, G., Benavides, J., and Romeo, E. (2000) Increased neurosteroids synthesis after brain and spinal cord injury in rats. *Neurosci. Lett.*, **284**, 65–68.
66. Meffre, D., Pianos, A., Liere, P., Eychenne, B., Cambourg, A., Schumacher, M., Stein, D.G., and Guennoun, R. (2007) Steroid profiling in brain and plasma of male and pseudopregnant female rats after traumatic brain injury: analysis by gas chromatography/mass spectrometry. *Endocrinology*, **148**, 2505–2517.
67. Marx, C.E., Trost, W.T., Shampine, L.J., Stevens, R.D., Hulette, C.M., Steffens, D.C., Ervin, J.F., Butterfield, M.I., Blazer, D.G., Massing, M.W., and Lieberman, J.A. (2006) The neurosteroid allopregnanolone is reduced in prefrontal cortex in Alzheimer's disease. *Biol. Psychiatry*, **60**, 1287–1294.
68. Weill-Engerer, S., David, J.P., Sazdovitch, V., Liere, P., Eychenne, B., Pianos, A., Schumacher, M., Delacourte, A., Baulieu, E.E., and Akwa, Y. (2002) Neurosteroid quantification in human brain regions: comparison between Alzheimer's and nondemented patients. *J. Clin. Endocrinol. Metab.*, **87**, 5138–5143.
69. Roglio, I., Bianchi, R., Gotti, S., Scurati, S., Giatti, S., Pesaresi, M., Caruso, D., Panzica, G.C., and Melcangi, R.C. (2008) Neuroprotective effects of dihydroprogesterone and progesterone in an experimental model of nerve crush injury. *Neuroscience*, **155**, 673–685.
70. Caruso, D., Scurati, S., Maschi, O., De Angelis, L., Roglio, I., Giatti, S., Garcia-Segura, L.M., and Melcangi, R.C. (2008) Evaluation of

neuroactive steroid levels by liquid chromatography-tandem mass spectrometry in central and peripheral nervous system: effect of diabetes. *Neurochem. Int.*, **52**, 560–568.

71. Pesaresi, M., Maschi, O., Giatti, S., Garcia-Segura, L.M., Caruso, D., and Melcangi, R.C. (2010) Sex differences in neuroactive steroid levels in the nervous system of diabetic and non-diabetic rats. *Horm. Behav.*, **57**, 46–55.

72. Caruso, D., Scurati, S., Roglio, I., Nobbio, L., Schenone, A., and Melcangi, R.C. (2008) Neuroactive steroid levels in a transgenic rat model of CMT1A Neuropathy. *J. Mol. Neurosci.*, **34**, 249–253.

73. Melcangi, R.C., Azcoitia, I., Ballabio, M., Cavarretta, I., Gonzalez, L.C., Leonelli, E., Magnaghi, V., Veiga, S., and Garcia-Segura, L.M. (2003) Neuroactive steroids influence peripheral myelination: a promising opportunity for preventing or treating age-dependent dysfunctions of peripheral nerves. *Prog. Neurobiol.*, **71**, 57–66.

74. Azcoitia, I., Leonelli, E., Magnaghi, V., Veiga, S., Garcia-Segura, L.M., and Melcangi, R.C. (2003) Progesterone and its derivatives dihydroprogesterone and tetrahydroprogesterone reduce myelin fiber morphological abnormalities and myelin fiber loss in the sciatic nerve of aged rats. *Neurobiol. Aging*, **24**, 853–860.

75. Melcangi, R.C., Magnaghi, V., and Martini, L. (2000) Aging in peripheral nerves: regulation of myelin protein genes by steroid hormones. *Prog. Neurobiol.*, **60**, 291–308.

76. Chavez-Delgado, M.E., Gomez-Pinedo, U., Feria-Velasco, A., Huerta-Viera, M., Castaneda, S.C., Toral, F.A., Parducz, A., Anda, S.L., Mora-Galindo, J., and Garcia-Estrada, J. (2005) Ultrastructural analysis of guided nerve regeneration using progesterone- and pregnenolone-loaded chitosan prostheses. *J. Biomed. Mater. Res. B Appl. Biomater.*, **74**, 589–600.

77. Islamov, R.R., Hendricks, W.A., Jones, R.J., Lyall, G.J., Spanier, N.S., and Murashov, A.K. (2002) 17 Beta-estradiol stimulates regeneration of sciatic nerve in female mice. *Brain Res.*, **943**, 283–286.

78. Gudemez, E., Ozer, K., Cunningham, B., Siemionow, K., Browne, E., and Siemionow, M. (2002) Dehydroepiandrosterone as an enhancer of functional recovery following crush injury to rat sciatic nerve. *Microsurgery*, **22**, 234–241.

79. Vita, G., Dattola, R., Girlanda, P., Oteri, G., Lo Presti, F., and Messina, C. (1983) Effects of steroid hormones on muscle reinnervation after nerve crush in rabbit. *Exp. Neurol.*, **80**, 279–287.

80. Yu, W.H. (1982) Effect of testosterone on the regeneration of the hypoglossal nerve in rats. *Exp. Neurol.*, **77**, 129–141.

81. Tanzer, L. and Jones, K.J. (2004) Neurotherapeutic action of testosterone on hamster facial nerve regeneration: temporal window of effects. *Horm. Behav.*, **45**, 339–344.

82. Jones, K.J., Brown, T.J., and Damaser, M. (2001) Neuroprotective effects of gonadal steroids on regenerating peripheral motoneurons. *Brain Res. Brain Res Rev.*, **37**, 372–382.

83. Huppenbauer, C.B., Tanzer, L., DonCarlos, L.L., and Jones, K.J. (2005) Gonadal steroid attenuation of developing hamster facial motoneuron loss by axotomy: equal efficacy of testosterone, dihydrotestosterone, and 17-beta estradiol. *J. Neurosci.*, **25**, 4004–4013.

84. Vinik, A.I., Park, T.S., Stansberry, K.B., and Pittenger, G.L. (2000) Diabetic neuropathies. *Diabetologia*, **43**, 957–973.

85. Sugimoto, K., Murakawa, Y., and Sima, A.A. (2000) Diabetic neuropathy–a continuing enigma. *Diabetes Metab. Res Rev.*, **16**, 408–433.

86. Yagihashi, S. (1997) Pathogenetic mechanisms of diabetic neuropathy: lessons from animal models. *J. Peripher. Nerv. Syst.*, **2**, 113–132.

87. Lauria, G., Lombardi, R., Borgna, M., Penza, P., Bianchi, R., Savino, C., Canta, A., Nicolini, G., Marmiroli, P., and Cavaletti, G. (2005) Intraepidermal nerve fiber density in rat foot

pad: neuropathologic-neurophysiologic correlation. *J. Peripher. Nerv. Syst.*, **10**, 202–208.
88. Biessels, G.J., Cristino, N.A., Rutten, G.J., Hamers, F.P., Erkelens, D.W., and Gispen, W.H. (1999) Neurophysiological changes in the central and peripheral nervous system of streptozotocin-diabetic rats. Course of development and effects of insulin treatment. *Brain*, **122** (Pt 4), 757–768.
89. Bianchi, R., Buyukakilli, B., Brines, M., Savino, C., Cavaletti, G., Oggioni, N., Lauria, G., Borgna, M., Lombardi, R., Cimen, B., Comelekoglu, U., Kanik, A., Tataroglu, C., Cerami, A., and Ghezzi, P. (2004) Erythropoietin both protects from and reverses experimental diabetic neuropathy. *Proc. Natl. Acad. Sci. U.S.A.*, **101**, 823–828.
90. Veiga, S., Leonelli, E., Beelke, M., Garcia-Segura, L.M., and Melcangi, R.C. (2006) Neuroactive steroids prevent peripheral myelin alterations induced by diabetes. *Neurosci. Lett.*, **402**, 150–153.
91. Yorek, M.A., Coppey, L.J., Gellett, J.S., Davidson, E.P., Bing, X., Lund, D.D., and Dillon, J.S. (2002) Effect of treatment of diabetic rats with dehydroepiandrosterone on vascular and neural function. *Am. J. Physiol. Endocrinol. Metab.*, **283**, E1067–E1075.
92. Leonelli, E., Bianchi, R., Cavaletti, G., Caruso, D., Crippa, D., Garcia-Segura, L.M., Lauria, G., Magnaghi, V., Roglio, I., and Melcangi, R.C. (2007) Progesterone and its derivatives are neuroprotective agents in experimental diabetic neuropathy: a multimodal analysis. *Neuroscience*, **144**, 1293–1304.
93. Roglio, I., Bianchi, R., Giatti, S., Cavaletti, G., Caruso, D., Scurati, S., Crippa, D., Garcia-Segura, L.M., Camozzi, F., Lauria, G., and Melcangi, R.C. (2007) Testosterone derivatives are neuroprotective agents in experimental diabetic neuropathy. *Cell. Mol. Life Sci.*, **64**, 1158–1168.
94. Windebank, A.J. and Grisold, W. (2008) Chemotherapy-induced neuropathy. *J. Peripher. Nerv. Syst.*, **13**, 27–46.
95. Quasthoff, S. and Hartung, H.P. (2002) Chemotherapy-induced peripheral neuropathy. *J. Neurol.*, **249**, 9–17.
96. Apfel, S.C. (2000) Managing the neurotoxicity of paclitaxel (Taxol) and docetaxel (Taxotere) with neurotrophic factors. *Cancer Invest.*, **18**, 564–573.
97. Cavaletti, G., Cavalletti, E., Montaguti, P., Oggioni, N., De Negri, O., and Tredici, G. (1997) Effect on the peripheral nervous system of the short-term intravenous administration of paclitaxel in the rat. *Neurotoxicology*, **18**, 137–145.
98. Sahenk, Z., Barohn, R., New, P., and Mendell, J.R. (1994) Taxol neuropathy. Electrodiagnostic and sural nerve biopsy findings. *Arch. Neurol.*, **51**, 726–729.
99. Roytta, M., Horwitz, S.B., and Raine, C.S. (1984) Taxol-induced neuropathy: short-term effects of local injection. *J. Neurocytol.*, **13**, 685–701.
100. Roglio, I., Bianchi, R., Camozzi, F., Carozzi, V., Cervellini, I., Crippa, D., Lauria, G., Cavaletti, G., and Melcangi, R.C. (2009) Docetaxel-induced peripheral neuropathy: protective effects of dihydroprogesterone and progesterone in an experimental model. *J. Peripher. Nerv. Syst.*, **14**, 36–44.
101. Ferzaz, B., Brault, E., Bourliaud, G., Robert, J.P., Poughon, G., Claustre, Y., Marguet, F., Liere, P., Schumacher, M., Nowicki, J.P., Fournier, J., Marabout, B., Sevrin, M., George, P., Soubrie, P., Benavides, J., and Scatton, B. (2002) SSR180575 (7-chloro-N,N,5-trimethyl-4-oxo-3-phenyl-3,5-dihydro-4H-pyridazino[4,5-b]i ndole-1-acetamide), a peripheral benzodiazepine receptor ligand, promotes neuronal survival and repair. *J. Pharmacol. Exp. Ther.*, **301**, 1067–1078.
102. Leonelli, E., Yague, J.G., Ballabio, M., Azcoitia, I., Magnaghi, V., Schumacher, M., Garcia-Segura, L.M., and Melcangi, R.C. (2005) Ro5-4864, a synthetic ligand of peripheral benzodiazepine receptor, reduces aging-associated myelin degeneration in the sciatic nerve of male rats. *Mech. Ageing Dev.*, **126**, 1159–1163.

8
Neuroprotective and Neurogenic Properties of Dehydroepiandrosterone and its Synthetic Analogs
Ioannis Charalampopoulos, Iakovos Lazaridis, and Achille Gravanis

8.1
Introduction

Dehydroepiandrosterone (DHEA) and dehydroepiandrosterone sulfate ester (DHEAS) are the most abundant steroids in humans. They are mainly produced by the zona reticularis of the adrenal cortex. In the early 1980s, both steroids were shown to be also produced within the nervous system [1, 2]. The enzymes required for the synthesis of DHEA were found to be expressed in neurons, astrocytes, and oligodendrocytes. The first step for the biosynthesis of DHEA is cholesterol's side chain cleavage by cytochrome P450scc and the production of pregnenolone, which is then metabolized to DHEA by the 17α-hydroxylase/c17 and 20-lyase activities of cytochrome P450c17. DHEA can be bidirectionally converted to the sulfated derivative DHEAS by hydroxysteroid sulfotransferase and back to DHEA by a sulfatase [3, 4]. During the development stages in mice, expression of P450c17 starts as early as E10.5 in neural crest cells and shortly after it is present in most of the spinal cord neural crest-derived tissues, including the peripheral nervous system (PNS) [5]. In the rat brain, cytochrome P450c17 is expressed during neonatal development and adulthood. The mRNA of P450c17 was found in the mesencephalon, cerebrum, diencephalon, and cerebellum with the expression levels being higher in the mesencephalon. High levels of P450c17 in association with lower levels of 3β-hydroxysteroid dehydrogenase/Δ^5-Δ^4-isomerase (3β-HSD), the enzyme that transforms DHEA to androstenedione, in the mesencephalon suggest that only DHEA – and not the other metabolites – is the main neurosteroid in this area [6]. Furthermore, P450c17 has been identified in hypothalamic and cortical neurons and astrocytes in culture [7, 8], as well as in neurons, astrocytes, and oligodendrocytes of the spinal cord of adult rats [9]. It is noteworthy that P450c17 is localized in the endoplasmic reticulum of the presynaptic and postsynaptic regions of pyramidal neurons in the hippocampal regions CA1–CA3, and in the granule neurons of the dentate gyrus, suggesting fast neuromodulatory actions of DHEA at the level of the synapse [10, 11]. The presence of P450c17 has been also described in neuronal and glial cells of the brain and pituitary of frog [12]. The importance of DHEA for neuronal function is supported by experimental evidence for an alternative Fe^{2+}-dependent P450c17-independent pathway of its production, although the concentration of iron needed is difficult to be attained *in vivo* [13, 14].

DHEA levels change profoundly throughout the life span. The tempo-spatial fluctuations in the concentration of DHEA and DHEAS within the developing nervous system suggest a significant regulatory role for these neurosteroids in the shaping of the developing nervous system. This hypothesis is further supported by recent findings associating DHEA to self-renewal or differentiation of neural progenitors and to the control of neuronal survival or apoptosis of mature neurons. Indeed, early in human neonatal life the concentration of DHEA starts to rise and remains at high levels throughout adulthood. It is postulated that elevated levels of DHEA protect the brain from endogenous and exogenous neurotoxic challenges, while promoting neurogenesis, in parallel. Furthermore, the progressive decline of DHEA levels during human aging [15–17] has been associated with neuronal loss and the appearance of neurodegenerative processes [18]. It is now well established that some of these effects of DHEA are mediated through modulation of various neurotransmitter receptors [19]. Indeed, DHEA and DHEAS are important regulators of neural function and fate, through their modulatory effects on $GABA_A$ [20], NMDA [21], or sigma-1 receptors [22]. Additionally, DHEA may affect neuronal fate interacting with cytoskeleton protein microtubule-associated protein 2 (MAP2) [23, 24], or with nerve growth factor (NGF) receptors [25]. This chapter reviews the experimental findings associating DHEA with neuroprotection. The molecular mechanisms by which this important neurosteroid affects neuronal cell fate throughout the life span and its potential therapeutic applications in neurodegeneration are also discussed.

8.2
Neuroprotective and Neurogenic Effects of DHEA in Hippocampal Neurons

Age-related dementias including Alzheimer's disease (AD) are characterized by progressive loss of memory and global cognitive decline. The histological hallmarks of AD are protein aggregates that form senile plaques from the disposition of misfolded amyloid-beta (Aβ), and hyperphosphorylated tau in neurofibrillary tangles, apoptotic loss of hippocampal and cortical neurons, and the appearance of neuroinflammation. The mechanisms underlying neural loss include oxidative stress, energy depletion, excitotoxicity, and damage due to neuroinflammation. During aging there is a progressive decline in the levels of circulating DHEA to an extent where at the age of 70 the circulating levels have reached 10–20% of their peak values. It is noteworthy that reduced levels of DHEA and DHEAS have been found in brain regions of patients with AD, compared with age matched nondemented controls [26, 27], suggesting a neuroprotective role for endogenous DHEA. Recent experimental findings support this hypothesis. Indeed, DHEA dose-dependently attenuates the loss of newborn hippocampal neurons induced by Aβ (25–35)-infusion in mice, and this effect is blocked by the sigma-1 receptor antagonist NE100 and is mimicked by the sigma-1 receptor agonist PRE084 [28]. These protective actions of DHEA are sensitive to the PI3K inhibitor LY294002. Aβ (25–35) decreases the levels of Akt phosphorylation, in a sigma-1 receptor-dependent manner. DHEA was shown to reverse the Aβ (25–35)-induced decrease of the dendritic density and length of doublecortin positive cells in the dentate gyrus. It appears that DHEA prevents the Aβ (25–35)-impaired survival and dendritic growth of newborn hippocampal neurons through a sigma-1 receptor-mediated modulation of PI3K kinase signaling. DHEA may also prevent AβP neurotoxicity, mediated by elevated levels of $Ca^{(2+)}$. AβP forms $Ca^{(2+)}$-permeable pores on membranes of rat hippocampal neurons, causing a marked increase in intracellular calcium level, and neuronal death. Preadministration of DHEA or DHEAS significantly inhibits the increase of intracellular calcium levels, induced by AβP. These findings suggest that supplementation of reduced levels in the elderly may be effective in preventing AβP neurotoxicity [29]. Oligodendrocytes and astrocytes may produce DHEA via a P450c17 alternative, Fe^{2+}-dependent pathway. Exposure of oligodendrocytes to β-amyloid increases DHEA formation [14], suggesting that DHEA synthesis can be regulated by neurotoxic β-amyloids, possibly as a reactive rescue response to the β-amyloid challenge.

DHEA also appears to protect hippocampal neurons against oxidative stress, by decreasing the levels of lipid peroxidation products in the hippocampus of both 12- and 22 month old rats, with the decrease being higher in the older group [30]. Peripheral benzodiazepine receptors are known to participate in the protection of neural cells from reactive oxygen species and their levels are reduced with aging. DHEA effectively maintains high levels of peripheral benzodiazepine receptors in mitochondria of cerebral cortex and reverses the impairment in learning/memory ability in the animal model of aging with D-galactose-induced neurodegeneration [31]. DHEA is also effective in rescuing hippocampal neurons from other oxidative challenges, including H_2O_2 and sodium nitroprusside [32].

DHEA and DHEAS have been shown to protect hippocampal neurons in culture against excitotoxic NMDA, AMPA, and kainic acid [33, 34]. It appears that DHEA exerts these protective effects by affecting major neurotoxic, prodeath signaling pathways. Indeed, pretreatment of mouse hippocampal HT22 cells with DHEA for 24 h completely reversed the glutamate-mediated neurotoxic effect of glucocorticoids by inhibiting the translocation of glucocorticoid receptors (GRs) into the nucleus [35]. Furthermore, DHEA inhibited the translocation of death-related stress-activated protein kinase 3 into the nucleus of primary hippocampal neurons [36]. DHEA reversed the NMDA-induced suppression of the PI3K/Akt pathway in P19-N neuronal cells [37], and inhibited the staurosporine-induced toxicity in human neuroblastoma SH-SY5Y cells by reducing the activity of caspase 3, thus blocking apoptotic cell loss. The increased viability produced by DHEA and DHEAS on staurosporine treated cells is abolished by inhibitors of prosurvival kinases PI3K and ERK/MAPK [38]. DHEA is proven equally effective in protecting hippocampal CA1/2 neurons from NMDA *in vivo* [33], significantly increasing the dose of NMDA necessary to induce seizures in mice [39].

Besides its cytoprotective effects in the hippocampus, DHEA appears to enhance memory in a number of animal studies and in humans [40]. Indeed, enhancing effects of DHEA have been reported in spatial learning in the model of chronic mild stress in mice [41]. Furthermore, DHEA improved memory retention in the foot-shock active avoidance test in mice; DHEA reversed the amnesic effects of muscarinic cholinergic antagonist scopolamine, via sigma-1 receptors [42, 43]. Moreover, DHEA reduced the neurodegeneration in CA1 area of the hippocampus and the learning deficits due to hypoxic insult with repeated exposure to carbon monoxide. However, these effects were not affected by the sigma-1 antagonist NE100 [22]. Finally, DHEAS enhanced working memory in a win-shift task in aged mice, using water escape motivation [44].

There is also strong experimental evidence that DHEA affects the physiological processes of memory through its actions on synapse function. Chronic treatment with DHEAS lowered the threshold pulse number required for induction of activity-dependent long-term potentiation (LTP) in the hippocampal Schaffer collateral CA1 synapses of rats [45]. These effects are abolished with coadministration of the sigma-1 antagonist NE100, although acute administration of NE100 does not seem to affect DHEAS actions, suggesting that sigma-1 receptors play a role in the chronic but not in the acute effects of DHEAS on LTP facilitation. Subthreshold high-frequency stimulation in the presence of DHEAS triggered the phosphorylation of Src and ERK2 kinases, while Src kinase inhibitor reversed DHEAS facilitating effects, suggesting that Src and ERK2 activation is required for the induction of LTP-facilitating effects of DHEAS [46, 47]. Administration of DHEA in gonadectomized male and female rats strongly increases the number of hippocampal CA1 spine synapses [48, 49]. Experiments with hippocampal slice preparations have shown that DHEAS enhanced the synaptic evoked population spike in the somatic region of CA1 neurons without affecting the dendritic synaptic potential [50, 51]. DHEAS reduced the paired pulse inhibition in the dentate gyrus and CA1 area of the hippocampus, via muscarinic M2 receptors [52]. DHEA may also play a role in the development and connectivity of the hippocampus. Indeed, DHEA increased immunoreactivity of the neuronal dendritic marker MAP2 in the hippocampus and nucleus accumbens of neonatal rats, and synapsin I and NPY (neuropeptide Y)-positive neurons in the hippocampus of postpuberty rats [53].

DHEA was also shown to increase neurogenesis in the dentate gyrus of the hippocampus and to protect existing neurons, antagonizing the neurotoxic effects of glucocorticoids [54]. Increased numbers of glial fibrillary acidic protein (GFAP)-positive astrocytes were also found in the hippocampus in DHEA-treated pre- and postpuberty rats [55].

8.3
Neuroprotective Effects of DHEA in Nigrostriatal Dopaminergic Neurons

Parkinson's disease (PD) is a very common neurodegenerative disorder with increasing prevalence in the elderly population. Although, for the initiation of the disease, many factors both genetic and environmental may play a role, the final outcome is the apoptotic loss of dopaminergic neurons in the substantia nigra pars compacta (SNc) that leads to extensive loss of striatal dopamine (DA) concentrations, the projection area of the SNc neurons. This loss of dopaminergic neurons leads to the clinical hallmarks of PD such as tremor, bradykinesia, and posture instability [56].

A chemical compound, the neurotoxin 1-methyl-4-phenyl-1,2,3,6-tetrahydropyridine (MPTP) reproduces (or produces) the pathology of PD when administered to experimental animals. Systemic administration of MPTP leads to severe apoptotic damage in the nigrostriatal dopaminergic system with similar behavioral consequences as in PD [57–59]. DHEA treatment of MPTP mice resulted in maintaining the concentrations of DA, and its metabolites dihydroxyphenylacetic acid (DOPAC), and homovanillic acid (HVA), to levels comparable to those of intact animals. Furthermore, DHEA prevented the decline of the mRNA levels of DA transporter and tyrosine hydroxylase resulting from MPTP toxicity [60]. In MPTP monkeys, DHEA improved the mean clinical score and potentiated locomotor activity in both moderately and severely impaired subjects [61, 62]. Administration of DHEA together with methyl-4-phenylpyridinium (MPP^+, the effector metabolite of MPTP) in the striatum of rats resulted in reduced loss of DA concentration and tyrosine hydroxylase or acetylcholinesterase-positive fibers in the striatum [63]. DHEAS was also protective, reducing MPP^+-induced apoptosis in cultured cerebellar granule cells [64]. These findings were corroborated by *in vitro* data, showing that DHEA and DHEAS increased acutely (peak effect between 10 and 30 min) and dose-dependently (EC_{50} in the nanomolar range) the release of catecholamines from dopaminergic PC12 cells [65]. It appears that the acute effect of these steroids involves actin depolymerization and submembrane actin filament disassembly, a fast-response cellular system regulating trafficking of catecholamine vesicles. DHEAS but not DHEA also affected catecholamine synthesis, increasing both the mRNA and protein levels of tyrosine hydroxylase.

DHEAS may also affect brain areas associated with nonmotor behavioral manifestations of PD, like the prelimbic cortex. Recent experimental findings have shown that DHEAS enhanced the excitatory postsynaptic currents in the prelimbic cortex, increasing glutamate release via D1 receptors [66]. On the other hand, DHEA induced the formation of synaptic inputs into the striatum, increasing the expression of synapsin I and DA production in the nigrostriatal system [67].

8.4
Neuroprotective Effects of DHEA in Autoimmune Neurodegenerative Processes

Multiple sclerosis (MS) is a major autoimmune disease of the central nervous system. The responsible component for the initiation and progression of the disease is adaptive immunity and the breakdown of tolerance, with autoreactive T cells and autoantibodies against protein and lipid components of myelin sheath. This attack results in degeneration of oligodendrocytes that produce the myelin and the underlying axons, leading to degeneration of neural cell bodies in the gray matter, and as the disease progresses, results in the significant cerebral atrophy that characterizes MS.

In human clinical studies, low serum levels of DHEA and DHEAS have been associated with the severity of fatigue, one of the most limiting symptoms of MS [68]. Furthermore, a large number of studies have demonstrated strong immunoregulatory actions of DHEA [69]. Administration of DHEA in the model of experimental autoimmune neuritis (EAN) in rats, resulted in a significant delay in the onset of disease with decreased infiltration of inflammatory cells in the PNS, decreased numbers of IFN-γ and TNF-α expressing cells, reduced levels of IFN-γ and TNF-α, and lower proliferation rates of autoreactive T cells in the spleen [70]. DHEA was also effective in reducing the incidence and severity of experimental autoimmune encephalomyelitis (EAE) in mice, reducing the Th1-mediated response [71]. In fact, DHEA significantly decreased the activation and proliferation of T lymphocytes as well as activation of NF-κB and the subsequent production of proinflammatory cytokines TNF-α, IFN-γ, and interleukin 12 (IL-12). The synthetic fluorinated DHEA analog fluasterone significantly delayed the onset, reduced the peak clinical score and cumulative disease index of EAE mice, and prevented or significantly attenuated relapses [72]. Moreover, T cells from treated EAE mice had significantly reduced proteolipid protein 139–151-specific T-cell proliferation responses and reduced numbers of TNF-α- and IFN-γ-producing cells in the CNS. This DHEA analog, which has been reported to have weak androgenic or estrogenic side effects, appears to have a potent inhibitory activity in EAE.

DHEA also appears to protect the other cellular components of autoimmune neurodegenerative processes, microglia cells. Indeed, DHEA-reduced protein and mRNA levels of nitric oxide synthase (iNOS) in LPS-stimulated BV2 microglial cells [73], and the production of proinflammatory TNF-α in LPS-stimulated astrocytes and microglia cells [74]. Finally, DHEA reduced the proliferation of astrocytes after stimulation with myelin basic protein [75, 76].

8.5
Neuroprotective Effects of DHEA against Brain Ischemia and Trauma

Damage of neural tissue, from ischemia or trauma, results in a secondary propagation of the lesion with degeneration of the surrounding area and distal neuronal axons. This secondary phase is driven by mechanisms such as the blood–brain barrier disruption, edema with infiltrating immune cells and neuroinflammation, degeneration of myelin, excitotoxicity, and oxidative stress [77]. The therapeutic strategy for ischemic and traumatic brain injury involves amelioration of these secondary events, reducing sequential injury, together with enhancement of regenerating mechanisms in order to promote tissue rebuilding.

There is now strong experimental evidence that DHEA exerts neuroprotective effects in a number of animal models of brain ischemia. In the model of transient global cerebral ischemia in rats, administration of DHEA 3–48 h after the induction of ischemia decreases the levels of cell death in the hippocampal CA1 neurons and improves spatial learning performance [78]. However, administration of DHEA 1 h before or after the induction of ischemia provokes an exacerbation of neuronal death and a decline of spatial learning performance, most probably due to activation of NMDA and sigma-1 receptors. Indeed, administration of NMDA or sigma-1 receptor antagonists prevented the deleterious effects of DHEA. On the other hand, DHEAS decreased the incidence of permanent paraplegia, provoked by reversible spinal cord ischemia in rabbits, in a $GABA_A$ receptor–dependent mechanism [79]. In the model of transient forebrain ischemia, administration of DHEAS reversed the impairment of LTP in hippocampal CA1 neurons, via sigma-1 receptors [78]. Furthermore, DHEAS prevented the biochemical changes in the ischemic injury of rat retina [80], and decreased the threshold shift of the action potential after transient cochlear ischemia [81]. Finally, DHEA restored the reactive oxygen species, the lactate dehydrogenase release, and the activity of Na/K-ATPase after ischemia with bilateral carotid artery occlusion in healthy and diabetic rats [82]. The above *in vivo* studies were further supported by *in vitro* data showing that DHEAS supplementation of rat cerebellar granule cell cultures decreases the levels of apoptosis, induced by oxygen–glucose deprivation [64], and this effect is blocked by pentobarbital, a $GABA_A$ receptor agonist.

DHEA and DHEAS were proven effective in protecting against brain trauma. In the model of focal cortical cold lesion, DHEAS significantly decreased the area of the lesion when administered either before or after the induction of the injury [83]. DHEA improved performance in the passive avoidance test and in the forced swimming test in mild traumatic brain injury, reversing the cognitive and behavioral damages induced by brain trauma [84]. DHEAS facilitated the recovery of the cortices in the surrounding area of focal cold brain injury and enhanced the amplitude of the cortical evoked responses [85]. Furthermore, DHEA facilitated the recovery of motor control and coordination after spinal cord injury in mice, increasing white matter and reducing gliosis in the surrounding area of the lesion [86]. DHEA was equally effective even seven days after traumatic brain injury, improving the performance in both sensory-motor and cognitive tasks [87]. DHEA minimized the damage caused after transection of the sciatic nerve, by increasing the number and the diameter of myelinated axons, and reducing atrophy of the gastrocnemius muscle [88]. Additionally, DHEA treatment in the model of penetrating brain trauma resulted in a significant decrease of gliotic tissue formation, decreasing the accumulation of astrocytes near the wounded area and reducing proliferation of reactive astrocytes [89]. DHEA inhibited the activation of astrocytes due to increased levels of potassium in hippocampal slice cultures [90]. The fluorinated DHEA analog, fluasterone, facilitated functional recovery of traumatic brain injury in rats, improving beam walk performance, neurological reflexes, and declarative memory [91]. Furthermore, both DHEA and its fluorinated analog inhibited IL-1β-induced cyclooxygenase-2 (COX2) mRNA and prostaglandin (PGE2) production in cultured mesangial cells.

8.6
Signaling Pathways Involved in the Effects of DHEA on Neuronal Cell Fate

The decline of circulating concentrations of DHEA and DHEAS during human aging has been associated with age-related degenerative processes [15–17], leading to the hypothesis that DHEA supplementation may improve the course of age-related health conditions, including neuronal loss [18]. Some of the potential antiaging properties of DHEA are attributed to its cytoprotective actions. Indeed, *in vitro* studies have shown that DHEA exerts potent antiapoptotic and prosurvival effects in various cells, including keratinocytes, lymphocytes, thymocytes, and endothelial cells [92–98]. In most of these cell systems, DHEA transmits its prosurvival signals via specific membrane binding sites of unknown nature, preventing cell apoptosis, through induction of antiapoptotic proteins.

DHEA was shown to activate similar signaling pathways and protect neuronal cells against various apoptotic challenges [38, 99, 100]. Indeed, DHEA protects neural crest–derived cells against serum or NGF deprivation–induced apoptosis at nanomolar concentrations [100]. The antiapoptotic effects of DHEA are mediated via binding to specific plasma membrane receptors (K_d: 0.9 nM) [101]; rapid activation of prosurvival kinases Shc/MEK1/2/ERK1/2, and PI3K/Akt; the induction of transcription factors CREB and NF-κB; and the subsequent production of antiapoptotic Bcl-2 proteins [18]. In fact, the prosurvival signaling pathways induced by DHEA in neuronal cells are initiated at the plasma membrane level, and have major common characteristics with those activated by NGF, the key neurotrophic factor of neural crest–derived cells. Neurotrophin NGF binds with high affinity (K_d: 0.1 nM) to the transmembrane tyrosine kinase (TrkA) receptor and with lower affinity (K_d: 1.0 nM) to the p75^{NTR} receptor, a membrane protein belonging to TNF receptor superfamily [102]. In the presence of TrkA receptors, p75^{NTR} participates in the formation of high-affinity binding sites, leading to cell survival signals. In the absence of TrkA, p75^{NTR} generates cell death signals. Docking of TrkA by NGF initiates receptor dimerization, and phosphorylation of tyrosine residues on the receptor, interacting with Shc and other adaptor proteins, resulting in the activation of PI3K/Akt and MEK/ERK signaling kinase pathways. These signals lead to the activation of prosurvival transcription factors CREB and NF-κB, the subsequent production of antiapoptotic Bcl-2 proteins and prevention of apoptotic neuronal death. In fact, DHEA can mimic NGF in inducing TrkA phosphorylation, and regulating the interaction of p75^{NTR} with its known cytoplasmic interactors TRAF-6, RIP2, or RhoGDI [25]. This rapid, within minutes, activation of NGF receptors by DHEA is followed by the sequential activation of prosurvival kinases Shc/ERK1/2 and Akt, leading to the production of antiapoptotic Bcl-2 proteins (Figure 8.1). The interaction of DHEA with NGF receptors is supported by *in vitro* experiments showing that deletion of TrkA receptor with specific siRNAs completely reverses its antiapoptotic actions, and *in vivo* experiments in NGF null mice, where systemic administration of DHEA during embryonic development significantly decreases the apoptotic loss of TrkA-positive sympathetic neurons [25]. These findings suggest that DHEA exerts its neurotrophic actions interacting with major

8.6 Signaling Pathways Involved in the Effects of DHEA on Neuronal Cell Fate | 145

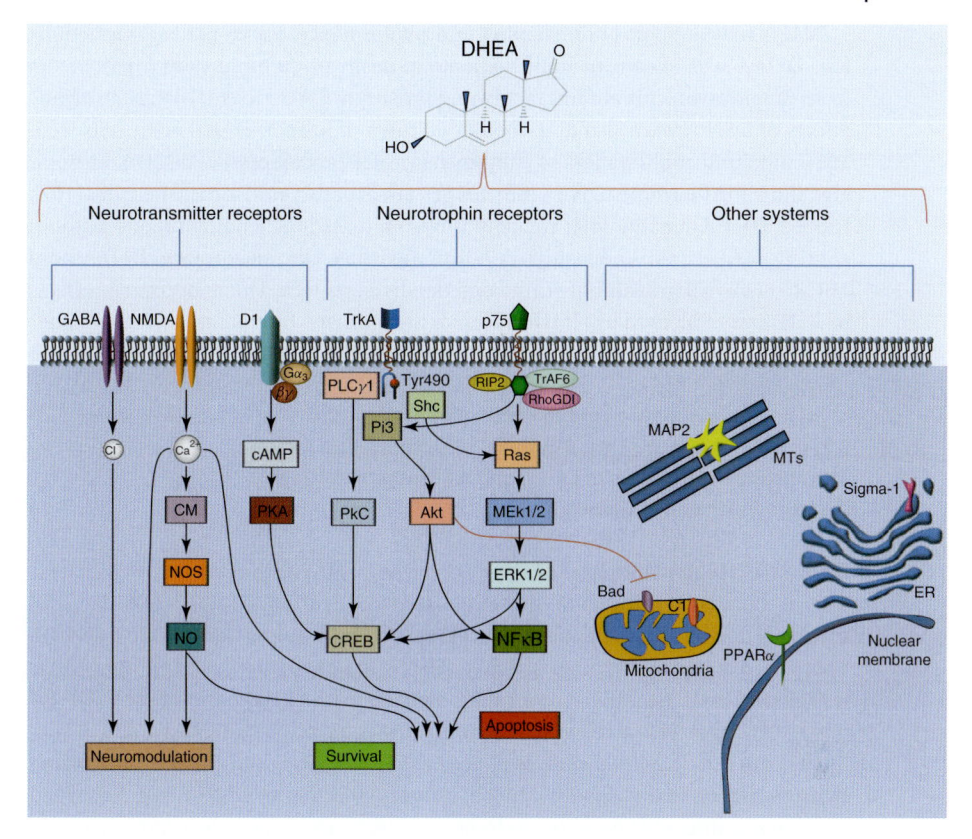

Figure 8.1 Signaling pathways involved in the neuroprotective effects of DHEA. DHEA is a multifaceted neurohormone, affecting neural tissues with multiple mechanisms of action. DHEA and DHEAS are important regulators of neural function through their modulatory effects on $GABA_A$, NMDA, or sigma-1 receptors. Additionally DHEA may affect neuronal cell fate interacting with cytoskeleton protein MAP2, or with dopamine-1 (D1) or nerve growth factor (NGF) receptors.

regulators of neuronal survival, NGF receptors. This intriguing hypothesis was first brought to light 15 years ago by Compagnone and Mellon, showing colocalized staining of P450c17, the rate-limiting enzyme of DHEA biosynthesis, and NGF receptors in mouse embryonic dorsal root ganglias (DRGs) [5]. About one-fifth of P450c17-immunopositive DRG neurons in the mouse were also found to be TrkA immunopositive. Among the TrkA-expressing cells, about one-third also express P450c17, being able to respond to both DHEA and NGF. The potential interaction of DHEA with NGF was further suggested by recent findings showing that axonal growth from sensory neurons was promoted by keratinocytes when the two cell types are cocultured [103]. The neuritogenic effect of keratinocytes is suppressed when the activity of NGF receptors was blocked with the TrkA inhibitor K252a or by inhibitors of steroidogenesis, and it was mimicked by DHEA.

DHEA may support brain shaping and maintenance, via its neurogenic actions too. DHEA was shown to inhibit apoptotic death of cultured neural precursors from rat embryonic forebrains, activating prosurvival Akt kinase [104]. Neurogenic effects of DHEA were also described in cultures of human neural stem cells [105, 106], inducing proliferation of human neuronal precursors via NMDA receptor signaling. Furthermore, DHEA induces growth of embryonic cortical neurites [107]. Indeed, DHEA at nanomolar concentrations increased the length of axons, without altering MAP-2-immunopositive dendritic neurites. On the contrary, DHEAS did not affect axonal growth but stimulated dendritic growth. The neuritogenic effects of DHEA were blocked by NMDA receptor antagonists. It remains unclear whether DHEA may also affect neural stem cell niches of the adult nervous system.

8.7
Therapeutic Perspectives of DHEA and its Synthetic Analogs in Neurodegenerative Diseases

DHEA replacement therapy has attracted extensive attention in the last two decades. This is because, in humans, the circulating concentrations of DHEA decrease markedly during aging, and have been associated with age-related cognitive decline. This has led to the hypothesis that DHEA supplementation during aging may improve mental health by preventing neuronal loss [108]. In rodents, DHEA possesses important neuroprotective and neurogenic properties *in vitro* and *in vivo*, but it is unclear whether these effects are mediated indirectly through its conversion to estrogens or to androgens. Indeed, naturally occurring DHEA is metabolized in humans into estrogens or androgens, which are known to exert important generalized endocrine side effects, including hormone-dependent neoplasias, thus, making the long-term clinical use of DHEA dubious. Furthermore, the broad effects of DHEA itself limits its therapeutic use. To circumvent the side effects of DHEA, one of the promising ways is to develop DHEA derivatives deprived of its endocrine moieties and with a greater specificity.

Fluasterone is a 17-fluorinated DHEA analog with weak estrogenic and androgenic effects. Fluasterone facilitated functional recovery of traumatic brain injury rats, relaxing middle cerebral artery, improving beam walk performance, neurological reflexes, and declarative memory [91]. Fluasterone was also proven effective in significantly delaying the onset, reduced the peak clinical score and cumulative disease index of EAE mice, and prevented or significantly attenuated relapses [72]. Two 7β-amino analogs of DHEA were recently evaluated on memory. DHEA and its two analogs improved retention memory in passive avoidance test in mice, without effect in the spatial working memory task [109]. The two synthetic compounds failed to reverse scopolamine-induced deficit in spontaneous alternation. Additionally, new 17 spiro analogs of DHEA were reported to have strong antiapoptotic and neuroprotective properties in various *in vitro* assays [110]. These compounds were proven highly effective (IC$_{50}$ at nanomolar level) in protecting cultured neural crest–derived cells against apoptosis, being devoid of any estrogenic or androgenic actions. The *in vivo* efficacy of these molecules is now under investigation in animal models of neurodegenerative diseases, such as EAE mice, rat diabetic retinal degeneration, and MPTP mice. All the above mentioned DHEA synthetic analogs might prove to be lead molecules for the synthesis of novel and safe neuroprotective and neurogenic agents, with therapeutic applications in neurodegenerative diseases.

References

1. Corpechot, C., Robel, P., Axelson, M., Sjovall, J., and Baulieu, E.E. (1981) Characterization and measurement of dehydroepiandrosterone sulfate in rat brain. *Proc. Natl. Acad. Sci. U.S.A.*, **78**, 4704–4707.
2. Baulieu, E.E. (1997) Neurosteroids: of the nervous system, by the nervous system, for the nervous system. *Recent Prog. Horm. Res.*, **52**, 1–32.
3. Miller, W.L. (2002) Androgen biosynthesis from cholesterol to DHEA. *Mol. Cell. Endocrinol.*, **198**, 7–14.
4. Mellon, S.M. (2001) Biosynthesis of neurosteroids and regulation of each synthesis. *Int. Rev. Neurobiol.*, **46**, 33–78.
5. Compagnone, N.A., Bulfone, A., Rubenstein, J.L., and Mellon, S.H. (1995) Steroidogenic enzyme P450c17 is expressed in the embryonic central nervous system. *Endocrinology*, **136**, 5212–5223.
6. Kohchi, C., Ukena, K., and Tsutsui, K.B. (1998) Age- and region-specific expressions of the messenger RNAs encoding for steroidogenic enzymes P450scc, P450c17 and 3b-HSD in the postnatal rat brain. *Brain Res.*, **801**, 233–238.
7. Zwain, I. and Yen, S. (1999) Neurosteroidogenesis in astrocytes, oligodendrocytes, and neurons of cerebral cortex of rat brain. *Endocrinology*, **140**, 3843–3852.
8. Zwain, I. and Yen, S. (1999) Dehydroepiandrosterone: biosynthesis and metabolism in the brain. *Endocrinology*, **140**, 880–887.
9. Kibaly, K., Patte-Mensah, C., and Mensah-Nyagan, A. (2005) Molecular and neurochemical evidence for the biosynthesis of dehydroepiandrosterone in the adult rat spinal cord. *J. Neurochem.*, **93**, 1220–1230.
10. Mukai, H., Takata, N., Ishii, H., Tanabe, N., Hojo, Y., Furukawa, A., Kimoto, T., and Kawato, S. (2006) Hippocampal synthesis of estrogens and androgens which are paracrine modulators of synaptic plasticity: synaptocrinology. *Neuroscience*, **138**, 757–764.
11. Hojo, Y., Hattori, T.A., Enami, T., Furukawa, A., Suzuki, K., Ishii, H.-T., Mukai, H., Morrison, J.-H., Janssen, W.-G., Kominami, S., Harada, N., Kimoto, T., and Kawato, S. (2004) Adult male rat hippocampus synthesizes estradiol from pregnenolone by cytochromes P45017 alpha and P450 aromatase localized in neurons. *Proc. Natl. Acad. Sci. U.S.A.*, **101**, 865–870.
12. Do Rego, J.L., Tremblay, Y., Luu-The, V., Repetto, E., Castel, H., Vallarino, M., Belanger, A., Pelletier, G., and Vaudry, H. (2007) Immunohistochemical localization and biological activity of the steroidogenic enzyme cytochrome P450 17α-hydroxylase/C17, 20-lyase (P450C17) in the frog brain and pituitary. *J. Neurochem.*, **100**, 251–268.
13. Cascio, C., Brown, R.C., Liu, Y., Han, Z., Hales, D.B., and Papadopoulos, V. (2000) Pathways of dehydroepiandrosterone formation in rat brain glia. *J. Steroid Biochem. Mol. Biol.*, **75**, 177–186.
14. Brown, R.-C., Cascio, C., and Papadopoulos, V. (2000) Pathways of neurosteroid biosynthesis in cell lines from human brain: regulation of dehydroepiandrosterone formation by oxidative stress and beta-amyloid peptide. *J. Neurochem.*, **74**, 847–859.

15. Belanger, A., Candas, B., Dupont, A., Cusan, L., Diamond, P., Gomez, J.L., and Labrie, F. (1994) Changes in serum concentrations of conjugated and unconjugated steroids in 40- to 80-year-old men. *J. Clin. Endocrinol. Metab.*, **79**, 1086–1090.
16. Migeon, C.J., Keller, A.R., Lawrence, B., and Shepart, T. (1957) Dehydroepiandrosterone and androsterone levels in human placenta. Effects of age and sex: day-to-day and diurnal variations. *J. Clin. Endocrinol. Metab.*, **17**, 1051–1062.
17. Vermeulen, A., Deslypene, J.P., Schelthout, W., Verdonck, L., and Rubens, R. (1982) Adrenocortical function in old age: response to acute adrenocorticotropin stimulation. *J. Clin. Endocrinol. Metab.*, **54**, 187–191.
18. Charalampopoulos, I., Margioris, A., and Gravanis, A. (2008) Neurosteroid dehydroepiandrosterone exerts anti-apoptotic effects by membrane-mediated, integrated genomic and non-genomic pro-survival signaling pathways. *J. Neurochem.*, **107**, 1457–1469.
19. Compagnone, N.A. and Mellon, S.H. (2000) Neurosteroids: biosynthesis and function of these novel neuromodulators. *Front. Neuroendocrinol.*, **21**, 1–56.
20. Majewska, M.D., Demirgören, S., Spivak, C.E., and London, E. (1990) The neurosteroid dehydroepiandrosterone sulfate is an allosteric antagonist of the GABAA receptor. *Brain Res.*, **526**, 143–146.
21. Johansson, T. and Le Grevès, P. (2005) The effect of dehydroepiandrosterone sulfate and allopregnanolone sulfate on the binding of [(3)H]ifenprodil to the N-methyl-d-aspartate receptor in rat frontal cortex membrane. *J. Steroid Biochem. Mol. Biol.*, **94**, 263–266.
22. Maurice, T., Roman, F., Privat, A. and (1996) Modulation by neurosteroids of the in vivo (+)-[3H]SKF-10,047 binding to sigma 1 receptors in the mouse forebrain. *J. Neurosci. Res.*, **46**, 734–743.
23. Laurine, E., Lafitte, D., Grégoire, C., Sérée, E., Loret, E., Douillard, S., Michel, B., Briand, C., and Verdier, J.M. (2003) Specific binding of dehydroepiandrosterone to the N terminus of the microtubule-associated protein MAP2. *J. Biol. Chem.*, **278**, 29979–29986.
24. Pérez-Neri, I., Montes, S., Ojeda-López, C., Ramírez-Bermúdez, J., and Ríos, C. (2008) Modulation of neurotransmitter systems by dehydroepiandrosterone and dehydroepiandrosterone sulfate: Mechanism of action and relevance to psychiatric disorders. *Prog. Neuropsychopharmacol. Biol. Psychiatry*, **32**, 1118–1130.
25. Lazaridis, I., Charalampopoulos, I., Vergou, V., Avlonitis, N., Calogeropoulou, T., and Gravanis, A. (2010) Neurosteroid dehydroepiandrosterone directly binds to NGF receptors, rescuing neuronal cells from apoptosis. FENS Abstracts 5, 112.7.
26. Weill-Engerer, S., David, J.P., Sazdovitch, V., Liere, P., Eychenne, B., Pianos, A., Schumacher, M., Delacourte, A., Baulieu, E.E., and Akwa, Y. (2002) Neurosteroid quantification in human brain regions: comparison between Alzheimer's and nondemented patients. *J. Clin. Endocrinol. Metab.*, **87** (11), 5138–5143.
27. Schumacher, M., Weill-Engerer, S., Liere, P., Robert, F., Franklin, R.J., Garcia-Segura, L.M., Lambert, J.J., Mayo, W., Melcangi, R.C., Parducz, A., Suter, U., Carelli, C., Baulieu, E.E., and Akwa, Y. (2003) Steroid hormones and neurosteroids in normal and pathological aging of the nervous system. *Prog. Neurobiol.*, **71**, 3–29.
28. Li, L., Xu, B., Zhu, Y., Chen, L., and Sokabe, M. (2010) DHEA prevents Abeta(25-35)-impaired survival of newborn neurons in the dentate gyrus through a modulation of PI3K-Akt-mTOR signaling. *Neuropharmacology*, **59**, 323–333.
29. Kato-Negishi, M. and Kawahara, M. (2008) Neurosteroids block the increase in intracellular calcium level induced by Alzheimer's beta-amyloid protein in long-term cultured rat hippocampal neurons. *Neuropsychiatr. Dis. Treat.*, **4**, 209–218.

30. Sinha, N., Baquer, N.-Z., and Sharma, D. (2005) Anti-lipidperoxidative role of exogenous dehydroepiendrosterone (DHEA) administration in normal ageing rat brain. *Indian J. Exp. Biol.*, **43**, 420–424.
31. Chen, C., Lang, S., Zuo, P., Yang, N., and Wang, X. (2008) Treatment with dehydroepiandrosterone increases peripheral benzodiazepine receptors of mitochondria from cerebral cortex in d-galactose-induced aged rats. *Basic Clin. Pharmacol. Toxicol.*, **103**, 493–501.
32. Bastianetto, S., Ramassamy, C., Poirier, J., and Quirion, R. (1999) Dehydroepiandrosterone (DHEA) protects hippocampal cells from oxidative stress-induced damage. *Brain Res. Mol. Brain Res.*, **66**, 35–41.
33. Kimonides, V.G., Khatibi, N.H., Svendsen, C.N., Sofroniew, M.V., and Herbert, J. (1998) Dehydroepiandrosterone (DHEA) and DHEA-sulfate (DHEAS) protect hippocampal neurons against excitatory amino acid-induced neurotoxicity. *Neurobiology*, **95**, 1852–1857.
34. Kurata, K., Takebayashi, M., Morinobu, S., and Yamawaki, S. (2004) beta-estradiol, dehydroepiandrosterone, and dehydroepiandrosterone sulfate protect against N-methyl-D-aspartate-induced neurotoxicity in rat hippocampal neurons by different mechanisms. *J. Pharmacol. Exp. Ther.*, **311**, 237–245.
35. Cardounel, A., Regelson, W., and Kalimi, M. (1999) Dehydroepiandrosterone protects hippocampal neurons against neurotoxin-induced cell death: mechanism of action. *Proc. Soc. Exp. Biol. Med.*, **222**, 145–149.
36. Kimonides, V.G., Spillantini, M.G., Sofroniew, M.V., Fawcett, J.W., and Herbert, J. (1999) Dehydroepiandrosterone antagonizes the neurotoxic effects of corticosterone and translocation of stress-activated protein kinase 3 in hippocampal primary cultures. *Neuroscience*, **89** (2), 429–436.
37. Xilouri, M. and Papazafiri, P. (2008) Induction of Akt by endogenous neurosteroids and calcium sequestration in P19 derived neurons. *Neurotox. Res.*, **13**, 209–219.
38. Leskiewicz, M., Regulska, M., Budziszewska, B., Jantas, D., Jaworska-Feil, L., Basta-Kaim, A., Kubera, M., Jagla, G., Nowak, W. and Lason, W. (2008) Effects of neurosteroids on hydrogen peroxide- and staurosporine-induced damage of human neuroblastoma SH-SY5Y Cells 1. *J. Neurosci. Res.*, **86**, 1361–1370.
39. Budziszewska, B., Siwanowicz, J., Leokiewicz, M., Jaworska-Feil, L., and Lasoń, W. (1998) Protective effects of neurosteroids against NMDA-induced seizures and lethality in mice. *Eur. Neuropsychopharmacol.*, **8**, 7–12.
40. Alhaj, H.A., Massey, A.E., and McAllister-Williams, R.H. (2006) Effects of DHEA administration on episodic memory, cortisol and mood in healthy young men: a double-blind, placebo-controlled study. *Psychopharmacology*, **188**, 541–551.
41. Zhang, X., Dong, Y., Yang, N., Liu, Y., Gao, R. and Zuo, P. (2007) Effects of ning shen ling granule and dehydroepiandrosterone on cognitive function in mice undergoing chronic mild stress. *Chin. J. Integr. Med.*, **13**, 46–49.
42. Flood, J.F., Smith, G.E., and Roberts, E. (1988) Dehydroepiandrosterone and its sulfate enhance memory retention in mice. *Brain Res.*, **447**, 269–278.
43. Urani, A., Privat, A., and Maurice, T. (1998) The modulation by neurosteroids of the scopolamine-induced learning impairment in mice involves an interaction with sigma 1 receptors. *Brain Res.*, **799**, 64–77.
44. Markowski, M., Ungeheuer, M., Bitran, D., and Locurto, C. (2001) Memory-enhancing effects of DHEAS in aged mice on a win-shift water escape task. *Physiol. Behav.*, **72** (4), 521–525.
45. Chen, L., Dai, X.N., and Sokabe, M. (2006) Chronic administration of dehydroepiandrosterone sulfate (DHEAS) primes for facilitated induction of long-term potentiation via sigma 1 receptor: optical imaging study in rat

hippocampal slices. *Neuropharmacology,* **50,** 380–392.

46. Chen, L., Miyamoto, Y., Furuya, K., Dai, X.N., Mori, N., and Sokabe, M. (2006) Chronic DHEAS administration facilitates hippocampal long-term potentiation via an amplification of Src-dependent NMDA receptor signaling. *Neuropharmacology,* **51,** 659–670.

47. Kaminska, M., Harris, J., Gijsbers, K., and Dubrovsky, B. (2000) Dehydroepiandrosterone sulfate (DHEAS) counteracts decremental effects of corticosterone on dentate gyrus LTP. Implications for depression. *Brain Res. Bull.,* **52,** 229–234.

48. Hajszan, T., MacLusky, N.J., and Leranth, C. (2004) Dehydroepiandrosterone increases hippocampal spine synapse density in ovariectomized female rats. *Endocrinology,* **145,** 1042–1045.

49. MacLusky, N.J., Hajszan, T., and Leranth, C. (2004) Effects of dehydroepiandrosterone and flutamide on hippocampal CA1 spine synapse density in male and female rats: implications for the role of androgens in maintenance of hippocampal structure. *Endocrinology,* **145,** 4154–4161.

50. Meyer, J., Lee, S., Wittenberg, G., Randall, R., and Gruol, D. (1999) Neurosteroid regulation of inhibitory synaptic transmission in the rat hippocampus in vitro. *Neuroscience,* **90,** 1177–1183.

51. Meyer, J. and Gruol, D. (1994) Dehydroepiandrosterone sulfate alters synaptic potentials in area CA1 of the hippocampal slice. *Brain Res.,* **633,** 253–261.

52. Steffensen, S., Jones, M., Hales, K., and Allison, D. (2006) Dehydroepiandrosterone sulfate and estrone sulfate reduce GABA-recurrent inhibition in the hippocampus via muscarinic acetylcholine receptors. *Hippocampus,* **16,** 1080–1090.

53. Iwata, M., Muneoka, K.T., Shirayama, Y., Yamamoto, A., and Kawahara, R. (2005) A study of a dendritic marker, microtubule-associated protein 2 (MAP-2), in rats neonatally treated neurosteroids, pregnenolone and dehydroepiandrosterone (DHEA). *Neurosci. Lett.,* **386,** 145–149.

54. Karishma, K.K. and Herbert, J. (2002) Dehydroepiandrosterone (DHEA) stimulates neurogenesis in the hippocampus of the rat, promotes survival of newly formed neurons and prevents corticosterone-induced suppression. *Eur. J. Neurosci.,* **16,** 445–453.

55. Shirayama, Y., Muneoka, K., Iwata, M., Ishida, H., Hazama, G., and Kawahara, R. (2005) Pregnenolone and dehydroepiandrosterone administration in neonatal rats alters the immunoreactivity of hippocampal synapsin I, neuropeptide Y and glial fibrillary acidic protein at post-puberty. *Neuroscience,* **133,** 147–157.

56. Lewis, S.J., and Barker, R.A. (2009) Understanding the dopaminergic deficits in Parkinson's disease: insights into disease heterogeneity. *J. Clin. Neurosci.,* **16,** 620–625.

57. Tatton, N. and Kish, J. (1997) In situ detection of apoptotic nuclei in the substantia nigra compacta of MPTP-treated mice using terminal deoxynucleotidyl transferase labelling and acridine orange staining. *Neuroscience,* **77,** 1037–1048.

58. Kuhn, K., Wellen, J., Link, N., Maskri, L., Lubbert, H. and Stickel, C.C. (2003) The mouse MPTP model: gene expression changes to dopaminergic neurons. *Eur. J. Neurosci.,* **17,** 1–12.

59. Schmidt, N. and Ferger, B. (2001) Neurochemical findings in the MPTP model of Parkinson's disease. *J. Neural Transm.,* **108,** 1263–1282.

60. D'Astous, M., Morissette, M., Tanguay, B., Callier, S., and Di Paolo, T. (2003) Dehydroepiandrosterone (DHEA) such as 17beta-estradiol prevents MPTP-induced dopamine depletion in mice. *Synapse,* **47,** 10–14.

61. Belanger, N., Gregoire, L., Bedard, P., and Di Paolo, T. (2006) DHEA improves symptomatic treatment of moderately and severely impaired MPTP monkeys. *Neurobiol. Aging,* **27,** 1684–1693.

62. Bélanger, N., Grégoire, L., Bédard, J., and Di Paolo, T. (2003) Estradiol and dehydroepiandrosterone potentiate levodopa-induced locomotor activity in 1-methyl-4-phenyl-1,2,3,6 tetrahydropyridine monkeys. *Endocrine*, **21**, 97–101.

63. Tomas-Camardiel, M., Sanchez-Hidalgo, M., Sanchez del Pino, M., Navarro, A., Machado, A., and Cano, J. (2002) Comparative study of the neuroprotective effect of dehydroepiandrosterone and 17 beta-estradiol against 1-methyl-4-phenylpyridinium toxicity on rat striatum. *Neuroscience*, **109**, 569–584.

64. Kaasik, A., Kalda, A., Jaako, K., and Zharkowski, A. (2001) DHEAS prevents oxygen-glucose deprivation-induced injury in cerebellar granule cell cultures. *Neuroscience*, **102**, 427–432.

65. Charalampopoulos, I., Dermitzaki, E., Vardouli, L., Tsatsanis, C., Stournaras, C., Margioris, A., and Gravanis, A. (2005) Dehydroepiandrosterone sulfate and allopregnanolone directly stimulate catecholamine production via induction of tyrosine hydroxylase and secretion by affecting actin polymerization. *Endocrinology*, **146**, 3309–3318.

66. Dong, L., Chenga, Z., Fua, Y.-M., Wanga, Z., Zhua, Y.-H., Suna, J., Donga, Y., and Zheng, P. (2007) Neurosteroid dehydroepiandrosterone sulfate enhances spontaneous glutamate release in rat prelimbic cortex through activation of dopamine D1 and sigma-1 receptor. *Neuropharmacology*, **52**, 966–974.

67. Muneoka, K., Iwata, M. and Shirayama, Y. (2009) Altered levels of synapsin I, dopamine transporter, dynorphin A, and neuropeptide Y in the nucleus accumbens and striatum at post-puberty in rats treated neonatally with pregnenolone or DHEA. *Int. J. Dev. Neurosci.*, **27**, 575–581.

68. Téllez, N., Comabella, M., Julià, E., Río, J., Tintoré, M., Brieva, L., Nos, C., and Montalban, X. (2006) Fatigue in progressive multiple sclerosis is associated with low levels of dehydroepiandrosterone. *Mult. Scler.*, **12**, 487–494.

69. Regelson, W., Loria, R. and Kalimi, M. (1994) Dehydroepiandrosterone (DHEA) – the "mother steroid". I. Immunologic action. *Ann. N. Y. Acad. Sci.*, **19**, 553–563.

70. Tan, X.D., Dou, Y., Shi, C., Duan, R. and Sun, R. (2009) Administration of dehydroepiandrosterone ameliorates experimental autoimmune neuritis in Lewis rats. *J. Neuroimmunol.*, **207**, 39–44.

71. Du, C., Khalil, M.-W., and Sriram, S. (2001) Administration of dehydroepiandrosterone suppresses experimental allergic encephalomyelitis in SJL/J mice. *J. Immunol.*, **167**, 7094–7101.

72. Offner, H., Zamora, A., Drought, H., Matejuk, A., Auci, D.-L., Morgan, E.-E., Vandenbark, A.-A., and Reading, C.-L. (2002) A synthetic androstene derivative and a natural androstene metabolite inhibit relapsing–remitting EAE. *J. Neuroimmunol.*, **130**, 113–128.

73. Wang, M., Huang, H., Chen, H., Kuo, J., and Jeng, K. (2001) Dehydroepiandrosterone inhibits lipopolysaccharide-induced nitric oxide production in BV-2 microglia. *J. Neurochem.*, **77**, 830–838.

74. Di Santo, E., Foddi, M.C., Ricciardi-Castagnoli, P., Mennini, T., and Ghezzi, P. (1996) DHEAS inhibits TNF production in monocytes, astrocytes and microglial cells. *Neuroimmunomodulation*, **3**, 285–288.

75. Muntwyler, R. and Bologa, L. (1989) In vitro hormonal regulation of astrocyte proliferation. *Schweiz. Arch. Neurol. Psychiatr.*, **140**, 29–33.

76. Bologa, L. and Sharma, J. (1987) Dehydroepiandrosterone and its sulfate derivative reduce neuronal death and enhance astrocytic differentiation in brain cell cultures. *J. Neurosci. Res.*, **17**, 225–234.

77. Greve, M.W., and Zink, B.J. (2009) Pathophysiology of traumatic brain injury. *Mt. Sinai J. Med.*, **76**, 97–104.

78. Li, Z., Cui, S., Zhou, R., Ge, Y., Sokabe, M., and Chen, L. (2009) DHEA-neuroprotection and -neurotoxicity after transient

79. Hsu, Y., Lapchak, P., Chapman, D., Nunez, S., Zivin, J., and Chung, P. (2000) Possible involvement of GABAA receptors ischemia model: possible involvement of GABAA receptors editorial dehydroepiandrosterone sulfate is neuroprotective in a reversible spinal cord. *Stroke*, **31**, 1953–1957.

80. Bucolo, C. and Drago, F. (2004) Effects of neurosteroids on ischemia-reperfusion injury in the rat retina: role of sigma1 recognition sites. *Eur. J. Pharmacol.*, **498**, 111–114.

81. Tabuchi, K., Oikawa, K., Uemaetomari, I., Tsuji, S., Wada, T., and Hara, A. (2003) Glucocorticoids and dehydroepiandrosterone sulfate ameliorate ischemia-induced injury of the cochlea. *Hear. Res.*, **180**, 51–56.

82. Aragno, M., Parola, S., Brignardello, E., Mauro, M., Tamagno, E., Manti, R., Danni, O., and Boccuzzi, G. (2000) Dehydroepiandrosterone prevents oxidative injury induced by transient ischemia/reperfusion in the brain of diabetic rats. *Diabetes*, **49**, 1924–1931.

83. Juhasz-Vedres, G., Rozsa, E., Rakos, G., Dobszay, M.-B., Kis, Z., Wolfling, J., Toldi, J., Parducz, R., and Farkas, F. (2006) Dehydroepiandrosterone sulfate is neuroprotective when administered either before or after injury in a focal cortical cold lesion model. *Endocrinology*, **147**, 683–686.

84. Milman, A., Zohar, O., Maayan, R., Weizman, R., and Pick, C. (2008) DHEAS repeated treatment improves cognitive and behavioral deficits after mild traumatic brain injury. *Eur. Neuropsychopharmacol.*, **18**, 181–187.

85. Lür, G., Rákos, G., Juhász-Vedres, G., Farkas, T., Kis, Z., and Toldi, J. (2006) Effects of dehydroepiandrosterone sulfate on the evoked cortical activity of controls and of brain-injured rats. *Cell. Mol. Neurobiol.*, **26**, 1505–1519.

86. Fiore, C., Inman, D.M., Hirose, S., Noble, L.-J., Igarashi, T. and Compagnone, N.A. (2004) Treatment with the neurosteroid dehydroepiandrosterone promotes recovery of motor behavior after moderate contusive spinal cord injury in the mouse. *J. Neurosci. Res.*, **75**, 391–400.

87. Hoffman, S.W., Virmani, S., Simkins, R.M., and Stein, D.G. (2003) The delayed administration of dehydroepiandrosterone sulfate improves recovery of function after traumatic brain injury in rats. *J. Neurotrauma.*, **20**, 859–870.

88. Ayhan, S., Markal, N., Siemionow, K., Araneo, B., and Siemionow, M. (2003) Effect of subepineurial dehydroepiandrosterone treatment on healing of transected nerves repaired with the epineurial sleeve technique. *Microsurgery*, **23**, 49–55.

89. García-Estrada, J., Luquín, S., Fernández, A.M. and Garcia-Segura, L.M. (1999) Dehydroepiandrosterone, pregnenolone and sex steroids down-regulate reactive astroglia in the male rat brain after a penetrating brain injury. *Int. J. Dev. Neurosci.*, **17**, 145–151.

90. Del Cerro, S., Garcia-Estrada, J., and Garcia-Segura, L.M. (1996) Neurosteroids modulate the reaction of astroglia to high extracellular potassium levels. *Glia*, **18** (4), 293–305.

91. Malik, A., Narayan, R., Wendling, W., Cole, R., Pashko, L., Schwartz, A., and Strauss, K. (2003) A novel dehydroepiandrosterone analog improves functional recovery in a rat traumatic brain injury model. *J. Neurotrauma.*, **20**, 463–476.

92. Liu, D. and Dillon, J.S. (2002) Dehydroepiandrosterone activates endothelial cell nitric-oxide synthase by a specific plasma membrane receptor coupled to Galpha(i2,3). *J. Biol. Chem.*, **277**, 21379–21388.

93. Liu, D., Ren, M., Bing, X., Stotts, C., Deorah, S., Love-Homan, L., and Dillon, J.S. (2006) Dehydroepiandrosterone inhibits intracellular calcium release in beta-cells by a plasma membrane-dependent mechanism. *Steroids*, **71**, 691–699.

94. Liu, D., Si, H., Reynolds, K., Zhen, W., Jia, Z., and Dillon, J.S. (2007) Dehydroepiandrosterone protects vascular endothelial cells against apoptosis

through a Galphai protein-dependent activation of phosphatidylinositol 3-kinase/Akt and regulation of antiapoptotic Bcl-2 expression. *Endocrinology*, **148**, 3068–3076.
95. Yan, C., Jiang, X., Pei, X., and Dai, Y.R. (1999) The in vitro antiapoptotic effect of dehydroepiandrosterone sulfate in mouse thymocytes and its relation to caspase-3/caspase-6. *Cell. Mol. Life Sci.*, **56**, 543–547.
96. Takahashi, H., Nakajima, A., and Sekihara, H. (2004) Dehydroepiandrosterone (DHEA) and its sulfate (DHEAS) inhibit the apoptosis in human peripheral blood lymphocytes. *J. Steroid Biochem. Mol. Biol.*, **88**, 261–264.
97. Liang, J., Yao, G., Yang, L., and Hou, Y. (2004) Dehydroepiandrosterone induces apoptosis of thymocyte through Fas/Fas-L pathway. *Int. Immunopharmacol.*, **4**, 1467–1475.
98. Alexaki, V.I., Charalampopoulos, I., Panayotopoulou, M., Kampa, M., Gravanis, A., and Castanas, E. (2009) Dehydroepiandrosterone protects human keratinocytes against apoptosis through membrane binding sites. *Exp. Cell Res.*, **315**, 2275–2283.
99. Gil-ad, I., Shtaif, B., Eshet, R., Maayan, R., Rehavi, M., and Weizman, A. (2001) Effect of dehydroepiandrosterone and its sulfate metabolite on neuronal cell viability in culture. *Isr. Med. Assoc. J.*, **3**, 639–643.
100. Charalampopoulos, I., Tsatsanis, C., Dermitzaki, E., Alexaki, I., Castanas, E., Margioris, A.-N., and Gravanis, A. (2004) Dehydroepiandrosterone and allopregnanolone protect sympathoadrenal cells against apoptosis, via Bcl-2 antiapoptotic proteins. *Proc. Natl. Acad. Sci. U.S.A.*, **101**, 8209–8221.
101. Charalampopoulos, I., Alexaki, I.-V., Lazaridis, I., Dermitzaki, E., Avlonitis, N., Tsatsanis, C., Calogeropoulou, T., Margioris, A., Castanas, E., and Gravanis, A. (2006) G-protein-associated membrane binding sites mediate the neuroprotective effect of Dehydroepiandrosterone. *FASEB J.*, **20**, 577–579.
102. Reichardt, L.F. (2006) Neurotrophin-regulated signaling pathways. *Phil. Trans. R. Soc. B.*, **361**, 1545–1564.
103. Ulmann, L., Rodeau, J., Danoux, L., Contet-Audonneau, J.L., Pauly, G., and Schlichter, R. (2009) Dehydroepiandrosterone and neurotrophins favor axonal growth in a sensory neuron-keratinocyte coculture model. *Neuroscience*, **159**, 514–525.
104. Zhang, B., Li, W., Ma, J., Barker, Y., Chang, W., Zhao, W., and Rubinow, D.R. (2002) Dehydroepiandrosterone (DHEA) and its sulfate derivative (DHEAS) regulate apoptosis during neurogenesis by triggering the Akt signaling pathway in opposing ways. *Brain Res. Mol. Brain Res.*, **98**, 58–66.
105. Suzuki, M., Wright, L., Marwah, P., Lardy, H., and Svendsen, C. (2004) Mitotic and neurogenic effects of dehydroepiandrosterone (DHEA) on human neural stem cell cultures derived from the fetal cortex. *Proc. Natl. Acad. Sci. U.S.A.*, **101**, 3202–3207.
106. Azizi, H., Mehrjardi, N.Z., Shahbazi, E., Hemmesi, K., Bahmani, M.K., and Baharvand, H. (2010) Dehydroepiandrosterone stimulates neurogenesis in mouse embryonal carcinoma cell- and human embryonic stem cell-derived neural progenitors and induces dopaminergic neurons. *Stem Cells Dev*, **19**, 809–819.
107. Compagnone, N.A. and Mellon, S.H. (1998) Dehydroepiandrosterone: a potential signalling molecule for neocortical organization during development. *Proc. Natl. Acad. Sci. U.S.A.*, **95**, 4678–4683.
108. Maninger, N., Wolkowitz, O.M., Reus, V.I., Epel, E.S., and Mellon, S.H. (2009) Neurobiological and neuropsychiatric effects of dehydroepiandrosterone (DHEA) and DHEA sulfate (DHEAS). *Front. Neuroendocrinol.*, **30**, 65–91.
109. Bazin, M.A., El Kihel, L., Boulouard, M., Bouët, V., and Rault, S. (2009) The effects of DHEA, 3 beta-hydroxy-5 alpha-androstane-6,17-dione, and 7-amino-DHEA analogues on short term and long term

memory in the mouse. *Steroids*, **74**, 931–937.

110. Calogeropoulou, T., Avlonitis, N., Minas, V., Pantzou, A., Alexi, X., Charalampopoulos, I., Zervou, M., Vergou, V., Lazaridis, I., Alexis, M.-N., and Gravanis, A. (2009) Novel dehydroepiandrosterone derivatives with anti-apoptotic, neuroprotective activity. *J. Med. Chem.*, **52**, 6569–6587.

9
Neurosteroids and Pain

Christine Patte-Mensah, Laurence Meyer, Véronique Schaeffer, Cherkaouia Kibaly, and Ayikoe G. Mensah-Nyagan

9.1
Introduction

Several investigations conducted in humans and animals have clearly demonstrated that the nervous system is capable of synthesizing bioactive steroids, also called *neurosteroids*. The process of neurosteroidogenesis occurs either in glial cells, in neurons, or within a cross talk between both cell types. The high conservation of neurosteroidogenesis through the vertebrate phylum reasonably suggests that neurosteroid production might be a pivotal mechanism for life. Using the brain as the exclusive center for neurosteroidogenesis, pharmacological and behavioral studies revealed that neurosteroids regulate neurobiological mechanisms such as cognition, stress, anxiety, depression, and neuroprotection. For a long time, the spinal cord (SC) has received little attention although this structure of the central nervous system (CNS) crucially controls neurophysiological processes including nociceptive and pain transmission, motor activities, reflexes, and neurovegetative functions. Intensive studies were recently performed using the rodent SC as a model to investigate the occurrence of neurosteroidogenesis in spinal neural networks. These studies revealed that the SC's dorsal horn (DH), which pivotally regulates pain transmission, contains several steroid-synthesizing enzymes, and actively produces various neurosteroids. The present chapter recapitulates current knowledge on neurosteroidogenesis in the SC and discusses the regulatory action exerted on pain sensation by neurosteroids endogenously produced in DH neural networks.

9.2
General Background on Neurosteroids

The term *neurosteroid* designates bioactive steroids endogenously synthesized in neurons and/or glial cells [1, 2]. The chemical structure of neurosteroids is not necessarily different from that of hormonal steroids. However, the main criterion

---

Hormones in Neurodegeneration, Neuroprotection, and Neurogenesis.
Edited by Achille G. Gravanis and Synthia H. Mellon
Copyright © 2011 WILEY-VCH Verlag GmbH & Co. KGaA, Weinheim
ISBN: 978-3-527-32627-3

required to consider an endogenous steroid as a neurosteroid is its production in the CNS or peripheral nervous system (PNS), independently from the activity of endocrine glands such as the adrenals and gonads. In addition to the genomic action generally used by all steroidal hormones, neurosteroids modulate the nervous system's activity in a paracrine or autocrine manner by acting through various membrane receptors including the GABA$_A$, NMDA, P2X, and sigma receptors [1–4]. Different categories of members can be distinguished in the neurosteroid family [5]. The nonexclusive neurosteroids such as pregnenolone (PREG), progesterone (PROG), or dehydroepiandrosterone (DHEA) are steroidal hormones that can also be synthesized by neurons or glial cells. Semiexclusive neurosteroids such as tetrahydroprogesterone (3α,5α-THP) also called *allopregnanolone* are mainly synthesized in the nervous system even if substantial amounts can be produced in the endocrine glands. A recent work has suggested the existence of exclusive neurosteroids such as epiallopregnanolone which may only be synthesized in nerve cells [6]. The demonstration of the ability of a neural structure to produce neurosteroids requires the localization of active forms of key steroidogenic enzymes such as cytochrome P450side-chain-cleavage (P450scc), cytochrome P450c17 (P450c17), 3α-hydroxysteroid oxidoreductase (3α-HSOR), 3β-hydroxysteroid dehydrogenase, and 5α-reductase (5α-R) in that structure [1, 7–9]. Among these enzymes, P450scc plays a crucial role since it catalyzes cholesterol (CHOL) conversion into PREG, the first pivotal step for neurosteroid biosynthesis [1, 10]. Occurrence of neurosteroid formation has been investigated in several animal species and it appeared that the process of neurosteroidogenesis is well conserved through the vertebrate phylum [1–3, 10]. This observation, which suggests that neurosteroidogenesis might be crucial for life, raises a great hope for the development of novel therapies based on the use of neurosteroids to improve the treatment of various neural disorders.

9.3
Overview on Pain

Pain is the first reason for medical consultation in many diseases. The International Association for the Study of Pain (IASP) defines pain as an unpleasant sensory and emotional experience associated with actual or potential tissue damage, or described in terms of such damage. This unpleasant sensation of hurt, discomfort, or distress may be a useful response of the organism like an early warning system that promotes survival in a dangerous environment. Pain could also be an expression of pathological changes in the nervous system. The former pain is beneficial while the latter is responsible for persistent suffering in millions of patients with a substantial cost to society due to disability, medical expenses, loss of work, decrease of productivity, and so on. Although pain is not homogeneous, three categories are generally considered in the literature: physiological, inflammatory, and neuropathic pain (for reviews see [11–14]). The conventional or classical anatomical circuit of pain extensively described in dozens of reviews and books implies peripheral nociceptors which are essential for the perception of pain. These terminals belong to primary sensory neurons whose cell bodies are located in the dorsal root ganglia. Central axons of primary afferents terminate in the SC DH and second-order spinal neurons projecting to the brain often have convergent inputs from different sensory fibers and tissues. The SC appears to be a pivotal structure in pain transmission and it is also well known that different brain areas are associated with various aspects of pain. For instance, the somatosensory areas 1 and 2 integrate the location and intensity of pain stimuli, whereas the anterior cingulate cortex, frontal cortex, and anterior insula regions are involved in cognitive and emotional components of pain. While anatomical structures and neural centers involved in pain control are well identified, molecular and neurochemical components of pain modulation deserve further clarification even though substantial progress has been made over the two last decades [14–24]. In fact, molecular processes involved in pain are only partially identified because pain, especially pathological pains resistant to the currently available analgesics, is probably characterized by multiple and sophisticated cellular mechanisms [25–30]. However, it is noteworthy that most of the pain processes which have already been identified seem to be subjected to an expression of neural plasticity or to the capacity of neurons to change their function, chemical profile, structure, or to trigger apoptotic processes, particularly in chronic pain states [31]. Therefore, preventing neural modifications responsible for chronic pain has now become a real challenge for biomedical research. Several families of endogenous compounds modulating neural activity and plasticity are suspected to play a crucial role in the mechanisms determining pain sensation. For instance, it is well documented that various neurotransmitters such as glutamate, substance P, serotonin, gamma-aminobutyric acid (GABA) are involved in the regulation of nociceptive processes leading to pain sensation [14, 18]. It has also been demonstrated that inflammatory and/or neuropathic pain may depend on the action of various cytokines and other molecules, including eicosanoids, endorphins, calcitonin gene–related peptide, free radicals, and transcription factors [32–34]. Because steroids control the development, activities, and plasticity of the nervous system, these compounds are of great interest in the modulation of pain [35–37]. In particular, several data have recently indicated that neurosteroids endogenously synthesized in the SC may crucially regulate pain sensation.

9.4
Involvement of Endogenous Neurosteroids in the Control of Pain

9.4.1
Evidence for the Local Production of Neurosteroids in the Spinal Circuit

The production of endogenous steroids in the SC has been proved by various studies which demonstrated the presence and biological activity of several key steroid-synthesizing enzymes in the rat SC. Among these enzymes are cytochrome P450scc, P450c17, 3β-hydroxysteroid dehydrogenase (3β-HSD), 5α-R, and 3α-HSOR (Figure 9.1). P450scc catalyzes the conversion of CHOL to PREG, the

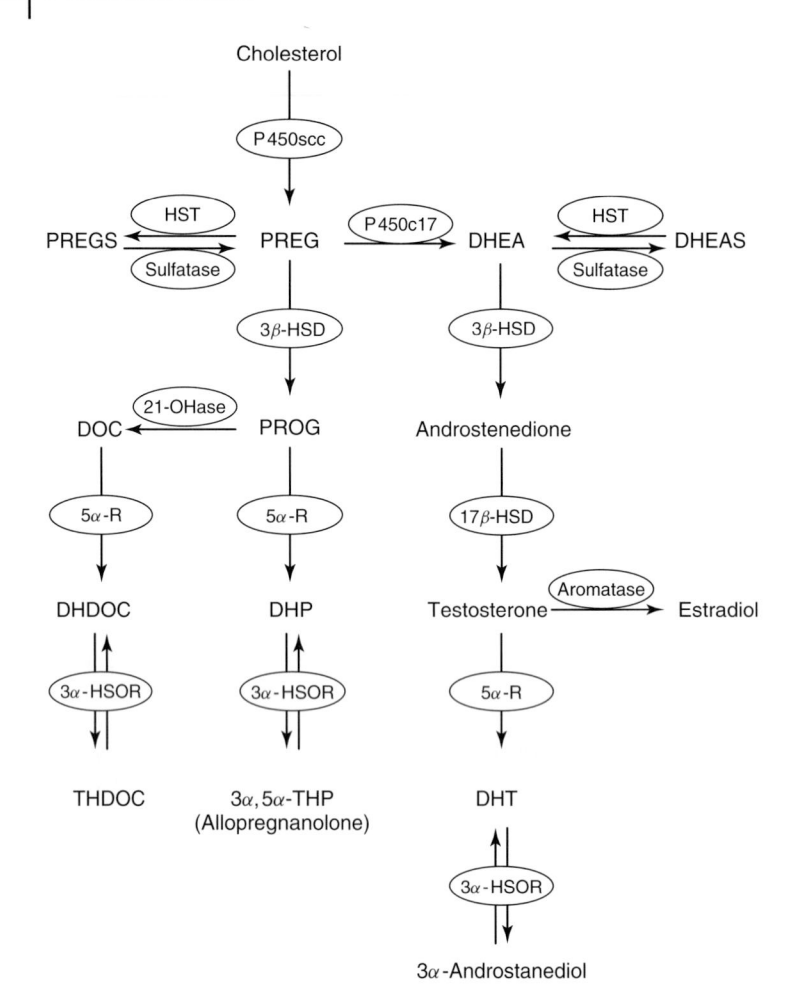

Figure 9.1 Biosynthetic pathways of neurosteroids. P450scc, cytochrome P450side-chain cleavage; P450c17, cytochrome P450c17 or 17α-hydroxylase/17,20 lyase; 3β-HSD, 3β-hydroxysteroid dehydrogenase; 21-OHase, 21-hydroxylase; 5α-R, 5α-reductase; 3α-HSOR, 3α-hydroxysteroid oxidoreductase; 17β-HSD, 17β-hydroxysteroid dehydrogenase; HST, hydroxysulfotransferase; PREG, pregnenolone; PREGS, pregnenolone sulfate; DHEA, dehydroepiandrosterone; DHEAS, dehydroepiandrosterone sulfate; PROG, progesterone; DOC, deoxycorticosterone; DHDOC, dihydrodeoxycorticosterone; THDOC, tetrahydrodeoxycorticosterone; DHP, dihydroprogesterone; 3α,5α-THP, 3α,5α-tetrahydroprogesterone or allopregnanolone; DHT, dihydrotestosterone.

first and rate-limiting step in the biosynthesis of all classes of steroid hormones. P450c17, also called *17α-hydroxylase/17,20 lyase*, converts PREG successively into 17-hydroxy-PREG and DHEA. P450c17 is also responsible for the transformation of PROG into 17-hydroxy-PROG and androstenedione, successively. The enzyme 3β-HSD catalyzes the conversion of Δ^5-3β-hydroxysteroids (PREG, 17-hydroxy-PREG, DHEA) into Δ^4-3-ketosteroids (PROG, 17-hydroxy-PROG, androstenedione). 5α-R is responsible for the transformation of testosterone (T), PROG, and deoxycorticosterone (DOC) into dihydrotestosterone (DHT), dihydroprogesterone (DHP), and dihydrodeoxycorticosterone (DHDOC), respectively. 3α-HSOR also called *3α-hydroxysteroid dehydrogenase* converts in a reversible manner DHT, DHP, and DHDOC into the respective neuroactive steroids 3α-androstanediol, allopregnanolone or 3α,5α-THP, and tetrahydrodeoxycorticosterone (THDOC).

The first anatomical and cellular distribution of P450scc in the adult rat SC was provided by immunohistochemical studies using two different antibodies against P450scc. One of these antisera was raised in rabbit against purified P450scc from bovine adrenocortical mitochondria [38–42]. The other antiserum was generated in rabbit against the carboxy-terminal amino acids 509–526 of rat P450scc [43]. The same anatomical and cellular distribution of P450scc-immunoreactivity was observed in the rat SC with both antisera. The highest density of P450scc-immunolabeling was found in superficial layers laminae I and II of DH where sensory neurons are located [14, 44–46]. Double-labeling experiments revealed that most of the P450scc-positive fibers in the DH also expressed immunoreactivity for microtubule-associated protein-2, a specific marker for neuronal fibers [45, 47, 48]. Motoneurons of the ventral horn (VH) also expressed immunoreactivity for P450scc, suggesting a possible role of the enzyme or its steroid products in the control of motor activity [44, 45]. Moreover, P450scc-immunostaining was detected in ependymal glial cells bordering the central canal in the SC, an observation which suggests a possible release of neurosteroids in the cerebrospinal fluid and their involvement in volume transmission mechanisms in the CNS [45, 49].

Well-validated biochemical experiments, which showed that homogenates from the adult rat SC are capable of converting CHOL into PREG (Figure 9.1), indicated that P450scc-like immunoreactivity detected in the spinal tissue corresponds to an active form of the enzyme [7, 8, 45, 50–52].

The demonstration of the presence and activity of P450c17 in the CNS has long remained controversial (for reviews see [1, 2]). Therefore, we combined molecular, immunohistochemical, and neurochemical approaches for a solid investigation of P450c17's existence and biological activity in adult rodent SC. This multitechnique study allowed the first anatomical and cellular mapping of a biologically active form of P450c17 in the adult rat SC [53]. Significant amounts of P450c17 mRNA were detected in all regions of the SC using the real-time polymerase chain reaction approach after reverse transcription (RT-rtPCR). By taking advantage of the availability of an antibody against P450c17, we revealed the presence of a specific protein in total homogenates and microsomal fractions from the rat SC and testis. The P450c17 antiserum used in our studies was also efficient in previous investigations which localized the enzyme in Leydig cells [54, 55]. This antiserum also allowed the anatomical and cellular localization of P450c17 throughout the white and gray matters of the SC when we used immunohistochemical approach combined with confocal laser microscope analysis. P450c17-immunostaining was found in both neurons and glial cells: in the white matter, the enzyme was mainly detected in astrocytes while in the gray matter, P450c17 was essentially found in neurons and oligodendrocytes [53]. The presence of P450c17 in the DH and VH suggested its potential involvement in the modulation of sensory or motor functions [14, 44, 46, 53]. Pulse-chase experiments, which revealed that SC slices converted [^{3}H]PREG into [^{3}H]DHEA (Figure 9.1), indicated that P450c17-like immunoreactivity detected in the adult rat SC corresponds to an active form of the enzyme [53]. The occurrence of P450c17 enzymatic activity was further demonstrated with biochemical experiments using ketoconazole, a selective inhibitor of the enzyme [56, 57]. A significant decrease was observed in the conversion of [^{3}H]PREG to [^{3}H]DHEA by SC slices when the pulse-chase experiments were performed in the presence of ketoconazole, a result which unambiguously confirms the existence of P450c17 activity in the adult rat SC [53].

The first isolation of 3β-HSD mRNA in the SC was performed in rat using the RT-PCR approach [58]. However, the anatomical and cellular distribution of 3β-HSD mRNA in the SC was provided by Coirini et al. [59] utilizing the in situ hybridization technique. This study revealed that the DH laminae I–III exhibited the highest density of 3β-HSD mRNAs which were also detected in layer X around the central canal, in the VH and in the lateral as well as ventral funiculi. At the cellular level, 3β-HSD mRNAs were found mainly in sensory neurons of the DH and in motoneurons of the VH throughout the cervical, thoracic, lumbar, and sacral segments of the SC [59]. Moreover, evidence for the existence of 3β-HSD protein and enzymatic activity in the SC was provided by Western blot analysis and gas chromatography/mass spectrometry assays which revealed that the concentrations of PREG and PROG were higher in the SC than in the plasma [59]. Recent studies have also confirmed the presence and activity of 3β-HSD in the rat SC by using real-time polymerase chain reaction and pulse-chase experiments combined with HPLC-Flo/one analysis of steroids newly synthesized from a radioactive precursor in spinal tissue [60].

The expression of 5α-R in the brain, but not the SC, has extensively been studied [61–66]. It has been suggested that the isoenzyme 5α-R type 1 (5α-R1) essentially plays a catabolic and neuroprotective role, whereas the isoform 2 or 5α-R2 participates in sexual differentiation of the CNS. However, the neurophysiological significance of these two isoenzymes remains a matter of speculation [67–70]. The first demonstration of 5α-R gene expression in the SC was provided by a recent study which revealed that, unlike to what is observed in the brain, the quantity of 5α-R2 mRNAs extracted from the whole adult rat SC is higher than that of 5α-R1 [71]. This work also indicated that mRNAs encoding 5α-R2 are expressed by motoneurons of the VH but did not provide any information about the presence or absence of the enzyme in the DH where sensory networks are located [14, 46, 72]. Therefore, a detailed immunohistochemical study was performed to determine the regional and cellular distribution of 5α-R1 and 5α-R2 in the adult rat SC [73]. The study was possible due to the availability of highly specific antisera against 5α-R1 and 5α-R2, which were previously used to successfully localize these enzymes in various steroidogenic tissues [74, 75]. Immunoreactivities for 5α-R1 and 5α-R2 were detected in the white matter of the SC from the cervical to sacral regions.

However, the intensity of 5α-R1-immunostaining was low and cell bodies as well as fibers containing this isoenzyme were observed mainly in the white matter of the cervical and thoracic segments. The 5α-R2 immunofluorescence, which was moderate in the white matter, was intense in the DH and VH of the gray matter [73]. Double-labeling identification with specific markers for nerve cells revealed that the 5α-R1 immunostaining was mainly expressed in oligodendrocytes and astrocytes of the white matter, whereas 5α-R2-immunolabeling colocalized with neurons and glial cells in the gray and white matters [47, 48, 73, 76, 77]. The observation of a restricted localization of 5α-R1 to the SC white matter is in agreement with previous studies indicating that the type 1 isoform of 5α-R is the most relevant isoenzyme present in myelinated structures of the male and female rat brain [68, 78–80].

There are four human 3α-HSOR isozymes, but, to date, only one isoform has been cloned in rats [81–84]. The enzymatic activity and mRNA encoding 3α-HSOR have been detected in the brain but the immunocytochemical mapping of the protein in the CNS has long remained unexplored [66, 85–87]. Taking advantage of the availability of a specific antiserum against the rat liver 3α-HSOR, we have recently determined the anatomical and cellular distribution of the enzyme in the rodent SC [73]. Relative titers, specificity, and effectiveness of the 3α-HSOR antibody have been shown by various biochemical and histochemical studies [81–83, 88, 89]. Intense immunoreactivity for 3α-HSOR was detected in white and gray matters of all SC segments. However, the highest density of 3α-HSOR-immunostaining was found in sensory areas of the DH [73]. Our study also revealed that 45% of 3α-HSOR-immunofluorescence was localized in oligodendrocytes, 35% in neurons, and 20% in astrocytes. A comparative analysis of 5α-R1-, 5α-R2-, and 3α-HSOR-positive elements in the SC made it possible to observe three different but interesting situations: (i) cell bodies and fibers containing both 3α-HSOR and 5α-R were identified; (ii) cells labeled only with the 5α-R1 or 5α-R2 antiserum were localized; and (iii) positive cell bodies expressing only 3α-HSOR-immunostaining were found [73]. Consequently, it appears that certain glial cells and neurons of the SC contain both 5α-R and 3α-HSOR enzymatic proteins which could catalyze biochemical reductions required for the biosynthesis of 3α,5α-reduced steroids such as 3α-androstanediol, 3α,5α-THP, and THDOC which control, through allosteric modulation of GABA$_A$ receptors, neurobiological mechanisms including stress, anxiety, analgesia, locomotion, and lordosis [90–94]. The production of neuroactive 3α,5α-reduced steroids may also involve collaboration among neurons, astrocytes, and oligodendrocytes which contain only one of the two enzymes; that is, 3α-HSOR or 5α-R. This collaboration may be done within the context of the cross talk between glial and neuronal elements in normal physiological or during pathological situations [95]. These suggestions could not rule out the possibility that 5α-R1 or 5α-R2 alone may convert, in the SC, PROG or T from peripheral sources into DHP or DHT that act via genomic receptors, the existence of which has been demonstrated in spinal tissues [96–99]. In a similar manner, 3α-HSOR alone may also convert, in the SC, peripheral DHP, DHT, or DHDOC into 3α,5α-THP, 3α-androstanediol, or THDOC, respectively, for the modulation of GABA$_A$ receptors [90, 94, 100, 101].

The fact that the rat spinal tissue homogenates are capable of converting [^{3}H]CHOL into various metabolites including 3α,5α-THP clearly indicates that 5α-R1, 5α-R2, and 3α-HSOR detected in the SC correspond to active forms of these enzymes [45, 73].

9.4.2
Endogenous Neurosteroids and Pain Modulation

On the basis of the principle that neurosteroids act mainly through autocrine or paracrine mechanisms, endogenous neurosteroid involvement in the regulation of a neurobiological process is plausible when neurosteroids are locally synthesized in the neural circuit controlling this process. It is true that neurosteroids modulate GABA$_A$, NMDA, and P2X receptors which are expressed in the SC and play a crucial role in the regulation of pain [1, 14, 18, 37]. However, the local synthesis of neurosteroids near their sites of action in pain neural centers is a prerequisite to render credible the possible involvement of endogenous neurosteroids in pain modulation. Therefore, the demonstration that the SC, which pivotally controls pain transmission [14, 18, 44], also contains the enzymatic machinery to locally synthesize neurosteroids (see Section 9.4.1) was extremely important to show that neurosteroids are produced and released near their sites of actions in the pain neural circuit. In addition, we observed that substance P, a major nociceptive neuropeptide secreted by primary afferents, inhibited in a dose-dependent manner allopregnanolone (3α,5α-THP) biosynthesis in the DH [102]. As the neurosteroid 3α,5α-THP is a potent allosteric stimulator of GABA$_A$ receptors, our observation suggested that substance P, by reducing 3α,5α-THP production, may indirectly decrease the spinal inhibitory tone and therefore facilitate noxious signal transmission.

To further investigate the possible role of neurosteroids endogenously produced in the DH in pain modulation, we performed a multidisciplinary study using the rat experimental model of neuropathic pain generated by sciatic nerve ligatures [103]. Molecular and biochemical investigations (quantitative real-time polymerase chain reaction after reverse transcription, Western blot, radioimmunoassay, pulse-chase experiments, high performance liquid chromatography, and continuous flow scintillation detection) revealed an up-regulation of enzymatic pathways (P450scc and 3α-HSOR) leading to 3α,5α-THP biosynthesis in the DH (Figure 9.2) [73, 104, 105]. In contrast, the biosynthetic pathway (P450c17) producing DHEA was down-regulated in neuropathic rat DH (Figure 9.2) [53, 106]. Behavioral studies using the plantar test (thermal nociceptive threshold) and the von Frey filament test (mechanical nociceptive threshold) showed that intrathecal administration of 3α,5α-THP in the lumbar SC induced analgesia in neuropathic-pain rats by suppressing the thermal hyperalgesia and mechanical allodynia characterizing these animals. Unlike 3α,5α-THP, intrathecal injection of Provera (3α-HSOR inhibitor) potentiated both thermal hyperalgesia and mechanical allodynia in neuropathic rats [105]. Acute DHEA treatment exerted a rapid pronociceptive and a delayed antinociceptive action. Inhibition of DHEA biosynthesis in the DH by intrathecally

9.4 Involvement of Endogenous Neurosteroids in the Control of Pain

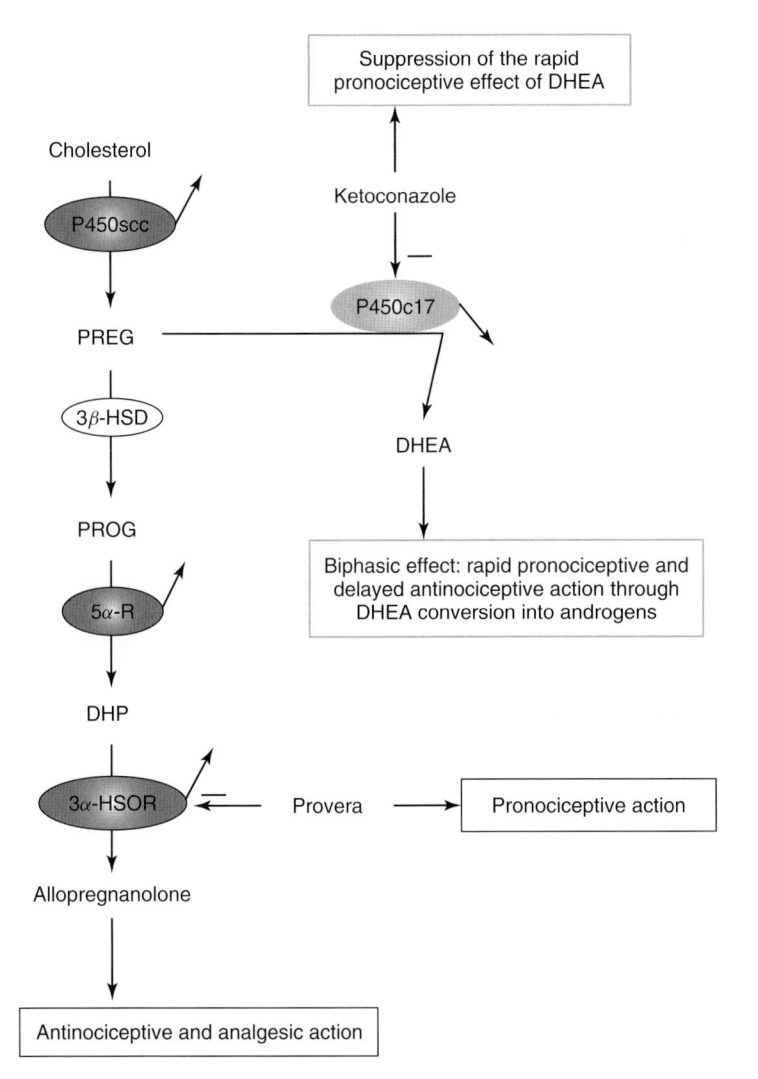

Figure 9.2 Regulation of neurosteroidogenic pathways in the spinal cord of neuropathic rats. During chronic neuropathic pain state induced by sciatic nerve injury, molecular expression and/or biological activities of the enzymes P450scc, 5α-reductase, and 3α-HSOR increased in the spinal cord. This up-regulation led to the production of increased level of allopregnanolone which exerted an antinociceptive action contributing to the coping of neuropathic rats with the chronic pain state. Intrathecal injection of Provera (a pharmacological inhibitor of 3α-HSOR) blocked the endogenous production of allopregnanolone in the spinal cord and exacerbated pain symptoms in neuropathic rats. The expression and enzymatic activity of P450c17 (DHEA-synthesizing enzyme) were repressed in the spinal cord during the chronic pain state. DHEA exerted a biphasic action on nociception: a rapid pronociceptive effect and a delayed antinociceptive action after its conversion into androgens such as testosterone. The blockade of endogenous DHEA production in the spinal cord with intrathecal administration of ketoconazole (P450c17 inhibitor) suppressed the rapid pronociceptive effect of DHEA.

administered ketoconazole (P450c17 inhibitor) induced analgesia in neuropathic rats. Chronic treatment of DHEA increased and maintained the elevated basal pain thresholds in neuropathic and control rats, suggesting that androgenic metabolites generated from daily injected DHEA exerted analgesic effects while DHEA itself (before being metabolized) induced a rapid pronociceptive action [106].

In agreement with our findings showing endogenous neurosteroid involvement in pain modulation, various other investigations using synthetic analogs of $3\alpha,5\alpha$-THP also revealed antinociceptive properties of neurosteroids in humans and animals [107–111]. Furthermore, it has clearly been demonstrated that 5α-reduced neurosteroids induce a potent peripheral analgesia which is mediated by both T-type calcium and $GABA_A$ channels [112].

9.5
Conclusion

Pain is a complex process which involves multiple biological and psychosocial events. The data reviewed in the present chapter show that endogenous neurosteroids and their synthetic analogs crucially control various mechanisms determining pain sensation. Future investigations aiming to develop effective strategies against chronic pain may consider the fact that endogenous neurosteroids produced in spinal sensory circuits interfere with nociceptive transmission and neurochemical mechanisms regulating pain sensation.

Acknowledgments

The authors want to thank their main sponsors, Université de Strasbourg (France) and Association Ti'toine (Normandie, France).

References

1. Baulieu, E. E., Robel, P., and Schumacher, M. (eds) (1999) *Contemporary Endocrinology*, Humana Press, Totowa.
2. Mensah-Nyagan, A.G., Do-Rego, J.L., Beaujean, D., Luu-The, V., Pelletier, G., and Vaudry, H. (1999) Neurosteroids: expression of steroidogenic enzymes and regulation of steroid biosynthesis in the central nervous system. *Pharmacol. Rev.*, 51, 63–81.
3. Mellon, S.H. and Griffin, L.D. (2002) Neurosteroids: biochemistry and clinical significance. *Trends Endocrinol. Metab.*, 13, 35–43.
4. Belelli, D. and Lambert, J.J. (2005) Neurosteroids: endogenous regulators of the GABA(A) receptor. *Nat. Rev. Neurosci.*, 6, 565–575.
5. Patte-Mensah, C. and Mensah-Nyagan, A.G. (2008) Peripheral neuropathy and neurosteroid formation in the central nervous system. *Brain Res. Rev.*, 57, 454–459.
6. Higashi, T., Nishio, T., Yokoi, H., Ninomiya, Y., and Shimada, K. (2007) Studies on Neurosteroids XXI: an improved liquid chromatography-tandem mass spectrometric method for determination of

5alpha-androstane-3alpha,17beta-diol in rat brains. *Anal. Sci.*, **23**, 1015–1019.
7. Mensah-Nyagan, A.G., Do-Rego, J.L., Feuilloley, M., Marcual, A., Lange, C., Pelletier, G., and Vaudry, H. (1996) In vivo and in vitro evidence for the biosynthesis of testosterone in the telencephalon of the female frog. *J. Neurochem.*, **67**, 413–422.
8. Mensah-Nyagan, A.G., Feuilloley, M., Do-Rego, J.L., Marcual, A., Lange, C., Tonon, M.C., Pelletier, G. et al. (1996) Localization of 17beta-hydroxysteroid dehydrogenase and characterization of testosterone in the brain of the male frog. *Proc. Natl. Acad. Sci. U.S.A.*, **93**, 1423–1428.
9. Compagnone, N.A. and Mellon, S.H. (2000) Neurosteroids: biosynthesis and function of these novel neuromodulators. *Front. Neuroendocrinol.*, **21**, 1–56.
10. Le Goascogne, C., Robel, P., Gouezou, M., Sananes, N., Baulieu, E.E., and Waterman, M. (1987) Neurosteroids: cytochrome P-450scc in rat brain. *Science*, **237**, 1212–1215.
11. Willis, W.D. (1985) *The Pain System: The Neural Basis of Nociceptive Transmission in the Mammalian Nervous System*, Karger, Basel.
12. Besson, J.M. and Chaouch, A. (1987) Peripheral and spinal mechanisms of nociception. *Physiol. Rev.*, **67**, 67–186.
13. Merskey, H. and Bogduk, N. (1994) *Classification of Chronic Pain: Descriptions of Chronic Pain Syndromes and Definition of Pain Terms*, 2nd edn, IASP Press, Seattle.
14. Millan, M.J. (1999) The induction of pain: an integrative review. *Prog. Neurobiol.*, **57**, 1–164.
15. Willis, W.D. Jr. (1995) Neurobiology. Cold, pain and the brain. *Nature*, **373**, 19–20.
16. Furst, S. (1999) Transmitters involved in antinociception in the spinal cord. *Brain Res. Bull.*, **48**, 129–141.
17. Nichols, M.L., Allen, B.J., Rogers, S.D., Ghilardi, J.R., Honore, P., Luger, N.M., Finke, M.P. et al. (1999) Transmission of chronic nociception by spinal neurons expressing the substance P receptor. *Science*, **286**, 1558–1561.
18. Millan, M.J. (2002) Descending control of pain. *Prog. Neurobiol.*, **66**, 355–474.
19. Cummins, T.R., Dib-Hajj, S.D., and Waxman, S.G. (2004) Electrophysiological properties of mutant Nav1.7 sodium channels in a painful inherited neuropathy. *J. Neurosci.*, **24**, 8232–8236.
20. Woolf, C.J. (2004) Pain: moving from symptom control toward mechanism-specific pharmacologic management. *Ann. Intern. Med.*, **140**, 441–451.
21. Coull, J.A., Beggs, S., Boudreau, D., Boivin, D., Tsuda, M., Inoue, K., Gravel, C. et al. (2005) BDNF from microglia causes the shift in neuronal anion gradient underlying neuropathic pain. *Nature*, **438**, 1017–1021.
22. Cox, J.J., Reimann, F., Nicholas, A.K., Thornton, G., Roberts, E., Springell, K., Karbani, G. et al. (2006) An SCN9A channelopathy causes congenital inability to experience pain. *Nature*, **444**, 894–898.
23. McMahon, S.B. and Bennett, D. (2007) Pain mechanisms. *Nat. Rev. Neurosci.*, **8**, 241. http://www.nature.com/nrn/poster/pain (accessed 2007).
24. Kawasaki, Y., Xu, Z.Z., Wang, X., Park, J.Y., Zhuang, Z.Y., Tan, P.H., Gao, Y.J. et al. (2008) Distinct roles of matrix metalloproteases in the early- and late-phase development of neuropathic pain. *Nat. Med.*, **14**, 331–336.
25. Arner, S. and Meyerson, B.A. (1988) Lack of analgesic effect of opioids on neuropathic and idiopathic forms of pain. *Pain*, **33**, 11–23.
26. Cherny, N.I. and Portenoy, R.K. (1994) The management of cancer pain. *CA Cancer J. Clin.*, **44**, 263–303.
27. Kingery, W.S. (1997) A critical review of controlled clinical trials for peripheral neuropathic pain and complex regional pain syndromes. *Pain*, **73**, 123–139.
28. Benedetti, F., Vighetti, S., Amanzio, M., Casadio, C., Oliaro, A., Bergamasco, B., and Maggi, G. (1998) Dose-response relationship of opioids in nociceptive and neuropathic postoperative pain. *Pain*, **74**, 205–211.

29. Koltzenburg, M. (1998) Painful neuropathies. *Curr. Opin. Neurol.*, **11**, 515–521.
30. Woolf, C.J. and Mannion, R.J. (1999) Neuropathic pain: aetiology, symptoms, mechanisms, and management. *Lancet*, **353**, 1959–1964.
31. Woolf, C.J. and Salter, M.W. (2000) Neuronal plasticity: increasing the gain in pain. *Science*, **288**, 1765–1769.
32. Basbaum, A.I. and Fields, H.L. (1984) Endogenous pain control systems: brainstem spinal pathways and endorphin circuitry. *Annu. Rev. Neurosci.*, **7**, 309–338.
33. White, F.A., Bhangoo, S.K., and Miller, R.J. (2005) Chemokines: integrators of pain and inflammation. *Nat. Rev. Drug Discov.*, **4**, 834–844.
34. Benemei, S., Nicoletti, P., Capone, J.G., and Geppetti, P. (2009) CGRP receptors in the control of pain and inflammation. *Curr. Opin. Pharmacol.*, **9**, 9–14.
35. Patte-Mensah, C., Kibaly, C., Boudard, D., Schaeffer, V., Béglé, A., Saredi, S., Meyer, L. et al. (2006) Neurogenic pain and steroid synthesis in the spinal cord. *J. Mol. Neurosci.*, **28**, 17–32.
36. Melcangi, R.C. and Mensah-Nyagan, A.G. (2008) Neurosteroids: measurement and pathophysiologic relevance. *Neurochem. Int.*, **52**, 503–505.
37. Mensah-Nyagan, A.G., Kibaly, C., Schaeffer, V., Venard, C., Meyer, L., and Patte-Mensah, C. (2008) Endogenous steroid production in the spinal cord and potential involvement in neuropathic pain modulation. *J. Steroid Biochem. Mol. Biol.*, **109**, 286–293.
38. Suhara, K., Gomi, T., Sato, H., Itagaki, E., Takemori, S., and Katagiri, M. (1978) Purification and immunochemical characterization of the two adrenal cortex mitochondrial cytochrome P-450-proteins. *Arch. Biochem. Biophys.*, **190**, 290–299.
39. Ikushiro, S., Kominami, S., and Takemori, S. (1992) Adrenal P-450scc modulates activity of P-45011 beta in liposomal and mitochondrial membranes. Implication of P-450scc in zone specificity of aldosterone biosynthesis in bovine adrenal. *J. Biol. Chem.*, **267**, 1464–1469.
40. Tsutsui, K. and Yamazaki, T. (1995) Avian neurosteroids. I. Pregnenolone biosynthesis in the quail brain. *Brain Res.*, **678**, 1–9.
41. Usui, M., Yamazaki, T., Kominami, S., and Tsutsui, K. (1995) Avian neurosteroids. II. Localization of a cytochrome P450scc-like substance in the quail brain. *Brain Res.*, **678**, 10–20.
42. Ukena, K., Usui, M., Kohchi, C., and Tsutsui, K. (1998) Cytochrome P450 side-chain cleavage enzyme in the cerebellar Purkinje neuron and its neonatal change in rats. *Endocrinology*, **139**, 137–147.
43. Roby, K.F., Larsen, D., Deb, S., and Soares, M.J. (1991) Generation and characterization of antipeptide antibodies to rat cytochrome P-450 side-chain cleavage enzyme. *Mol. Cell. Endocrinol.*, **79**, 13–20.
44. Haines, D.E., Mihailoff, G.A., and Yezierski, R.P. (1997) in *Fundamental Neuroscience* (ed. D.E. Haines), Churchill Livingstone Inc., New York, pp. 129–141.
45. Patte-Mensah, C., Kappes, V., Freund-Mercier, M.J., Tsutsui, K., and Mensah-Nyagan, A.G. (2003) Cellular distribution and bioactivity of the key steroidogenic enzyme, cytochrome P450 side chain cleavage, in sensory neural pathways. *J. Neurochem.*, **86**, 1233–1246.
46. Willis, W.D., Westlund, K.N., and Carlton, S.M. (1995) in *The Rat Nervous System*, 2nd edn (ed. G. Paxinos), Academic Press Inc., Sydney, pp. 725–750.
47. Kennedy, P.G. (1982) Neural cell markers and their applications to neurology. *J. Neuroimmunol.*, **2**, 35–53.
48. Nagle, R.B. (1988) Intermediate filaments: a review of the basic biology. *Am. J. Surg. Pathol.*, **12**(Suppl 1), 4–16.
49. Fuxe, K. and Agnati, L.F. (1992) *Volume Transmission in the Brain: Novel Mechanisms for Neuronal Transmission*, Raven, New York.
50. Mensah-Nyagan, A.G., Feuilloley, M., Dupont, E., Do-Rego, J.L., Leboulenger, F., Pelletier, G., and

Vaudry, H. (1994) Immunocytochemical localization and biological activity of 3 beta-hydroxysteroid dehydrogenase in the central nervous system of the frog. *J. Neurosci.*, **14**, 7306–7318.
51. Mensah-Nyagan, A.G., Do-Rego, J.L., Beaujean, D., Luu-The, V., Pelletier, G., and Vaudry, H. (2001) Regulation of neurosteroid biosynthesis in the frog diencephalon by GABA and endozepines. *Horm. Behav.*, **40**, 218–225.
52. Mensah-Nyagan, A.G., Beaujean, D., Luu-The, V., Pelletier, G., and Vaudry, H. (2001) Anatomical and biochemical evidence for the synthesis of unconjugated and sulfated neurosteroids in amphibians. *Brain Res. Brain Res. Rev.*, **37**, 13–24.
53. Kibaly, C., Patte-Mensah, C., and Mensah-Nyagan, A.G. (2005) Molecular and neurochemical evidence for the biosynthesis of dehydroepiandrosterone in the adult rat spinal cord. *J. Neurochem.*, **93**, 1220–1230.
54. Hales, D.B. (1992) Interleukin-1 inhibits Leydig cell steroidogenesis primarily by decreasing 17 alpha-hydroxylase/C17-20 lyase cytochrome P450 expression. *Endocrinology*, **131**, 2165–2172.
55. Hales, D.B., Sha, L.L., and Payne, A.H. (1987) Testosterone inhibits cAMP-induced de Novo synthesis of Leydig cell cytochrome P-450 (17 alpha) by an androgen receptor-mediated mechanism. *J. Biol. Chem.*, **262**, 11200–11206.
56. Kuhn-Velten, W.N. and Lessmann, M. (1992) Ketoconazole inhibition of the bifunctional cytochrome P450c17 does not affect androgen formation from the endogenous lyase substrate. The catalytic site remains refractory in the course of intermediary hydroxyprogesterone processing. *Biochem. Pharmacol.*, **44**, 2371–2378.
57. Swart, P., Swart, A.C., Waterman, M.R., Estabrook, R.W., and Mason, J.I. (1993) Progesterone 16 alpha-hydroxylase activity is catalyzed by human cytochrome P450 17 alpha-hydroxylase. *J. Clin. Endocrinol. Metab.*, **77**, 98–102.
58. Sanne, J.L. and Krueger, K.E. (1995) Expression of cytochrome P450 side-chain cleavage enzyme and 3 beta-hydroxysteroid dehydrogenase in the rat central nervous system: a study by polymerase chain reaction and in situ hybridization. *J. Neurochem.*, **65**, 528–536.
59. Coirini, H., Gouezou, M., Liere, P., Delespierre, B., Pianos, A., Eychenne, B., Schumacher, M. et al. (2002) 3 Beta-hydroxysteroid dehydrogenase expression in rat spinal cord. *Neuroscience*, **113**, 883–891.
60. Saredi, S., Patte-Mensah, C., Melcangi, R.C., and Mensah-Nyagan, A.G. (2005) Effect of streptozotocin-induced diabetes on the gene expression and biological activity of 3beta-hydroxysteroid dehydrogenase in the rat spinal cord. *Neuroscience*, **135**, 869–877.
61. Saitoh, H., Hirato, K., Yanaihara, T., and Nakayama, T. (1982) A study of 5 alpha-reductase in human fetal brain. *Endocrinol. Jpn.*, **29**, 461–467.
62. Melcangi, R.C., Celotti, F., Castano, P., and Martini, L. (1993) Differential localization of the 5 alpha-reductase and the 3 alpha-hydroxysteroid dehydrogenase in neuronal and glial cultures. *Endocrinology*, **132**, 1252–1259.
63. Pelletier, G., Luu-The, V., and Labrie, F. (1994) Immunocytochemical localization of 5 alpha-reductase in rat brain. *Mol. Cell. Neurosci.*, **5**, 394–399.
64. Martini, L., Celotti, F., and Melcangi, R.C. (1996) Testosterone and progesterone metabolism in the central nervous system: cellular localization and mechanism of control of the enzymes involved. *Cell. Mol. Neurobiol.*, **16**, 271–282.
65. Negri-Cesi, P., Poletti, A., and Celotti, F. (1996) Metabolism of steroids in the brain: a new insight into the role of 5alpha-reductase and aromatase in brain differentiation and functions. *J. Steroid Biochem. Mol. Biol.*, **58**, 455–466.
66. Stoffel-Wagner, B., Watzka, M., Steckelbroeck, S., Ludwig, M., Clusmann, H., Bidlingmaier, F., Casarosa, E. et al. (2003) Allopregnanolone serum levels and

expression of 5 alpha-reductase and 3 alpha-hydroxysteroid dehydrogenase isoforms in hippocampal and temporal cortex of patients with epilepsy. *Epilepsy Res.*, **54**, 11–19.

67. Celotti, F., Melcangi, R.C., and Martini, L. (1992) The 5 alpha-reductase in the brain: molecular aspects and relation to brain function. *Front. Neuroendocrinol.*, **13**, 163–215.

68. Poletti, A., Coscarella, A., Negri-Cesi, P., Colciago, A., Celotti, F., and Martini, L. (1998) 5 alpha-reductase isozymes in the central nervous system. *Steroids*, **63**, 246–251.

69. Poletti, A., Negri-Cesi, P., Rabuffetti, M., Colciago, A., Celotti, F., and Martini, L. (1998) Transient expression of the 5alpha-reductase type 2 isozyme in the rat brain in late fetal and early postnatal life. *Endocrinology*, **139**, 2171–2178.

70. Torres, J.M. and Ortega, E. (2003) Differential regulation of steroid 5alpha-reductase isozymes expression by androgens in the adult rat brain. *FASEB J.*, **17**, 1428–1433.

71. Pozzi, P., Bendotti, C., Simeoni, S., Piccioni, F., Guerini, V., Marron, T.U., Martini, L. *et al.* (2003) Androgen 5-alpha-reductase type 2 is highly expressed and active in rat spinal cord motor neurones. *J. Neuroendocrinol.*, **15**, 882–887.

72. Julius, D. and Basbaum, A.I. (2001) Molecular mechanisms of nociception. *Nature*, **413**, 203–210.

73. Patte-Mensah, C., Penning, T.M., and Mensah-Nyagan, A.G. (2004) Anatomical and cellular localization of neuroactive 5 alpha/3 alpha-reduced steroid-synthesizing enzymes in the spinal cord. *J. Comp. Neurol.*, **477**, 286–299.

74. Andersson, S. and Russell, D.W. (1990) Structural and biochemical properties of cloned and expressed human and rat steroid 5 alpha-reductases. *Proc. Natl. Acad. Sci. U.S.A.*, **87**, 3640–3644.

75. Thigpen, A.E., Cala, K.M., and Russell, D.W. (1993) Characterization of Chinese hamster ovary cell lines expressing human steroid 5 alpha-reductase isozymes. *J. Biol. Chem.*, **268**, 17404–17412.

76. Raff, M.C., Mirsky, R., Fields, K.L., Lisak, R.P., Dorfman, S.H., Silberberg, D.H., Gregson, N.A. *et al.* (1978) Galactocerebroside is a specific cell-surface antigenic marker for oligodendrocytes in culture. *Nature*, **274**, 813–816.

77. Matus, A. (1990) Microtubule-associated proteins. *Curr. Opin. Cell. Biol.*, **2**, 10–14.

78. Melcangi, R.C., Celotti, F., Ballabio, M., Castano, P., Poletti, A., Milani, S., and Martini, L. (1988) Ontogenetic development of the 5 alpha-reductase in the rat brain: cerebral cortex, hypothalamus, purified myelin and isolated oligodendrocytes. *Brain Res. Dev. Brain Res.*, **44**, 181–188.

79. Melcangi, R.C., Celotti, F., Ballabio, M., Poletti, A., Castano, P., and Martini, L. (1988) Testosterone 5 alpha-reductase activity in the rat brain is highly concentrated in white matter structures and in purified myelin sheaths of axons. *J. Steroid Biochem.*, **31**, 173–179.

80. Poletti, A., Celotti, F., Rumio, C., Rabuffetti, M., and Martini, L. (1997) Identification of type 1 5alpha-reductase in myelin membranes of male and female rat brain. *Mol. Cell. Endocrinol.*, **129**, 181–190.

81. Pawlowski, J.E., Huizinga, M., and Penning, T.M. (1991) Cloning and sequencing of the cDNA for rat liver 3 alpha-hydroxysteroid/dihydrodiol dehydrogenase. *J. Biol. Chem.*, **266**, 8820–8825.

82. Jez, J.M., Bennett, M.J., Schlegel, B.P., Lewis, M., and Penning, T.M. (1997) Comparative anatomy of the aldo-keto reductase superfamily. *Biochem. J.*, **326** (Pt 3), 625–636.

83. Jez, J.M., Flynn, T.G., and Penning, T.M. (1997) A new nomenclature for the aldo-keto reductase superfamily. *Biochem. Pharmacol.*, **54**, 639–647.

84. Penning, T.M., Jin, Y., Heredia, V.V., and Lewis, M. (2003) Structure-function relationships in 3alpha-hydroxysteroid dehydrogenases: a comparison of the rat and human isoforms. *J. Steroid Biochem. Mol. Biol.*, **85**, 247–255.

85. Krieger, N.R. and Scott, R.G. (1984) 3 alpha-Hydroxysteroid oxidoreductase in rat brain. *J. Neurochem.*, **42**, 887–890.
86. Krieger, N.R. and Scott, R.G. (1989) Nonneuronal localization for steroid converting enzyme: 3 alpha-hydroxysteroid oxidoreductase in olfactory tubercle of rat brain. *J. Neurochem.*, **52**, 1866–1870.
87. Khanna, M., Qin, K.N., and Cheng, K.C. (1995) Distribution of 3 alpha-hydroxysteroid dehydrogenase in rat brain and molecular cloning of multiple cDNAs encoding structurally related proteins in humans. *J. Steroid Biochem. Mol. Biol.*, **53**, 41–46.
88. Smithgall, T.E. and Penning, T.M. (1988) Electrophoretic and immunochemical characterization of 3 alpha-hydroxysteroid/dihydrodiol dehydrogenases of rat tissues. *Biochem. J.*, **254**, 715–721.
89. Jez, J.M., Schlegel, B.P., and Penning, T.M. (1996) Characterization of the substrate binding site in rat liver 3alpha-hydroxysteroid/dihydrodiol dehydrogenase. The roles of tryptophans in ligand binding and protein fluorescence. *J. Biol. Chem.*, **271**, 30190–30198.
90. Majewska, M.D. (1992) Neurosteroids: endogenous bimodal modulators of the GABAA receptor. Mechanism of action and physiological significance. *Prog. Neurobiol.*, **38**, 379–395.
91. Paul, S.M. and Purdy, R.H. (1992) Neuroactive steroids. *FASEB J.*, **6**, 2311–2322.
92. Frye, C.A., Duncan, J.E., Basham, M., and Erskine, M.S. (1996) Behavioral effects of 3 alpha-androstanediol. II: Hypothalamic and preoptic area actions via a GABAergic mechanism. *Behav. Brain Res.*, **79**, 119–130.
93. Frye, C.A., Rhodes, M.E., Rosellini, R., and Svare, B. (2002) The nucleus accumbens as a site of action for rewarding properties of testosterone and its 5alpha-reduced metabolites. *Pharmacol. Biochem. Behav.*, **74**, 119–127.
94. Reddy, D.S. (2003) Is there a physiological role for the neurosteroid THDOC in stress-sensitive conditions? *Trends Pharmacol. Sci.*, **24**, 103–106.
95. Melcangi, R.C., Poletti, A., Cavarretta, I., Celotti, F., Colciago, A., Magnaghi, V., Motta, M. et al. (1998) The 5alpha-reductase in the central nervous system: expression and modes of control. *J. Steroid Biochem. Mol. Biol.*, **65**, 295–299.
96. Lumbroso, S., Sandillon, F., Georget, V., Lobaccaro, J.M., Brinkmann, A.O., Privat, A., and Sultan, C. (1996) Immunohistochemical localization and immunoblotting of androgen receptor in spinal neurons of male and female rats. *Eur. J. Endocrinol.*, **134**, 626–632.
97. Matsumoto, A. (1997) Hormonally induced neuronal plasticity in the adult motoneurons. *Brain Res. Bull.*, **44**, 539–547.
98. Kastrup, Y., Hallbeck, M., Amandusson, A., Hirata, S., Hermanson, O., and Blomqvist, A. (1999) Progesterone receptor expression in the brainstem of the female rat. *Neurosci. Lett.*, **275**, 85–88.
99. Labombarda, F., Guennoun, R., Gonzalez, S., Roig, P., Lima, A., Schumacher, M., and De Nicola, A.F. (2000) Immunocytochemical evidence for a progesterone receptor in neurons and glial cells of the rat spinal cord. *Neurosci. Lett.*, **288**, 29–32.
100. Frye, C.A. (2001) The role of neurosteroids and non-genomic effects of progestins and androgens in mediating sexual receptivity of rodents. *Brain Res. Brain Res. Rev.*, **37**, 201–222.
101. Lambert, J.J., Belelli, D., Peden, D.R., Vardy, A.W., and Peters, J.A. (2003) Neurosteroid modulation of GABAA receptors. *Prog. Neurobiol.*, **71**, 67–80.
102. Patte-Mensah, C., Kibaly, C., and Mensah-Nyagan, A.G. (2005) Substance P inhibits progesterone conversion to neuroactive metabolites in spinal sensory circuit: a potential component of nociception. *Proc. Natl. Acad. Sci. U.S.A.*, **102**, 9044–9049.
103. Bennett, G.J. and Xie, Y.K. (1988) A peripheral mononeuropathy in rat that produces disorders of pain sensation like those seen in man. *Pain*, **33**, 87–107.

104. Patte-Mensah, C., Li, S., and Mensah-Nyagan, A.G. (2004) Impact of neuropathic pain on the gene expression and activity of cytochrome P450side-chain-cleavage in sensory neural networks. *Cell. Mol. Life Sci.*, **61**, 2274–2284.
105. Meyer, L., Venard, C., Schaeffer, V., Patte-Mensah, C., and Mensah-Nyagan, A.G. (2008) The biological activity of 3alpha-hydroxysteroid oxido-reductase in the spinal cord regulates thermal and mechanical pain thresholds after sciatic nerve injury. *Neurobiol. Dis.*, **30**, 30–41.
106. Kibaly, C., Meyer, L., Patte-Mensah, C., and Mensah-Nyagan, A.G. (2008) Biochemical and functional evidence for the control of pain mechanisms by dehydroepiandrosterone endogenously synthesized in the spinal cord. *FASEB J.*, **22**, 93–104.
107. Goodchild, C.S., Guo, Z., and Nadeson, R. (2000) Antinociceptive properties of neurosteroids I. Spinally-mediated antinociceptive effects of water-soluble aminosteroids. *Pain*, **88**, 23–29.
108. Goodchild, C.S., Robinson, A., and Nadeson, R. (2001) Antinociceptive properties of neurosteroids IV: pilot study demonstrating the analgesic effects of alphadolone administered orally to humans. *Br. J. Anaesth.*, **86**, 528–534.
109. Nadeson, R. and Goodchild, C.S. (2000) Antinociceptive properties of neurosteroids II. Experiments with Saffan and its components alphaxalone and alphadolone to reveal separation of anaesthetic and antinociceptive effects and the involvement of spinal cord GABA(A) receptors. *Pain*, **88**, 31–39.
110. Nadeson, R. and Goodchild, C.S. (2001) Antinociceptive properties of neurosteroids III: experiments with alphadolone given intravenously, intraperitoneally, and intragastrically. *Br. J. Anaesth.*, **86**, 704–708.
111. Gambhir, M., Mediratta, P.K., and Sharma, K.K. (2002) Evaluation of the analgesic effect of neurosteroids and their possible mechanism of action. *Indian J. Physiol. Pharmacol.*, **46**, 202–208.
112. Pathirathna, S., Brimelow, B.C., Jagodic, M.M., Krishnan, K., Jiang, X., Zorumski, C.F., Mennerick, S. et al. (2005) New evidence that both T-type calcium channels and $GABA_A$ channels are responsible for the potent peripheral analgesic effects of 5alpha-reduced neuroactive steroids. *Pain*, **114**, 429–443.

Part III
Polypeptide Hormones and Neuroprotection

Hormones in Neurodegeneration, Neuroprotection, and Neurogenesis.
Edited by Achille G. Gravanis and Synthia H. Mellon
Copyright © 2011 WILEY-VCH Verlag GmbH & Co. KGaA, Weinheim
ISBN: 978-3-527-32627-3

10
The Insulin/IGF-1 System in Neurodegeneration and Neurovascular Disease

Przemyslaw (Mike) Sapieha and Lois Smith

10.1
Introduction

Neurodegenerative diseases are multifactorial, yet one common variable relates to perturbations in energy metabolism. Simply put, neurodegenerative diseases present varying degrees of compromised utilization of substrates that ensure adequate cellular function. Perhaps the best example is the inefficient catabolism of glucose during cellular respiration to generate ATP. It is therefore pertinent to consider the involvement of hormones that influence cellular glucose uptake and utilization in the progression of neurodegenerative diseases. Insulin and peptides such as insulin-like growth factors (IGFs) are particularly relevant in this context and an emerging body of evidence is pointing toward their involvement in neuronal homeostasis. Interestingly, neurons resemble insulin-producing islet cells more closely than any other cell type, commencing with the fact that islet cells evolved from insulin-producing neurons [1].

The role of the insulin/IGF system in the central nervous system (CNS) reaches far beyond that of glucose uptake and has been receiving increasing attention as its deregulation may exacerbate neurodegenerative disease. Although a direct link between insulin/IGF systems and neurodegeneration has yet to be established, a number of neurodegenerative genetic disorders such as Alzheimer's disease (AD), Parkinson's disease (PD), ataxia–telangiectasia, Huntington disease, Prader–Willi syndrome, and Werner syndrome have been associated with insulin resistance [2]. Moreover, direct evidence for a role of the insulin system in the nervous system comes from rodent studies that demonstrate compromised regulation of energy metabolism in animals that are deficient in the insulin receptor (IR), specifically in nervous tissue [3].

Beyond direct action on neurons, the insulin/IGF axis plays a notable role in ensuring the sound development and preservation of vascular beds within the nervous system. It is therefore conceivable that an impaired insulin/IGF axis could be a contributor to conditions such as vascular dementia. The involvement of insulin/IGF in neurovascular disease is perhaps best illustrated by work done in the retina (a direct extension of the CNS) where IGF levels correlate inversely

Hormones in Neurodegeneration, Neuroprotection, and Neurogenesis.
Edited by Achille G. Gravanis and Synthia H. Mellon
Copyright © 2011 WILEY-VCH Verlag GmbH & Co. KGaA, Weinheim
ISBN: 978-3-527-32627-3

with vessel dropout in pathological conditions such as retinopathy of prematurity (ROP) [4].

In this chapter, we summarize the current evidence for involvement of the insulin/IGF pathways in neurodegenerative disease and reexamine the classic concept of CNS insensitivity to insulin. As neuronal survival and homeostasis is extremely reliant on vascular integrity, we expand on the vasoprotective role of these molecules in neuronal tissue.

10.2
Insulin and Insulin Growth Factors

Insulin and related peptides are primal players in cellular biology that have been acquired early (600 million years ago) in multicellular eukaryote evolution [5, 6]. Evidence for the primitive origin of the insulin/IGF system stems from the conservation of signaling pathways between *Caenorhabditis elegans*, *Drosophila melanogaster*, rodents, and humans [7]. IGF-1 is a single-chain polypeptide that is structurally related to insulin. It is synthesized predominantly by the liver in response to the growth hormone. Over 32 members of the IGF family have been identified in *C. elegans* [7], while in mammals only 3 are known. While the actions of insulin in adult systems are thought to be primarily at the level of metabolic regulation, the biological actions of IGF are part of a more complex system.

The cellular actions of IGF-1 are mainly described as occurring through the activation of a type-I IGF tyrosine–kinase membrane receptor (IGF-R). The affinity of IGF-1 to its receptor is modulated by a family of insulin-like growth factor binding proteins (IGFBPs) as the binding affinity toward these carrier proteins is higher than that of its receptor. Systemically, the vast majority of IGF-1 (up to 98%) is bound to one of six IGFBPs, of which IGFBP3 is the most abundant [8]. IGFBPs prolong the circulating half-life of IGF and thus augment tissue delivery [9]. Once in the tissue, IGFBPs can either potentiate IGF signaling by releasing them in the proximity of their receptors or, conversely, hinder signaling by sequestering them. The binding of IGF to the IGFBPs is therefore a key regulatory step in IGF signaling.

10.3
Local versus Systemic Actions

The direct actions of insulin and IGF-1 on the CNS remained elusive largely because of the low levels of the hormones produced in neural tissue. In fact, even in the late 1980s, the brain was dubbed as "insulin insensitive" in medical textbooks. This was perhaps due to the fact that only the developing CNS produces detectable amounts of these hormones and these levels drop considerably in the mature brain. In fact IGF-1 mRNA production is both low and restricted to a few structures in the mature CNS, while the local expression of insulin in the CNS is debatable [10–12].

The first evidence for insulin sensitivity in the CNS comes from the anatomical distribution of the receptors for these hormones. They are present abundantly throughout the nervous system on both neurons and glia of the cortex, retina, substantia nigra, hypothalamus, and hippocampus [12–14]. Importantly, both IGF-1 and insulin can readily cross the blood–brain barrier via receptor-mediated processes and can, therefore, feasibly exert their biological activity via their neuronal-resident receptors [15–17]. Moreover, raising peripheral insulin levels directly augments concentrations in the brain and CSF (cerebrospinal fluid).

An alternative scenario would call for a contribution from local, tissue-specific sources of the hormones. While the role of systemic insulin is widely accepted, the relevance of local production remains to be elucidated regardless of the demonstration of neuronal production. Conversely, the systemic role of IGF-1 is not recognized. This is in part due to prominent local generation of IGF-1 from neuronal sources (in both the brain and retina), brain endothelium, and reactive infiltrating microglia [12, 18–22] confounding distinctions between local and systemic sources.

10.4
Insulin/IGF Signaling Pathway

The converging role of insulin and IGF-1 in neurophysiology and neuropathology remains to be fully explored. Both IR and insulin-like growth factor 1 receptor (IGF-1R) are tyrosine kinase receptors each having two extracellular subunits and two transmembrane domains. As they share more than 50% of the amino acid sequence homology, both IGF-1 and insulin bind to both IR and IGF1R promiscuously, although the affinity for their respective receptor remains 100-fold greater [23]. This is particularly pertinent in conditions with chronically elevated levels of insulin where excess insulin will be available to cross-react with IGF-R. The entwined nature of these pathways is further complicated by the fact that their receptors form heterodimers that can be activated by either ligand. On the basis of the above, it is likely that insulin/IGF signaling is reliant on the tissue distribution of both receptors. IRs are expressed throughout the CNS, with the highest levels being in the olfactory bulb, hypothalamus, cerebral cortex, cerebellum, and hippocampus [13, 24]. Upon hormone binding to IR, the autophosphorylation of the receptors leads to the recruitment and phosphorylation of insulin receptor substrate (IRS) proteins which induce the activation of the phosphatidylinositol 3-kinase/Akt and the Ras/MAPK pathways.

The molecular targets of insulin in the CNS are mainly described for the hypothalamus (see below). Although several other regions of the brain and CNS respond to insulin and receptors for insulin are present throughout, the exact significance with respect to neuronal systems is presently less well characterized. Conversely, modes of IGF function in the CNS are considerably better understood. As alluded to below, it is thought that the neurotrophic actions of the insulin system are carried out by IGF-1 and its activation of IGF-1R. IGF-1R is ubiquitously expressed throughout the CNS, including neurons, glia, oligodendrocytes, microglia, endothelial cells, and pericytes. The downstream effectors of these receptors are intertwined, complex, and emerging. As with IR, of the multiple cascades stimulated by IGF-1/IGF-R, the Akt pathway blocks apoptotic cascades via canonical pathways [25] and the Ras/MAPK pathways provokes growth. A simplified view of this complex signaling system suggests that insulin's actions are predominantly metabolic while IGF-1 is preferentially mitogenic [26].

10.5
The Insulin/IGF Axis in the Brain

The involvement of the insulin/IGF axis in cellular homeostasis, cell metabolism, and growth has been well defined. The regional effect of insulin on glucose metabolism in the brain is thought to be mediated via the insulin-sensitive glucose transporters GLUT4 and GLUT8. Evidence for overlapping expression of insulin, IR, and the glucose transporters (GLUTs) in the hippocampus and hypothalamus is suggestive of such a mechanism [27, 28].

Other than its effect on glucose uptake, insulin plays a role in regulating nutrient homeostasis by acting on the arcuate nucleus of the hypothalamus [29]. Spikes of insulin in the brain suppress appetite (anorexigenic) while blocking insulin signaling has the opposite effect (orexigenic). This is thought to be the result of insulin's action on neuropeptide expression. For example, the anorexigenic neuropeptides proopiomelanocortin (POMC) and α-melanocyte-stimulating hormone (α-MSH) are increased by insulin while the orexigenic neuropeptide Y (NPY) and the agouti-related peptide (AgRP) are decreased [1].

10.6
Insulin/IGF and Neuroprotection

Although the neuroactive and neurometabolic properties of insulin and its derivatives have been recognized for decades, they have only been accepted as a prototypical neuronal survival factor since the late 1990s [30]. Subsequent to this work, insulin and insulin-like factors were perceived as possible neurotrophic factors playing key roles in neuronal health and disease (Figure 10.1). Key findings were provided from transgenic studies where mice engineered to systemically overexpress IGF-1 showed increases in brain weight that were of greater proportions compared to their gain in body weight [31]. Similarly, transgenic mice designed to overproduce plasma-IGF-1 specifically in the adult showed significant increases in brain size [32], further illustrating the important neurotrophic support provided by this factor.

10.6 Insulin/IGF and Neuroprotection | 177

(a) Insulin/IGF-1 as survival factors

MAPK (growth)
PI3K (survival)
Neuron

(b) Insulin/IGF-1 in AD

Insulin insensitivity
Tau hyperphosphorylation
βA

(c) Insulin/IGF-1 in PD

Dopaminergic neurons

- Insulin
- IGF-1
- VEGF
- IR
- IGF-1R
- VEGFR
- Dopamine transporter
- Insulin degrading enzyme
- Insulin-resistant IR

(d) Insulin/IGF-1 in neurovascular homeostasis

↑NO
↓ET-1
↑VEGF signaling
Vessel

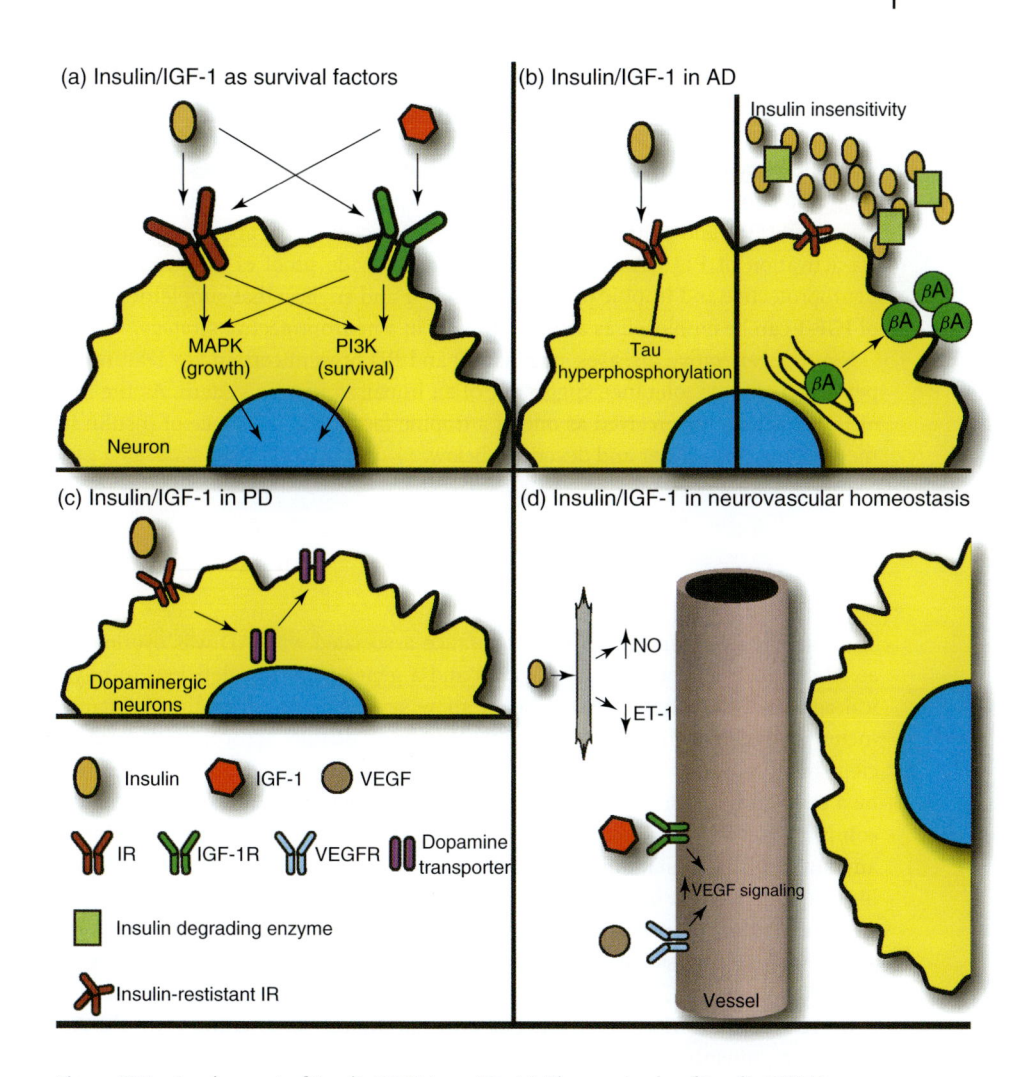

Figure 10.1 Involvement of insulin/IGF-1 in neuronal health. (a) IGF-1 and insulin can act directly as neuronal survival factors by activating their receptors and consequently stimulating MAPK (growth) and PI3K (survival) pathways. (b) In AD, insulin can be protective by preventing tau hyperphosphorylation. In the context of insulin resistance, Aβ precursor protein secretion is augmented. Moreover, Aβ competes with *insulin degrading enzyme* (IDE) and further contributes to the elevated levels of multimeric aggregates of amyloid protein observed in AD. (c) The exact role of insulin/IGF-1 in PD remains to be determined. Thus far it is known that insulin augments levels of dopamine transporter in dopaminergic neurons. (d) The insulin/IGF-1 plays key roles in neurovascular health, which is paramount to neuronal health. Insulin increases nitric oxide (NO) and decreases endothelin-1 (ET-1) and therefore directly affects vasodilation and regional blood flow. Moreover, IGF-1 potentiates VEGF signaling, further contributing to vascular homeostasis and adequate neuronal perfusion and metabolic supply.

Relevantly, IGF-1 can also stimulate neuronal metabolism [33] and excitability [34]. Both of these properties can enhance the cell's propensity to survive following insult. Direct evidence for the neuroprotective properties of IGF-1 on CNS neurons comes from numerous studies modeling various CNS pathologies. Examples illustrate the protective effects of IGF-1 against ischemic injuries of the CNS [35], protection against oxidative stress [36], hyperglycemic events [37], glutamate excitoxicity [38], and on transected retinal ganglion cells where apoptosis is blocked via activation of PI3K/Akt [39]. These studies provide proof of principle for the neuroprotective and trophic properties of IGF-1 and suggest that supplementation of IGF-1 can be employed as a survival factor in a neuroprotective strategy.

Neurodegenerative diseases such as AD and PD present certain key features of perturbed glucose tolerance suggestive of an impaired insulin system. As the role of IGF-1 is largely perceived as one of a trophic factor, the influence of insulin in these diseases is subtler and described below.

10.7
Alzheimer's Disease

Traditionally, the impaired glucose tolerance associated with AD was overlooked as an offshoot of dietary insufficiencies and a general lifestyle deficient in physical activity. Interestingly, AD patients show several signs of insulin resistance including high insulin levels following glucose challenge, accompanied by deficient insulin-mediated glucose uptake. In line, AD progression is accompanied by a reduction in IR number and tyrosine kinase activity [40]. Remarkably, this compromised response occurs in the preliminary phases of the disease, prior to any physical manifestations and independent of dietary intake [41] and therefore likely stems from an initial cellular deregulation. Presently, a direct link between AD and syndromes presenting elements of insulin resistance such as type-2 diabetes mellitus is less clear. Many of the associations have been made based on epidemiological studies, yet concrete biochemical proof remains to be obtained.

In AD, the multimeric aggregates of amyloid protein (originating from the $A\beta$ precursor protein) are a cardinal neurological feature of the disease. Relevant to insulin homeostasis, these aggregates have a pancreatic analog consisting of islet amyloid polypeptide (IAPP) [42, 43]. IAPP and the $A\beta$ precursor protein share 90% structural homology and may be a contributing factor, explaining the predisposition to insulin resistance and hypersecretion of insulin noted in people suffering from Alzheimer's [3, 44].

The tissue content of $A\beta$ is directly reliant on its rate of synthesis, release, and proteolytic degradation. *In vitro* studies have shown that insulin can influence the release of intracellular $A\beta$ from neuronal cultures by accelerating their trafficking from the Golgi networks outward to the plasma membrane [45]. Therefore, high concentrations of circulating insulin may increase accumulation of extracellular deposits of $A\beta$ in the brain. Moreover, elevated insulin levels may also directly affect the intracellular breakdown of $A\beta$ via substrate competition with the metalloprotease *insulin degrading enzyme* (IDE) [46]. As its name states, IDE's main role is thought to be the catabolism of insulin but it can also mediate the degradation of Aβ. Evidence for the implication of IDE in AD is supported by findings that show an overall reduction in the expression and activity of this enzyme in the brains of patients with AD [47, 48]. This is corroborated by findings of reduced Aβ degradation and high insulin levels in mice knocked out for IDE.

Insulin's link to AD may also be made with respect to *tau*-enriched neurofibrillary tangles, which are another hallmark of AD. Tau proteins are microtubule-associated proteins that, when hyperphosphorylated, form inclusions of tangled neurofilaments. Insulin can influence the extent of tau phosphorylation via its regulation of glycogen synthase kinase 3β (GSK3β) [49]. In line, mice deficient in IRS2 show more hyperphosphorylated tau and the ensuing neurofibrillar tangles [50].

10.8
Parkinson's Disease

As with AD, a link between PD and insulin resistance can be made. The prominent pathological manifestation of PD is the progressive degeneration of dopaminergic neurons in the substantia nigra, affecting inhibitory GABA pathways. Although there is presently a lack of well-controlled human studies unequivocally linking PD to insulin insensitivity, there exists a body of evidence suggestive of this association. The first qualitative evidence of involvement of insulin in PD comes from the dense expression of IR in dopaminergic neurons within the substantia nigra. Importantly, the level of IR is severely reduced in subjects affected by PD [14, 51]. The exact role of IR in dopamininergic neurons remains to be fully elucidated; however, it can potentially be linked to PD as intracerebroventricular delivery of insulin augments levels of dopamine transporter mRNA within the substantia nigra and increases its activity [52, 53].

From the clinical side, data points to compromised insulin-mediated glucose uptake as an early clinical feature of PD [54]. In fact, perturbed glucose tolerance affects up to 80% of patients suffering from PD [55]. Relevantly, drugs used to slow PD such as levodopa provoke both hyperinsulinemia and hyperglycemia [54]. Similarly, other pharmacological frontlines against PD, such as bromocriptine, are thought to increase insulin sensitivity [56]. Taken together, it is tempting to speculate that insulin plays an important role in regulating the function of dopaminergic neurons and is suggestive of a link to PD.

10.9
Vascular Dementia

Vascular dementia loosely encompasses syndromes that lead to vascular lesions in the brain. They can be caused by general vascular disease attaining the brain or by focal lesions such as strokes. Adequate neuronal function requires a tight coupling of cerebral blood circulation to supply the metabolic demand neurons. It is well known that both insulin resistance and hyperinsulinemia represent severe risk factors for vascular disease. Perturbations in insulin signaling can have direct negative effects on vasculature. For example, under normal circumstances, insulin has vasoactive effects that directly affect vasodilation, and consequently, regional blood flow. This is caused in part by insulin-provoked increases in nitric oxide (NO) and decreases in endothelin-1 (ET-1). Conversely, in a context of insulin resistance, responsiveness decreases and there is a drop in NO and augmentation in ET-1, resulting in vasoconstriction and poor capillary recruitment. Ultimately, this further deregulates tissue glucose levels and exacerbates endothelial dysfunction. Within the brain, this scenario is highly deleterious and will impede neuronal function [57].

Epidemiological evidence for insulin deregulation partaking in vascular dementia comes from studies demonstrating an increased risk in patients with type-2 diabetes or hypertension [58–60]. In line, a direct link can be drawn between insulin resistance and hypertension in that 50% of individuals who have hypertension have varying degrees of insulin resistance [61]. Moreover, type-2 diabetes mellitus and insulin resistance in general are positively correlated with increased risk of stroke as well as a poorer prognosis for recovery [52]. It is also becoming clear that the progression of AD is accompanied by vascular injury with microvascular decay being a prominent feature [62]. As both AD and vascular disease share common deficits in insulin signaling, the contribution of insulin resistance to the progression of both pathologies invites further investigation.

10.10
Neurovascular Degeneration

Perhaps the best arena in which to study the influence of the insulin/IGF axis on neurovascular health is the neurovascular interface of the retina in the disease context of ROP. Neuronal survival is heavily reliant on adequate perfusion from the neighboring vasculature. The retina is undoubtedly the most accessible structure of the CNS. Moreover, it presents a highly stereotyped vascular bed, amenable to investigation.

The development of human retinal vasculature commences upon the fourth month of gestation and only concludes at term (40 weeks) [63]. Hence, when an infant is born prematurely, its retinal blood supply is incomplete and highly vulnerable to decay. This vascular deficiency makes the retina prone to complications, birth weight and gestational age being the most significant susceptibility factors making the insulin/IGF-1 pathway a potential culprit.

Paradoxically, the incidence of ROP increased (predominantly in affluent countries) with the advent of oxygen supplementation to overcome the inefficient O_2 absorption of premature infant lungs. A direct link was later drawn between O_2 therapy and ROP [64, 65]. ROP manifests as a biphasic disease; phase 1 is associated with vascular dropout and growth cessation yielding a hypoxic retina while phase 2 is characterized by a profuse deregulated vascular growth initiated by the hypoxic state of the tissue.

Oxygen-dependent factors are well defined in the context of endothelial survival in the relative hyperoxia encountered in the extrauterine environment, as well as the supplemental oxygen given to premature infants. These augmented oxygen levels suppress vascular endothelial growth factor (VEGF) and consequently stunt normal vessel formation (phase I of ROP development). As the retina becomes functionally active, the lack of adequate vascularization leads to tissue hypoxia and triggers a staunch rise in VEGF levels [66–68] which subsequently drives the exaggerated and aberrant preretinal vasoproliferation associated with the secondary phase of the disease.

However, currently adapted oxygenation protocols (which limit high hemoglobin saturation with oxygen) for premature subjects have led to a reduction in the incidence of ROP [69–71]. Yet, in light of these adjustments, ROP persists as more premature infants are saved and remains a major cause of blindness in the industrialized world, an observation that highlights the importance of nonoxygen-regulated factors. Of all contributors, prematurity remains the greatest risk for ROP [71], suggesting that certain key factors present *in utero* that partake in normal fetal development are lacking in preterm infants [72–75].

IGF-1 is essential for fetal development at all stages of pregnancy [76]. Plasma levels of IGF-1 rise with gestational age and are considerably augmented during the third trimester of pregnancy (when premature infants predisposed to ROP are born) [77, 78]. Importantly, IGF-1 decreases postnatally, thus underscoring the placental and amniotic source of the factor [79].

Evidence for an implication of IGF-1 paucity in the vascular phenotype associated with ROP was demonstrated in animal models of IGF-1 deficient mice, where retinal vessel growth was retarded, akin to patterns noted in premature babies with ROP [72]. Similarly, in human subjects, low IGF-1 serum levels directly correlate with the severity of ROP, and, interestingly, may also account for abnormal brain development [80].

In neonatal mice with decreased IGF-1, it was observed that VEGF levels remain elevated, yet these animals have less retinal vascular development [74]. These findings point to a mechanism of action where IGF-1 does not directly modulate VEGF levels. In fact, IGF-1 controls the maximal VEGF-induced activation of Akt in endothelial cells and therefore modulates vessel survival [72], an essential event in preventing phase I of ROP. Moreover, IGF-1 is required for maximum VEGF-induced activation of p44/42 MAPK – essential for endothelial cell proliferation and thus the neovascularization observed in phase II of ROP [75]. In this regard, IGF-1 acts as a permissive factor for VEGF-dependent endothelial growth and survival such that VEGF alone would be insufficient to induce the exaggerated angiogenesis associated with ROP and other proliferative retinopathies.

Increased IGFBP3 has also been shown to augment vessel survival (in an IGF-1 independent manner) in a mouse model of oxygen-induced retinopathy and consequently reduces the severity of the disease. Importantly, in premature infants of 30–35 weeks postmenstrual age, IGFBP3 levels were found to be significantly diminished in infants with ROP as compared to those without ROP [73].

The importance of an obligate presence of IGF-1 in adequate concentrations as described above raises the possibility of pharmacological restoration of IGF-1 (and IGFBP3) to *in utero* levels as a strategy for countering the vascular degeneration associated with the first phase of ROP. Clinical trials are being undertaken to address the merits of this strategy in the premature population [81].

10.11
Conclusion

There is emerging evidence for the involvement of the insulin/IGF-1 axis in neurodegerative diseases, highlighting the existence of a common mechanism in these pathologies. As dysglycemia is becoming a true epidemic, it will be important to integrate the notion of preserving sound energy metabolism in future therapeutic approaches to counter neurodegenerative disease. The insulin/IGF-1 system therefore becomes an attractive axis for studies to help curb neurodegeneration.

References

1. Benoit, S.C., Air, E.L., Coolen, L.M., Strauss, R., Jackman, A., Clegg, D.J., Seeley, R.J., and Woods, S.C. (2002) The catabolic action of insulin in the brain is mediated by melanocortins. *J. Neurosci.*, **22**, 9048–9052.
2. Ristow, M. (2004) Neurodegenerative disorders associated with diabetes mellitus. *J. Mol. Med.*, **82**, 510–529.
3. Razay, G. and Wilcock, G.K. (1994) Hyperinsulinaemia and Alzheimer's disease. *Age Ageing*, **23**, 396–399.
4. Smith, L.E. (2003) Pathogenesis of retinopathy of prematurity. *Semin. Neonatol.*, **8**, 469–473.
5. Chan, S.J., Cao, Q.P., and Steiner, D.F. (1990) Evolution of the insulin superfamily: cloning of a hybrid insulin/insulin-like growth factor cDNA from amphioxus. *Proc. Natl. Acad. Sci. U.S.A.*, **87**, 9319–9323.
6. McRory, J.E. and Sherwood, N.M. (1997) Ancient divergence of insulin and insulin-like growth factor. *DNA Cell Biol.*, **16**, 939–949.
7. Pierce, S.B., Costa, M., Wisotzkey, R., Devadhar, S., Homburger, S.A., Buchman, A.R., Ferguson, K.C., Heller, J., Platt, D.M., Pasquinelli, A.A., Liu, L.X., Doberstein, S.K., and Ruvkun, G. (2001) Regulation of DAF-2 receptor signaling by human insulin and ins-1, a member of the unusually large and diverse C. elegans insulin gene family. *Genes Dev.*, **15**, 672–686.
8. Frystyk, J., Bek, T., Flyvbjerg, A., Skjaerbaek, C., and Orskov, H. (2003) The relationship between the circulating IGF system and the presence of retinopathy in Type 1 diabetic patients. *Diabet. Med.*, **20**, 269–276.
9. Firth, S.M. and Baxter, R.C. (2002) Cellular actions of the insulin-like growth factor binding proteins. *Endocr. Rev.*, **23**, 824–854.
10. Bach, M.A., Shen-Orr, Z., Lowe, W.L. Jr., Roberts, C.T., and LeRoith, D. (1991) Insulin-like growth factor I mRNA levels are developmentally regulated in specific regions of the rat brain. *Brain Res. Mol. Brain Res.*, **10**, 43–48.

11. Bartlett, W.P., Li, X.S., Williams, M., and Benkovic, S. (1991) Localization of insulin-like growth factor-1 mRNA in murine central nervous system during postnatal development. *Dev. Biol.*, **147**, 239–250.
12. Lofqvist, C., Willett, K.L., Aspegren, O., Smith, A.C., Aderman, C.M., Connor, K.M., Chen, J., Hellstrom, A., and Smith, L.E. (2009) Quantification and localization of the IGF/insulin system expression in retinal blood vessels and neurons during oxygen-induced retinopathy in mice. *Invest. Ophthalmol. Vis. Sci.*, **50**, 1831–1837.
13. Havrankova, J., Roth, J., and Brownstein, M. (1978) Insulin receptors are widely distributed in the central nervous system of the rat. *Nature*, **272**, 827–829.
14. Unger, J.W., Livingston, J.N., and Moss, A.M. (1991) Insulin receptors in the central nervous system: localization, signalling mechanisms and functional aspects. *Prog. Neurobiol.*, **36**, 343–362.
15. Frank, H.J., Pardridge, W.M., Morris, W.L., Rosenfeld, R.G., and Choi, T.B. (1986) Binding and internalization of insulin and insulin-like growth factors by isolated brain microvessels. *Diabetes*, **35**, 654–661.
16. Pardridge, W.M. (1993) Transport of insulin-related peptides and glucose across the blood-brain barrier. *Ann. N. Y. Acad. Sci.*, **692**, 126–137.
17. Reinhardt, R.R. and Bondy, C.A. (1994) Insulin-like growth factors cross the blood-brain barrier. *Endocrinology*, **135**, 1753–1761.
18. Aguado, F., Sanchez-Franco, F., Rodrigo, J., Cacicedo, L., and Martinez-Murillo, R. (1994) Insulin-like growth factor I-immunoreactive peptide in adult human cerebellar Purkinje cells: co-localization with low-affinity nerve growth factor receptor. *Neuroscience*, **59**, 641–650.
19. Beilharz, E.J., Russo, V.C., Butler, G., Baker, N.L., Connor, B., Sirimanne, E.S., Dragunow, M., Werther, G.A., Gluckman, P.D., Williams, C.E., and Scheepens, A. (1998) Co-ordinated and cellular specific induction of the components of the IGF/IGFBP axis in the rat brain following hypoxic-ischemic injury. *Brain Res. Mol. Brain Res.*, **59**, 119–134.
20. Craner, M.J., Klein, J.P., Black, J.A., and Waxman, S.G. (2002) Preferential expression of IGF-I in small DRG neurons and down-regulation following injury. *Neuroreport*, **13**, 1649–1652.
21. Garcia-Segura, L.M., Perez, J., Pons, S., Rejas, M.T., and Torres-Aleman, I. (1991) Localization of insulin-like growth factor I (IGF-I)-like immunoreactivity in the developing and adult rat brain. *Brain Res.*, **560**, 167–174.
22. Guthrie, K.M., Nguyen, T., and Gall, C.M. (1995) Insulin-like growth factor-1 mRNA is increased in deafferented hippocampus: spatiotemporal correspondence of a trophic event with axon sprouting. *J. Comp. Neurol.*, **352**, 147–160.
23. Pandini, G., Frasca, F., Mineo, R., Sciacca, L., Vigneri, R., and Belfiore, A. (2002) Insulin/insulin-like growth factor I hybrid receptors have different biological characteristics depending on the insulin receptor isoform involved. *J. Biol. Chem.*, **277**, 39684–39695.
24. van Houten, M., Posner, B.I., Kopriwa, B.M., and Brawer, J.R. (1979) Insulin-binding sites in the rat brain: in vivo localization to the circumventricular organs by quantitative radioautography. *Endocrinology*, **105**, 666–673.
25. Peruzzi, F., Prisco, M., Dews, M., Salomoni, P., Grassilli, E., Romano, G., Calabretta, B., and Baserga, R. (1999) Multiple signaling pathways of the insulin-like growth factor 1 receptor in protection from apoptosis. *Mol. Cell Biol.*, **19**, 7203–7215.
26. Kim, J.J. and Accili, D. (2002) Signalling through IGF-I and insulin receptors: where is the specificity? *Growth Horm. IGF Res.*, **12**, 84–90.
27. Apelt, J., Mehlhorn, G., and Schliebs, R. (1999) Insulin-sensitive GLUT4 glucose transporters are colocalized with GLUT3-expressing cells and demonstrate a chemically distinct neuron-specific localization in rat brain. *J. Neurosci. Res.*, **57**, 693–705.
28. Schulingkamp, R.J., Pagano, T.C., Hung, D., and Raffa, R.B. (2000) Insulin receptors and insulin action in the

28. brain: review and clinical implications. Neurosci. Biobehav. Rev., 24, 855–872.
29. Porte, D. Jr., Baskin, D.G., and Schwartz, M.W. (2005) Insulin signaling in the central nervous system: a critical role in metabolic homeostasis and disease from C. elegans to humans. Diabetes, 54, 1264–1276.
30. Dudek, H., Datta, S.R., Franke, T.F., Birnbaum, M.J., Yao, R., Cooper, G.M., Segal, R.A., Kaplan, D.R., and Greenberg, M.E. (1997) Regulation of neuronal survival by the serine–threonine protein kinase Akt. Science, 275, 661–665.
31. Mathews, L.S., Hammer, R.E., Behringer, R.R., D'Ercole, A.J., Bell, G.I., Brinster, R.L., and Palmiter, R.D. (1988) Growth enhancement of transgenic mice expressing human insulin-like growth factor I. Endocrinology, 123, 2827–2833.
32. Reiss, K., Cheng, W., Ferber, A., Kajstura, J., Li, P., Li, B., Olivetti, G., Homcy, C.J., Baserga, R., and Anversa, P. (1996) Overexpression of insulin-like growth factor-1 in the heart is coupled with myocyte proliferation in transgenic mice. Proc. Natl. Acad. Sci. U.S.A., 93, 8630–8635.
33. Bondy, C.A. and Cheng, C.M. (2002) Insulin-like growth factor-1 promotes neuronal glucose utilization during brain development and repair processes. Int. Rev. Neurobiol., 51, 189–217.
34. Carro, E., Nunez, A., Busiguina, S., and Torres-Aleman, I. (2000) Circulating insulin-like growth factor I mediates effects of exercise on the brain. J. Neurosci., 20, 2926–2933.
35. Gluckman, P., Klempt, N., Guan, J., Mallard, C., Sirimanne, E., Dragunow, M., Klempt, M., Singh, K., Williams, C., and Nikolics, K. (1992) A role for IGF-1 in the rescue of CNS neurons following hypoxic-ischemic injury. Biochem. Biophys. Res. Commun., 182, 593–599.
36. Heck, S., Lezoualc'h, F., Engert, S., and Behl, C. (1999) Insulin-like growth factor-1-mediated neuroprotection against oxidative stress is associated with activation of nuclear factor kappaB. J. Biol. Chem., 274, 9828–9835.
37. Russell, J.W. and Feldman, E.L. (1999) Insulin-like growth factor-I prevents apoptosis in sympathetic neurons exposed to high glucose. Horm. Metab. Res., 31, 90–96.
38. Vincent, A.M., Mobley, B.C., Hiller, A., and Feldman, E.L. (2004) IGF-I prevents glutamate-induced motor neuron programmed cell death. Neurobiol. Dis., 16, 407–416.
39. Kermer, P., Klocker, N., Labes, M., and Bahr, M. (2000) Insulin-like growth factor-I protects axotomized rat retinal ganglion cells from secondary death via PI3-K-dependent Akt phosphorylation and inhibition of caspase-3 In vivo. J. Neurosci., 20, 2–8.
40. Fehm, H.L., Perras, B., Smolnik, R., Kern, W., and Born, J. (2000) Manipulating neuropeptidergic pathways in humans: a novel approach to neuropharmacology? Eur. J. Pharmacol., 405, 43–54.
41. Craft, S., Newcomer, J., Kanne, S., Dagogo-Jack, S., Cryer, P., Sheline, Y., Luby, J., Dagogo-Jack, A., and Alderson, A. (1996) Memory improvement following induced hyperinsulinemia in Alzheimer's disease. Neurobiol. Aging, 17, 123–130.
42. Hoppener, J.W., Ahren, B., and Lips, C.J. (2000) Islet amyloid and type 2 diabetes mellitus. N. Engl. J. Med., 343, 411–419.
43. Mosselman, S., Hoppener, J.W., Zandberg, J., van Mansfeld, A.D., Geurts van Kessel, A.H., Lips, C.J., and Jansz, H.S. (1988) Islet amyloid polypeptide: identification and chromosomal localization of the human gene. FEBS Lett., 239, 227–232.
44. Janson, J., Laedtke, T., Parisi, J.E., O'Brien, P., Petersen, R.C., and Butler, P.C. (2004) Increased risk of type 2 diabetes in Alzheimer disease. Diabetes, 53, 474–481.
45. Gasparini, L., Gouras, G.K., Wang, R., Gross, R.S., Beal, M.F., Greengard, P., and Xu, H. (2001) Stimulation of beta-amyloid precursor protein trafficking by insulin reduces intraneuronal beta-amyloid and requires mitogen-activated protein kinase signaling. J. Neurosci., 21, 2561–2570.

46. Qiu, W.Q., Walsh, D.M., Ye, Z., Vekrellis, K., Zhang, J., Podlisny, M.B., Rosner, M.R., Safavi, A., Hersh, L.B., and Selkoe, D.J. (1998) Insulin-degrading enzyme regulates extracellular levels of amyloid beta-protein by degradation. *J. Biol. Chem.*, **273**, 32730–32738.
47. Farris, W., Mansourian, S., Chang, Y., Lindsley, L., Eckman, E.A., Frosch, M.P., Eckman, C.B., Tanzi, R.E., Selkoe, D.J., and Guenette, S. (2003) Insulin-degrading enzyme regulates the levels of insulin, amyloid beta-protein, and the beta-amyloid precursor protein intracellular domain in vivo. *Proc. Natl. Acad. Sci. U.S.A.*, **100**, 4162–4167.
48. Ho, L., Qin, W., Pompl, P.N., Xiang, Z., Wang, J., Zhao, Z., Peng, Y., Cambareri, G., Rocher, A., Mobbs, C.V., Hof, P.R., and Pasinetti, G.M. (2004) Diet-induced insulin resistance promotes amyloidosis in a transgenic mouse model of Alzheimer's disease. *FASEB J.*, **18**, 902–904.
49. Hong, M. and Lee, V.M. (1997) Insulin and insulin-like growth factor-1 regulate tau phosphorylation in cultured human neurons. *J. Biol. Chem.*, **272**, 19547–19553.
50. Schubert, M., Brazil, D.P., Burks, D.J., Kushner, J.A., Ye, J., Flint, C.L., Farhang-Fallah, J., Dikkes, P., Warot, X.M., Rio, C., Corfas, G., and White, M.F. (2003) Insulin receptor substrate-2 deficiency impairs brain growth and promotes tau phosphorylation. *J. Neurosci.*, **23**, 7084–7092.
51. Figlewicz, D.P., Szot, P., Chavez, M., Woods, S.C., and Veith, R.C. (1994) Intraventricular insulin increases dopamine transporter mRNA in rat VTA/substantia nigra. *Brain Res.*, **644**, 331–334.
52. Baird, T.A., Parsons, M.W., Barber, P.A., Butcher, K.S., Desmond, P.M., Tress, B.M., Colman, P.G., Jerums, G., Chambers, B.R., and Davis, S.M. (2002) The influence of diabetes mellitus and hyperglycaemia on stroke incidence and outcome. *J. Clin. Neurosci.*, **9**, 618–626.
53. Liu, Z., Wang, Y., Zhao, W., Ding, J., Mei, Z., Guo, L., Cui, D., and Fei, J. (2001) Peptide derived from insulin with regulatory activity of dopamine transporter. *Neuropharmacology*, **41**, 464–471.
54. Van Woert, M.H. and Mueller, P.S. (1971) Glucose, insulin, and free fatty acid metabolism in Parkinson's disease treated with levodopa. *Clin. Pharmacol. Ther.*, **12**, 360–367.
55. Sandyk, R. (1993) The relationship between diabetes mellitus and Parkinson's disease. *Int. J. Neurosci.*, **69**, 125–130.
56. Sirtori, C.R., Bolme, P., and Azarnoff, D.L. (1972) Metabolic responses to acute and chronic L-dopa administration in patients with parkinsonism. *N. Engl. J. Med.*, **287**, 729–733.
57. Cersosimo, E. and DeFronzo, R.A. (2006) Insulin resistance and endothelial dysfunction: the road map to cardiovascular diseases. *Diabetes Metab. Res. Rev.*, **22**, 423–436.
58. Meyer, J.S., McClintic, K.L., Rogers, R.L., Sims, P., and Mortel, K.F. (1988) Aetiological considerations and risk factors for multi-infarct dementia. *J. Neurol. Neurosurg. Psychiatr.*, **51**, 1489–1497.
59. Ott, A., Stolk, R.P., Hofman, A., van Harskamp, F., Grobbee, D.E., and Breteler, M.M. (1996) Association of diabetes mellitus and dementia: the Rotterdam Study. *Diabetologia*, **39**, 1392–1397.
60. Ott, A., Stolk, R.P., van Harskamp, F., Pols, H.A., Hofman, A., and Breteler, M.M. (1999) Diabetes mellitus and the risk of dementia: The Rotterdam Study. *Neurology*, **53**, 1937–1942.
61. Meigs, J.B. (2003) Epidemiology of the insulin resistance syndrome. *Curr. Diab. Rep.*, **3**, 73–79.
62. Taguchi, A. (2009) Vascular factors in diabetes and Alzheimer's disease. *J. Alzheimer's Dis.*, **16**, 859–864.
63. Roth, A.M. (1977) Retinal vascular development in premature infants. *Am. J. Ophthalmol.*, **84**, 636–640.
64. Michaelson, I. (1948) The mode of development of the vascular system of the retina with some observations on its significance for certain retinal disorders. *Trans. Ophthalmol. Soc. U.K.*, **68**, 137–180.
65. Ashton, N., Ward, B., and Serpell, G. (1954) Effect of oxygen on developing

retinal vessels with particular reference to the problem of retrolental fibroplasia. Br. J. Ophthalmol., **38**, 397–432.

66. Adamis, A.P., Shima, D.T., Yeo, K.T., Yeo, T.K., Brown, L.F., Berse, B., D'Amore, P.A., and Folkman, J. (1993) Synthesis and secretion of vascular permeability factor/vascular endothelial growth factor by human retinal pigment epithelial cells. Biochem. Biophys. Res. Commun., **193**, 631–638.

67. Aiello, L.P., Avery, R.L., Arrigg, P.G., Keyt, B.A., Jampel, H.D., Shah, S.T., Pasquale, L.R., Thieme, H., Iwamoto, M.A., Park, J.E. et al. (1994) Vascular endothelial growth factor in ocular fluid of patients with diabetic retinopathy and other retinal disorders. N. Engl. J. Med., **331**, 1480–1487.

68. Aiello, L.P., Pierce, E.A., Foley, E.D., Takagi, H., Chen, H., Riddle, L., Ferrara, N., King, G.L., and Smith, L.E. (1995) Suppression of retinal neovascularization in vivo by inhibition of vascular endothelial growth factor (VEGF) using soluble VEGF-receptor chimeric proteins. Proc. Natl. Acad. Sci. U.S.A., **92**, 10457–10461.

69. Kinsey, V.E., Arnold, H.J., Kalina, R.E., Stern, L., Stahlman, M., Odell, G., Driscoll, J.M. Jr., Elliott, J.H., Payne, J., and Patz, A. (1977) PaO2 levels and retrolental fibroplasia: a report of the cooperative study. Pediatrics, **60**, 655–668.

70. Lucey, J.F. and Dangman, B. (1984) A reexamination of the role of oxygen in retrolental fibroplasia. Pediatrics, **73**, 82–96.

71. Simons, B.D. and Flynn, J.T. (1999) Retinopathy of prematurity and associated factors. Int. Ophthalmol. Clin., **39**, 29–48.

72. Hellstrom, A., Perruzzi, C., Ju, M., Engstrom, E., Hard, A.L., Liu, J.L., Albertsson-Wikland, K., Carlsson, B., Niklasson, A., Sjodell, L., LeRoith, D., Senger, D.R., and Smith, L.E. (2001) Low IGF-I suppresses VEGF-survival signaling in retinal endothelial cells: direct correlation with clinical retinopathy of prematurity. Proc. Natl. Acad. Sci. U.S.A., **98**, 5804–5808.

73. Lofqvist, C., Chen, J., Connor, K.M., Smith, A.C., Aderman, C.M., Liu, N., Pintar, J.E., Ludwig, T., Hellstrom, A., and Smith, L.E. (2007) IGFBP3 suppresses retinopathy through suppression of oxygen-induced vessel loss and promotion of vascular regrowth. Proc. Natl. Acad. Sci. U.S.A., **104**, 10589–10594.

74. Smith, L. (1997) Essential role of growth hormone in ischemia-induced retinal neovascularization. Science, **276**, 1706–1709.

75. Smith, L.E., Shen, W., Perruzzi, C., Soker, S., Kinose, F., Xu, X., Robinson, G., Driver, S., Bischoff, J., Zhang, B., Schaeffer, J.M., and Senger, D.R. (1999) Regulation of vascular endothelial growth factor-dependent retinal neovascularization by insulin-like growth factor-1 receptor. Nat. Med., **5**, 1390–1395.

76. Langford, K., Nicolaides, K., and Miell, J.P. (1998) Maternal and fetal insulin-like growth factors and their binding proteins in the second and third trimesters of human pregnancy. Hum. Reprod., **13**, 1389–1393.

77. Lassarre, C., Hardouin, S., Daffos, F., Forestier, F., Frankenne, F., and Binoux, M. (1991) Serum insulin-like growth factors and insulin-like growth factor binding proteins in the human fetus. Relationships with growth in normal subjects and in subjects with intrauterine growth retardation. Pediatr. Res., **29**, 219–225.

78. Reece, E.A., Wiznitzer, A., Le, E., Homko, C.J., Behrman, H., and Spencer, E.M. (1994) The relation between human fetal growth and fetal blood levels of insulin-like growth factors I and II, their binding proteins, and receptors. Obstet. Gynecol., **84**, 88–95.

79. Lineham, J.D., Smith, R.M., Dahlenburg, G.W., King, R.A., Haslam, R.R., Stuart, M.C., and Faull, L. (1986) Circulating insulin-like growth factor I levels in newborn premature and full-term infants followed longitudinally. Early Hum. Dev., **13**, 37–46.

80. Lofqvist, C., Engstrom, E., Sigurdsson, J., Hard, A.L., Niklasson, A., Ewald, U., Holmstrom, G., Smith, L.E., and Hellstrom, A. (2006) Postnatal head

growth deficit among premature infants parallels retinopathy of prematurity and insulin-like growth factor-1 deficit. *Pediatrics*, **117**, 1930–1938.

81. Chen, J. and Smith, L. (2007) Retinopathy of prematurity. *Angiogenesis*, **10**, 133–140.

11
Leptin Neuroprotection in the Central Nervous System

Feng Zhang, Suping Wang, Armando P. Signore, Zhongfang Weng, and Jun Chen

11.1
Introduction

11.1.1
Origin, Source, and Structure of Leptin

In 1949, a line of unusually fat mice was observed by Ingalls in the Jackson Laboratory. The mutant mice were obese, with body weights up to three times that of normal animals [1]. The gene responsible for the mutation was identified in 1994 [2] and was named *obese* (*ob*) [1]. The *ob* gene product is a protein made primarily in the adipose tissue. This protein participates in mammalian energy homeostasis by raising energy expenditure and reducing food consumption, leading to weight loss [2, 3]. In 1995, Halaas *et al.* named this adipose protein *leptin*, the name derived from the Greek root *leptos*, meaning "thin" or "slender" [3].

Human leptin consists of 146 amino acids with a molecular weight of approximately 16 kDa [2], and it shares 70% of its sequence homology with leptin from other species. Leptin is a member of the long-chain class-I helical cytokine family [4], which also includes the growth hormone, prolactin, and erythropoietin. The structure of leptin consists of a four-helix bundle with one very short strand segment and two longer interconnected loops. The vast majority is synthesized by white adipose tissue, although other organs including the stomach [5], placenta [6], skeletal muscle [7], heart [8], and brain [9] also produce leptin at much lower levels. There is still much speculation as to the functions of nonadipose-derived leptin, especially in the nervous system.

11.1.2
Functions of Leptin

The primary known effects of leptin on mammalian physiology are suppression of food intake, stimulation of thermogenesis, inhibition of feeding, and regulation of neuroendocrine secretion [10–12]. Leptin binds to receptors on the cell membrane of hypothalamic neurons and regulates their activity. The immediate effects of

Hormones in Neurodegeneration, Neuroprotection, and Neurogenesis.
Edited by Achille G. Gravanis and Synthia H. Mellon
Copyright © 2011 WILEY-VCH Verlag GmbH & Co. KGaA, Weinheim
ISBN: 978-3-527-32627-3

leptin signaling are inhibition of neuropeptide Y and anandamide signaling, which stimulate feeding; and promotion of the effects of the α-melanin stimulating hormone, which represses the desire for food, leading to inhibition of food intake and promotion of catabolic pathways. Leptin also stimulates the secretion of gonadotropin-releasing hormone from the anterior pituitary, regulating puberty and reproduction [10–12].

In addition to its role as a modulator of diet intake, a number of new roles have been recently proposed for leptin: two of the most promising include neurotrophic and neuroprotective effects [13–15]. These new functions implicate leptin in more widespread impacts on the nervous system, including plasticity and maintenance, and repair of nervous system control. These effects will be examined in detail later in the chapter.

11.1.3
Leptin Receptors

Leptin receptor (ObR or LEPR), also known as *CD295*, is the product of the diabetes (*db*) or *Lepr* gene [16], encoding a single-transmembrane domain receptor that belongs to the cytokine receptor class-I superfamily [16, 17]. The *Lepr* gene encodes six alternatively spliced forms of LEPRs, ObR-a through ObR-f [10, 14, 16]. Of the six forms, ObR-b is the only full-length protein product of the *Lepr* gene, consisting of 1162 amino acid residuals, of which 303 form the intracellular signaling motif [11, 14]. It is also the only full-length functional receptor. All other LEPRs are collectively called *short forms* because their intracellular domains are truncated to various degrees and their signaling function is largely lost [11, 14]. The short forms of the LEPR are important for the transport and regulation of the bioavailability of leptin in the central nervous system. The highly truncated ObR-e isoform is a singular case in that it lacks even the membrane-binding portions that all other ObR contain and is instead secreted into the bloodstream, where it can bind circulating leptin, thus regulating the circulating levels of the hormone [16, 18].

The long form of LEPR is most abundantly expressed in the hypothalamus, with the greatest concentration occurring in the arcuate, ventromedial, dorsomedial, and lateral hypothalamic nuclei [19–21]. It is also expressed at lower levels in other regions of the brain including the hippocampus, cerebral cortex, substantia nigra pars compacta, and cerebellum [22, 23]. In contrast, the short forms of LEPR are highly expressed in the choroid plexus, blood–brain barrier (BBB) endothelial cells, and hypothalamus. Much broader and lower levels of short forms also exist throughout most of the other regions of the brain [14, 19, 22, 24].

Once leptin binds to its receptor, it induces homodimerization, which in turn recruits Janus tyrosine kinase 2 (JAK2). Activation of JAK2 occurs via autophosphorylation at Tyr1007/Tyr1008 [14, 25]. The activated JAK2 then phosphorylates tyrosine residues of the intracellular domain of Obr-b and kinases. This mechanism amplifies the leptin signal to provide activation of multiple effector kinases.

Phosphorylation of Obr-b occurs at up to three sites. These phosphorylated tyrosines then become docking sites for further downstream signaling components.

The established docking sites are Tyr985, Tyr1027, and Tyr1138. Phosphorylation of Tyr985 leads to the recruitment and phosphorylation of SH2 domain-containing protein-tyrosine phosphatase (SHP2), which consequently binds and activates growth factor receptor-bound protein 2 (GRB2). GRB2, together with SHP2, activates several signaling pathways including the mitogen-activated protein kinase (MEK) and extracellular signal-related kinase (ERK) signaling pathways [10, 14]. ERK has dual effects on cellular activity including altered gene expression such as c-fos and brain-derived neurotrophic factor (BDNF) via cAMP-response element binding protein (CREB) [14, 22, 26]. ERK also phosphorylates and inhibits a number of proapoptotic proteins such as BAD (binding alcohol dehydrogenase) and Bim-EL [14, 27]. Parallel to MEK/ERK activation, phosphorylation of Tyr1027 recruits the insulin receptor substrate (IRS), leading to activation of phosphoinositide-3 kinase (PI3K) and its downstream effector, Akt [14, 28]. Akt is a critical kinase that modulates neuronal viability with both prosurvival and antiapoptotic activity [29].

A very different but particularly important downstream pathway for leptin action is the signal transducer and activator of transcription 3 (STAT3). Monomers of STAT3 are recruited to phosphorylated Tyr1138. Upon phosphorylation by JAK2, STAT3 dimerizes and is activated [14, 26]. Dimerized STAT3 mediates a series of transcriptional events including several associated with food intake and energy expenditure such as the synthesis of proopiomelanocortin [30]. Expression of a negative feedback regulator of ObR-b phosphorylation by JAK2 is also enhanced by STAT3, the suppressor of cytokine signaling 3 (SOCS3) [14, 31]. STAT3 may also mediate transcription of the antiapoptotic protein Bcl-xL [32].

11.1.4
Leptin Transport across the Blood–Brain Barrier

In order for it to have any biological activity in the brain, leptin released from adipocytes must infiltrate or bypass the BBB from the bloodstream. Leptin can indeed concentrate in the brain, most notably in the region containing the highest concentrations of LEPRs – the hypothalamus. Lower levels of leptin have been shown in other brain structures including the cortex, striatum, and hippocampus [20, 33].

Transport rates across the BBB for leptin have been measured using ^{125}I-leptin and autoradiography. The rate of leptin transport across the BBB is about 20 times faster than that of albumin, with more than half the injected leptin entering the brain within 20 min. The hypothalamus and choroid plexus display the highest radioactivity, while the rest of the parenchyma has relatively low radioactivity. The transport of ^{125}I-leptin across the BBB is inhibited by unlabeled leptin in a dose-dependent manner, indicating that the transport is competitive and saturable. The transport system is one-way, as leptin is cleared from the brain via the cerebrospinal fluid (CSF) [20].

The leptin transport system is composed of two subsystems: a rapid, high-affinity system and a slower, low-affinity system [34]. The high-affinity system is concentrated in the hypothalamus and choroid plexus, with K_m values of 0.2 ng ml^{-1} and 1.1 ng ml^{-1}, respectively. The low-affinity system serves much of the rest of the brain, with K_m values of 130 ng ml^{-1}, 345 ng ml^{-1}, and 88 ng ml^{-1} measured in the cerebral cortex, striatum, and hippocampus, respectively [34]. As a result, leptin uptake in the hypothalamus is about 11–38 times faster than in other brain structures.

The actual mechanisms responsible for leptin transport across the BBB are still not well understood, as a leptin-specific transporter is yet to be discovered. Several hypotheses have been proposed to account for leptin transport [33]. First, leptin may leak through the fenestrated capillaries located in the median eminence and then enter the hypothalamus. Second, the short forms of the LEPR, especially ObR-a, are highly expressed in brain endothelial cells and in choroid plexus capillaries [24, 35]. The short LEPRs may facilitate leptin's entrance into the central nervous system [33]. This hypothesis is supported by a report showing that a mutation in the short LEPR results in a decrease in leptin transport [36]. Alternatively, leptin may use cocarriers for its transport across the BBB. For example, megalin (gp330), or low-density lipoprotein receptor-related protein, is an endocytotic receptor that may mediate leptin transport. It is expressed in the choroid plexus and in the proximal tubules of the kidney, where it seems to be involved in leptin excretion [37].

11.2
Mutation of Leptin or Leptin Receptors

Mutations in both leptin and its receptors occur and manifest as specific phenotypes. Mice that are homozygous for a mutation in the obese gene (*ob/ob*), resulting in the complete lack of any normal leptin protein, are characterized by extreme obesity, hyperglycemia, glucose intolerance, and diminished fertility [1, 2]. The mutation itself is a C to T conversion in the *ob* gene, resulting in a change of an arginine to a stop codon TGA [2]. Patients with the leptin mutation demonstrate morbid obesity and hypogonadism [38].

Several types of mutation in the *Lepr* gene have also been identified. In 1966, Hummel discovered a new mutation in C57Bl/Ks mice that developed diabetes and obesity [39]. The gene responsible was named *diabetes* (*db*), and the homozygous mutation in mouse *db* gene is the cause of the *db/db* phenotype in mice. Later, the *db* gene was found to be identical to the LEPR gene, *Lepr* [16]. The mutant protein has a truncated cytoplasmic region, producing LEPRs that are defective in signal transduction [40]. The Zucker fatty (*fa/fa*) rat is characterized by a different mutation: a missense mutation in the extracellular domain of the LEPR leading to obesity, diabetes, and hypertension [41]. The *fa* mutation of Q269P results in a dramatic structural change in the LEPR, disrupting its leptin-binding function.

11.3
Neurotrophic Role of Leptin

Neurons require continuous albeit very low concentrations of neurotrophic factors for survival. Withdrawal of neurotrophins results in altered neuronal function and eventually, in neuronal demise. For example, neurotrophic factor withdrawal from neuronal cultures leads to neuronal apoptosis [32, 42], and serum withdrawal from neuroblastoma SH-SY5Y cell culture leads to cell death [43]. When leptin is added to the medium as serum withdrawal takes place, neuronal death is attenuated [32, 42, 43]. This suggests that leptin has neurotrophic activity and can at least partially substitute for neurotrophic factors following neurotrophic withdrawal *in vitro*.

Neuronal development requires the interplay between many neurotrophic elements, and leptin may be involved in normal neuronal and glial maturation in the brain [13]. Leptin deficiency in *ob/ob* or *db/db* mice results in decreased brain weight, impaired myelination, and reduced levels of several neuronal and glial marker proteins [13, 44]. Leptin treatment can increase brain weight and protein content, and increase locomotor activity in *ob/ob* mice [13], further supporting leptin's neurotrophic role.

Leptin may also contribute to neurogenesis. In the adult brain, neurogenesis is restricted to only two regions: the hippocampal dentate gyrus and the subventricular zone. These areas contain both immature neuronal precursors and immature neurons. Recent studies have demonstrated the expression of LEPRs on these nondifferentiated neuronal populations in murine and monkey brain [45, 46]. The administration of exogenous leptin can enhance the neurogenesis of hippocampal progenitor cells, indicating that functional LEPRs are a part of their signaling repertoire [46].

Leptin also has a potent angiogenic effect [47, 48]. Administration of leptin has induced neovascularization in corneas of normal rats but not in corneas from *fa/fa* Zucker rats that do not have LEPRs capable of signal transduction [49]. The angiogenic effects of leptin have been shown to be quite robust, inducing more significant increases in vessel length and tortuosity than that caused by the vascular endothelial growth factor [47].

11.4
Leptin Neuroprotection against Disorders of the Central Nervous System

Leptin protects CNS from acute and chronic disorders, which is revealed with both *in vivo* and *in vitro* models as summarized in Tables 11.1 and 11.2.

11.4.1
Acute Neurological Disorders

Brain ischemia is a disorder induced by thrombosis or embolism in the cerebrovascular system, leading to a rapid reduction in the blood supply to specific

Table 11.1 Leptin neuroprotection *in vitro*.

Insult	Neuron	Effect	Mechanism	References
Neurotrophin withdrawal	Hippocampal	Reduces death	STAT3, PI3K	[32]
	DA	Reduces death	MAPK, Akt, NF-κB	[42]
	SH-SY5Y	Reduces death	STAT3, Akt, ERK1/2	[43]
6-OHDA	DA	Reduces death	MAPK, Akt, NF-κB	[42]
	MN9D	Reduces death	STAT, ERK1/2, CREB, BDNF	[22]
NMDA	Cortical and hippocampal	Reduces death	JAK2/STAT3, Akt, Bcl-xL	[32, 50]
OGD	Mouse cortical	Reduces death	NF-κB/c-Rel, Bcl-xL	[9]
	Rat cortical	Reduces death	ERK1/2, STAT3, BDNF	[23]
Retinoic acid-AD	SH-SY5Y and cortical	Reduces tau phosphorylation	AMPK	[51, 52]
Aβ expression-AD	SH-SY5Y and cortical	Reduces Aβ level	AMPK	[51]
Fe^{2+}	Hippocampal	Reduces death	JAK2/STAT3, Akt, Bcl-xL	[32]
Electronic stimulation	Hippocampal slices	Inhibits synaptic responses in CA1	JAK2, PI3K	[53]

brain regions. Subsequent damage to the brain can be somewhat mild to very severe including neuronal dysfunction, loss of neuronal populations, and even death. Stroke is particularly difficult to treat due to its unpredictability, the extreme vulnerability of neurons to ischemia, and the limitations of available medication.

The major causes of ischemic neuronal death are glutamate excitotoxicity, oxidative stress, and apoptosis [29, 56]. Several lines of evidence demonstrate that leptin can target these processes and consequently reduce neuronal injury after ischemia. The antiexcitotoxicity ability of leptin was initially demonstrated *in vitro* by Dicou *et al.* using murine neurons [50]. They showed that leptin pretreatment of primary neuronal culture prevented excitotoxic neuronal death induced by N-methyl-D-aspartate (NMDA), and that this neuroprotective effect of leptin could be compromised by AG490, a JAK2 inhibitor. This was confirmed later by Guo *et al.* [32], who reported that leptin pretreatment of rat hippocampal neurons increased neuronal survival after exposure to NMDA.

Apoptosis is a type of programmed cell death characterized by a variety of morphological changes including membrane blebbing, cell shrinkage, DNA fragmentation,

Table 11.2 Leptin neuroprotection *in vivo*.

Model	Species	Effect	Mechanism	References
Ischemia	Rat-global	Reduces CA1 neuronal death	PI3K/Akt, ERK1/2	[54]
	Mouse-tMCAO	Reduces infarct	ERK1/2, CREB, STAT3	[23]
	Mouse-pMCAO	Reduces infarct	NF-κB, Bcl-xL	[9]
Seizure	Mouse	Reduces CA3 neuronal death	STAT3, PI3K	[32, 53]
	Rat	Anticonvulsive	JAK2, PI3K	[53]
PD	Mouse	Reduces DA neuronal death	STAT3, ERK 1/2, CREB, BDNF	[22]
AD	Mouse	Reduce Aβ level	AMPK	[51, 55]

Abbreviation: AD, Alzheimer's disease; AMPK, AMP-activated protein kinase; BDNF, brain-derived neurotrophic factor; CREB, cAMP-response element binding; ERK, extracellular signal-related kinase; JAK2, Janus tyrosine kinase 2; MCAO; middle cerebral artery occlusion; NF-κB, nuclear factor-kappa B; PD, Parkinson's disease; PI3K, phosphoinositide-3 kinase; and STAT3, signal transducer and activators of transcription 3.

and chromatin condensation. Apoptosis is the major form of neuronal death induced by hypoxia ischemia in rat pups and in global ischemia in adult rats; it also plays a critical role in the development of the infarct following focal cerebral ischemia, especially in the penumbral region [29, 57]. Leptin has been shown to attenuate apoptosis induced by the removal of serum in a dose-dependent manner, in cultured SH-SY5Y neuroblastoma cells [43]. In similar studies, leptin has reduced neuronal apoptosis caused by neurotrophic factor withdrawal, excitotoxic and oxidative insults [32], and oxygen–glucose deprivation [9]. The antiapoptotic mechanisms of leptin involve the JAK2/STAT3, PI3k/Akt, and MEK/ERK signaling pathways, and up-regulation of the antiapoptotic protein Bcl-xL in neurons [9, 32, 43].

Leptin is also effective in attenuating ischemic injury in animal models. Global brain ischemia causes selective and delayed hippocampal CA1 neuronal death, and mimics brain injury in cardiac arrest and resuscitation. Leptin has been shown to markedly increase the number of surviving CA1 neurons following experimental global brain ischemia [54]. The same study also demonstrated that leptin neuroprotection is mediated via its receptor, as the infusion of a specific leptin antagonist 10 min prior to ischemia abolished the prosurvival effects of leptin [54]. Thrombotic and embolic stroke models have also been tested. Exogenous leptin administration can reduce infarct volumes in focal brain ischemia induced by middle cerebral artery occlusion in mice [9, 23]. This neuroprotection is dose-dependent and also mediated by the LEPR, as the protection requires activation of downstream signaling pathways including MAPK (mitogen-activated protein kinase), STAT3, GRB2, and nuclear factor-kappa B (NF-κB) [9, 23]. These findings make leptin a promising prospective agent for the treatment of stroke. One possible approach is to use leptin in conjunction with tPA; such a combination treatment could allow leptin to extend the practical time window for successful tPA treatment.

Recently, an anticonvulsive effect has been postulated for leptin [58] based on the reports that leptin induced NMDA receptor–mediated long-term depression [58] and inhibited electrical activities in the hippocampus [59, 60]. In support of this hypothesis, it has been demonstrated that a ketogenic diet in rat increases serum levels of leptin and suppresses seizure activities [61, 62]. More directly, in animal models, leptin can reduce the duration and frequency of focal seizures induced by neocortical injections of 4-aminopyridine as seen in rats; it can also delay the onset of generalized seizures induced by pentylenetetrazole as seen in mice. The underlying mechanism is associated with an inhibitory effect of leptin on synaptic transmission [53]. This inhibition is mediated by LEPRs, as leptin has failed to inhibit synaptic activities in hippocampal slices prepared from db/db mice [53].

In the kainate-induced seizure model, injection of kainate into the murine hippocampus causes neuronal degeneration in the CA3 and CA1 subregions. This hippocampal neuron loss then induces seizures in most animals, with db/db mice showing much greater seizure activity [32]. Leptin infusion into the cerebral ventricle has reduced kainate-induced neuronal death. The activation of STAT3 and PI3K signaling pathways by leptin plays a key role in leptin-mediated inhibition of seizure [32, 53]. Taken together, these reports demonstrate that leptin can suppress the initiation and spread of seizures, suggesting that leptin may have therapeutic potential in the clinical management of epilepsy.

11.4.2
Neurodegenerative Diseases and Other Disorders

The pathological hallmarks of Alzheimer's disease (AD) include senile plaques and neurofibrillary tangles (NFTs). The plaques are extracellular, consisting primarily of amyloid beta (Aβ) peptides. The NFT are intracellular and result from hyperphosphorylation of a microtubule-interacting protein known as *tau*, leading to its aggregation. Both Aβ and tau are considered neurotoxic, and excessive accumulation leads to neuronal demise and subsequently, to cognitive decline.

Cognitive loss in the elderly is inversely associated with serum levels of leptin [63]. Patients with higher leptin concentrations demonstrate less cognitive decline compared to those with lower serum leptin levels. In keeping with the lower levels of CNS leptin, individuals with midlife obesity have an increased tendency to develop dementia later in life when compared to the general population [64]. These correlations suggest that leptin may have a neurotrophic effect by preventing neuronal loss and dementia in the AD population group.

Experiments using AD animal models have shown that leptin confers beneficial effects. One line of AD-transgenic mice has lower serum levels of leptin, and leptin supplementation has reduced the brain Aβ levels of these mice [51, 55].

Additionally, leptin has been very effective in reducing tau phosphorylation and aggregation in a concentration- and time-dependent manner [52]. AMP-activated protein kinase (AMPK) also plays an important role in leptin's neuroprotective effects against AD [51, 52, 65]. Following its activation by leptin, AMPK can decrease neuronal β-secretase activity and regulate the phosphorylation of tau [51, 55]. Thus, leptin holds promise as an approach for the treatment of AD.

Parkinson's disease (PD) is a neurodegenerative movement disorder characterized by chronic and progressive impairment of the control of movement. Its primary clinical symptoms, such as muscle rigidity, tremor, and bradykinesia, are caused by the loss of dopaminergic (DA) neurons in the substantia nigra pars compacta. These DA neurons form the nigrostriatal pathway to the striatum and are essential for the modulation of movement, including limiting excessive movements while at rest [14, 66].

Leptin may play a role in the homeostatic regulation of dopamine transport and storage. In Zucker *fa/fa* rats, DA transporters have been shown to be up-regulated in the substantia nigra, suggesting that there may be increased DA clearance, and perhaps decreased DA signaling in this leptin-deficient rat [67]. In line with this suggestion, *ob/ob* mice exhibit decreased overall dopamine stores in their midbrain dopamine neurons leading to diminished neurotransmission capacity [68].

A direct neuroprotective effect of leptin has been reported using the 6-hydroxydopamine (6-OHDA) PD model, using both *in vitro* and *in vivo* studies [22]. Leptin-treated DA neurons have been shown to be less susceptible to 6-OHDA toxicity, and leptin has preserved the function of the nigrostriatal tract [22]. Leptin receptor-dependent neuroprotective mechanisms include activation of the JAK2–STAT and MEK/ERK signaling pathways. Furthermore, another vital effector is activation of ERK1/2, resulting in enhanced nuclear localization of the transcription factor CREB [22]. Use of an LEPR antagonist or knockdown of the leptin adaptor signaling proteins JAK2 or GRB2 has resulted in loss of protection, suggesting that the protection is mediated by LEPR [22].

Leptin can also up-regulate the expression of BDNF [22, 69] and may play an important role in leptin-mediated neuroprotection against PD. A well-known prosurvival factor for DA neurons, BDNF is diminished in PD [70]. BDNF acts by binding to the TrkB receptor kinase and subsequently activating downstream signaling pathways that include PI3-K and MEK/ERK; thus, BDNF uses many of the same intermediary signaling molecules as leptin [14]. As both leptin and BDNF activate common signaling pathways, leptin may induce positive feedback by increasing BDNF expression that may, in turn, further activate those same signals [14].

Patients with Huntington's disease (HD) have decreased plasma levels of leptin in the CSF [71]. In a mouse model of HD the reduction of circulating leptin also occurs, starting at early stages and becoming more pronounced as the animal ages and the disease-like state progresses [72]. Although these results are suggestive, it remains to be tested whether leptin can prevent neuronal loss in HD.

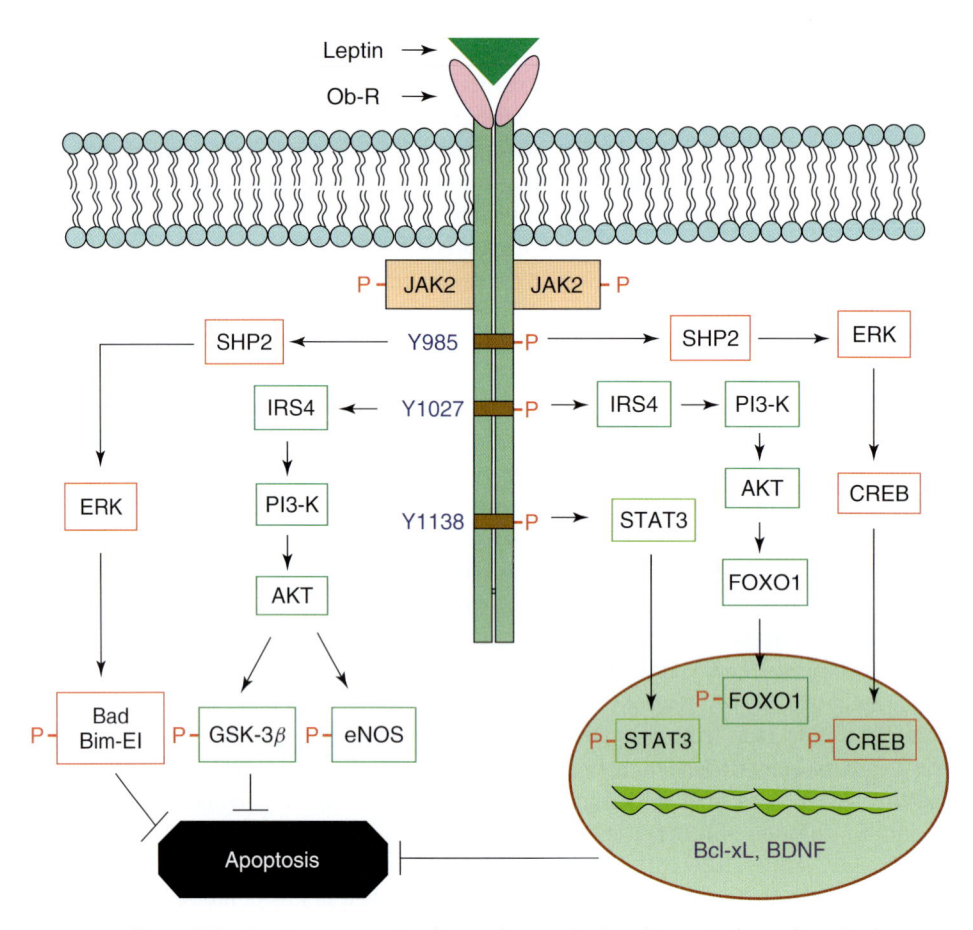

Figure 11.1 Leptin receptor signaling and neuroprotective pathways. The left half of the diagram demonstrates the posttranslational regulation of leptin-mediated neuroprotective effects that primarily occur in acute conditions such as stroke and seizures. The right half shows the translational regulation of leptin neuroprotection that might occur in chronic conditions such as Parkinson's disease. Once leptin binding activates its receptor, the second messenger kinase JAK2 contributes to three cascades. The phosphorylation of Y985 recruits SH2 domain-containing protein-tyrosine phosphatase PTPN11 (SHP2) to the intracellular domain of the leptin receptor. Activation of the SHP2 will then activate the MEK/ERK signaling pathway. One consequence of activating this particular pathway is the stimulation of BDNF production, which occurs via phosphorylation of the transcription factor cAMP-response binding element (CREB). JAK2 also recruits and directly activates STAT3 via Y1138. STAT3 dimerization allows it to translocate to the nucleus and affect the transcription of a number of genes that mediate neuronal activity and survival. The phosphorylation of Y1077 recruits IRS4, which sequentially activates PI3-K and Akt. Further downstream signaling by these cascades induces the inhibition of GSK-3β glycogen synthase kinase and transcriptional control via forkhead box 01 (FOXO1), which is involved in modulation of gluconeogenic responses.

11.4.3
Leptin Neuroprotective Mechanisms

Like other peptide hormones, leptin influences the nervous system by first binding to its receptor on the surface of target neurons. LEPR activation may also be required for leptin's neuroprotective effect. In support of this ligand-receptor and functional specificity, studies have shown that leptin neuroprotection can be largely blocked by a specific leptin antagonist made up of three mutant leptin proteins [22, 54]. These three mutants bind LEPRs with an affinity similar to that of genuine leptin but do not exert any biological effect because of the mutations (L39A/D40A/F41A) [73]. Furthermore, the neuroprotective effect of leptin can be attenuated by interrupting its downstream signaling pathways such as JAK2/STAT3, PI3K/Akt, and MEK/ERK with specific kinase inhibitors [14, 15], indicating the importance of LEPR and its downstream signaling activation.

Among these signaling pathways, the JAK2/STAT3 pathway primarily mediates new protein synthesis, while PI3K/Akt and MEK/ERK contribute to posttranslational modification and influence protein synthesis; in particular, protein phosphorylation [14, 29, 74]. Therefore, their roles in leptin-mediated neuroprotection may vary under different conditions (Figure 11.1). In acute stress, such as brain ischemia or seizure, posttranslational modification via ERK or Akt plays a major role in neuroprotection [23, 53, 54], while translational regulation may play only a small role because of protein synthesis inhibition and depletion of energy [75]. In subacute or chronic conditions such as neurotrophin withdrawal or neurodegenerative diseases, translational regulation, such as the up-regulation of BDGF and Bcl-xL, may become more important in leptin-mediated neuroprotection [22, 32, 43].

11.5
Significance

Human leptin *per se* is a natural peptide circulating in the blood and thus, it should be safe to administer exogenous leptin to humans. Indeed, it has been safely injected into human bodies to treat morbid obesity [38]. In addition to its role as a moderator of diet intake, leptin is now established as a neurotrophin and neuroprotective agent. Leptin seems a highly promising candidate for the treatment of a variety of neurological disorders including AD, PD, and ischemic stroke.

References

1. Ingalls, A.M., Dickie, M.M., and Snell, G.D. (1950) Obese, a new mutation in the house mouse. *J. Hered.*, **41**, 317–318.

2. Zhang, Y., Proenca, R., Maffei, M., Barone, M., Leopold, L., and Friedman, J.M. (1994) Positional cloning of the mouse obese gene and

its human homologue. *Nature*, **372**, 425–432.
3. Halaas, J.L., Gajiwala, K.S., Maffei, M., Cohen, S.L., Chait, B.T., Rabinowitz, D., Lallone, R.L., Burley, S.K., and Friedman, J.M. (1995) Weight-reducing effects of the plasma protein encoded by the obese gene. *Science*, **269**, 543–546.
4. Huising, M.O., Kruiswijk, C.P., and Flik, G. (2006) Phylogeny and evolution of class-I helical cytokines. *J. Endocrinol.*, **189**, 1–25.
5. Bado, A., Levasseur, S., Attoub, S., Kermorgant, S., Laigneau, J.P., Bortoluzzi, M.N., Moizo, L., Lehy, T., Guerre-Millo, M., Le Marchand-Brustel, Y., and Lewin, M.J. (1998) The stomach is a source of leptin. *Nature*, **394**, 790–793.
6. Hoggard, N., Hunter, L., Duncan, J.S., Williams, L.M., Trayhurn, P., and Mercer, J.G. (1997) Leptin and leptin receptor mRNA and protein expression in the murine fetus and placenta. *Proc. Natl. Acad. Sci. U.S.A.*, **94**, 11073–11078.
7. Wang, J., Liu, R., Hawkins, M., Barzilai, N., and Rossetti, L. (1998) A nutrient-sensing pathway regulates leptin gene expression in muscle and fat. *Nature*, **393**, 684–688.
8. Purdham, D.M., Zou, M.X., Rajapurohitam, V., and Karmazyn, M. (2004) Rat heart is a site of leptin production and action. *Am. J. Physiol. Heart Circ. Physiol.*, **287**, H2877–H2884.
9. Valerio, A., Dossena, M., Bertolotti, P., Boroni, F., Sarnico, I., Faraco, G., Chiarugi, A., Frontini, A., Giordano, A., Liou, H.C., Grazia De Simoni, M., Spano, P., Carruba, M.O., Pizzi, M., and Nisoli, E. (2009) Leptin is induced in the ischemic cerebral cortex and exerts neuroprotection through NF-kappaB/c-Rel-dependent transcription. *Stroke*, **40**, 610–617.
10. Hegyi, K., Fulop, K., Kovacs, K., Toth, S., and Falus, A. (2004) Leptin-induced signal transduction pathways. *Cell Biol. Int.*, **28**, 159–169.
11. Friedman, J.M. and Halaas, J.L. (1998) Leptin and the regulation of body weight in mammals. *Nature*, **395**, 763–770.
12. Friedman, J.M. (2009) Leptin at 14 y of age: an ongoing story. *Am. J. Clin. Nutr.*, **89**, 973S–979S.
13. Ahima, R.S., Bjorbaek, C., Osei, S., and Flier, J.S. (1999) Regulation of neuronal and glial proteins by leptin: implications for brain development. *Endocrinology*, **140**, 2755–2762.
14. Signore, A.P., Zhang, F., Weng, Z., Gao, Y., and Chen, J. (2008) Leptin neuroprotection in the CNS: mechanisms and therapeutic potentials. *J. Neurochem.*, **106**, 1977–1990.
15. Tang, B.L. (2008) Leptin as a neuroprotective agent. *Biochem. Biophys. Res. Commun.*, **368**, 181–185.
16. Tartaglia, L.A., Dembski, M., Weng, X., Deng, N., Culpepper, J., Devos, R., Richards, G.J., Campfield, L.A., Clark, F.T., Deeds, J., Muir, C., Sanker, S., Moriarty, A., Moore, K.J., Smutko, J.S., Mays, G.G., Wool, E.A., Monroe, C.A., and Tepper, R.I. (1995) Identification and expression cloning of a leptin receptor, OB-R. *Cell*, **83**, 1263–1271.
17. Baumann, H., Morella, K.K., White, D.W., Dembski, M., Bailon, P.S., Kim, H., Lai, C.F., and Tartaglia, L.A. (1996) The full-length leptin receptor has signaling capabilities of interleukin 6-type cytokine receptors. *Proc. Natl. Acad. Sci. U.S.A.*, **93**, 8374–8378.
18. Gallardo, N., Arribas, C., Villar, M., Ros, M., Carrascosa, J.M., Martinez, C., and Andres, A. (2005) ObRa and ObRe are differentially expressed in adipose tissue in aged food-restricted rats: effects on circulating soluble leptin receptor levels. *Endocrinology*, **146**, 4934–4942.
19. Fei, H., Okano, H.J., Li, C., Lee, G.H., Zhao, C., Darnell, R., and Friedman, J.M. (1997) Anatomic localization of alternatively spliced leptin receptors (Ob-R) in mouse brain and other tissues. *Proc. Natl. Acad. Sci. U.S.A.*, **94**, 7001–7005.
20. Banks, W.A., Kastin, A.J., Huang, W., Jaspan, J.B., and Maness, L.M. (1996) Leptin enters the brain by a saturable system independent of insulin. *Peptides*, **17**, 305–311.
21. Mercer, J.G., Hoggard, N., Williams, L.M., Lawrence, C.B., Hannah, L.T., and Trayhurn, P. (1996) Localization

of leptin receptor mRNA and the long form splice variant (Ob-Rb) in mouse hypothalamus and adjacent brain regions by in situ hybridization. *FEBS Lett.*, **387**, 113–116.
22. Weng, Z., Signore, A.P., Gao, Y., Wang, S., Zhang, F., Hastings, T., Yin, X.M., and Chen, J. (2007) Leptin protects against 6-hydroxy-dopamine-induced dopaminergic cell death via mitogen-activated protein kinase signaling. *J. Biol. Chem.*, **282**, 34479–34491.
23. Zhang, F., Wang, S., Signore, A.P., and Chen, J. (2007) Neuroprotective effects of leptin against ischemic injury induced by oxygen-glucose deprivation and transient cerebral ischemia. *Stroke*, **38**, 2329–2336.
24. Golden, P.L., Maccagnan, T.J., and Pardridge, W.M. (1997) Human blood–brain barrier leptin receptor. Binding and endocytosis in isolated human brain microvessels. *J. Clin. Invest.*, **99**, 14–18.
25. Taga, T. and Kishimoto, T. (1997) Gp130 and the interleukin-6 family of cytokines. *Annu. Rev. Immunol.*, **15**, 797–819.
26. Banks, A.S., Davis, S.M., Bates, S.H., and Myers, M.G. Jr. (2000) Activation of downstream signals by the long form of the leptin receptor. *J. Biol. Chem.*, **275**, 14563–14572.
27. Cheung, E.C. and Slack, R.S. (2004) Emerging role for ERK as a key regulator of neuronal apoptosis. *Sci. STKE*, **2004**, PE45.
28. Wauman, J., De Smet, A.S., Catteeuw, D., Belsham, D., and Tavernier, J. (2008) Insulin receptor substrate 4 couples the leptin receptor to multiple signaling pathways. *Mol. Endocrinol.*, **22**, 965–977.
29. Zhang, F., Yin, W., and Chen, J. (2004) Apoptosis in cerebral ischemia: executional and regulatory signaling mechanisms. *Neurol. Res.*, **26**, 835–845.
30. Cowley, M.A., Smart, J.L., Rubinstein, M., Cerdan, M.G., Diano, S., Horvath, T.L., Cone, R.D., and Low, M.J. (2001) Leptin activates anorexigenic POMC neurons through a neural network in the arcuate nucleus. *Nature*, **411**, 480–484.
31. Bjorbaek, C., Elmquist, J.K., Frantz, J.D., Shoelson, S.E., and Flier, J.S. (1998) Identification of SOCS-3 as a potential mediator of central leptin resistance. *Mol. Cell*, **1**, 619–625.
32. Guo, Z., Jiang, H., Xu, X., Duan, W., and Mattson, M.P. (2008) Leptin-mediated cell survival signaling in hippocampal neurons mediated by JAK STAT3 and mitochondrial stabilization. *J. Biol. Chem.*, **283**, 1754–1763.
33. Ziylan, Y.Z., Baltaci, A.K., and Mogulkoc, R. (2009) Leptin transport in the central nervous system. *Cell Biochem. Funct.*, **27**, 63–70.
34. Zlokovic, B.V., Jovanovic, S., Miao, W., Samara, S., Verma, S., and Farrell, C.L. (2000) Differential regulation of leptin transport by the choroid plexus and blood–brain barrier and high affinity transport systems for entry into hypothalamus and across the blood-cerebrospinal fluid barrier. *Endocrinology*, **141**, 1434–1441.
35. Merino, B., Diez-Fernandez, C., Ruiz-Gayo, M., and Somoza, B. (2006) Choroid plexus epithelial cells co-express the long and short form of the leptin receptor. *Neurosci. Lett.*, **393**, 269–272.
36. Kastin, A.J., Pan, W., Maness, L.M., Koletsky, R.J., and Ernsberger, P. (1999) Decreased transport of leptin across the blood–brain barrier in rats lacking the short form of the leptin receptor. *Peptides*, **20**, 1449–1453.
37. Dietrich, M.O., Spuch,C., Antequera, D., Rodal, I., de Yebenes, J.G., Molina, J.A., Bermejo, F., and Carro, E. (2008) Megalin mediates the transport of leptin across the blood-CSF barrier. *Neurobiol. Aging*, **29**, 902–912.
38. Licinio, J., Caglayan, S., Ozata, M., Yildiz, B.O., de Miranda, P.B., O'Kirwan, F., Whitby, R., Liang, L., Cohen, P., Bhasin, S., Krauss, R.M., Veldhuis, J.D., Wagner, A.J., DePaoli, A.M., McCann, S.M., and Wong, M.L. (2004) Phenotypic effects of leptin replacement on morbid obesity, diabetes mellitus, hypogonadism, and behavior in

leptin-deficient adults. *Proc. Natl. Acad. Sci. U.S.A.*, **101**, 4531–4536.

39. Hummel, K.P., Dickie, M.M., and Coleman, D.L. (1966) Diabetes, a new mutation in the mouse. *Science*, **153**, 1127–1128.
40. Lee, G.H., Proenca, R., Montez, J.M., Carroll, K.M., Darvishzadeh, J.G., Lee, J.I., and Friedman, J.M. (1996) Abnormal splicing of the leptin receptor in diabetic mice. *Nature*, **379**, 632–635.
41. Takaya, K., Ogawa, Y., Isse, N., Okazaki, T., Satoh, N., Masuzaki, H., Mori, K., Tamura, N., Hosoda, K., and Nakao, K. (1996) Molecular cloning of rat leptin receptor isoform complementary DNAs--identification of a missense mutation in Zucker fatty (fa/fa) rats. *Biochem. Biophys. Res. Commun.*, **225**, 75–83.
42. Oldreive, C.E., Harvey, J., and Doherty, G.H. (2008) Neurotrophic effects of leptin on cerebellar Purkinje but not granule neurons in vitro. *Neurosci. Lett.*, **438**, 17–21.
43. Russo, V.C., Metaxas, S., Kobayashi, K., Harris, M., and Werther, G.A. (2004) Antiapoptotic effects of leptin in human neuroblastoma cells. *Endocrinology*, **145**, 4103–4112.
44. van der Kroon, P.H., and Speijers, G.J. (1979) Brain deviations in adult obese-hyperglycemic mice (ob/ob). *Metabolism*, **28**, 1–3.
45. Tonchev, A.B. (2008) Expression of leptin receptor in progenitor cell niches of adult monkey brain. *Comptes Rendus De L Academie Bulgare Des. Sci.*, **61**, 1219–1224.
46. Garza, J.C., Guo, M., Zhang, W., and Lu, X.Y. (2008) Leptin increases adult hippocampal neurogenesis in vivo and in vitro. *J. Biol. Chem.*, **283**, 18238–18247.
47. Talavera-Adame, D., Xiong, Y., Zhao, T., Arias, A.E., Sierra-Honigmann, M.R., and Farkas, D.L. (2008) Quantitative and morphometric evaluation of the angiogenic effects of leptin. *J. Biomed. Opt.*, **13**, 064017.
48. Segal-Lieberman, G., Bradley, R.L., Kokkotou, E., Carlson, M., Trombly, D.J., Wang, X., Bates, S., Myers, M.G., Flier, J.S., and Maratos-Flier, E. Jr. (2003) Melanin-concentrating hormone is a critical mediator of the leptin-deficient phenotype. *Proc. Natl. Acad. Sci. U.S.A*, **100**, 10085–10090.
49. Sierra-Honigmann, M.R., Nath, A.K., Murakami, C., García-Cardeña, G., Papapetropoulos, A., Sessa, W.C., Madge, L.A., Schechner, J.S., Schwabb, M.B., Polverini, P.J., and Flores-Riveros, J.R. (1998) Biological action of leptin as an angiogenic factor. *Science*, **281**, 1683–1686.
50. Dicou, E., Attoub, S., and Gressens, P. (2001) Neuroprotective effects of leptin in vivo and in vitro. *Neuroreport*, **12**, 3947–3951.
51. Greco, S.J., Sarkar, S., Johnston, J.M., and Tezapsidis, N. (2009) Leptin regulates tau phosphorylation and amyloid through AMPK in neuronal cells. *Biochem. Biophys. Res. Commun.*, **380**, 98–104.
52. Greco, S.J., Sarkar, S., Johnston, J.M., Zhu, X., Su, B., Casadesus, G., Ashford, J.W., Smith, M.A., and Tezapsidis, N. (2008) Leptin reduces Alzheimer's disease-related tau phosphorylation in neuronal cells. *Biochem. Biophys. Res. Commun.*, **376**, 536–541.
53. Xu, L., Rensing, N., Yang, X.F., Zhang, H.X., Thio, L.L., Rothman, S.M., Weisenfeld, A.E., Wong, M., and Yamada, K.A. (2008) Leptin inhibits 4-aminopyridine- and pentylenetetrazole-induced seizures and AMPAR-mediated synaptic transmission in rodents. *J. Clin. Invest.*, **118**, 272–280.
54. Zhang, F. and Chen, J. (2008) Leptin protects hippocampal CA1 neurons against ischemic injury. *J. Neurochem.*, **107**, 578–587.
55. Fewlass, D.C., Noboa, K., Pi-Sunyer, F.X., Johnston, J.M., Yan, S.D., and Tezapsidis, N. (2004) Obesity-related leptin regulates Alzheimer's Abeta. *FASEB J.*, **18**, 1870–1878.
56. Lipton, P. (1999) Ischemic cell death in brain neurons. *Physiol. Rev.*, **79**, 1431–1568.
57. Doyle, K.P., Simon, R.P., and Stenzel-Poore, M.P. (2008) Mechanisms of ischemic brain damage. *Neuropharmacology*, **55**, 310–318.

58. Durakoglugil, M., Irving, A.J., and Harvey, J. (2005) Leptin induces a novel form of NMDA receptor-dependent long-term depression. *J. Neurochem.*, **95**, 396–405.
59. Shanley, L.J., Irving, A.J., Rae, M.G., Ashford, M.L., and Harvey, J. (2002) Leptin inhibits rat hippocampal neurons via activation of large conductance calcium-activated K^+ channels. *Nat. Neurosci.*, **5**, 299–300.
60. Shanley, L.J., O'Malley, D., Irving, A.J., Ashford, M.L., and Harvey, J. (2002) Leptin inhibits epileptiform-like activity in rat hippocampal neurones via PI 3-kinase-driven activation of BK channels. *J. Physiol.*, **545**, 933–944.
61. Thio, L.L., Erbayat-Altay, E., Rensing, N., and Yamada, K.A. (2006) Leptin contributes to slower weight gain in juvenile rodents on a ketogenic diet. *Pediatr. Res.*, **60**, 413–417.
62. Murphy, P. (2005) Use of the ketogenic diet as a treatment for epilepsy refractory to drug treatment. *Expert. Rev. Neurother.*, **5**, 769–775.
63. Holden, K.F., Lindquist, K., Tylavsky, F.A., Rosano, C., Harris, T.B., and Yaffe, K. (2008) Serum leptin level and cognition in the elderly: findings from the Health ABC Study. *Neurobiol. Aging*, **30**, 1483–1489.
64. Whitmer, R.A., Gustafson, D.R., Barrett-Connor, E., Haan, M.N., Gunderson, E.P., and Yaffe, K. (2008) Central obesity and increased risk of dementia more than three decades later. *Neurology*, **71**, 1057–1064.
65. Uotani, S., Abe, T., and Yamaguchi, Y. (2006) Leptin activates AMP-activated protein kinase in hepatic cells via a JAK2-dependent pathway. *Biochem. Biophys. Res. Commun.*, **351**, 171–175.
66. Brotchie, J. and Fitzer-Attas, C. (2009) Mechanisms compensating for dopamine loss in early Parkinson disease. *Neurology*, **72**, S32–S38.
67. Figlewicz, D.P., Patterson, T.A., Johnson, L.B., Zavosh, A., Israel, P.A., and Szot, P. (1998) Dopamine transporter mRNA is increased in the CNS of Zucker fatty (fa/fa) rats. *Brain Res. Bull.*, **46**, 199–202.
68. Roseberry, A.G., Painter, T., Mark, G.P., and Williams, J.T. (2007) Decreased vesicular somatodendritic dopamine stores in leptin-deficient mice. *J. Neurosci.*, **27**, 7021–7027.
69. Komori, T., Morikawa, Y., Nanjo, K., and Senba, E. (2006) Induction of brain-derived neurotrophic factor by leptin in the ventromedial hypothalamus. *Neuroscience*, **139**, 1107–1115.
70. Nagatsu, T., Mogi, M., Ichinose, H., and Togari, A. (2000) Changes in cytokines and neurotrophins in Parkinson's disease. *J. Neural. Transm. Suppl.*, **60**, 277–290.
71. Popovic, V., Svetel, M., Djurovic, M., Petrovic, S., Doknic, M., Pekic, S., Miljic, D., Milic, N., Glodic, J., Dieguez, C., Casanueva, F.F., and Kostic, V. (2004) Circulating and cerebrospinal fluid ghrelin and leptin: potential role in altered body weight in Huntington's disease. *Eur. J. Endocrinol.*, **151**, 451–455.
72. Phan, J., Hickey, M.A., Zhang, P., Chesselet, M.F., and Reue, K. (2009) Adipose tissue dysfunction tracks disease progression in two Huntington's disease mouse models. *Hum. Mol. Genet.*, **18**, 1006–1016.
73. Gertler, A. (2006) Development of leptin antagonists and their potential use in experimental biology and medicine. *Trends Endocrinol. Metab.*, **17**, 372–378.
74. Li, W.X. (2008) Canonical and non-canonical JAK-STAT signaling. *Trends Cell Biol.*, **18**, 545–551.
75. DeGracia, D.J., Jamison, J.T., Szymanski, J.J., and Lewis, M.K. (2008) Translation arrest and ribonomics in post-ischemic brain: layers and layers of players. *J. Neurochem.*, **106**, 2288–2301.

12
Somatostatin and Neuroprotection in Retina
Kyriaki Thermos

12.1
Introduction

The neuropeptide somatostatin (somatotropin release inhibitory factor, SRIF) was originally identified as the major inhibitor of the release of growth hormone (GH) from the pituitary [1–3]. It is a cyclic tetradecapeptide, which is widely distributed in the peripheral and central nervous system (CNS). It mediates a diverse number of physiological actions by interacting with specific receptors belonging to the GPCR family [4–7]. Five SRIF receptor subtypes have been cloned, namely sst_{1-5}, with sst_2 existing as two-splice variants sst_{2A} and sst_{2B} [7] (for nomenclature see [8]). These receptors are expressed differentially in different tissues, coupled to diverse intracellular pathways, and mediate SRIF's actions in a plethora of physiological systems and disease states.

In the brain, the activation of somatostatin receptors influences a number of functions [9] such as locomotor activity [10–13] and cognition [14] and is implicated in the pathophysiology of various diseases such as epilepsy [15, 16] and Alzheimer's disease [17]. In addition, SRIF has been shown to have neuroprotective effects in different paradigms of neurotoxicity (focal ischemia [18] and excitotoxicity [19]).

In recent years, investigations have been directed toward the search of agents that ameliorate neuronal damage and protect the CNS from ischemia and neurotoxicity. The retina is part of the vertebrate CNS, and has been suggested to be "an approachable part of the brain" [20]. Owing to the simplicity of its anatomical organization, it has been used by many investigators as a model for the study of neural mechanisms and neurotransmitter interactions in the brain. This chapter provides an overview of the functional mapping of somatostatin receptors in the retina and summarizes the knowledge acquired to date on the neuroprotective role of somatostatin against ischemia-induced cell death and the therapeutic potential of somatostatin analogs in retinal disease.

Hormones in Neurodegeneration, Neuroprotection, and Neurogenesis.
Edited by Achille G. Gravanis and Synthia H. Mellon
Copyright © 2011 WILEY-VCH Verlag GmbH & Co. KGaA, Weinheim
ISBN: 978-3-527-32627-3

12.2
Somatostatin and Related Peptides

SRIF is derived from a longer N-terminally extended form of 28 amino acids, namely SRIF-28 [21]. Both peptides are synthesized in variable amounts by differential processing of a larger precursor molecule, preprosomatostatin [22]. SRIF is predominant in the CNS, pancreas, and stomach and SRIF-28 is predominant in the intestine [7]. In the late 1990s, a novel peptide cortistatin (CST) was identified in the cortex bearing a strong similarity to somastostatin. It is derived from a preproCST molecule that is cleaved to produce CST-14 and CST-29, two forms reminiscent of SRIF-14 and SRIF-28. CST-14 shares 11 of its 14 amino acids with SRIF [23, 24].

12.3
Somatostatin Receptors and Signaling

SRIF, SRIF-28, and CST mediate a diverse number of physiological actions by interacting with specific receptors [4, 5, 7]. As mentioned above, five SRIF receptor subtypes have been cloned, namely sst_{1-5}. SRIF and CST display high affinity for all receptor subtypes, whereas SRIF-28 binds preferentially to sst_5 [7, 25–27]. An orphan G protein–coupled receptor MrgX2 was reported as the human CST receptor [28]. However, its functional significance remains to be elucidated.

The cloning of the somatostatin receptors promoted the development of specific sst agonists and antagonists [29–34], specific antibodies [35, 36], and genetically modified mice [37–40], which aided our understanding of the signaling pathways involved in somatostatin's multiple actions.

SRIF was initially characterized to have pertussis toxin–sensitive actions, to inhibit adenylyl cyclase and to reduce cAMP levels [41]. Subsequent studies reported the differential involvement of ssts in the modulation of a number of effector systems including ion channels (K^+ and Ca^{2+}) and a variety of enzymes [7, 42–45]. These effector systems activate intracellular signaling pathways that lead to SRIF's physiological actions.

12.4
Somatostatin and its Receptors in Retina

The presence of immunoreactive SRIF in retina was first reported in rat [46]. Subsequently, SRIF was localized in retinas of many species, where it is localized primarily in wide-field amacrine cells (ACs) with processes that ramify in the inner plexiform layer (IPL), in cells of the ganglion cell layer (GCL), and in neurons in the inner nuclear layer (INL) (for reviews see [43, 47]).

Electrophysiological studies have supported the hypothesis that SRIF functions as a neurotransmitter, neuromodulator, or trophic factor in retina [48–51]. The presence of somatostatin receptors in discrete retinal neurons, as described below, substantiates the multifaceted role of SRIF in the retina. Visual information is transferred from photoreceptors to bipolar to ganglion cells (GCs) and to the brain via the optic nerve. This vertical visual pathway is gated by glutamate that is synthesized in the above mentioned neurons and represents the major neurotransmitter in the retina. Somatostatin receptor activation may influence glutamate levels either directly or indirectly via somatostatin's actions on other neurotransmitter systems found in the retina, such as dopamine (DA) and nitric oxide (NO) (for review see [45, 47, 52, 53]).

12.5
Localization of Somatostatin Receptors in Retinal Neurons

Subsequent to the cloning of the ssts, RT-PCR studies suggested the presence of the mRNA of different ssts in the retina [54–56]. However, immunohistochemical studies employing antibodies raised against the different ssts led to the functional mapping of these receptors in the retina (Figure 12.1) [47, 57, 58].

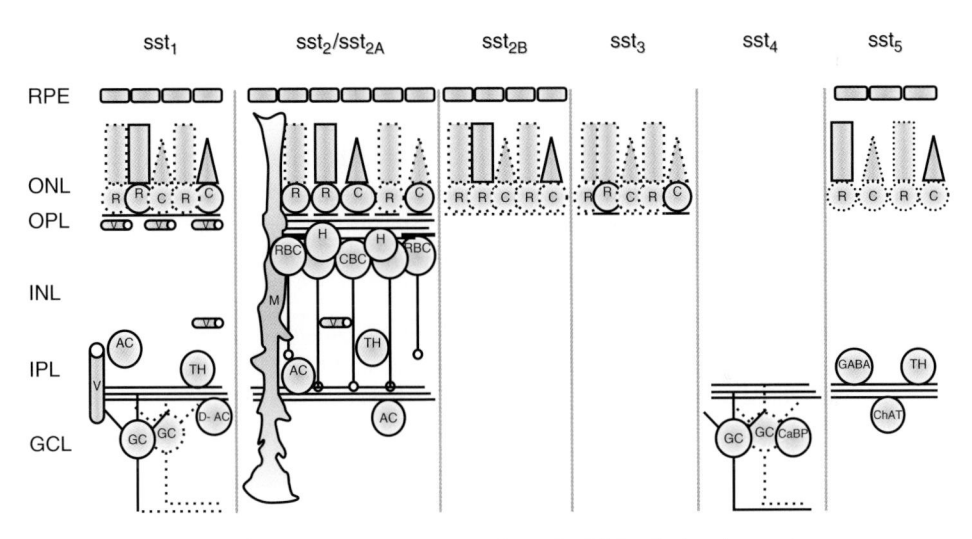

Figure 12.1 Mapping of somatostatin receptors (sst_1–sst_5) in vertebrate retina. RPE, retinal pigment epithelium; ONL, outer nuclear layer; OPL, outer plexiform layer; INL, inner nuclear layer; IPL, inner plexiform layer; GCL, ganglion cell layer; R, rods; C, cones; RBCs, rod bipolar cells; CBCs, cone bipolar cells; v, vessels; H, horizontal cells; ACs, amacrine cells; DACs, displaced amacrine cells; TH, tyrosine hydroxylase; GCs, ganglion cells; GABA, γ-amino butyric acid; ChAT, choline acetyl transferase; CaBP, calcium binding protein. Sst_1: [55, 56, 59–61], sst_2/sst_{2A}: [50, 55, 56, 58–60, 62, 63, 66], sst_{2B}: [63, 64], sst_3: [66], sst_4: [60, 67], and sst_5: [57, 58].

12.5.1
Sst$_1$

Sst$_1$ immunoreactivity (IR) was first reported from rat retinal studies, to be located in SRIF-expressing ACs located in the INL, in displaced amacrine cells (DACs) in the GCL and in a small number of GCs [59]. In addition, sst$_1$ IR was also detected in cell processes lengthening throughout the IPL in sublamina S1, in blood vessels of the inner retina, and the retinal pigment epithelium (RPE) of the rat [60]. In the rabbit, sst$_1$ IR was found to be located in non-tyrosine hydroxylase (TH) and TH-immunoreactive ACs, displaced amacrines, and the GCL [61]. Sst$_1$ IR was also reported in human retina to be localized in retinal GCs, individual cells in the outer nuclear layer (ONL) and INL, in retinal blood vessels, and the RPE (cultured cells and tissue) [55, 56].

12.5.2
Sst$_2$

Sst$_2$ receptors have been localized on medium-sized ACs in the INL, large ACs of the IPL coexpressing TH, and inner segments of cone photoreceptors [59]. In a recent report, sst$_2$ receptors were also found on cholinergic neurons [58]. The use of specific antibodies raised against the two isoforms of the sst$_2$ receptor, namely sst$_{2A}$ and sst$_{2B}$, depicted the presence of the sst$_{2A}$ in rod bipolar cells (RBCs), horizontal cells, and cone photoreceptors, as well as TH-containing ACs in rat [62, 63]. Sst$_{2B}$ receptors were found to be localized only in the outer segments of photoreceptors [63] and in hRPE cells [64]. Both sst$_{2A}$ and sst$_{2B}$ receptors were found to be colocalized with NADPH-diaphorase, a marker for nitric oxide synthase (NOS) [65] in retinal neurons [60, 63] and hRPE cells [64].

In human ocular tissue, sst$_2$ immunostain was reported in the outer and inner segments of rods and cones and individual cells in the ONL, INL, and GCL, RPE cells in culture and in endothelial cells of blood vessels [56]. Sst$_{2A}$ IR was also reported in the inner and outer nuclear and plexiform layers of the human retina and the RPE [55].

In lower vertebrates, sst$_{2A}$ IR was localized to both inner and outer retina of the salamander [50]. Specifically, sst$_{2A}$ IR was observed in rod and cone photoreceptors with prominent staining throughout the inner segment and synaptic terminals, in bipolar and amacrine cell bodies, in the outer plexiform layer (OPL), and in all laminae of the IPL. Diffuse immunostain was reported in the GCL. In retinas of the adult newt *Pleurodeles waltl*, sst$_{2A}$ IR was also observed in rod bipolar, inner segments of the cone photoreceptors, and in the region corresponding to connecting cilia of rods [66].

12.5.3
Sst$_3$

The presence of sst$_3$ mRNA was detected by RT-PCR in rat [54] and human retina [55, 56]; yet, sst$_3$ IR has not been localized in mammalian retina. Sst$_3$ immunostain was reported in the lower vertebrate newt (*Pl. waltl*) to be intensely present in the inner segments of cones and in cilia of rods [66].

12.5.4
Sst$_4$

Sst$_4$ IR was confined to multistratified processes of the IPL, the plasma membrane, the cytoplasm of cell bodies, clusters of long processes in the GCL, and optic nerve fibers of the rat retina. Colocalization of sst$_4$ and the ganglion cell marker microtubule associated protein 1A (MAP 1A) was observed in both cell bodies and processes of the GCL and in processes in the IPL. This colocalization signifies that the sst$_4$ expression is restricted in ganglion cell bodies, the dendritic field and axons, and not on amacrine cell processes in the IPL or cell bodies of DACs in the GCL [60]. Sst$_4$ IR was also localized to calbindin (calcium binding protein, CaBP)-IR and CaBP-non-IR cells in the GCL of the mouse retina [67].

12.5.5
Sst$_5$

While low sst$_5$ mRNA levels were detected by RT-PCR in rat [54], mouse [67], and human retina [56], the presence and location of the sst$_5$ subtype in the retina was only recently reported. Sst$_5$ receptors were located on GABA-ergic, dopaminergic, and cholinergic ACs in rat retina [57]. In agreement with these data, it was also shown that sst$_5$ IR is located on neuronal processes and some cell somata in the IPL and OPL in rat retina, as well as in the outer segments of photoreceptors and the RPE [58].

The above mentioned studies provide conclusive evidence to support the differential localization of sst$_1$–sst$_5$ receptors in the vertebrate retina and suggest that somatostatin plays an important role in retinal circuitry and vision.

12.6
Somatostatin Receptor Function in Retinal Circuitry

The presence of somatostatin receptors in photoreceptors, rod bipolar, and GCs suggests that somatostatin synthesized and released by a subcategory of ACs may act at a distance in a paracrine fashion, and may influence the release of glutamate, the major neurotransmitter of these retinal cell types. In addition, the presence of ssts in TH-, NADPH-, or SRIF-containing cells suggested a role of somatostatin in the regulation of DA and NO release, as well as of its own release.

12.6.1
Effects on Glutamate Release

Somatostatin modulated voltage-gated K^+ and Ca^{2+} currents through activation of sst_{2A} receptors present in the photoreceptors of the salamander retina [50]. These results suggested a possible role for somatostatin in the regulation of glutamate transmitter release from photoreceptors. Studies performed on isolated RBCs of the rabbit retina showed SRIF and octreotide ($sst_{2,3,5}$ analog) to modulate Ca^{2+}-voltage-dependent K^+ channels by activating sst_2 receptors [51]. Somatostatin was also shown to inhibit calcium influx into rat bipolar cell axonal terminals [68]. The sst_2 receptor couples to $G_{o\alpha}$, localized in the IPL in the rabbit retina [69]. This relationship may be instrumental in somatostatin's modulation of the ion channels found on RBCs and the subsequent neurotransmitter release. Earlier electrophysiological studies have shown that $G_{o\alpha 2}$ couples SRIF receptors to Ca^{2+} channels [70]. These studies suggested that SRIF modulates glutamate release in a negative manner.

The direct effects of somatostatin and sst_2 activation on glutamate release were subsequently reported. Activation of sst_2 receptors in mouse retinal explants led to the inhibition of K^+-induced glutamate release [71]. Similarly, SRIF-induced inhibition of glutamate release was significantly more evident in retinas of mice that lack the sst_1 receptor and overexpress the sst_2 subtype [72].

12.6.2
Effects on Dopamine Release

The presence of sst_1 and sst_2 on TH-expressing ACs in rat retina suggested a role for somatostatin in the regulation of DA release. This was substantiated in an *ex vivo* study, where activation of sst_1 and sst_2 receptors in rat retinal explants led to increases in DA release [73]. DA is one of the major neuromodulators of retinal circuitry [74]. It influences the physiology of outer and inner retina, being a very important messenger for light adaptation, but is also involved in trophic functions [74]. SRIF may influence DA release directly via sst_1 and sst_2 receptors present on TH-containing neurons (as described above) or indirectly via its actions on the release of glutamate from RBCs and by altering the glutamatergic input to the dopaminergic cell [72]. Activation of D_2 DA receptors in the retina has been shown to inhibit the release of aspartate and glutamate [75]. Therefore, this pathway may also be involved in SRIF's inhibition of glutamate levels.

12.6.3
Effects on Nitric Oxide/GMP

NO is another important neuromodulator of retinal circuitry [76]. Most retinal cell types express NOS [77] or NADPH-diaphorase [65, 78, 79]. Colocalization of ssts with NADPH-diaphorase [60, 63, 64] suggested a role of SRIF as a modulator of NO release. Indeed, SRIF was shown to increase NO_x^- basal levels in a concentration-dependent manner by activating solely the sst_2 subtype in the retina [60] and hRPE [64]. NO is believed to play a variety of roles in ocular physiology [80], including maintenance of tight junction integrity [81], development of immune and inflammatory responses [82], and blood flow regulation [83]. Therefore, SRIF may influence the abovementioned functions by regulating NO production.

As in other systems, NO activates a soluble guanylyl cyclase (sGC), leading to an increase in cGMP production in the retina [84]. NO/cGMP is an important signaling pathway [85, 86] and an important regulator of retinal physiology [87]. cGMP influences three main targets, namely protein kinase G (PKG), cyclic nucleotide gated channels (CNGCs), and the cyclic nucleotide phosphodiesterase (PDE). Ion channels contain PKG consensus phosphorylation sites and many are under the control of cGMP/PKG.

Somatostatin was reported to inhibit a neuronal Ca^{2+} current via a cGMP-dependent protein kinase [88]. In rat retina, cGMP was shown to regulate light-induced changes in glutamate release [89]. In a recent study, SRIF increased cGMP levels in rat retinal explants in a concentration-dependent manner via an SRIF/sst_2/NO mechanism involving neuronal NOS [90]. These data suggested that NO/cGMP may be one of the signaling pathways via which SRIF's actions are mediated in the retina [45].

The triad SRIF, NO, and DA have been classified as retinal neuromodulators. SRIF appears to regulate NO [60, 64], cGMP [90], and DA levels [73]. More recent studies support that DA, NO, and cGMP also modulate SRIF levels [91]. All three neuromodulators influence the release of glutamate. Therefore, a reciprocal feedback regulation may exist to provide the fine-tuning of retinal circuitry and light–dark adaptation processes.

12.6.4
Effects on Somatostatin Release

Sst_1 was found to be located in SRIF-expressing cells in the INL [59] suggesting that sst_1 may function as an autoreceptor in the retina. The conduction of functional studies *ex vivo* in retinal explants showed that the sst_1 selective agonist CH275, and not the $sst_{2/5}$ selective agonist MK678, reduced SRIF levels. These findings suggested that the sst_1 receptor is the functional autoreceptor for SRIF in the retina [92, 93]. In retinas of sst_1 KO mice, SRIF levels were significantly increased [94], thus substantiating the autoreceptor nature of the sst_1 subtype. Recently, Ke and Zhong [57] reported the presence of sst_5 receptors in retinal ACs that express SRIF, and suggested that the sst_5 may also play an autoreceptor role. However, this has to be substantiated with functional studies.

12.7
Neuroprotection by Somatostatin Analogs

In the brain, SRIF has been shown to have neuroprotective effects in different paradigms of neurotoxicity. Glutamate-mediated excitotoxicity is an accepted mechanism via which ischemic neuronal death is mediated in different models of ischemia, including permanent middle cerebral artery occlusion (MCAO). SRIF reduced NMDA-dependent excitotoxicity in cultured embryonic neurons of the cerebral cortex [19]. CST also protected against kainate-induced neurotoxicity in rat brain [95], and somatostatin, octreotide, and CST protected in a dose-dependent manner against ischemic neuronal brain damage following MCAO [18].

The retina is part of the vertebrate CNS. Owing to the simplicity of its anatomical organization, it has been used by many investigators as a model for the study of neural mechanisms and neurotransmitter interactions in the brain. During the last decade, the functional mapping of somatostatin receptors and their involvement in retinal disease have been investigated [45, 47, 52, 53, 96]. The neuroprotective properties of SRIF in brain [18, 19, 95] provided the impetus for the investigation of the putative neuroprotective effects of SRIF and its selective sst analogs in different models of retinal ischemia.

12.7.1
Retinal Ischemia and Excitotoxicity

Ischemia is characterized by lack of cellular energy and metabolic failure, alteration of normal neuronal membrane processes, and ionic homeostasis which lead to membrane depolarization, increase in extracellular glutamate levels, and activation of ionotropic glutamate receptors [97]. A subsequent cascade of events involving the rise in intracellular calcium levels and the activation of NO formation are believed to be important in cell death [98]. Retinal ischemia leads to an increase in apoptosis and a marked loss of ganglion and ACs leading to vision impairment and blindness [98]. It is implicated in many ocular diseases such as glaucoma [99, 100] and diabetic retinopathy (DR) [101, 102]. The availability of *in vivo* and *in vitro* animal models of retinal ischemia has aided the study of the neuroprotective properties and therapeutic potential of various agents.

12.7.2
Anti-Ischemic Actions of SRIF

12.7.2.1 *Ex vivo* Studies
In the past decade, we have focused our studies on the investigation of the possible neuroprotective effects of SRIF and its analogs against retinal ischemic and excitotoxic insults. Initially, an *ex vivo* chemical model of rat retinal ischemia was employed [103]. This model involves the blockade of oxidative phosphorylation (NaCN) and glycolysis (iodoacetic acid) and is believed to be useful in the understanding of the early events underlying the pathophysiology of ischemia [104, 105].

sst_2 receptor activation reversed the chemical ischemia-induced decrease of PKC-IR in rod bipolar and choline acetyl transferase (ChAT-), TH-, and bNOS-IR in ACs. It was also efficient in reversing retinal cell death as observed with the TUNEL assay [103]. Comparable results were also observed in an earlier study where octreotide reduced the increase of lipid peroxidation levels observed subsequent to ischemia/reperfusion retinal injury [106]. In agreement with these data, the increased presence of functional sst_2 receptors, in retinas of sst_1 k.o. mice [94], reduced retinal cell death and markedly lowered the ischemia-induced increase of glutamate release in an *ex vivo* model of mouse retinal ischemia involving hypoxia and iodoacetic acid [107].

12.7.2.2 *In vivo* Studies

To prove the therapeutic significance of SRIF's neuroprotective properties mentioned above, *in vivo* evidence was essential and, hence, an *in vivo* model of excitotoxicity was employed [108]. The excitatory amino acid (RS)-α-amino-3-hydroxy-5-methyl-4-isoxazolepropionic acid hydrobromide (AMPA) was intravitreally injected and retinal cell loss was examined. AMPA differentially influenced retinal cell viability. The degree of loss observed was directly coupled to the presence of the AMPA receptor subtypes. AMPA receptors belong to the non-NMDA family of ionotropic receptors. They are composed of four subunits, $GluR_1$–$GluR_4$ that are differentially expressed in the retina [109]. AMPA receptors lacking $GluR_2$ have an increased permeability to intracellular calcium ions, making the neurons that express these receptors less vulnerable [110]. AMPA eliminated ChAT-IR and bNOS-IR, in a dose-dependent manner, 24 h after its injection in the rat eye. No other retinal markers were affected by AMPA treatment. Intravitreal coinjection of AMPA with SRIF or the sst_2 selective analog L-779,976 [29] or lanreotide ($sst_{2/5}$ analog) protected the retina from the AMPA-induced toxic effects. Sst_1 and sst_4 selective analogs had no effect [108]. Recently, sst_5 receptors were found to be expressed in the retina [57]. *In vivo* evidence suggested that these receptors are functional and that their activation by intravitreally administered sst_5 specific ligands (L-817,818) [29] protected the retina from AMPA excitotoxicity [58]. Therefore, both sst_2 and sst_5 ligands are useful neuroprotectants against excitotoxicity in the retina.

12.8 Mechanisms of SRIF's Neuroprotection

12.8.1 Involvement of NO/cGMP

The mechanisms via which SRIF-ergic ligands prevent the damage produced by chemical ishemia or other neurotoxic insults are not known. However, SRIF's ability to inhibit voltage-gated Ca^{2+} channels [111] may be responsible for the lowering of the intracellular calcium ion concentration responsible for the toxic effects [98].

As mentioned earlier, SRIF is known to inhibit neuronal calcium currents via a mechanism involving a cGMP-dependent protein kinase [88]. Also, cGMP was shown to be important in SRIF's protective actions against NMDA-induced neuronal death in cortical cultures [19]. These findings suggested that cGMP may be an intermediate signal for SRIF's neuroprotection. cGMP is the catalysis product of sGC upon its activation by NO. SRIF increased NO [60] and cGMP [90] levels in rat retinal explants via an sst_2/SHP-1/NO/cGMP-mediated mechanism. One can conjecture that NO may trigger the synthesis of cGMP in the same cell expressing the sst_2 receptor or neighboring cells, and provide neuroprotection. Therefore, the question was raised as to whether NO has a dual role in the retina; in particular, is it involved in both toxicity and neuroprotection, as was earlier suggested [112].

In order to investigate whether a NO/cGMP mechanism is involved in the neuroprotection properties of SRIF, the neuroprotective properties of NO and/or cGMP in retina had to be established first. To target the latter, Mastrodimou et al. [113] employed a wide range of concentrations of arginine, the substrate of NOS, and NO donors (sodium nitroprusside, 3-morpholinosydnonimine or SIN-1, and NONOate, the slow NO releaser), as well as the membrane-permeable cGMP analog 8-Br-cGMP in the model of chemical ischemia. The results of this study showed that SIN-1 and NONOate offered neuroprotection against ischemia-induced ChAT-IR depletion and TUNEL assay–assessed cell death in a concentration-dependent manner. 8-Br-cGMP also protected the retina from chemical ischemia. The use of peroxynitrite scavenger, L-cysteine, partially reduced the SIN-1 protection suggesting that a NO/peroxynitrite/cGMP pathway is involved in the neuroprotection. Both peroxynitrite and cGMP have been shown to influence ion channel physiology and the subsequent release of neurotransmitters [88, 114]. In addition, NO mediates the down-regulation of excitatory amino acid receptors (AMPA and NMDA) via NO cGMP-dependent and -independent (S-nitrosylation) pathways [86, 112, 115–119]. The results presented suggest that NO has a dual role in the retina. Further studies are essential in order to elucidate whether the above-mentioned mechanisms apply to the neuroprotective role of NO in the retina.

12.8.2
NO/cGMP Mediates SRIF's Neuroprotective Effects

To investigate the involvement of the NO/cGMP pathway in the neuroprotection offered by the SRIF analogs, NOS and sGC blockade was established and the effects of the sst_2 selective ligands were examined in the chemical ischemia model. The NOS inhibitor N(gamma)-monomethyl-L-arginine (NMMA) and the sGC inhibitors 1-H[1,2,4]oxadiazolol[4,3,a]quinoxalin-1-one (ODQ) and NS 2028 reversed the protective effect of lanreotide, thus implicating NO/sGC and cGMP in the SRIF-mediated neuroprotective mechanisms [113].

Interestingly, the neuroprotective effects of NO against NMDA-mediated retinal excitotoxicity may also be associated with the induction of TH expression and increase in DA levels [120]. Therefore, it is becoming more apparent that the retinal neuromodulators NO, DA, and SRIF not only influence retinal circuitry and vision but are also involved in the protection of the retina from ischemia or excitotoxicity.

The mechanism involved in the sst_2-mediated protection against AMPA toxicity *in vivo* may also involve the NO/cGMP pathway. Blockade of either NOS or sGC reduced the sst_2-mediated retinal neuroprotection, while 8-Br-cGMP protected the retina from the excitotoxic effects of AMPA administration [121]. In the retina, ionotropic AMPA/kainic acid receptors have been shown to be negatively regulated by cGMP. In retinal horizontal cells, the function of kainate receptors was depressed by cGMP, and by NO in an sGC–PKG-dependent manner [122]. In addition, it has been shown that the redox states of NO (NO^-, NO^+) down-regulate the NMDA channel activity by reacting with critical cysteine residues (S-nitrosylation), thus contributing to neuroprotection [112, 118]. Therefore, there is ample support for the role of the NO/cGMP–PKG and NO/S-nitrosylation cascades in negatively regulating excitatory amino acid receptors and in decreasing their neurototoxic effects (Figure 12.2).

The above-mentioned mechanism is not a plausible one for the sst_5 neuroprotective effects in the AMPA excitotoxicity model. Sst_5 activation did not lead to increases in NO levels in the retina, as was shown with sst_2 activation [60]. The

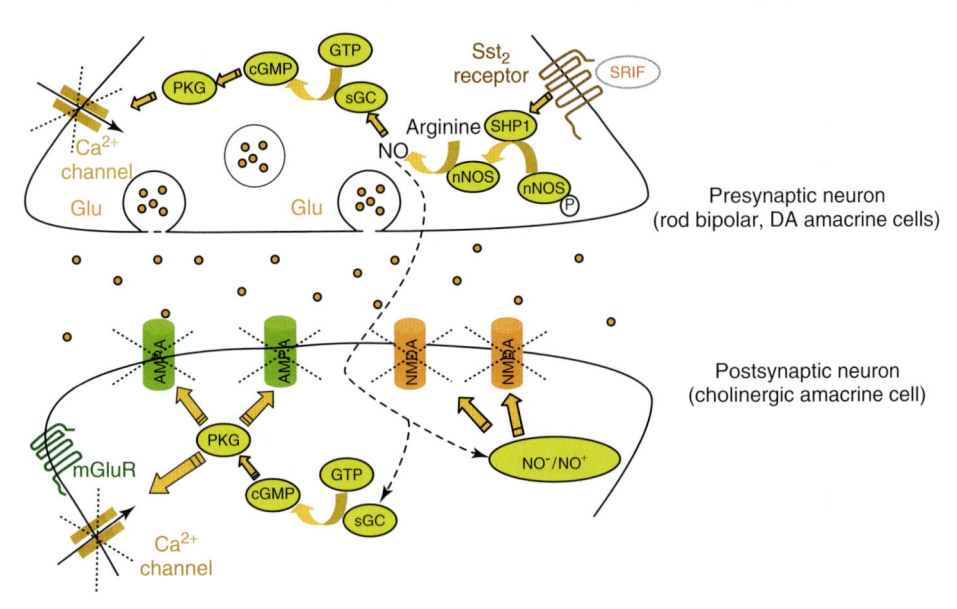

Figure 12.2 Putative mechanism of sst_2 neuroprotection in the retina via NO/cGMP signaling. SRIF increases NO and cGMP levels in rat retinal explants via an sst_2/SHP-1/NO/cGMP-mediated mechanism [60, 90]. The NO/peroxynitrite/cGMP pathway is involved in retinal neuroprotection against chemical ischemia [113]. Both peroxynitrite and cGMP have been shown to influence ion channel physiology and the subsequent release of neurotransmitters [88, 114]. In addition, NO mediates the down-regulation of excitatory amino acid receptors (AMPA and NMDA) via NO cGMP-dependent and -independent (S-nitrosylation) pathways [86, 112, 115–119].

inhibitory neurotransmitter GABA, however, may be involved in the sst_5-mediated protection. Sst_5 expression was reported on somatodendritic compartments of GABA-ergic ACs [57]. Activation of these receptors by SRIF may lead to the increase of GABA levels and neuroprotection via the activation of GABA-ergic receptors [98].

In conclusion, SRIF and its analogs (sst_2/sst_5) may provide neuroprotection by activating ssts located on retinal neurons and influencing the release of excitatory or inhibitory amino acids, intracellular calcium ions, NO/cGMP or DA signaling. Therefore, the pharmacological profile of the somatostatinergic agents renders them promising therapeutics in retinal diseases whose pathophysiology involves ischemic and excitotoxic insults.

12.9
Therapeutic Potential of Somatostatin Agents

DR (proliferative and nonproliferative), a complication of diabetes, is the leading cause of blindness in working-age people in industrialized countries. It is defined as a vascular disease and characterized by neovascularization. Neovascularization (angiogenesis) is due to the secretion of growth factors in response to ischemia [123–125]. The ability of SRIF and analogs to inhibit the release of growth factors, such as GH, insulin-like growth factor (IGF), basic fibroblast growth factor (bFGF) [126–128], and vascular endothelial growth factor (VEGF) [107, 129], has implicated SRIF as a retinal antiangiogenic agent. However, the mechanisms via which somatostatinergic ligands act as antiangiogenics are still under investigation (for review see [130, 131]).

The SRIF analog lanreotide was shown to have a direct antiangiogenic and an indirect GH inhibitory effect on patients with proliferative DR [132]. A number of clinical studies support the opinion that antiangiogenic efficacy of SRIF analogs lies not only in systemic inhibition of GH but also in the local modulation of the effects of growth factors (for a recent review see [96]). A deficit of SRIF was observed in the vitreous fluid of diabetic patients [133].

In addition to its neovascular characteristics, there are data suggesting that DR is a neurodegenerative disease of the eye. A number of studies indicate that diabetes, like ischemia, increases the concentration of glutamate in the vitreous fluid [134–136]. Several lines of evidence designate that glutamate excitotoxicity is responsible for the neuronal loss in DR and suggest that the cause of neuronal apoptosis in the retina during diabetes may be chronic glutamate excitotoxicity (for review see [137]).

Functional changes in vision were identified in DR before and independent of abnormal vessel formation [138]. These changes suggested that diabetes has direct effects on neural retina, which lead to apoptosis. Retinas of rats employed in the streptozotocin model of Type I diabetes were characterized by increases in apoptotic TUNEL-positive neuronal and/or glial cells, but not vascular cells [101]. In the same study, it was also observed that in human retinas with DR, TUNEL-positive cells are not limited to areas with retinal vascular lesions. These data indicated that retinal apoptosis is a generalized response of retinal neural cells to diabetes.

Lower SRIF expression in the retina, possibly associated with retinal neurodegeneration, was reported to be among the early events in DR [139]. More recently, it was shown that the mRNA and IR levels of the neuropeptide CST were also significantly lower in the neuroretina and the RPE of diabetic donors [140]. In addition, increases in apoptotic cells and neuroglial activation were observed in diabetic versus nondiabetic retinas and associated with the lower expression of SRIF-28 [139] and CST [140]. These results led the authors to suggest that SRIF and CST deficiency may be involved in diabetic retinal neurodegeneration. In agreement with these studies, it was previously shown that CST protected the rat retina against chemical ischemia *ex vivo*, supporting the neuropeptide's neuroprotective role [103].

12.10
Conclusions

Over the last decade, the functional mapping of somatostatin receptors (sst_1, sst_2, sst_4, sst_5) in the mammalian retina has given us substantial information on the role of somatostatin in retinal physiology. Somatostatin and selective agonists for each receptor subtype differentially influence retinal circuitry and vision processes. Sst_2 and sst_5 receptor activation provides neuroprotection to the retina and specific ligands for these receptors show promise in the treatment of retinal ischemia. Experimental and clinical evidence have been increasing in recent years in support of this tenet. However, further prospective clinical trials are needed to assess SRIF's role in the treatment of DR and in other retinopathies, which were mentioned but not discussed in this chapter. Sst_2 and sst_5 selective ligands should be further examined with respect to their stability and their ability to cross the retinal blood barrier. This information will provide answers as to the most appropriate route of administration and dose regimen for optimum efficacy. Agents that could be administered intravitreally may prove most successful in clinical studies.

It is becoming more evident that the wisest strategy for the treatment of retinopathies is to focus on more than one target systems. SRIF analogs could be combined with laser photocoagulation (primary therapy) or the new and efficacious antineovascular (anti-VEGF) agents in a multidrug treatment to ensure the most efficacious therapy in retinopathies whose pathophysiology involves ischemia-induced neurodegeneration and/or ischemia-induced neovascularization.

Acknowledgments

I would like to thank Ms Foteini Kiagiadaki for the artwork and her comments on the manuscript and to acknowledge the support of the European Social Fund and National Resources and the Greek Ministry of Development-GSRT (Programs

Pythagoras and PENED). This chapter is based on a more extensive earlier review by Vasilaki and Thermos.

Abbreviations

bFGF	basic fibroblast growth factor
CNS	central nervous system
CST	cortistatin
DA	dopamine
DR	diabetic retinopathy
GCL	ganglion cell layer
GH	growth hormone
IGF	insulin-like growth factor
INL	inner nuclear layer
IPL	inner plexiform layer
MCAO	middle cerebral artery occlusion
NO	nitric oxide
ONL	outer nuclear layer
OPL	outer plexiform layer
PKG	protein kinase G
RPE	retinal pigment epithelium
SRIF	somatotropin release inhibitory factor
TH	tyrosine hydroxylase
VEGF	vascular endothelial growth factor

References

1. Krulich, L., Dhariwal, A.-P., and McCann, S.-M. (1968) Stimulatory and inhibitory effects of purified hypothalamic extracts on growth hormone release from rat pituitary in vitro. *Endocrinology*, **83**, 783–790.
2. Brazeau, P., Vale, W., Burgus, R., Ling, N., Butcher, M., Rivier, J., and Guillemin, R. (1973) Hypothalamic polypeptide that inhibits the secretion of immunoreactive pituitary growth hormone. *Science*, **179**, 77–79.
3. Guillemin, R. (2008) Somatostatin: the beginnings, 1972. *Mol. Cell. Endocrinol.*, **286**, 3–4.
4. Thermos, K. and Reisine, T. (1988) Somatostatin receptor subtypes in the clonal anterior pituitary cell lines AtT-20 and GH3. *Mol. Pharmacol.*, **33**, 370–377.
5. Thermos, K., He, H.-T., Wang, H.-L., Margolis, N., and Reisine, T. (1989) Biochemical properties of brain somatostatin receptors. *Neuroscience*, **31**, 131–141.
6. Thermos, K., Meglasson, M.-D., Nelson, J., Lounsbury, K.-M., and Reisine, T. (1990) Pancreatic beta-cell somatostatin receptors. *Am. J. Physiol.*, **259**, 216–224.
7. Olias, G., Viollet, C., Kusserow, H., Epelbaum, J., and Meyerhof, W. (2004) Regulation and function of somatostatin receptors. *J. Neurochem.*, **89**, 1057–1091.

8. Hoyer, D., Bell, G.-I., Berelowitz, M., Epelbaum, J., Feniuk, W., Humphrey, P.-P., O'Carroll, A.-M. et al. (1995) Classification and nomenclature of somatostatin receptors. *Trends Pharmacol. Sci.*, **16**, 86–88.
9. Viollet, C., Lepousez, G., Loudes, C., Videau, C., Simon, A., and Epelbaum, J. (2008) Somatostatinergic systems in brain: networks and functions. *Mol. Cell. Endocrinol.*, **286**, 75–87.
10. Raynor, K., Lucki, I., and Reisine, T. (1993) Somatostatin receptors in the nucleus accumbens selectively mediate the stimulatory effect of somatostatin on locomotor activity in rats. *J. Pharmacol. Exp. Ther.*, **265**, 67–73.
11. Marazioti, A., Kastellakis, A., Antoniou, K., Papasava, D., and Thermos, K. (2005) Somatostatin receptors in the ventral pallidum/substantia innominata modulate rat locomotor activity. *Psychopharmacology*, **181**, 319–326.
12. Marazioti, A., Pitychoutis, P.-M., Papadopoulou-Daifoti, Z., Spyraki, C., and Thermos, K. (2008) Activation of somatostatin receptors in the globus pallidus increases rat locomotor activity and dopamine release in the striatum. *Psychopharmacology*, **201**, 413–422.
13. Santis, S., Kastellakis, A., Kotzamani, D., Pitarokoili, K., Kokona, D., and Thermos, K. (2009) Somatostatin increases rat locomotor activity by activating sst(2) and sst(4) receptors in the striatum and via glutamatergic involvement. *N. S. Arch. Pharmacol.*, **379**, 181–189.
14. Baraban, S.-C. and Tallent, M.-K. (2004) Interneuron diversity series: interneuronal neuropeptides – endogenous regulators of neuronal excitability. *Trends Neurosci.*, **27**, 135–142.
15. Vezzani, A. and Hoyer, D. (1999) Brain somatostatin: a candidate inhibitory role in seizures and epileptogenesis. *Eur. J. Neurosci.*, **11**, 3767–3776.
16. Tallent, M.-K. and Qiu, C. (2008) Somatostatin: an endogenous antiepileptic. *Mol. Cell. Endocrinol.*, **286**, 96–103.
17. Burgos-Ramos, E., Hervás-Aguilar, A., Aguado-Llera, D., Puebla-Jiménez, L., Hernández-Pinto, A.-M., Barrios, V., and Arilla-Ferreiro, E. (2008) Somatostatin and Alzheimer's disease. *Mol. Cell. Endocrinol.*, **286**, 104–111.
18. Rauca, C., Schäfer, K., and Höllt, V. (1999) Effects of somatostatin, octreotide and cortistatin on ischaemic neuronal damage following permanent middle cerebral artery occlusion in the rat. *N. S. Arch. Pharmacol.*, **360**, 633–638.
19. Forloni, G., Lucca, E., Angeretti, N., Chiessa, R., and Vezzani, A. (1997) Neuroprotective effect of somatostatin on nonapoptotic NMDA-induced neuronal death: role of cyclic GMP. *J. Neurochem.*, **68**, 319–327.
20. Dowling, J.E. (1987) *The Retina*, Harvard University Press.
21. Pradayrol, L., Jörnvall, H., Mutt, V., and Ribet, A. (1980) N-terminally extended somatostatin: the primary structure of somatostatin-28. *FEBS Lett.*, **109**, 55–58.
22. Tostivint, H., Lihrmann, I., and Vaudry, H. (2008) New insight into the molecular evolution of the somatostatin family. *Mol. Cell. Endocrinol.*, **286**, 5–17.
23. de Lecea, L., Criado, J.-R., Prospero-Garcia, O., Gautvik, K.-M., Schweitzer, P., Danielson, P.-E., Dunlop, C.-L. et al. (1996) A cortical neuropeptide with neuronal depressant and sleep-modulating properties. *Nature*, **381**, 242–245.
24. de Lecea, L. (2008) Cortistatin – functions in the central nervous system. *Mol. Cell. Endocrinol.*, **286**, 88–95.
25. Patel, Y.-C. and Srikant, C.-B. (1994) Subtype selectivity of peptide analogs for all five cloned human somatostatin receptors (hsstr 1–5). *Endocrinology*, **135**, 2814–2817.
26. Panetta, R., Greenwood, M.-T., Warszynska, A., Demchyshyn, L.-L., Day, R., Niznik, H.-B., Srikant, C.-B., and Patel, Y.-C. (1994) Molecular cloning, functional characterization, and chromosomal localization of a human somatostatin receptor (somatostatin receptor type 5) with preferential affinity for somatostatin-28. *Mol. Pharmacol.*, **45**, 417–427.

27. Møller, L.-N., Stidsen, C.-E., Hartmann, B., and Holst, J.-J. (2003) Somatostatin receptors. *Biochim. Biophys. Acta*, **1616**, 1–84.
28. Robas, N., O'Reilly, M., Katugampola, S., and Fidock, M. (2003) Maximizing serendipity: strategies for identifying ligands for orphan G-protein-coupled receptors. *Curr. Opin. Pharmacol.*, **3**, 121–126.
29. Rohrer, S.-P., Birzin, E.-T., Mosley, R.-T., Berk, S.-C., Hutchins, S.-M., Shen, D.-M., Xiong, Y. et al. (1998) Rapid identification of subtype-selective agonists of the somatostatin receptor through combinatorial chemistry. *Science*, **282**, 737–740.
30. Bass, R.-T., Buckwalter, B.-L., Patel, B.-P., Pausch, M.-H., Price, L.-A., Strnad, J., and Hadcock, J.-R. (1996) Identification and characterization of novel somatostatin antagonists. *Mol. Pharmacol.*, **50**, 709–715.
31. Reubi, J.-C., Schaer, J.-C., Wenger, S., Hoeger, C., Erchegyi, J., Waser, B., and Rivier, J. (2000) SST3-selective potent peptidic somatostatin receptor antagonists. *Proc. Natl. Acad. Sci.*, **97**, 13973–13978.
32. Hoyer, D., Nunn, C., Hannon, J., Schoeffter, P., Feuerbach, D., Schuepbach, E., Langenegger, D. et al. (2004) SRA880, in vitro characterization of the first non-peptide somatostatin sst(1) receptor antagonist. *Neurosci. Lett.*, **361**, 132–135.
33. Tulipano, G., Soldi, D., Bagnasco, M., Culler, M.-D., Taylor, J.-E., Cocchi, D., and Giustina, A. (2002) Characterization of new selective somatostatin receptor subtype-2 (sst2) antagonists, BIM-23627 and BIM-23454. Effects of BIM-23627 on GH release in anesthetized male rats after short-term high-dose dexamethasone treatment. *Endocrinology*, **143**, 1218–1224.
34. Martin, R.-E., Green, L.-G., Guba, W., Kratochwil, N., and Christ, A. (2007) Discovery of the first nonpeptidic, small-molecule, highly selective somatostatin receptor subtype 5 antagonists: a chemogenomics approach. *J. Med. Chem.*, **50**, 6291–6294.
35. Helboe, L., Møller, M., Nørregaard, L., SciØdt, M., and Stidsen, C.-E. (1997) Development of selective antibodies against the human somatostatin receptor subtypes sst1-sst5. *Brain Res. Mol. Brain Res.*, **49**, 82–88.
36. Schulz, S., Handel, M., Schreff, M., Schmidt, H., and Hollt, V. (2000) Localisation of five somatostatin receptors in the rat central nervous system using subtype-specific antibodies. *J. Physiol.*, **94**, 259–264.
37. Kreienkamp, H.-J., Akgün, E., Baumeister, H., Meyerhof, W., and Richter, D. (1999) Somatostatin receptor subtype 1 modulates basal inhibition of growth hormone release in somatotrophs. *FEBS Lett.*, **462**, 464–466.
38. Viollet, C., Vaillend, C., Videau, C., Bluet-Pajot, M.-T., Ungerern, A., L'Héritier, A., Kopp, C. et al. (2000) Involvement of sst2 somatostatin receptor in locomotor, exploratory activity and emotional reactivity in mice. *Eur. J. Neurosci.*, **12**, 3761–3770.
39. Allen, J.-P., Hathway, G.-J., Clarke, N.-J., Jowett, M.-I., Topps, S., Kendrick, K.-M., Humphrey, P.-P. et al. (2003) Somatostatin receptor 2 knockout/lacZ knockin mice show impaired motor coordination and reveal sites of somatostatin action within the striatum. *Eur. J. Neurosci.*, **17**, 1881–1895.
40. Strowski, M.-Z., Kohler, M., Chen, H.-Y., Trumbauer, M.-E., Li, Z., Szalkowski, D., Gopal-Truter, S. et al. (2003) Somatostatin receptor subtype 5 regulates insulin secretion and glucose homeostasis. *Mol. Endocrinol.*, **17**, 93–106.
41. Mahy, N., Woolkalis, M., Thermos, K., Carlson, K., Manning, D., and Reisine, T. (1988) Pertussis toxin modifies the characteristics of both the inhibitory GTP binding proteins and the somatostatin receptor in anterior pituitary tumor cells. *J. Pharmacol. Exp. Ther.*, **246**, 779–785.
42. Csaba, Z. and Dournaud, P. (2001) Cellular biology of somatostatin receptors. *Neuropeptides*, **35**, 1–23.

43. Cervia, D. and Bagnoli, P. (2007) An update on somatostatin receptor signaling in native systems and new insights on their pathophysiology. *Pharmacol. Ther.*, **116**, 322–341.
44. Florio, T. (2008) Molecular mechanisms of the antiproliferative activity of somatostatin receptors (SSTRs) in neuroendocrine tumors. *Front. Biosci.*, **13**, 822–840.
45. Thermos, K. (2008) Novel signals mediating the functions of somatostatin: The emerging role of NO/cGMP. *Mol. Cell. Endocrinol.*, **286**, 49–57.
46. Shapiro, B., Kronheim, S., and Pimstone, B. (1979) The presence of immunoreactive somatostatin in rat retina. *Horm. Metab. Res.*, **11**, 79–80.
47. Thermos, K. (2003) Functional mapping of somatostatin receptors in the retina: a review. *Vision Res.*, **43**, 1805–1815.
48. Ferriero, D.-M. and Sagar, S.-M. (1987) Development of somatostatin immunoreactive neurons in rat retina. *Dev. Brain Res.*, **34**, 207–214.
49. Zalutsky, R.-A. and Miller, R.-F. (1990) The physiology of somatostatin in the rabbit retina. *J. Neurosci.*, **10**, 383–393.
50. Akopian, A., Johnson, J., Gabriel, R., Brecha, N., and Witkovsky, P. (2000) Somatostatin modulates voltage-gated K^+ and Ca^{++} currents in rod and cone photoreceptors of the salamander retina. *J. Neurosci.*, **20**, 929–936.
51. Petrucci, C., Resta, V., Fieni, F., Bigiani, A., and Bagnoli, P. (2001) Modulation of potassium current and calcium influx by somatostatin in rod bipolar cells isolated from the rabbit retina via sst2 receptors. *N. S. Arch. Pharmacol.*, **363**, 680–694.
52. Casini, G., Catalani, E., Dal Monte, M., and Bagnoli, P. (2005) Functional aspects of the somatostatinergic system in the retina and the potential therapeutic role of somatostatin in retinal disease. *Histol. Histopathol.*, **20**, 615–632.
53. Cervia, D., Casini, G., and Bagnoli, P. (2008) Physiology and pathology of somatostatin in the mammalian retina: A current view. *Mol. Cell. Endocrinol.*, **286**, 112–122.
54. Mori, M., Aihara, M., and Shimizu, T. (1997) Differential expression of somatostatin receptors in the rat eye: sstR4 is intensely expressed in the iris/ciliary body. *Neurosci. Lett.*, **223**, 185–188.
55. van Hagen, P.-M., Baarsma, G.-S., Mooy, C.-M., Ercoskan, E.-M., ter Averst, E., Hofland, L.-J., Lamberts, S.-W., and Kuijpers, R.-W. (2000) Somatostatin and somatostatin receptors in retinal diseases. *Eur. J. Endocrinol.*, **143** (Suppl 1), S43–S51.
56. Klisovic, D.-D., O'Dorisio, M.-S., Katz, S.-E., Sall, J.-W., Balster, D., O'Dorisio, T.-M., Craig, E., and Lubow, M. (2001) Somatostatin receptor gene expression in human ocular tissues: RT-PCR and immunohistochemical study. *Invest. Ophthalmol. Vis. Sci.*, **42**, 2193–2201.
57. Ke, J.B. and Zhong, Y.M. (2007) Expression of somatostatin receptor subtype 5 in rat retinal amacrine cells. *Neuroscience*, **144**, 1025–1032.
58. Kiagiadaki, F., Savvaki, M., and Thermos, K. (2010) Activation of somatostatin receptor (sst$_5$) protects the rat retina from AMPA-induced neurotoxicity. *Neuropharmacology*, **58**, 297–303.
59. Helboe, L. and Moller, M. (1999) Immunohistochemical localization of somatostatin receptor subtypes sst$_1$ and sst$_2$ in the rat retina. *Invest. Ophthalmol. Vis. Sci.*, **40**, 2376–2382.
60. Vasilaki, A., Mouratidou, M., Schulz, S., and Thermos, K. (2002) Somatostatin influences nitric oxide production in the rat retina. *Neuropharmacology*, **43**, 899–909.
61. Cristiani, R., Fontanesi, G., Casini, G., Petrucci, C., Viollet, C., and Bagnoli, P. (2000) Expression of somatostatin subtype 1 receptor in the rabbit retina. *Invest. Ophthalmol. Vis. Sci.*, **41**, 3191–3199.
62. Johnson, J., Wu, V., Wong, H., Walsh, J.-H., and Brecha, N.-C. (1999) Somatostatin receptor subtype 2A expression in the rat retina. *Neuroscience*, **94**, 675–683.
63. Vasilaki, A., Gardette, R., Epelbaum, J., and Thermos, K. (2001) NADPH-diaphorase colocalization

with somatostatin receptor subtypes sst2A and sst2B in the retina. *Invest. Ophthalmol. Vis. Sci.*, **42**, 1600–1609.
64. Vasilaki, A., Papadaki, T., Notas, G., Kolios, G., Mastrodimou, N., Hoyer, D., Tsilimbaris, M. et al. (2004) Effect of somatostatin on nitric oxide production in human retinal pigment epithelium cell cultures. *Invest. Ophthalmol. Vis. Sci.*, **45**, 1499–1506.
65. Dawson, T.-M., Bredt, D.-S., Fotuhi, M., Hwang, P.-M., and Snyder, S.-H. (1991) Nitric oxide synthase and neuronal NADPH diaphorase are identical in brain and peripheral tissues. *Proc. Nat. Acad. Sci. U.S.A.*, **88**, 7797–7801.
66. Grigoryan, E.-N., Vasilaki, A., Mastrodimou, N., and Thermos, K. (2003) Somatostatin receptor immunoreactivity in the eye of the adult newt (Pleurodeles waltlii Michan). *Neurosci. Lett.*, **337**, 143–146.
67. Cristiani, R., Petrucci, C., Dal Monte, M., and Bagnoli, P. (2002) Somatostatin (SRIF) and SRIF receptors in the mouse retina. *Brain Res.*, **936**, 1–14.
68. Johnson, J., Caravelli, M.-L., and Brecha, N. (2001) Somatostatin inhibits calcium influx into rat rod bipolar cell axonal terminals. *Vis. Neurosci.*, **18**, 101–108.
69. Vasilaki, A., Georgoussi, Z., and Thermos, K. (2003) Somatostatin receptors (sst2) are coupled to Go and modulate GTPase activity in the rabbit retina. *J. Neurochem.*, **84**, 625–632.
70. Kleuss, C., Scherübl, H., Hescheler, J., Schultz, G., and Wittig, B. (1992) Different beta-subunits determine G-protein interaction with transmembrane receptors. *Nature*, **358**, 424–426.
71. Dal Monte, M., Petrucci, C., Cozzi, A., Allen, J.-P., and Bagnoli, P. (2003) Somatostatin inhibits potassium-evoked glutamate release by activation of the sst(2) somatostatin receptor in the mouse retina. *N. S. Arch. Pharmacol.*, **367**, 188–192.
72. Bigiani, A., Petrucci, C., Ghiaroni, V., Dal Monte, M., Cozzi, A., Kreienkamp, H.J., Richter, D., and Bagnoli, P. (2004) Functional correlates of somatostatin receptor 2 overexpression in the retina of mice with genetic deletion of somatostatin receptor 1. *Brain. Res.*, **1025**, 177–185.
73. Kouvidi, E., Papadopoulou-Daifoti, Z., and Thermos, K. (2006) Somatostatin modulates dopamine release in rat retina. *Neurosci. Lett.*, **391**, 82–86.
74. Witkovsky, P. (2004) Dopamine and retinal function. *Doc. Ophthalmologica*, **108**, 17–40.
75. Kamisaki, Y., Hamahashi, T., Mita, C., and Itoh, T. (1991) D-2 dopamine receptors inhibit release of aspartate and glutamate in rat retina. *J. Pharmacol. Exp. Ther.*, **256**, 634–638.
76. Bredt, D.-S. and Snyder, S.-H. (1992) Nitric oxide, a novel neuronal messenger. *Neuron*, **8**, 3–11.
77. Alderton, W.-K., Cooper, C.-E., and Knowles, R.-G. (2001) Nitric oxide synthases: structure, function and inhibition. *Biochem. J.*, **357**, 593–615.
78. Haverkamp, S. and Eldred, W.-D. (1988) Localization of nNOS in photoreceptor, bipolar and horizontal cells in turtle and rat retinas. *Neuroreport*, **9**, 2231–2235.
79. Roufail, E., Stringer, M., and Rees, S. (1995) Nitric oxide synthase immunoreactivity and NADPH diaphorase staining are co-localised in neurons closely associated with the vasculature in rat and human retina. *Brain Res.*, **684**, 36–46.
80. Goldstein, I.-M., Ostwald, P., and Roth, S. (1996) Nitric oxide: a review of its role in retinal function and disease. *Vision Res.*, **36**, 2979–2994.
81. Zech, J.-C., Pouvreau, I., Cotinet, A., Goureau, O., Le Varlet, B., and de Kozak, Y. (1998) Effect of cytokines and nitric oxide on tight junctions in cultured rat retinal pigment epithelium. *Invest. Ophthalmol. Vis. Sci.*, **39**, 1600–1608.
82. Holtkamp, G.-M., Kijlstra, A., Peek, R., and de Vos, A.-F. (2001) Retinal pigment epithelium-immune system interactions: cytokine production and cytokine-induced changes. *Prog. Retinal Eye Res.*, **20**, 29–48.
83. Schmetterer, L. and Polak, K. (2001) Role of nitric oxide in the control of

ocular blood flow. *Prog. Retin. Eye Res.*, **20**, 823–847.
84. Koch, K.-W., Lambrecht, H.-G., Haberecht, M., Redburn, D., and Schmidt, H.-H.-H.-W. (1994) Functional coupling of a Ca^{2+}/calmodulin-dependent nitric oxide synthase and a soluble guanylyl cyclase in vertebrate photoreceptor cells. *EMBO J.*, **13**, 3312–3320.
85. Koesling, D. and Friebe, A. (1999) Soluble guanylyl cyclase: structure and regulation. *Rev. Physiol. Biochem. Pharmacol.*, **135**, 41–65.
86. Ahern, G.-P., Klyachko, V.-A., and Jackson, M.-B. (2002) cGMP and S-nitrosylation: two routes for modulation of neuronal excitability by NO. *Trends Neurosci.*, **25**, 510–517.
87. Ding, J.-D. and Weinberg, R.-J. (2007) Distribution of soluble guanylyl cyclase in rat retina. *J. Comp. Neurol.*, **502**, 734–745.
88. Meriney, S.-D., Gray, D.-B., and Pilar, G.-R. (1994) Somatostatin-induced inhibition of neuronal Ca^{2+} current modulated by cGMP-dependent protein kinase. *Nature*, **369**, 336–339.
89. Barabás, P., Kovács, I., Kovács, R., Pálhalmi, J., Kardos, J., and Schousboe, A. (2002) Light-induced changes in glutamate release from isolated rat retina is regulated by cyclic guanosine monophosphate. *J. Neurosci. Res.*, **67**, 149–155.
90. Mastrodimou, N., Kiagiadaki, F., Hodjarova, M., Karagianni, E., and Thermos, K. (2006) Somatostatin receptors (sst2) regulate cGMP production in rat retina. *Regul. Pept.*, **133**, 41–46.
91. Kiagiadaki, F., Koulakis, E., and Thermos, K. (2008) Dopamine (D1) receptor activation and nitrinergic agents influence somatostatin levels in rat retina. *Exp. Eye Res.*, **86**, 18–24.
92. Mastrodimou, N. and Thermos, K. (2004) The somatostatin receptor (sst1) modulates the release of somatostatin in rat retina. *Neurosci. Lett.*, **356**, 13–16.
93. Thermos, K., Bagnoli, P., Epelbaum, J., and Hoyer, D. (2006) The somatostatin sst1 receptor: an autoreceptor for somatostatin in brain and retina? *Pharmacol. Ther.*, **110**, 455–464.
94. Dal Monte, M., Petrucci, C., Vasilaki, A., Cervia, D., Grouselle, D., Epelbaum, J., Kreienkamp, H.J. et al. (2003) Genetic deletion of somatostatin receptor 1 alters somatostatinergic transmission in the mouse retina. *Neuropharmacology*, **45**, 1080–1092.
95. Braun, H., Schulz, S., Becker, A., Schröder, H., and Höllt, V. (1998) Protective effects of cortistatin (CST-14) against kainate-induced neurotoxicity in rat brain. *Brain Res.*, **803**, 54–60.
96. Vasilaki, A. and Thermos, K. (2009) Somatostatin analogues as therapeutics in retinal disease. *Pharmacol. Ther.*, **122**, 324–333.
97. Lipton, S.-A. and Rosenberg, P.-A. (1994) Excitatory amino acids as a final common pathway in neurologic disorders. *N. Engl. J. Med.*, **330**, 613–622.
98. Osborne, N.-N., Casson, R.-J., Wood, J.-P., Chidlow, G., Graham, M., and Melena, J. (2004) Retinal ischemia: mechanisms of damage and potential therapeutic strategies. *Prog. Retin. Eye Res.*, **23**, 91–147.
99. Osborne, N.-N. (2008) Pathogenesis of ganglion "cell death" in glaucoma and neuroprotection: focus on ganglion cell axonal mitochondria. *Prog. Brain Res.*, **173**, 339–352.
100. Seki, M. and Lipton, S.-A. (2008) Targeting excitotoxic/free radical signaling pathways for therapeutic intervention in glaucoma. *Prog. Brain Res.*, **173**, 495–510.
101. Barber, A.-J., Lieth, E., Khin, S.-A., Antonetti, D.-A., Buchanan, A.-G., and Gardner, T.-W. (1998) Neural apoptosis in the retina during experimental and human diabetes. Early onset and effect of insulin. *J. Clin. Invest.*, **102**, 783–791.
102. Antonetti, D.-A., Barber, A.-J., Bronson, S.-K., Freeman, W.-M., Gardner, T.-W., Jefferson, L.-S., Kester, M. et al. (2006) Diabetic retinopathy: seeing beyond glucose-induced microvascular disease. *Diabetes*, **55**, 2401–2411.
103. Mastrodimou, N., Lambrou, G.-N., and Thermos, K. (2005) Effect of somatostatin analogues on chemically induced

ischaemia in the rat retina. *N. S. Arch. Pharmacol.*, **371**, 44–53.

104. Reiner, P.-B., Laycock, A.-G., and Doll, C.-J. (1990) A pharmacological model of ischemia in the hippocampal slice. *Neurosci. Lett.*, **119**, 175–178.

105. Ferreira, I.-L., Duarte, C.-B., Neves, A.-R., and Carvalho, A.-P. (1998) Culture medium components modulate retina cell damage induced by glutamate, kainate or "chemical ischemia". *Neurochem. Int.*, **32**, 387–396.

106. Celiker, U., Ilhan, N., Ozercan, I., Demir, T., and Celiker, H. (2002) Octreotide reduces ischaemia-reperfusion injury in the retina. *Acta Ophthalmol. Scand.*, **80**, 395–400.

107. Catalani, E., Cervia, D., Martini, D., Bagnoli, P., Simonetti, E., Timperio, A.M., and Casini, G. (2007) Changes in neuronal response to ischemia in retinas with genetic alterations of somatostatin receptor expression. *Eur. J. Neurosci.*, **25**, 1447–1459.

108. Kiagiadaki, F. and Thermos, K. (2008) Effect of intravitreal administration of somatostatin and sst2 analogues on AMPA-induced neurotoxicity in rat retina. *Invest. Ophthalmol. Vis. Sci.*, **49**, 3080–3089.

109. Peng, Y.-W., Blackstone, C.-D., Huganir, R.-L., and Yau, K.-W. (1995) Distribution of glutamate receptor subtypes in the vertebrate retina. *Neuroscience*, **66**, 483–497.

110. Geiger, J.-R., Melcher, T., Koh, D.-S., Sakmann, B., Seeburg, P.-H., Jonas, P., and Monyer, H. (1995) Relative abundance of subunit mRNAs determines gating and Ca^{2+} permeability of AMPA receptors in principal neurons and interneurons in rat CNS. *Neuron*, **15**, 193–204.

111. Tallent, M., Liapakis, G., O'Carroll, A.M., Lolait, S.-J., Dichter, M., and Reisine, T. (1996) Somatostatin receptor subtypes SSTR2 and SSTR5 couple negatively to an L-type Ca^{2+} current in the pituitary cell line AtT-20. *Neuroscience*, **71**, 1073–1081.

112. Lipton, S.-A., Choi, Y.-B., Pan, Z.-H., Lei, S.-Z., Chen, H.-S., Sucher, N.-J., Loscalzo, J. *et al.* (1993) A redox-based mechanism for the neuroprotective and neurodestructive effects of nitric oxide and related nitroso-compounds. *Nature*, **364**, 626–632.

113. Mastrodimou, N., Kiagiadaki, F., and Thermos, K. (2008) The role of nitric oxide and cGMP in somatostatin's protection of retinal ischemia. *Invest. Ophthalmol. Vis. Sci.*, **49**, 342–349.

114. Ohkuma, S., Katsura, M., Higo, A., Shirotani, K., Hara, A., Tarumi, C., and Ohgi, T. (2001) Peroxynitrite affects $Ca2+$ influx through voltage-dependent calcium channels. *J. Neurochem.*, **76**, 341–350.

115. Manzoni, O., Prezeau, L., Marin, P., Deshager, S., Bockaert, J., and Fagni, L. (1992) Nitric oxide-induced blockade of NMDA receptors. *Neuron*, **8**, 653–662.

116. Kim, Y.-M., Chung, H.-T., Kim, S.-S., Han, J.-A., Yoo, Y.-M., Kim, K.-M., Lee, G.-H. *et al.* (1999) Nitric oxide protects PC12 cells from serum deprivation-induced apoptosis by cGMP-dependent inhibition of caspase signaling. *J. Neurosci.*, **19**, 6740–6747.

117. Vidwans, A., Kim, S., Coffin, D.-O., Wink, D.-A., and Hewet, S.-J. (1999) Analysis of the neuroprotective effects of various nitric oxide donor compounds in murine mixed cortical cell culture. *J. Neurochem.*, **72**, 1843–1852.

118. Kim, W.-K., Choi, Y.-B., Rayudu, P.V., Das, P., Asaad, W., Arnelle, D.R., Stamler, J.S., and Lipton, S.A. (1999) Attenuation of NMDA receptor activity and neurotoxicity by nitroxyl anion, NO^-. *Neuron*, **24**, 461–469.

119. Nelson, E.-J., Connoly, J., and McArthur, P. (2003) Nitric oxide and S-nitrosylation: excitotoxic and cell signaling mechanism. *Biol. Cell*, **95**, 3–8.

120. Kitaoka, Y., Kumai, T., Isenoumi, K., Kitaoka, Y., Motoki, M., Kobayashi, S., and Ueno, S. (2003) Neuroprotective effect of nitric oxide against NMDA-induced neurotoxicity in the rat retina is associated with tyrosine hydroxylase expression. *Brain Res.*, **977**, 46–54.

121. Kiagiadaki, F. and Thermos, K. (2008) The role of nitric oxide and cGMP in the somatostatin analog induced protection of the retina from AMPA-induced excitotoxicity. 22nd Conference of the Hellenic Society for Neuroscience, p. 175.
122. McMahon, D.-G. and Ponomareva, L.-V. (1996) Nitric oxide and cGMP modulate retinal glutamate receptors. *J. Neurophysiol.*, **76**, 2307–2315.
123. Spranger, J. and Pfeiffer, A.-F. (2001) New concepts in pathogenesis and treatment of diabetic retinopathy. *Exp. Clin. Endocrinol. Diabetes*, **109**, 438–450.
124. Caldwell, R.-B., Bartoli, M., Behzadian, M.-A., El-Remessy, A.-E., Al-Shabrawey, M., Platt, D.-H., Liou, G.-I., and Caldwell, R.-W. (2005) Vascular endothelial growth factor and diabetic retinopathy: role of oxidative stress. *Curr. Drug Targets*, **6**, 511–524.
125. Afzal, A., Shaw, L.-C., Ljubimov, A.-V., Boulton, M.-E., Segal, M.-S., and Grant, M.-B. (2007) Retinal and choroidal microangiopathies: therapeutic opportunities. *Microvasc. Res.*, **74**, 131–144.
126. Grant, M.-B., Caballero, S., and Millard, W.-J. (1993) Inhibition of IGF-I and b-FGF stimulated growth of human retinal endothelial cells by the somatostatin analogue, octreotide: a potential treatment for ocular neovascularization. *Regul. Pept.*, **48**, 267–278.
127. Smith, L.-E., Kopchick, J.-J., Chen, W., Knapp, J., Kinose, F., Daley, D., Foley, E. et al. (1997) Essential role of growth hormone in ischemia-induced retinal neovascularization. *Science*, **276**, 1706–1709.
128. Sall, J.-W., Klisovic, D.-D., O'Dorisio, M.-S., and Katz, S.-E. (2004) Somatostatin inhibits IGF-1 mediated induction of VEGF in human retinal pigment epithelial cells. *Exp. Eye Res.*, **79**, 465–476.
129. Dal Monte, M., Ristori, C., Cammalleri, M., and Bagnoli, P. (2009) Somatostatin analogs affect retinal angiogenesis in a mouse model of oxygen-induced retinopathy: involvement of the somatostatin receptor subtype 2. *Invest. Ophthalmol. Vis. Sci*, **50**, 3596–3606.
130. García de la Torre, N., Wass, J.-H.-A., and Turner, H.-E. (2002) Antiangiogenic effects of somatostatin analogs. *Clin. Endocrinol.*, **57**, 425–441.
131. Dasgupta, P. (2004) Somatostatin analogues: Multiple roles in cellular proliferation, neoplasia, and angiogenesis. *Pharmacol. Ther.*, **10**, 61–85.
132. McCombe, M., Lightman, S., Eckland, D.-J., Hamilton, A.-M., and Lightman, S.-L. (1991) Effect of a long-acting somatostatin analogue (BIM23014) on proliferative diabetic retinopathy: a pilot study. *Eye*, **5**, 569–575.
133. Simó, R., Lecube, A., Sararols, L., Garcia-Arumi, J., Segura, R.M., and Hernandez, C. (2002) Deficit of somatostatin-like immunoreactivity in the vitreous fluid of diabetic patients. *Diabetes Care*, **25**, 2282–2286.
134. Ambati, J., Chalam, K.-V., Chawla, D.-K., D'Angio, C.-T., Guillet, E.-G., Rose, S.-J., Vanderlinde, R.-E., and Ambati, B.-K. (1997) Elevated gamma aminobutyric acid, glutamate, and vascular endothelial growth factor levels in the vitreous of patients with proliferative diabetic retinopathy. *Arch. Ophthalmol.*, **115**, 1161–1166.
135. Lieth, E., Barber, A.-J., Xu, B., Dice, C., Ratz, M.-J., Tanase, D., and Strother, J.-M. (1998) Glial reactivity and impaired glutamate metabolism in short-term experimental diabetic retinopathy. *Diabetes*, **47**, 815–820.
136. Kowluru, R.-A., Engerman, R.-L., Case, G.-L., and Kern, T.-S. (2001) Retinal glutamate in diabetes and effect of antioxidants. *Neurochem. Int.*, **38**, 385–390.
137. Barber, A.-J. (2003) A new view of diabetic retinopathy: a neurodegenerative disease of the eye. *Prog. Neuropsychopharmacol. Biol. Psychiatry*, **27**, 283–290.
138. Lieth, E., Gardner, T.-W., Barber, A.-J., Antonetti, D.-A., and Penn State Retina Research Group (2000) Retinal neurodegeneration: early pathology in diabetes. *Clin. Experiment. Ophthalmol.*, **28**, 3–8.

139. Carrasco, E., Hernández, C., Miralles, A., Huguet, P., Farrés, J., and Simó, R. (2007) Lower somatostatin expression is an early event in diabetic retinopathy and is associated with retinal neurodegeneration. *Diabetes Care*, **30**, 2902–2908.

140. Carrasco, E., Hernández, C., de Torres, I., Farrés, J., and Simó, R. (2008) Lowered cortistatin expression is an early event in the human diabetic retina and is associated with apoptosis and glial activation. *Mol. Vis.*, **14**, 1496–1502.

13
Neurotrophic Effects of PACAP in the Cerebellar Cortex

Anthony Falluel-Morel, Hubert Vaudry, Hitoshi Komuro, Dariusz C. Gorecki, Ludovic Galas, and David Vaudry

Pituitary adenylate cyclase-activating polypeptide (PACAP) is a 38-amino acid peptide that was originally isolated from the ovine hypothalamus on the basis of its ability to stimulate cAMP formation in rat anterior pituitary cells [1]. PACAP belongs to the vasoactive intestinal polypeptide (VIP)–secretin–GHRH–glucagon superfamily. The sequence of PACAP has been remarkably well conserved during evolution, which suggests that the peptide should be involved in the regulation of important biological functions [2]. PACAP acts through three distinct receptors named PAC1-R, VPAC1-R, and VPAC2-R. PACAP has a much higher affinity than VIP for PAC1-R while both peptides bind similarly to VPAC1-R and VPAC2-R. All three receptors are positively coupled to the adenylyl cyclase, but PAC1-R also stimulates the phospholipase C (PLC) and the extracellular regulated kinase (ERK) pathways [3]. PAC1-R has several variants that result mainly from the alternative splicing of the mRNA regions encoding the first extracellular domain and the third intracellular cytoplasmic loop [4], which are coupled to differential transduction mechanisms. For instance, both the *short* and *hop* variants strongly activate the protein kinase A (PKA) and PLC pathways, whereas the *hip* isoform does not signal through the PLC pathway at all [5]. In addition, the *hip/hop* form displays an intermediate phenotype with a reduced ability to activate both signal transduction pathways [5].

PACAP and its receptors are widely distributed in the brain and peripheral organs, and the peptide has been found to exert pleiotropic effects on hormone secretion, vasodilation, regulation of inflammation, and apoptosis [3]. Besides, increasing evidence indicates that PACAP exerts neurotrophic actions during development and neuroprotective activities in the adult brain. Here, we review the current knowledge regarding the role of PACAP during histogenesis of the cerebellar cortex.

13.1
Expression of PACAP and its Receptors in the Developing Cerebellum

In the brain, the highest concentrations of PACAP are found in the hypothalamus and cerebellum [6]. PACAP is detected in the rat brain as early as embryonic day 14

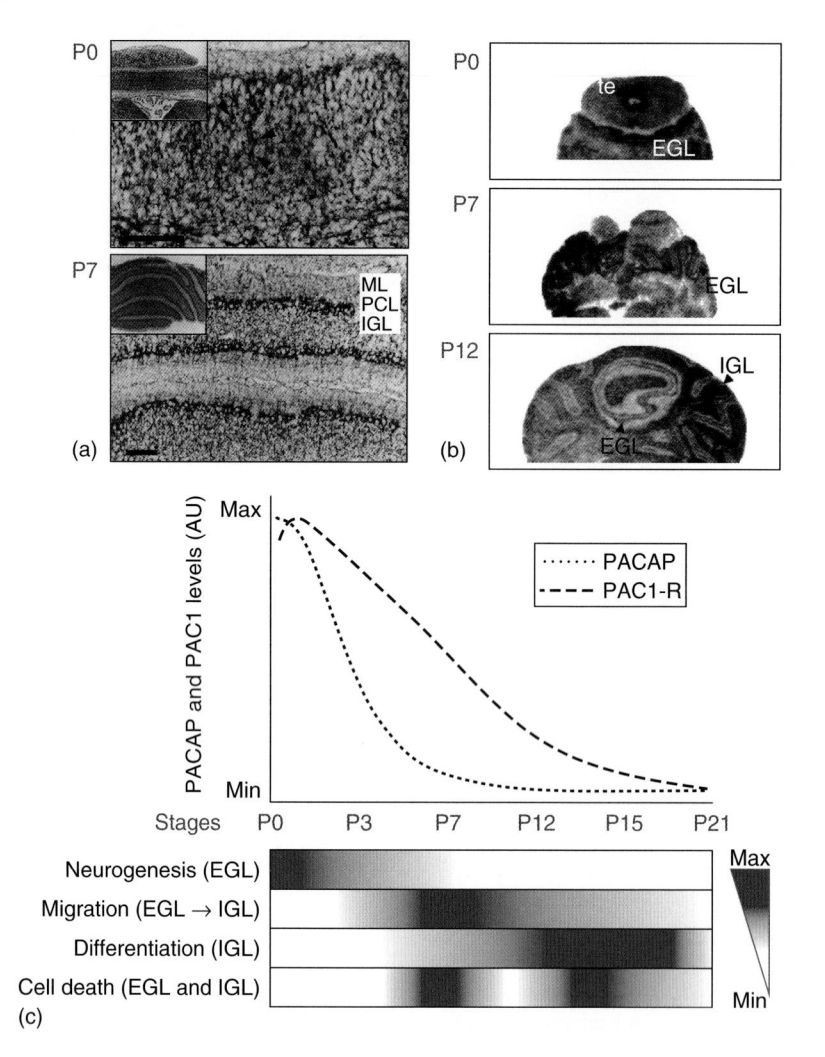

Figure 13.1 Expression and distribution of PACAP and its receptor PAC1-R during the postnatal cerebellar development. (a) Immunostaining for PACAP on frontal sections of the cerebellar cortex at birth (P0, top) and at postnatal day 7 (P7, bottom). In a P0 rat, PACAP is mainly found in Purkinje cell fibers and in a few cell bodies (arrowheads). At P7, PACAP immunoreactivity is visible in Purkinje cells throughout the cerebellar cortex and in nerve fibers of the internal granule cell layer. ML, molecular layer; PCL, Purkinje cell layer; IGL, internal granule cell layer. Scale bars: 100 µm. (Reprinted with permission from [9].) (b) Localization of PAC1-R mRNA by *in situ* hybridization in the cerebellum of postnatal rat at P0, P7, and P12. EGL, external granule cell layer; IGL, internal granule cell layer; tc, tectal neuroepithelium. Scale bar: 3 mm. (Reprinted with permission from [10].) (c) Developmental time-course of the expression of PACAP and PAC1-R in the rat cerebellum during the postnatal period. The chart below the main graph represents the time frame of the main steps of histogenesis (proliferation, migration, differentiation, and cell death).

(E14) and the concentration of the peptide gradually increases throughout the pre- and postnatal development periods [7]. In the cerebellum, the concentration of PACAP culminates immediately after birth and then gradually declines to reach adult levels within the next few weeks (Figure 13.1). In newborn rodents, PACAP-positive cells and nerve fibers are localized in the Purkinje cell layer [8, 9]. In the adult rat cerebellum, PACAP is mainly present in the soma and dendrites of Purkinje cells and in nerve fibers surrounding granule cells [9].

PACAP binding sites have been identified in both the developing and adult cerebellum (Figure 13.1). In the rat cerebellum, PACAP binding sites are transiently expressed from postnatal day 8 (P8) to P25 in a germinative matrix, in the external granule cell layer (EGL), and in the medulla [11, 12]. PACAP binding sites are also present in the internal granule cell layer (IGL) in both the developing and mature cerebellum [11]. Specifically, these binding sites are located on the membrane of the granule cells [12]. Molecular characterization of PACAP receptors revealed that, in the EGL, granule cells express both PAC1-R and VPAC1-R mRNA, but not VPAC2-R mRNA [10]. These receptors are functional, since treatment of granule cells with PACAP leads to the stimulation of both the adenylyl cyclase and PLC pathways [12–14].

13.2
Effects of PACAP on Granule Cell Proliferation

Studies that aimed at investigating the effect of PACAP on cell proliferation have shown that the peptide exhibits either promitotic or antimitotic activities. In fact, even on a single cell type, PACAP may exert differential effects depending on the types of receptor variants that are expressed. Such an effect was specifically described in neocortical and hindbrain precursors, where PACAP acts as a promitogenic factor through stimulation of the PAC1-R *hop* isoform while, at later stages of development when cells start to express the *short* PAC1-R isoform, PACAP becomes antimitogenic [15–18].

In the cerebellum, PACAP is generally described as an antimitogenic peptide, although this effect is the consequence of the prodifferentiating action of PACAP (see below). Indeed, most functional studies regarding the effect of PACAP during cerebellar corticogenesis in rodents have been conducted after birth [19, 20] within a time frame when cell proliferation is already declining. However, at earlier stages of development (E18–P1), PACAP may exert a promitotic effect in the EGL [20]. These observations indicate that a detailed investigation of the effect of PACAP throughout cerebellar development is now required.

13.3
Effects of PACAP on Granule Cell Migration

Real-time videomicroscopy observations of cultured granule cells revealed that PACAP reduces neuronal motility [8, 21]. Experiments performed on cerebellar

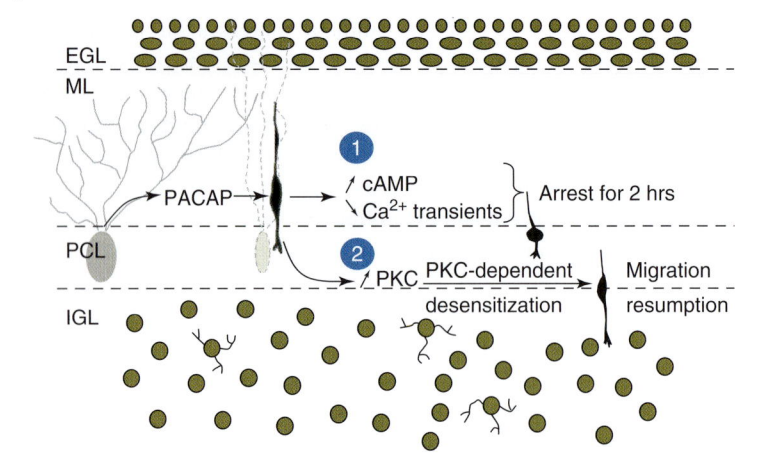

Figure 13.2 Schematic representation of the effects of PACAP on cerebellar granule cell migration. (1) When granule cells reach the Purkinje cell layer, they get exposed to PACAP which, upon receptor activation, induces an increase in intracellular cAMP and a subsequent inhibition of the amplitude of spontaneous Ca^{2+} transients. As shown using specific pharmacological agents, regulation of these second messengers is responsible for the arrest signal evoked by PACAP. (2) PACAP also induces protein kinase C activation in granule cells, which mediates the desensitization of PAC1-R in Purkinje cells, allowing granule cells to resume their migration toward the IGL. cAMP, cyclic adenosine mono phosphate; EGL, external granule cell layer; IGL, internal granule cell layer; PCL, Purkinje cell layer; PKC, protein kinase C.

tissue slices from P10 mice have shown that application of exogenous PACAP induces a strong inhibition of granule cell migration in the EGL and the molecular layer (ML) but has no effect on cell migration in the Purkinje cell layer and in the IGL [8]. In contrast, the PACAP antagonist PACAP(6-38) accelerates granule cell migration in the Purkinje cell layer but does not modify cell velocity in the EGL, ML, and IGL. These data provide strong evidence for the involvement of PACAP in the control of granule cell migration and suggest that endogenous PACAP is responsible for the reduction of the velocity of granule cell migration observed in the Purkinje cell layer during development (Figure 13.2). So far, PACAP is the first messenger that has been shown to regulate the migration of granule cells at the level of Purkinje neurons. The functional significance of this pause during development of the cerebellar cortex is not fully understood yet but the available results suggest that PACAP signaling is a key element in the final neuronal migration of granule cells in the absence of radial glia guidance within the IGL.

In acute slice preparations, PACAP stimulates cAMP production, which in turn decreases the amplitude of spontaneous calcium transients in granule cells undergoing migration (Figure 13.2) [8]. Using specific pharmacological agents, it has been shown that cAMP and calcium mediate the arrest signal evoked by PACAP. However, even though the effect of PACAP on cell migration is robust, it only lasts for about 2 h because of a desensitization mechanism mediated through the protein kinase C pathway [8]. The current hypothesis is that after an initial response to endogenous PACAP released by Purkinje cells, PACAP receptors borne by granule cells undergo desensitization to allow their migration within the IGL. This desensitization process is likely to be beneficial inasmuch as the IGL contains substantial amounts of PACAP that may, later on, promote cell differentiation.

13.4
Effects of PACAP on Granule Cell Survival

Numerous studies have shown that PACAP is a potent neuroprotective factor [3]. In the cerebellum, apoptotic cell death plays a critical role during histogenesis. Indeed, a large number of granule cells are eliminated in the premigratory zone of the EGL and later on in the IGL. The facts that PACAP is produced by Purkinje cells in the immature cerebellum and that PAC1-R is expressed in the EGL and IGL strongly suggest that PACAP may participate in the control of cell survival. Consistent with this hypothesis, *in vitro* studies have shown that PACAP promotes granule cell survival (Figure 13.3a). In addition, treatment of cultured granule cells with PACAP prevents cell death induced by different agents such as ceramides [22, 23], ethanol [24], H_2O_2 [25], and cisplatin [26].

Local administration of PACAP *in vivo* at the surface of the cerebellum of P8 rat pups increases the number of granule cells present in the IGL [27], suggesting that the prodifferentiating and trophic effects of PACAP observed *in vitro* may reflect a physiological role in the organism. More recent studies, conducted in knockout animals, provide functional evidence for a role of endogenous PACAP in the control of morphogenetic cell death. These studies show that disruption of PACAP or PAC1-R significantly reduces the thickness of the IGL and that this effect can be ascribed to an increase in apoptotic cell death (Figure 13.3b,c) [19, 20].

The mechanisms of action of PACAP have been widely investigated in cerebellar neurons [2, 3, 28]. PACAP acts as a potent inhibitor of caspase-3 activity through a blockade of the mitochondrial apoptotic pathway. This signaling cascade is also activated by cAMP stimulators or PACAP analogs and blocked by a dominant inhibitory mutant of the cAMP-dependent PKA. Downstream of PKA, PACAP induces the phosphorylation of ERK through Rap1 and Ras activation [29]. ERK activation is required for the long-lasting inhibition of caspase-3 activity and contributes to the neuroprotective effect of PACAP.

13.5
Effects of PACAP on Granule Cell Differentiation

It is difficult to distinguish between the protective and the prodifferentiating effects of PACAP as these two processes are interdependent and share common pathways. However, several studies indicate that PACAP promotes granule cell differentiation. Treatment of granule cells with PACAP *in vitro* induces the appearance of a dense network of long neurites (Figure 13.3a) [23]. This apparent differentiation

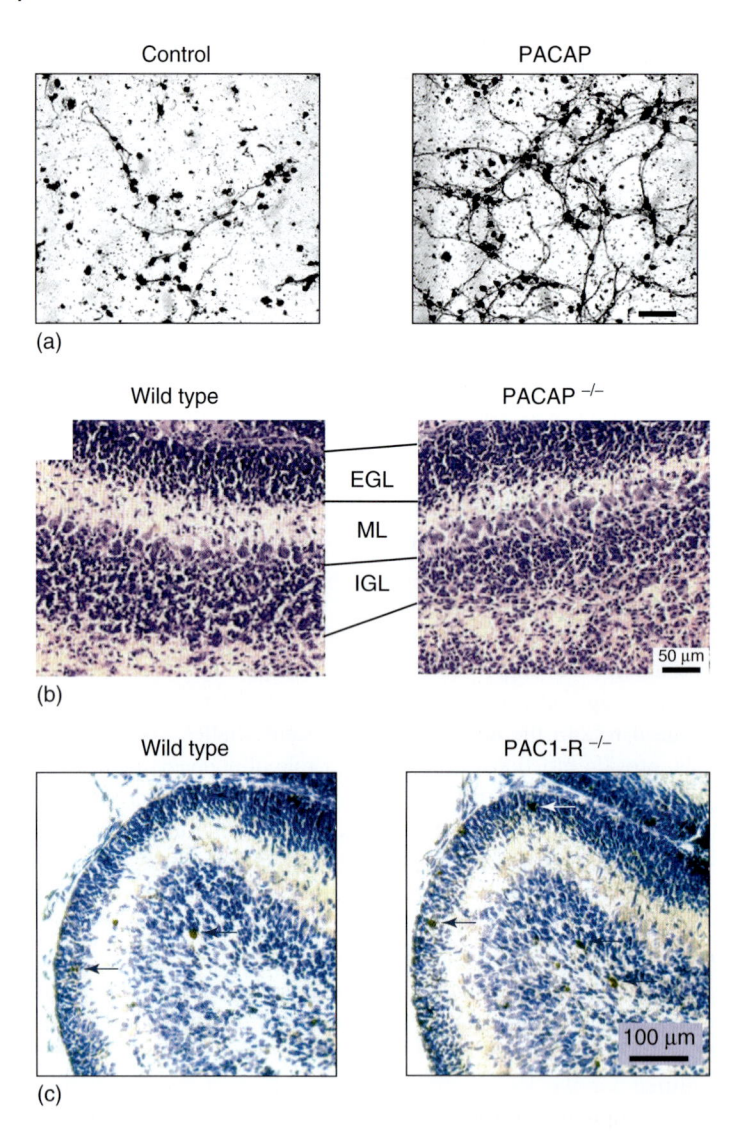

Figure 13.3 Neurotrophic effects of PACAP on cerebellar granule cells. (a) Microphotographs illustrating the effect of PACAP on cell survival and neurite outgrowth in rat cerebellar granule cells. Granule cells from 8 day old (P8) rats were cultured for 48 h in control conditions (left) or in the presence of 10^{-8} M PACAP (right). Scale bar: 25 μm. (Reprinted with permission from [23].) (b) Measurement of the thickness of the cerebellar layers in wild-type and PACAP knockout mice at P7. EGL, external granule cell layer; ML, molecular layer; IGL, internal granule cell layer. (Reprinted with permission from [19].) (c) Measurement of active caspase-3 immunoreactivity in P7 mouse cerebellum. In wild-type animals, very few caspase-3-positive cells (arrows) are found in the EGL and IGL, while a higher number is seen in the same regions of PAC1$^{-/-}$ animals. Scale bar: 100 μm. (Reprinted with permission from [20].)

process is associated with an accumulation of actin at the emergence cone and phosphorylation of the tau protein [21]. A recent study indicates that PACAP, acting through Epac, stimulates Rap and p38 MAPK leading to the mobilization of intracellular Ca^{2+} stores [30]. This new pathway may facilitate the maturation of granule precursors into excitable neurons. In support of this notion, PACAP has been shown to enhance the release of glutamate induced by granule cell depolarization [31].

The antimitotic effect of PACAP can also be associated with its prodifferentiating activity. Notably, in the EGL, granule cells proliferate actively and then stop to divide in order to migrate through the different layers of the cerebellar cortex [32]. Sonic hedgehog (Shh) is a key morphogen produced by Purkinje neurons, which stimulates granule cell proliferation [33]. PACAP has been found to markedly inhibit Shh-induced thymidine incorporation in both rat and mouse granule cells [18], suggesting that PACAP serves as a signal for granule cells to stop proliferating and initiate differentiation. Nevertheless, alternative mechanisms are probably also involved, particularly as PACAP has been shown to inhibit cell proliferation through stimulation of the tumor suppressor gene *Lot1* [34].

13.6
Functional Relevance

PACAP appears to play multiple functions during development of the cerebellum: PACAP modulates granule cell proliferation, transiently stops their migration in the Purkinje cell layer, contributes to their differentiation, and promotes their survival. The ability of PACAP to protect cultured granule cells from apoptosis induced by neurotoxic molecules suggests that endogenous PACAP might also defend the cerebellum against neuronal damages occurring during brain histogenesis, notably in the case of alcohol exposure or hypoxia. Consistent with this hypothesis, it has been shown that endogenous PACAP released into the culture medium can protect granule cells exposed to stressors such as ethanol or hydrogen peroxide [35].

Mutation of the Shh receptor gene Patched-1 is often responsible for the development of medulloblastoma (the most common malignant brain tumor in childhood), which is believed to arise from precursor cells from the EGL. Deletion of a single copy of the *PACAP* gene leads to a 2.5-fold increase in medulloblastoma incidence in Patched-1 knockout animals, demonstrating that PACAP exerts a powerful inhibitory control on the induction, growth, and/or survival of these tumors [36]. In fact, treatment of medulloblastoma cell lines with PACAP strongly reduces cell proliferation, indicating the possible involvement of PACAP in medulloblastoma pathogenesis. On the basis of these observations, PACAP is currently exciting interest as a possible candidate for development into a therapeutically valuable agent.

Importantly, as PACAP receptors exhibit a similar pattern of expression in the developing human cerebellum [37], it is likely that the neurotrophic effects of PACAP documented in rodents may also occur in humans. As a step toward testing this notion, it will now be possible to conduct preclinical therapeutic studies in nonhuman primate models, which have functional PACAP receptors [38, 39].

Acknowledgments

This work was supported by INSERM, the Institut de Recherches Scientifiques sur les Boissons (IREB), the Interreg 4A FEDER project AdMiN, and an ANR Jeune Chercheuse-Jeune Chercheur.

References

1. Miyata, A., Arimura, A., Dahl, R.R., Minamino, N., Uehara, A., Jiang, L., Culler, M.D., and Coy, D.H. (1989) Isolation of a novel 38 residue-hypothalamic polypeptide which stimulates adenylate cyclase in pituitary cells. *Biochem. Biophys. Res. Commun.*, **164**, 567–574.
2. Vaudry, D., Gonzalez, B.J., Basille, M., Yon, L., Fournier, A., and Vaudry, H. (2000) Pituitary adenylate cyclase-activating polypeptide and its receptors: from structure to functions. *Pharmacol. Rev.*, **52**, 269–324.
3. Vaudry, D., Falluel-Morel, A., Bourgault, S., Basille, M., Burel, D., Wurtz, O., Fournier, A., Chow, B.K.C., Hashimoto, H., and Vaudry, H. (2009) Pituitary adenylate cyclase-activating polypeptide and its receptors: 20 years after the discovery. *Pharmacol. Rev.* **3**, 283–357.
4. Ushiyama, M., Ikeda, R., Sugawara, H., Yoshida, M., Mori, K., Kangawa, K., Inoue, K., Yamada, K., and Miyata, A. (2007) Differential intracellular signaling through PAC1 isoforms as a result of alternative splicing in the first extracellular domain and the third intracellular loop. *Mol. Pharmacol.*, **72**, 103–111.
5. Spengler, D., Waeber, C., Pantaloni, C., Holsboer, F., Bockaert, J., Seeburg, P.H., and Journot, L. (1993) Differential signal transduction by five splice variants of the PACAP receptor. *Nature*, **365**, 170–175.
6. Kivipelto, L., Absood, A., Arimura, A., Sundler, F., Hakanson, R., and Panula, P. (1992) The distribution of pituitary adenylate cyclase-activating polypeptide-like immunoreactivity is distinct from helodermin- and helospectin-like immunoreactivities in the rat brain. *J. Chem. Neuroanat.*, **5**, 85–94.
7. Tatsuno, I., Somogyvari-Vigh, A., and Arimura, A. (1994) Developmental changes of pituitary adenylate cyclase activating polypeptide (PACAP) and its receptor in the rat brain. *Peptides*, **15**, 55–60.
8. Cameron, D.B., Galas, L., Jiang, Y., Raoult, E., Vaudry, D., and Komuro, H. (2007) Cerebellar cortical-layer-specific control of neuronal migration by pituitary adenylate cyclase-activating polypeptide. *Neuroscience*, **146**, 697–712.
9. Nielsen, H.S., Hannibal, J., and Fahrenkrug, J. (1998) Expression of pituitary adenylate cyclase activating polypeptide (PACAP) in the postnatal and adult rat cerebellar cortex. *Neuroreport*, **9**, 2639–2642.
10. Basille, M., Vaudry, D., Coulouarn, Y., Jegou, S., Lihrmann, I., Fournier, A., Vaudry, H., and Gonzalez, B. (2000) Comparative distribution of pituitary adenylate cyclase-activating polypeptide (PACAP) binding sites and PACAP receptor mRNAs in the rat brain during development. *J. Comp. Neurol.*, **425**, 495–509.
11. Basille, M., Gonzalez, B.J., Fournier, A., and Vaudry, H. (1994) Ontogeny of pituitary adenylate cyclase-activating polypeptide (PACAP) receptors in the rat cerebellum: a quantitative autoradiographic study. *Brain Res. Dev. Brain Res.*, **82**, 81–89.

12. Basille, M., Gonzalez, B.J., Leroux, P., Jeandel, L., Fournier, A., and Vaudry, H. (1993) Localization and characterization of PACAP receptors in the rat cerebellum during development: evidence for a stimulatory effect of PACAP on immature cerebellar granule cells. *Neuroscience*, **57**, 329–338.
13. Basille, M., Gonzalez, B.J., Desrues, L., Demas, M., Fournier, A., and Vaudry, H. (1995) Pituitary adenylate cyclase-activating polypeptide (PACAP) stimulates adenylyl cyclase and phospholipase C activity in rat cerebellar neuroblasts. *J. Neurochem.*, **65**, 1318–1324.
14. Favit, A., Scapagnini, U., and Canonico, P.L. (1995) Pituitary adenylate cyclase-activating polypeptide activates different signal transducing mechanisms in cultured cerebellar granule cells. *Neuroendocrinology*, **61**, 377–382.
15. Lelievre, V., Hu, Z., Byun, J.Y., Ioffe, Y., and Waschek, J.A. (2002) Fibroblast growth factor-2 converts PACAP growth action on embryonic hindbrain precursors from stimulation to inhibition. *J. Neurosci. Res.*, **67**, 566–573.
16. Lu, N., Zhou, R., and DiCicco-Bloom, E. (1998) Opposing mitogenic regulation by PACAP in sympathetic and cerebral cortical precursors correlates with differential expression of PACAP receptor (PAC1-R) isoforms. *J. Neurosci. Res.*, **53**, 651–662.
17. Nicot, A. and DiCicco-Bloom, E. (2001) Regulation of neuroblast mitosis is determined by PACAP receptor isoform expression. *Proc. Natl. Acad. Sci. U.S.A.*, **98**, 4758–4763.
18. Nicot, A., Lelievre, V., Tam, J., Waschek, J.A., and DiCicco-Bloom, E. (2002) Pituitary adenylate cyclase-activating polypeptide and sonic hedgehog interact to control cerebellar granule precursor cell proliferation. *J. Neurosci.*, **22**, 9244–9254.
19. Allais, A., Burel, D., Isaac, E.R., Gray, S.L., Basille, M., Ravni, A., Sherwood, N.M., Vaudry, H., and Gonzalez, B.J. (2007) Altered cerebellar development in mice lacking pituitary adenylate cyclase-activating polypeptide. *Eur. J. Neurosci.*, **25**, 2604–2618.
20. Falluel-Morel, A., Tascau, L.I., Sokolowski, K., Brabet, P., and DiCicco-Bloom, E. (2008) Granule cell survival is deficient in PAC1-/- mutant cerebellum. *J. Mol. Neurosci.*, **36**, 38–44.
21. Falluel-Morel, A., Vaudry, D., Aubert, N., Galas, L., Benard, M., Basille, M., Fontaine, M., Fournier, A., Vaudry, H., and Gonzalez, B.J. (2005) Pituitary adenylate cyclase-activating polypeptide prevents the effects of ceramides on migration, neurite outgrowth, and cytoskeleton remodeling. *Proc. Natl. Acad. Sci. U.S.A.*, **102**, 2637–2642.
22. Falluel-Morel, A., Aubert, N., Vaudry, D., Basille, M., Fontaine, M., Fournier, A., Vaudry, H., and Gonzalez, B.J. (2004) Opposite regulation of the mitochondrial apoptotic pathway by C2-ceramide and PACAP through a MAP-kinase-dependent mechanism in cerebellar granule cells. *J. Neurochem.*, **91**, 1231–1243.
23. Gonzalez, B.J., Basille, M., Vaudry, D., Fournier, A., and Vaudry, H. (1997) Pituitary adenylate cyclase-activating polypeptide promotes cell survival and neurite outgrowth in rat cerebellar neuroblasts. *Neuroscience*, **78**, 419–430.
24. Vaudry, D., Rousselle, C., Basille, M., Falluel-Morel, A., Pamantung, T.F., Fontaine, M., Fournier, A., Vaudry, H., and Gonzalez, B.J. (2002) Pituitary adenylate cyclase-activating polypeptide protects rat cerebellar granule neurons against ethanol-induced apoptotic cell death. *Proc. Natl. Acad. Sci. U.S.A.*, **99**, 6398–6403.
25. Vaudry, D., Pamantung, T.F., Basille, M., Rousselle, C., Fournier, A., Vaudry, H., Beauvillain, J.C., and Gonzalez, B.J. (2002) PACAP protects cerebellar granule neurons against oxidative stress-induced apoptosis. *Eur. J. Neurosci.*, **15**, 1451–1460.
26. Aubert, N., Vaudry, D., Falluel-Morel, A., Desfeux, A., Fisch, C., Ancian, P., de Jouffrey, S., Le Bigot, J.F., Couvineau, A., Laburthe, M., Fournier, A., Laudenbach, V., Vaudry, H., and Gonzalez, B.J. (2008) PACAP prevents toxicity induced by cisplatin in rat and primate neurons but not in proliferating

27. Vaudry, D., Gonzalez, B.J., Basille, M., Pamantung, T.F., Fournier, A., and Vaudry, H. (2000) PACAP acts as a neurotrophic factor during histogenesis of the rat cerebellar cortex. Ann. N. Y. Acad. Sci., 921, 293–299.
28. Vaudry, D., Falluel-Morel, A., Leuillet, S., Vaudry, H., and Gonzalez, B.J. (2003) Regulators of cerebellar granule cell development act through specific signaling pathways. Science, 300, 1532–1534.
29. Obara, Y., Horgan, A.M., and Stork, P.J. (2007) The requirement of Ras and Rap1 for the activation of ERKs by cAMP, PACAP, and KCl in cerebellar granule cells. J. Neurochem., 101, 470–482.
30. Ster, J., De Bock, F., Guerineau, N.C., Janossy, A., Barrere-Lemaire, S., Bos, J.L., Bockaert, J., and Fagni, L. (2007) Exchange protein activated by cAMP (Epac) mediates cAMP activation of p38 MAPK and modulation of Ca^{2+}-dependent K^+ channels in cerebellar neurons. Proc. Natl. Acad. Sci. U.S.A., 104, 2519–2524.
31. Aoyagi, K. and Takahashi, M. (2001) Pituitary adenylate cyclase-activating polypeptide enhances $Ca^{(2+)}$-dependent neurotransmitter release from PC12 cells and cultured cerebellar granule cells without affecting intracellular Ca(2+) mobilization. Biochem. Biophys. Res. Commun., 286, 646–651.
32. Altman, J. (1972) Postnatal development of the cerebellar cortex in the rat. 3. Maturation of the components of the granular layer. J. Comp. Neurol., 145, 465–513.
33. Dahmane, N. and Ruiz i Altaba, A. (1999) Sonic hedgehog regulates the growth and patterning of the cerebellum. Development, 126, 3089–3100.
34. Fila, T., Trazzi, S., Crochemore, C., Bartesaghi, R., and Ciani, E. (2009) Lot1 is a key element of the pituitary adenylate cyclase-activating polypeptide (PACAP)/cyclic AMP pathway that negatively regulates neuronal precursor proliferation. J. Biol. Chem., 284, 15325–15338.
35. Vaudry, D., Hamelink, C., Damadzic, R., Eskay, R.L., Gonzalez, B., and Eiden, L.E. (2005) Endogenous PACAP acts as a stress response peptide to protect cerebellar neurons from ethanol or oxidative insult. Peptides, 26, 2518–2524.
36. Lelievre, V., Seksenyan, A., Nobuta, H., Yong, W.H., Chhith, S., Niewiadomski, P., Cohen, J.R., Dong, H., Flores, A., Liau, L.M., Kornblum, H.I., Scott, M.P., and Waschek, J.A. (2008) Disruption of the PACAP gene promotes medulloblastoma in ptc1 mutant mice. Dev. Biol., 313, 359–370.
37. Basille, M., Cartier, D., Vaudry, D., Lihrmann, I., Fournier, A., Freger, P., Gallo-Payet, N., Vaudry, H., and Gonzalez, B. (2006) Localization and characterization of pituitary adenylate cyclase-activating polypeptide receptors in the human cerebellum during development. J. Comp. Neurol., 496, 468–478.
38. Aubert, N., Basille, M., Falluel-Morel, A., Vaudry, D., Bucharles, C., Jolivel, V., Fisch, C., De Jouffrey, S., Le Bigot, J.F., Fournier, A., Vaudry, H., and Gonzalez, B.J. (2007) Molecular, cellular, and functional characterizations of pituitary adenylate cyclase-activating polypeptide and its receptors in the cerebellum of New and Old World monkeys. J. Comp. Neurol., 504, 427–439.
39. Jolivel, V., Basille, M., Aubert, N., de Jouffrey, S., Ancian, P., Le Bigot, J.F., Noack, P., Massonneau, M., Fournier, A., Vaudry, H., Gonzalez, B.J., and Vaudry, D. (2009) Distribution and functional characterization of pituitary adenylate cyclase-activating polypeptide receptors in the brain of non-human primates. Neuroscience, 160, 434–451.

14
The Corticotropin-Releasing Hormone in Neuroprotection
Christian Behl and Angela Clement

Corticotropin-releasing hormone (CRH) is a small peptide with diverse biological functions. Initially identified as the central modulator of the mammalian hypothalamic-pituitary-adrenal (HPA) axis and playing a key role in mediating neuroendocrine effects in response to stress, CRH research of the last decade has uncovered additional roles of the CRH. On the basis of findings that the CRH system is disturbed in neurodegenerative disorders and that the receptors for CRH are also widely expressed in brain areas not directly related to the stress response, a potential role for CRH in neuroprotection has been proposed. Until now, various investigations have demonstrated a direct protective role of CRH and its related peptides at the cellular level and also *in vivo*. The molecular downstream targets of the neuroprotective CRH activity are just starting to evolve. CRH and CRH receptors are back on the map and are in the focus of many researchers working on novel approaches to prevention and treatment of neurodegenerative syndromes.

14.1
Introduction

CRH is *the* stress hormone of the body. CRH (also called corticotropin-releasing factor, CRF) regulates the activity of the HPA axis and is released upon stressful stimuli. It is secreted by the hypothalamus and mediates neuroendocrine effects [1]. CRH, an oligopeptide consisting of 41 amino acids, was discovered by Vale *et al.* in 1981 [2, 3] and has been found to regulate the stress response by turning on the HPA axis. As for other releasing hormones of the hypothalamus, CRH is derived from a larger precursor and is released following proteolytic processing. Neuroendocrine hypothalamic neurons are molecular interfaces that turn electrophysiological signals from the periphery into biochemical signals which then switch on the respective hormone axis, such as the HPA, activated by the molecular player CRH. Initially believed to be involved in just regulating the endocrine stress response, by studying the expression of CRH and its receptors it rapidly became clear that CRH has a rather broad spectrum of activities, including mediating the autonomic, behavioral, and immunological stress response. Owing to its central function in the regulation of the HPA axis, a dysfunction of CRH signaling has been linked to psychiatric disorders, mainly anxiety syndromes and depression [4–7]. CRH and CRH receptors along the HPA axis are therefore important pharmacological targets for a novel antidepressant treatment approach [8–10].

Over the last decade, the evidence that has accumulated proves that CRH has many extra-HPA activities. CRH can directly modulate neuronal activities related to memory and cognition and has been shown to possess neuroprotective properties. Recent work reveals the molecular details of the neuroprotective actions of CRH.

14.2
The CRH Family of Proteins and Molecular Signal Transduction

The CRH protein family consists of the four peptides: CRH, urocortin 1 (UCN1), urocortin 2 (UCN2), and urocortin 3 (UCN3). An additional component of the CRH system is the corticotropin-releasing hormone-binding protein (CRH-BP) that is highly conserved among vertebrates, shows overlapping expression with CRH and UCN1, and binds both peptides with high affinity thus modulating the access of CRH and UCN1 to their respective receptors [11, 12]. The actions of CRH and its related peptides are mediated by two distinct high-affinity transmembrane receptors, corticotropin-releasing hormone receptor 1 (CRH-R1) and corticotropin-releasing hormone receptor 2 (CRH-R2). The two mammalian CRH receptor subtypes share a 70% sequence homology and in both cases molecular splice variants are known, although only CRH-R2 splice variants appear to be functional [13, 14]. Consistent with the structural differences, both receptors display distinct pharmacological profiles. CRH and UCN1 are ligands for both receptors, CRH-R1 and CRH-R2. CRH binds with 10 times higher affinity to CRH-R1 than to CRH-R2, whereas UCN1 binds to both receptor subtypes with similar affinity [14] (Table 14.1). UCN2 and UCN3 bind selectively to CRH-R2. Therefore, the urocortins appear to be the natural ligands for CRH-R2. In the brain, CRH itself is found with the highest density in cell bodies of the parvocellular neurons of the hypothalamic paraventricular

Table 14.1 Overview of the CRH system.

CRH family member	Localization	Receptor affinity
CRH	PVN of the hypothalamus, CNA, cortex, hippocampus, brain stem, periphery	CRH-R1 (10-fold), CRH-R2
UCN1	EW nucleus of midbrain	CRH-R1=CRH-R2
UCN2	Hypothalamus, LC, brain stem, spinal cord	CRH-R2
UCN3	Hypothalamus, amygdala, periphery	CRH-R2

EW nucleus, Edinger–Westphal nucleus; PVN, paraventricular nucleus; CNA, central nucleus of the amygdala; and LC, locus ceruleus.

nucleus and the central nucleus of the amygdala, and to a lesser extent in other brain areas. CRH is also present in a variety of peripheral tissues [15, 16].

Interestingly, both receptors are widely distributed throughout the CNS, which is indicative of extra-HPA activities of the CRH and related neuropeptides. In the CNS, CRH-R1 appears to be more abundantly expressed than CRH-R2. The latter is mainly found in the peripheral tissue, for example, prominently in the heart. As expected, the density of CRH-R1 is highest in the anterior lobe of the pituitary. In addition, CRH-R1 is also found to be expressed in the cerebral cortex, cerebellum, hippocampus, amygdala, and the brain stem. CRH receptors belong to the group of G-protein-coupled 7 transmembrane receptors. Upon binding of CRH to CRH-R1, G-proteins and subsequently the cAMP-synthesizing enzyme adenylate cyclase are activated, leading to an increase in intracellular cAMP levels [17]. In addition to the initially described activation of the downstream signaling mediator protein kinase A (PKA), numerous other signaling pathways have been shown to be activated by CRH (for a review see [15]).

CRH receptors are pharmacological targets and both peptidergic and nonpeptidergic antagonists that can block the signal transduction through CRH-R1 have been described in the search for treatment of affective disorders associated with a deregulated HPA axis. Such antagonists can be either selective, such as antalarmin [18] R121919 [19] and the very recently synthesized potent CRH-R1 specific antagonist compound 13–15 [20], or nonselective, such as the frequently used α-helical peptide CRH9-41 [21].

14.3
From the Physiology to the Pathophysiology of CRH

On the basis of its discovery as central stress hormone of the mammalian HPA axis, it was found that CRH controls the adaptation of behavior and the autonomic and immunological stress response. CRH was found to exert anxiogenic properties in animal models, improve alertness, modulate feeding and ingestion, control body temperature, affect blood flow and blood pressure, and may have a proinflammatory effect [22]. All these activities are linked to the hypothalamus and the pituitary, where CRH receptors are abundantly expressed. CRH is clearly a stress hormone that is of vital importance for the mammalian body. Adequate response to harmful external stress, including the proper supply of energy, is essential since there is no survival without the possibility of coping with stress in an adequate manner. Therefore, CRH as central mediator of the HPA axis serves as an overall protective hormone for the organism. On the other hand, a dysregulation of the HPA axis can be detrimental to the organism and a number of psychiatric disorders are associated with high blood levels of CRH. For chronic depression and a number of anxiety-related disorders, a causal relationship between high CRH levels and these disorders has been found [4, 5, 11]. Continuously increased CRH levels and a dysregulated HPA axis can cause disease. Consequently, CRH-R1-selective antagonists have been developed to pharmacologically reduce an HPA axis overdrive and may indeed be beneficial in the treatment of depression and anxiety [23, 24].

A protective role for CRH was hypothesized based on the observation that CRH levels in the brains of patients with Alzheimer's disease (AD) were reduced [25–27]. Indeed, CRH was shown to display neuroprotective activities under oxidative stress in rat primary cerebellar neurons, human IMR32 neuroblastoma cells, mouse pituitary adenoma AtT20 cells, and HT22 cells (a mouse clonal hippocampal cell line with stable overexpression of CRH-R1 [28]). In addition, a potential neuroprotective activity of CRH has been suggested by various studies demonstrating decreased CRH levels and changes in CRH receptor expression in a number of neurodegenerative disorders, including stroke, ischemia, and AD [29, 30].

14.4
CRH and Neurodegenerative Conditions

Early work from the groups of DeSouza and Nemeroff showed that there are reciprocal changes in the immunoreactivity of CRH and CRH receptors in the cerebral cortex and striatum of patients suffering from AD [25, 26, 31, 32]. Reduced levels of CRH in the cerebrospinal fluid (CSF) of AD patients have recently been reported [33–35]. By including a battery of neuropsychological testing, a correlation of low CRH levels in the CSF with greater cognitive impairment has been detected [27]. Further, increased CRH receptor immunoreactivity was correlated with AD-associated amyloid β (Aβ) protein deposits in AD tissue [32].

Fueled by the *postmortem* observations of changes in the CRH system under disease condition, some initial *in vivo* studies focused on the effect of available CRH-R-antagonists and found protective effects of the antagonist α-helical CRH9–41 in rodent models of experimental ischemia [36, 37]. Other studies at the same time demonstrated that the activation (and not the block) of CRH receptors causes protection in hypoxia and ischemia-reperfusion models in cardiac myocytes and rat brain [38, 39]. This initial discrepancy in detecting a neuroprotective effect for the activation as well as for the block of CRH receptor signaling may be due in part to differences in the experimental *in vivo* models used. Moreover, it should be mentioned that the frequently used α-helical CRH9–41 antagonist is reported to exhibit partial agonist activity, particularly when used at higher doses (>25 µg) and may thus induce effects similar to CRH in the brain tissue [40, 41].

14.5
Protective Activities of CRH

Owing to the finding that lower CRH levels are associated with AD pathology, an experimental increase in CRH levels in the brain and its consequences with respect to cognitive function were investigated. Treatment of mice with CRH itself, as well as with other CRH-BP ligands that remove CRH from the CRH-BP but do not show binding to CRH receptors had interesting effects: increasing the levels of "free" CRH levels in the brain up to control levels by promoting its dissociation from the high-affinity CRH-BP had cognition-enhancing properties in models of learning and memory in mice, without the characteristic stress effects of CRH receptor agonists [42]. After this interesting study, many labs have investigated CRH as a neuroprotective neuropeptide in a variety of disease-related models in vitro and in vivo (Table 14.2). As a result of these studies, different downstream signaling pathways and potential mediators of the protective activity have been described. Our own lab has contributed data on molecular mediators of CRH's neuroprotective activity under conditions of oxidative stress. In 2003, by employing primary neuronal cultures from different brain regions (cerebellum, cerebral cortex, hippocampus), we found that CRH exerts brain region-specific neuroprotective effects against Aβ toxicity [43]. At low CRH concentrations, neuroprotective effects were observed only in cerebellar and hippocampal cultures, whereas at high concentrations protection of cortical neurons was also observed. Western blot analysis with phospho-specific antibodies directed against mitogen-activated protein kinase (MAPK), cAMP response element-binding protein (CREB), and glycogen synthase kinase (GSK)3β also resulted in brain region-specific differences regarding intracellular signaling. Low CRH concentrations were correlated with cell survival and activation of the CREB pathway and inactivation of GSK3β in cerebellar and hippocampal cultures, whereas in cortical cultures, higher concentrations of CRH were necessary to activate CREB and inactivate GSK3β. Importantly, MAPK activation has been observed only in cortical neurons. Differences in signaling were found to be independent of receptor expression levels because RT-PCR analysis indicated no region-specific differences in CRH-R1 mRNA expression [44]. Brain

Table 14.2 Neuroprotective effects of CRH and UCN1.

Protected cell type/cell model	Toxin	References
Primary neurons of cerebellum, hippocampus, cortex from rat	Aβ-peptide, glutamate, FeSO$_4$, 4-hydroxynonenal	[43, 45, 46]
Rat primary cerebellar neurons, AtT20 pituitary adenoma; IMR32 neuroblastoma, CRH-R1-overexpressing HT22 hippocampal cell clone	Aβ-peptide; H$_2$O$_2$	[28]
Hippocampal slice cultures from rat	Hypoxia; glutamate, H$_2$O$_2$, Aβ-peptide	[39, 47, 48]
Hippocampal neurons in mice with brain-specific CRH overexpression	Kainate injections (i.p.) – in vivo model of acute excitotoxic stress	[49]
Hippocampal neurons	Rat in vivo model of ischemia	[50]
Dopaminergic neurons of substantia nigra and striatum	Rat in vivo model of Parkinson's disease	[51, 52]

region-specific differences in the activation of the MAPK-pathway were also shown *in vivo* after intracerebroventricular injection of CRH into mouse brains [44].

In a follow-up study, we showed that the expression of the well-known endogenous neuroprotective factor, the brain-derived neurotrophic factor (BDNF), is enhanced by CRH [53]. This activity is mediated by CRH-R1 and inhibited by the activation of another G-protein-coupled receptor system, the endocannabinoid receptor 1 (CB1). The finding that BDNF may act as a downstream effector of CRH, mediating its neuroprotective activity, was recently confirmed *in vivo* as well. Employing mice overexpressing murine CRH in the CNS, we observed a differential response of CRH-overexpressing mice (CRH-COEhom-Nes) to acute oxidative stress-related excitotoxicity induced by kainate as compared to controls (CRH-COEcon-Nes) [49]. Interestingly, CRH-overexpression reduced the duration of epileptic seizures and prevented kainate-induced neurodegeneration and neuroinflammation in the hippocampus [49]. These findings further highlighted a neuroprotective action of CRH *in vivo*. Interestingly, this neuroprotective effect was accompanied by increased basal levels of BDNF in CRH-COEhom-Nes mice [49], suggesting a potential role for BDNF in mediating CRH-induced neuroprotective actions against acute excitotoxicity *in vivo*.

Very recently, by linking CRH signaling to the function of gap junctions, we detected a novel activity of CRH that may also contribute to its neuroprotective role. An enhanced gap junction communication has been reported to confer neuroprotection after various neurotoxic insults. In our study, we showed that CRH treatment up-regulates connexin43 (Cx43) expression as well as gap junctional communication in a CRH receptor-dependent manner in neuroblastoma cells, primary astrocytes, and organotypic hippocampal slice cultures. MAPK and PKA-CREB-coupled pathways are involved in the signaling cascade from CRH to enhanced Cx43 function. Inhibition of CRH-promoted gap junction communication by the gap junction inhibitor carbenoxolone could prevent neuroprotective actions of CRH in cell and tissue culture models, suggesting that gap junction molecules are involved in the neuroprotective effects of CRH [47]. Interestingly, coculture studies of primary neurons and astrocytes revealed that astrocytic Cx43 most probably contributes to the neuroprotective effects of CRH [47]. Thus, BDNF and the gap junction molecule Cx43 act as downstream targets of CRH signaling, mediating the neuroprotective effects of CRH toward experimentally induced oxidative stress (Figure 14.1).

In order to further dissect the signal transduction cascades activated by CRH-R1, a comparative proteome approach was performed *in vitro*, utilizing murine corticotroph AtT-20 cells [54]. Amongst others, they identified alterations in PRKAR1A, the regulatory subunit of PKA; in PGK1 and PGAM1, key regulators of glycolysis; and in proteins involved in proteosome-mediated proteolysis, PSMC2 and PSMA3.

Several studies revealed interactions of the CRH system with other G-protein-coupled receptor systems. Thus, it has been shown that CRH via CRH receptors stimulates the expression of the opioids β-endorphin, dynamin, and encephalin known to exert neuroprotective effects (summarized in [50]). In

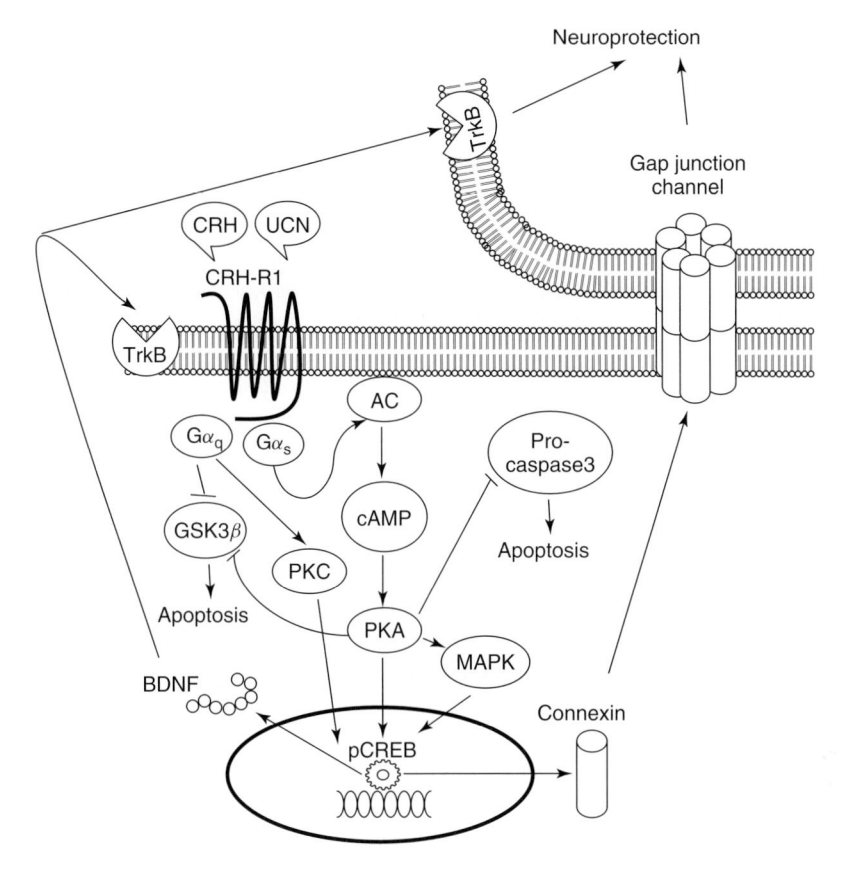

Figure 14.1 Downstream targets of CRH involved in neuroprotection. Binding of CRH or its related peptide UCN to the G-protein-coupled CRH-R1 leads to conformational changes of the receptor which are translated into cell type–specific stimulation of trimeric G-proteins (e.g., Gαs and Gαq). Subsequently, distinct signaling pathways involving protein kinases such as PKA, PKC, and MAPK are activated. Activation of these signaling molecules promotes phosphorylation of transcription factors like CREB. Activated transcription factors then regulate the expression of proteins, for example, BDNF and Cx43, that show neuroprotective properties in experimental in vitro and in vivo models. In addition, CRH-induced activation of Gαq and PKA leads to inactivation of GSK3β and thereby to an inhibition of apoptotic processes. Furthermore, PKA activation by CRH proved to be antiapoptotic by blocking procaspase 3 activation.

addition, a cross talk between the CRH system and the cannabinoid system has been described [53, 55, 56].

Moreover, other labs have intensively investigated the role of CRH as a neuroprotectant and confirmed the initial findings [45, 48]. The CRH-related UCN1 also protects against oxidative stress through the activation of CRH-R1. Employing subtype-specific CRH receptor antagonists, evidence was provided that the signaling pathways that mediate the neuroprotective effect of UCN1 involve PKA, protein kinase C (PKC), and MAPK. The discovery of a biological activity of UCN1 in hippocampal neurons suggests a role for this peptide in adaptive responses of hippocampal neurons to potentially lethal oxidative and excitotoxic insults [46]. Elliott-Hunt and colleagues reported that CRH in very low physiological concentrations (2 pM), prevented glutamate-induced neurotoxicity via receptor-mediated mechanisms in organotypic hippocampal cultures both during and, interestingly, after the glutamate-induced insult. With respect to downstream signaling it was found that activation of the adenylate cyclase pathway and induction of MAPK phosphorylation mediates the CRH action. Most importantly, this study showed that CRH can afford neuroprotection against neurotoxicity up to 12 h following the insult, suggesting that CRH is acting at a late stage in the neuronal death cycle, a finding which might be important in the development of novel neuroprotective agents in order to improve neuronal survival following the insult [48].

In addition to AD, in Parkinson's disease (PD) too the CRH system is severely affected and peptides of the CRH family have also been investigated in experimental models of PD. Initially it was described that UCN1 reverses key features of nigrostriatal damage in two paradigms of PD, the hemiparkinsonian 6-hydroxydopamine (6-OHDA)-lesioned rat and the lipopolysaccharide (LPS)-injected rat; the latter serving as a neuroinflammatory model for PD [51, 57, 58]. Furthermore, it was shown that UCN1 treatment prevented the loss of the dopamine that is usually observed in these models. Fourteen days after stereotactic injections of 6-OHDA or LPS and UCN1, extracellular dopamine levels in striata ipsilateral of injection sites of 6-OHDA/LPS- and UCN1-treated rats were comparable with sham injected control rats, while rats given 6-OHDA/LPS and vehicle had considerably lower dopamine levels. In addition, the dopamine metabolites dihydroxyphenylacetic acid and homovanillic acid were also preserved in dialysates from UCN1-treated rats. The effects of UCN1 were entirely blocked by the antagonists α-helical and NBI 27914. The selective CRH-R2 ligand UCN3 had no effect. This *in vivo* study strongly suggests that UCN1 is capable of maintaining adequate nigrostriatal function *in vivo* via CRH-R1 following neurotoxic challenge [52]. In addition to its neuroprotective activities, UCN1 is also known for its cardioprotective effects, which are mediated by CRH-R2 [59].

14.6
Lessons from the Heart

In recent years, powerful cytoprotective effects of the urocortins have been described against acute ischemia and reperfusion injury in the heart. Most studies focused on the effects of UCN1, and the signaling pathways following initial binding of UCN1 to CRH-R2 have been extensively investigated. In addition to activating several kinases including PKC, PKA, and MAPK, the urocortins target mitochondria by promoting the opening of mitochondrial ATP-sensitive potassium channels, for example, by stimulating expression of Kir6.1, a pore-forming subunit of an ATP-sensitive potassium channel [60]. These potassium channels also show prominent expression in certain regions of the brain and neural cell types [61] and have been shown to possess neuroprotective properties, for example, against epileptic seizures [62]. Thus, mitochondrial potassium channels might be additional targets of the CRH-mediated neuroprotective action in the brain and deserve further attention.

14.7
Outlook

As with other factors, the current labeling of CRH as a stress-inducing factor is not fully consistent with its wide range of activities. Being initially isolated as the mediator of the HPA axis and central modulator of the body's stress response, today, CRH is acknowledged as a neuropeptide with many different effects. It is clear that in addition to the important role of CRH in the hypothalamus and pituitary, CRH also acts on extrahypothalamic areas. An impressive role for CRH in neuroprotection is developing, and different downstream targets which may mediate CRH's neuroprotective activity are identified. Early work on CRH has already emphasized an important impact of CRH on synaptic plasticity, learning, and memory formation. It appears that a controlled activation of CRH receptors in the brain may stabilize neural function and with all the current knowledge on nonpeptidergic antagonists of CRH-R1, it is no science fiction to develop agonists of CRH-R1. In addition, initial *in vivo* studies on the memory promoting and protective effect of compounds that displace CRH from its natural *in vivo* binding protein (CRH-BP) should be revisited and extended in models of neurodegeneration in order to develop novel strategies for neurostabilization and neuroprotection under disease conditions.

References

1. Smith, S.M. and Vale, W.W. (2006) The role of the hypothalamic-pituitary-adrenal axis in neuroendocrine responses to stress. *Dialogues Clin. Neurosci.*, **8**(4), 383–395.
2. Spiess, J., Rivier, J., Rivier, C., and Vale, W. (1981) Primary structure of corticotropin-releasing factor from bovine hypothalamus. *Proc. Natl. Acad. Sci. U.S.A.*, **78**(10), 6517–6521.
3. Vale, W., Spiess, J., Rivier, C., and Rivier, J. (1981) Characterization of a 41-residue ovine hypothalamic peptide that stimulates secretion of corticotropin and beta-endorphin. *Science*, **213**(4514), 1394–1397.
4. de Kloet, E.R., Joels, M., and Holsboer, F. (2005) Stress and the brain: from adaptation to disease. *Nat. Rev. Neurosci.*, **6**(6), 463–475.
5. Holsboer, F., Von Bardeleben, U., Gerken, A., Stalla, G.K., and Muller, O.A. (1984) Blunted corticotropin and normal cortisol response to human corticotropin-releasing factor in depression. *N. Engl. J. Med.*, **311**(17), 1127.
6. Nemeroff, C.B. and Vale, W.W. (2005) The neurobiology of depression: inroads to treatment and new drug discovery. *J. Clin. Psychiatry*, **66**(Suppl 7), 5–13.
7. von Bardeleben, U. and Holsboer, F. (1988) Human corticotropin releasing

hormone: clinical studies in patients with affective disorders, alcoholism, panic disorder and in normal controls. *Prog. Neuropsychopharmacol. Biol. Psychiatry.*, Suppl. 12, S165–S187.
8. Holsboer, F. (1999) The rationale for corticotropin-releasing hormone receptor (CRH-R) antagonists to treat depression and anxiety. *J. Psychiatr. Res.*, **33**(3), 181–214.
9. Holsboer, F. and Ising, M. (2008) Central CRH system in depression and anxiety – evidence from clinical studies with CRH1 receptor antagonists. *Eur. J. Pharmacol.*, **583**(2–3), 350–357.
10. Tellew, J.E. and Luo, Z. (2008) Small molecule antagonists of the corticotropin releasing factor (CRF) receptor: recent medicinal chemistry developments. *Curr. Top. Med. Chem.*, **8**(6), 506–520.
11. Behan, D.P., Grigoriadis, D.E., Lovenberg, T., Chalmers, D., Heinrichs, S., Liaw, C., and Desouza, E.B. (1996) Neurobiology of corticotropin releasing factor (CRF) receptors and CRF-binding protein – implications for the treatment of CNS disorders. *Mol. Psychiatry*, **1**(4), 265–277 (Review).
12. Lovejoy, D.A., Aubry, J.M., Turnbull, A., Sutton, S., Potter, E., Yehling, J., Rivier, C., and Vale, W.W. (1998) Ectopic expression of the CRF-binding protein: minor impact on HPA axis regulation but induction of sexually dimorphic weight gain. *J. Neuroendocrinol.*, **10**(7), 483–491.
13. Hauger, R.L., Risbrough, V., Brauns, O., and Dautzenberg, F.M. (2006) Corticotropin releasing factor (CRF) receptor signaling in the central nervous system: new molecular targets. *CNS Neurol. Disord. Drug Targets*, **5**(4), 453–479.
14. Perrin, M.H., and Vale, W.W. (1999) Corticotropin releasing factor receptors and their ligand family. *Ann. N. Y. Acad. Sci.*, **885**, 312–328.
15. Dautzenberg, F.M. and Hauger, R.L. (2002) The CRF peptide family and their receptors: yet more partners discovered. *Trends Pharmacol. Sci.*, **23**(2), 71–77.
16. Steckler, T. and Holsboer, F. (1999) Corticotropin-releasing hormone receptor subtypes and emotion. *Biol. Psychiatry*, **46**(11), 1480–1508.
17. De Souza, E.B. (1995) Corticotropin-releasing factor receptors: physiology, pharmacology, biochemistry and role in central nervous system and immune disorders. *Psychoneuroendocrinology*, **20**(8), 789–819.
18. Webster, E.L., Lewis, D.B., Torpy, D.J., Zachman, E.K., Rice, K.C., and Chrousos, G.P. (1996) In vivo and in vitro characterization of antalarmin, a nonpeptide corticotropin-releasing hormone (crh) receptor antagonist – suppression of pituitary acth release and peripheral inflammation. *Endocrinology*, **137**(12), 5747–5750.
19. Keck, M.E., Welt, T., Wigger, A., Renner, U., Engelmann, M., Holsboer, F., and Landgraf, R. (2001) The anxiolytic effect of the CRH(1) receptor antagonist R121919 depends on innate emotionality in rats. *Eur. J. Neurosci.*, **13**(2), 373–380.
20. Gilligan, P.J., He, L., Clarke, T., Tivitmahaisoon, P., Lelas, S., Li, Y.W., Heman, K., Fitzgerald, L., Miller, K., Zhang, G., Marshall, A., Krause, C., McElroy, J., Ward, K., Shen, H., Wong, H., Grossman, S., Nemeth, G., Zaczek, R., Arneric, S.P., Hartig, P., Robertson, D.W., and Trainor, G. (2009) 8-(4-Methoxyphenyl)pyrazolo[1,5-a]-1,3,5-triazines: selective and centrally active corticotropin-releasing factor receptor-1 (CRF1) antagonists. *J. Med. Chem.*, **52**(9), 3073–3083.
21. Steckler, T. and Dautzenberg, F.M. (2006) Corticotropin-releasing factor receptor antagonists in affective disorders and drug dependence – an update. *CNS Neurol. Disord. Drug Targets*, **5**(2), 147–165.
22. Arborelius, L., Owens, M.J., Plotsky, P.M., and Nemeroff, C.B. (1999) The role of corticotropin-releasing factor in depression and anxiety disorders. *J. Endocrinol.*, **160**(1), 1–12.
23. Zobel, A.W., Nickel, T., Kunzel, H.E., Ackl, N., Sonntag, A., Ising, M., and Holsboer, F. (2000) Effects of the high-affinity corticotropin-releasing hormone receptor 1 antagonist R121919 in major depression: the first 20 patients treated. *J. Psychiatr. Res.*, **34**(3), 171–181.

24. Gilligan, P.J., Clarke, T., He, L., Lelas, S., Li, Y.W., Heman, K., Fitzgerald, L., Miller, K., Zhang, G., Marshall, A., Krause, C., McElroy, J.F., Ward, K., Zeller, K., Wong, H., Bai, S., Saye, J., Grossman, S., Zaczek, R., Arneric, S.P., Hartig, P., Robertson, D., and Trainor, G. (2009) Synthesis and structure-activity relationships of 8-(pyrid-3-yl)pyrazolo[1,5-a]-1,3,5-triazines: potent, orally bioavailable corticotropin releasing factor receptor-1 (CRF1) antagonists. *J. Med. Chem.*, **52**(9), 3084–3092.

25. Bissette, G., Reynolds, G.P., Kilts, C.D., Widerlov, E., and Nemeroff, C.B. (1985) Corticotropin-releasing factor-like immunoreactivity in senile dementia of the Alzheimer type. Reduced cortical and striatal concentrations. *J. Am. Med. Assoc.*, **254**(21), 3067–3069.

26. De Souza, E.B., Whitehouse, P.J., Kuhar, M.J., Price, D.L., and Vale, W.W. (1986) Reciprocal changes in corticotropin-releasing factor (CRF)-like immunoreactivity and CRF receptors in cerebral cortex of Alzheimer's disease. *Nature*, **319**(6054), 593–595.

27. Pomara, N., Singh, R.R., Deptula, D., LeWitt, P.A., Bissette, G., Stanley, M., and Nemeroff, C.B. (1989) CSF corticotropin-releasing factor (CRF) in Alzheimer's disease: its relationship to severity of dementia and monoamine metabolites. *Biol. Psychiatry*, **26**(5), 500–504.

28. Lezoualc'h, F., Engert, S., Berning, B., and Behl, C. (2000) Corticotropin-releasing hormone-mediated neuroprotection against oxidative stress is associated with the increased release of non-amyloidogenic amyloid beta precursor protein and with the suppression of nuclear factor-kappaB. *Mol. Endocrinol.*, **14**(1), 147–159.

29. Bayatti, N. and Behl, C. (2005) The neuroprotective actions of corticotropin releasing hormone. *Ageing Res. Rev.*, **4**(2), 258–270.

30. Whitehouse, P.J., Vale, W.W., Zweig, R.M., Singer, H.S., Mayeux, R., Kuhar, M.J., Price, D.L., and De Souza, E.B. (1987) Reductions in corticotropin releasing factor-like immunoreactivity in cerebral cortex in Alzheimer's disease, Parkinson's disease, and progressive supranuclear palsy. *Neurology*, **37**(6), 905–909.

31. Auchus, A.P., Green, R.C., and Nemeroff, C.B. (1994) Cortical and subcortical neuropeptides in Alzheimer's disease. *Neurobiol. Aging*, **15**(4), 589–595.

32. Behan, D.P., Khongsaly, O., Owens, M.J., Chung, H.D., Nemeroff, C.B., and De Souza, E.B. (1997) Corticotropin-releasing factor (CRF), CRF-binding protein (CRF-BP), and CRF/CRF-BP complex in Alzheimer's disease and control postmortem human brain. *J. Neurochem.*, **68**(5), 2053–2060.

33. Davis, K.L., Mohs, R.C., Marin, D.B., Purohit, D.P., Perl, D.P., Lantz, M., Austin, G., and Haroutunian, V. (1999) Neuropeptide abnormalities in patients with early Alzheimer disease. *Arch. Gen. Psychiatry*, **56**(11), 981–987.

34. May, R.C., Kelly, R.A., and Mitch, W.E. (1987) Mechanisms for defects in muscle protein metabolism in rats with chronic uremia. Influence of metabolic acidosis. *J. Clin. Invest.*, **79**(4), 1099–1103.

35. Mouradian, M.M., Farah, J.M. Jr., Mohr, E., Fabbrini, G., O'Donohue, T.L., and Chase, T.N. (1986) Spinal fluid CRF reduction in Alzheimer's disease. *Neuropeptides*, **8**(4), 393–400.

36. Lyons, M.K., Anderson, R.E., and Meyer, F.B. (1991) Corticotropin releasing factor antagonist reduces ischemic hippocampal neuronal injury. *Brain Res.*, **545**(1–2), 339–342.

37. Strijbos, P.J., Relton, J.K., and Rothwell, N.J. (1994) Corticotrophin-releasing factor antagonist inhibits neuronal damage induced by focal cerebral ischaemia or activation of NMDA receptors in the rat brain. *Brain Res.*, **656**(2), 405–408.

38. Brar, B.K., Stephanou, A., Okosi, A., Lawrence, K.M., Knight, R.A., Marber, M.S., and Latchman, D.S. (1999) CRH-like peptides protect cardiac myocytes from lethal ischaemic injury. *Mol. Cell Endocrinol.*, **158**(1–2), 55–63.

39. Fox, M.W., Anderson, R.E., and Meyer, F.B. (1993) Neuroprotection by corticotropin releasing factor during hypoxia in rat brain. *Stroke*, **24**(7), 1072–1075.
40. Winslow, J.T., Newman, J.D., and Insel, T.R. (1989) CRH and alpha-helical-CRH modulate behavioral measures of arousal in monkeys. *Pharmacol. Biochem. Behav.*, **32**(4), 919–926.
41. Chang, F.C. and Opp, M.R. (1998) Blockade of corticotropin-releasing hormone receptors reduces spontaneous waking in the rat. *Am. J. Physiol.*, **275**(3, Pt 2), R793–R802.
42. Behan, D.P., Heinrichs, S.C., Troncoso, J.C., Liu, X.J., Kawas, C.H., Ling, N., and De Souza, E.B. (1995) Displacement of corticotropin releasing factor from its binding protein as a possible treatment for Alzheimer's disease. *Nature*, **378**(6554), 284–287 [see comments].
43. Bayatti, N., Zschocke, J., and Behl, C. (2003) Brain region-specific neuroprotective action and signaling of corticotropin-releasing hormone in primary neurons. *Endocrinology*, **144**(9), 4051–4060.
44. Refojo, D., Echenique, C., Muller, M.B., Reul, J.M., Deussing, J.M., Wurst, W., Sillaber, I., Paez-Pereda, M., Holsboer, F., and Arzt, E. (2005) Corticotropin-releasing hormone activates ERK1/2 MAPK in specific brain areas. *Proc. Natl. Acad. Sci. U. S. A.*, **102**(17), 6183–6188.
45. Pedersen, W.A., McCullers, D., Culmsee, C., Haughey, N.J., Herman, J.P., and Mattson, M.P. (2001) Corticotropin-releasing hormone protects neurons against insults relevant to the pathogenesis of Alzheimer's disease. *Neurobiol. Dis.*, **8**(3), 492–503.
46. Pedersen, W.A., Wan, R., Zhang, P., and Mattson, M.P. (2002) Urocortin, but not urocortin II, protects cultured hippocampal neurons from oxidative and excitotoxic cell death via corticotropin-releasing hormone receptor type I. *J. Neurosci.*, **22**(2), 404–412.
47. Hanstein, R., Trotter, J., Behl, C., and Clement, A.B. (2009) Increased connexin 43 expression as a potential mediator of the neuroprotective activity of the corticotropin-releasing hormone. *Mol. Endocrinol.*, **23**(9), 1479–1493.
48. Elliott-Hunt, C.R., Kazlauskaite, J., Wilde, G.J., Grammatopoulos, D.K., and Hillhouse, E.W. (2002) Potential signalling pathways underlying corticotrophin-releasing hormone-mediated neuroprotection from excitotoxicity in rat hippocampus. *J. Neurochem.*, **80**(3), 416–425.
49. Hanstein, R., Lu, A., Wurst, W., Holsboer, F., Deussing, J.M., Clement, A.B., and Behl, C. (2008) Transgenic overexpression of corticotropin releasing hormone provides partial protection against neurodegeneration in an in vivo model of acute excitotoxic stress. *Neuroscience*, **156**(3), 712–721.
50. Charron, C., Frechette, S., Proulx, G., and Plamondon, H. (2008) In vivo administration of corticotropin-releasing hormone at remote intervals following ischemia enhances CA1 neuronal survival and recovery of spatial memory impairments: a role for opioid receptors. *Behav. Brain Res.*, **188**(1), 125–135.
51. Abuirmeileh, A., Harkavyi, A., Kingsbury, A., Lever, R., and Whitton, P.S. (2008) The CRF-like peptide urocortin produces a long-lasting recovery in rats made hemiparkinsonian by 6-hydroxydopamine or lipopolysaccharide. *J. Neurol. Sci.*, **271**(1–2), 131–136.
52. Abuirmeileh, A., Harkavyi, A., Kingsbury, A., Lever, R., and Whitton, P.S. (2009) The CRF-like peptide urocortin greatly attenuates loss of extracellular striatal dopamine in rat models of Parkinson's disease by activating CRF(1) receptors. *Eur. J. Pharmacol.*, **604**(1–3), 45–50.
53. Bayatti, N., Hermann, H., Lutz, B., and Behl, C. (2005) Corticotropin-releasing hormone-mediated induction of intracellular signaling pathways and brain-derived neurotrophic factor expression is inhibited by the activation of the endocannabinoid system. *Endocrinology*, **146**(3), 1205–1213.
54. Kronsbein, H.C., Jastorff, A.M., Maccarrone, G., Stalla, G., Wurst, W., Holsboer, F., Turck, C.W., and Deussing, J.M. (2008)

CRHR1-dependent effects on protein expression and posttranslational modification in AtT-20 cells. *Mol. Cell Endocrinol.*, **292**(1–2), 1–10.

55. Cota, D., Marsicano, G., Tschop, M., Grubler, Y., Flachskamm, C., Schubert, M., Auer, D., Yassouridis, A., Thone-Reineke, C., Ortmann, S., Tomassoni, F., Cervino, C., Nisoli, E., Linthorst, A.C., Pasquali, R., Lutz, B., Stalla, G.K., and Pagotto, U. (2003) The endogenous cannabinoid system affects energy balance via central orexigenic drive and peripheral lipogenesis. *J. Clin. Invest.*, **112**(3), 423–431.

56. Hermann, H. and Lutz, B. (2005) Co-expression of the cannabinoid receptor type 1 with the corticotropin-releasing hormone receptor type 1 in distinct regions of the adult mouse forebrain. *Neurosci. Lett.*, **375**(1), 13–18.

57. Abuirmeileh, A., Harkavyi, A., Lever, R., Biggs, C.S., and Whitton, P.S. (2007) Urocortin, a CRF-like peptide, restores key indicators of damage in the substantia nigra in a neuroinflammatory model of Parkinson's disease. *J. Neuroinflammation*, **4**, 19.

58. Abuirmeileh, A., Lever, R., Kingsbury, A.E., Lees, A.J., Locke, I.C., Knight, R.A., Chowdrey, H.S., Biggs, C.S., and Whitton, P.S. (2007) The corticotrophin-releasing factor-like peptide urocortin reverses key deficits in two rodent models of Parkinson's disease. *Eur. J. Neurosci.*, **26**(2), 417–423.

59. Davidson, S.M. and Yellon, D.M. (2009) Urocortin: a protective peptide that targets both the myocardium and vasculature. *Pharmacol. Rep.*, **61**(1), 172–182.

60. Lawrence, K.M., Chanalaris, A., Scarabelli, T., Hubank, M., Pasini, E., Townsend, P.A., Comini, L., Ferrari, R., Tinker, A., Stephanou, A., Knight, R.A., and Latchman, D.S. (2002) K(ATP) channel gene expression is induced by urocortin and mediates its cardioprotective effect. *Circulation*, **106**(12), 1556–1562.

61. Thomzig, A., Laube, G., Pruss, H., and Veh, R.W. (2005) Pore-forming subunits of K-ATP channels, Kir6.1 and Kir6.2, display prominent differences in regional and cellular distribution in the rat brain. *J. Comp. Neurol.*, **484**(3), 313–330.

62. Soundarapandian, M.M., Wu, D., Zhong, X., Petralia, R.S., Peng, L., Tu, W., and Lu, Y. (2007) Expression of functional Kir6.1 channels regulates glutamate release at CA3 synapses in generation of epileptic form of seizures. *J. Neurochem.*, **103**(5), 1982–1988.

15
Neuroprotective and Neurogenic Effects of Erythropoietin
Helmar C. Lehmann and Ahmet Höke

15.1
Introduction

Erythropoietin (EPO) represents the essential growth factor for the production of red blood cells. Its synthesis is tightly regulated by blood oxygen level; hypoxia induces the production of EPO by interstitial fibroblasts in the kidneys and liver. In recent years, it has become clear that erythropoietin and its receptor (EPOR) are not only expressed in tissues which are directly involved in hematopoiesis but can also be found in the brain, peripheral nervous system (PNS), reproductive tract, and other organs [1–4]. These discoveries raised the possibility that EPO may perform a variety of physiological functions for support and maintenance in these tissues. This is supported by experimental studies showing that EPO is able to prevent apoptosis in neurons and enhance axonal regeneration in a variety of different experimental paradigms of injury in the central nervous system (CNS) and PNS. In this chapter, we review the neuroprotective and neurogenic effects of EPO in different models of injury in the CNS and PNS.

15.2
EPO in Models of Neonatal Hypoxic-Ischemic Brain Injury

The potential of EPO as a neuroprotective agent has been extensively studied in various *in vivo* models of hypoxic-ischemic brain injury [5–8]. EPO appears to be an effective agent although the exact mechanism of action is still unclear. Multiple mechanisms have been proposed including decreased susceptibility to glutamate toxicity [9, 10], neurotrophic factor-like activity [9, 11], prevention of apoptosis [12–15], and decrease of inflammation that often accompanies ischemia [16, 17].

In a mouse model of human periventricular leukomalacia, it has been demonstrated that EPO can reduce excitotoxic brain injury [18]. In this study a single injection of EPO was sufficient to significantly reduce the infarct size when given

Hormones in Neurodegeneration, Neuroprotection, and Neurogenesis.
Edited by Achille G. Gravanis and Synthia H. Mellon
Copyright © 2011 WILEY-VCH Verlag GmbH & Co. KGaA, Weinheim
ISBN: 978-3-527-32627-3

within a therapeutic window of 4 h. This beneficial effect correlated with an increase in EPO mRNA. Similarly, in the neonatal rat brain, prenatal treatment with EPO was found to significantly reduce white matter injury induced by intrauterine exposure to lipopolysaccharides [19]. The neuroprotective effect of EPO was found to be associated with a significant reduction in the proinflammatory cytokines tumor necrosis factor-α (TNF-α) and interleukin-6 (IL-6). Another study evaluated the long-term effects of recombinant EPO during neonatal development in the absence of any injury [20]. The authors found that newborn rats that were exposed to repeated doses of recombinant EPO of up to 5000 U kg^{-1} did not show an increase in adverse effects and instead, demonstrated improvements in development. This and other studies have established the relative safety of recombinant EPO in various animal models.

The pharmacokinetics of EPO has been an important issue regarding its use in models of neonatal CNS injury. EPO is a relatively large molecule (molecular weight over 30 kDa) and it is unclear if it crosses the blood–brain barrier in significant amounts, even if there is a temporary or permanent breakdown of the blood–brain barrier during ischemia. Experimental data indicate that only a small proportion of recombinant EPO crosses the blood–brain barrier in a dose-dependent fashion. Thus, neuroprotective levels of EPO require administration of high doses of recombinant EPO. After administration of a high dose intraperitoneally, EPO reaches a peak in the brain after 10 h and more importantly, relatively high levels can still be detected 20 h after injection [21]. The same group also determined the optimal dose in a model of hypoxic brain injury (unilateral carotid ligation plus 90 min of 8% hypoxia) in postnatal seven day old rats and found that in this model, three doses of 5000 U kg^{-1} had the most beneficial effect [22]. These and other encouraging results from preclinical trials provide a rationale for the use of recombinant EPO in human newborns. Recently, the results from a randomized, double-blinded trial in preterm infants were reported [23]. In this trial, infants with gestational ages between 24 and 31 weeks were randomized to receive five doses of recombinant human erythropoietin (rhEPO) or a placebo. Although this study failed to detect significant differences in the percentage of infants who survived without brain injury, rhEPO was not associated with a higher rate of adverse events. This relative safety of rhEPO in newborns may justify a larger trial with more statistical power to detect a potentially beneficial effect of EPO in this group of patients.

Interestingly, EPO seems to not only be protective in hypoxic-ischemic brain injury but also in oxygen-induced brain injury. Newborn rats that are exposed to prolonged hyperoxia show increased apoptosis in the hippocampus and this effect can be prevented by intraperitoneal injection of rhEPO at a dose of 1000 U kg^{-1} per day [24]. Similarly, in another study, systemic treatment with a higher dose (20 000 IE kg^{-1}) of rhEPO reduced hyperoxia-induced neuronal apoptosis in rodent pup brains. Proteomic analysis revealed that rhEPO inhibited a high proportion of hyperoxia-induced proteomic changes, suggesting a potential mechanism of action [25].

15.3
EPO in Models of Ischemic Stroke in Adults

Ischemic stroke, caused by abrupt disruption of the brain blood flow that results in impaired neurological function, represents one of the most common causes of permanent disability and a major socioeconomic burden. The oxygen deprivation caused by a reduction of the blood flow initiates a complex cascade of events that eventually results in neuronal (and glial) death. Despite major efforts, therapeutic approaches that were aimed to protect neurons from cell death have not been successful in clinical trials, partly due to the complexity of the pathological processes that contribute to neuronal death following ischemia. Owing to its various effector mechanisms, EPO is of particular interest as a potential treatment for stroke. This view is supported by various preclinical studies that used EPO and its derivatives in different stroke models (Table 15.1).

Intracerebral and intraperitoneal administration of EPO has been reported to be beneficial in the rodent model of permanent middle cerebral artery occlusion (MCAO) [26, 27, 33]. In this model of focal ischemia, the therapeutic window was quite prolonged; recombinant EPO resulted in significant reduction in infarct volume when given within 6 h after vessel occlusion [27]. Subsequent studies determined different doses of systemic EPO treatment and found that rhEPO given at doses of 500, 1150, or 5000 IU kg^{-1} can significantly reduce cortical infarct volume [28]. Paradoxically, and in contrast to those encouraging results, transgenic mice that overexpress human EPO in neuronal cells in the CNS were found to be more susceptible to ischemia in this model of permanent occlusion of the middle cerebral artery [34]. Likewise, transgenic mice that systemically overexpress human EPO, which results in hematocrit levels of 80%, were found to have elevated numbers of Mac-1 immunoreactive cells in infarcted tissue and increased infarct volume [34]. These studies indicate that moderately increased levels of EPO in the brain are probably not sufficient to provide significant tissue protection or that EPO overexpression has to be in glial cells rather than neuronal cells so as to be effective. Furthermore, the systemic overexpression study suggests that chronically high levels of EPO may reverse potential beneficial effects due to an elevated hematocrit.

Consequently, more recent studies evaluated the use of nonerythropoietic EPO derivatives in models of focal ischemia. Carbamylated erythropoietin (CEPO) does not bind to EPOR but has been shown to exert neuroprotective effects similar to EPO [17]. Rats that were treated with CEPO after focal ischemia showed a better outcome in behavioral and histological measures (i.e., perifocal microglial activation, polymorphomonuclear cell infiltration, and white matter damage) as compared to control animals. Likewise, other nonerythropoietic derivatives (carbamylated darbepoetin-α and mutant EPO-S100E) were also found to be neuroprotective [15]. Another study compared different doses of rhEPO and rhCEPO in the rat model of embolic MCAO and found that both treatments substantially reduced the number of apoptotic cells and activated microglia in the infarcted region [28]. In order to reduce systemic side effects of rhEPO, another approach has involved the intranasal delivery of rhEPO [35, 36]. Intranasal administration of rhEPO in

Table 15.1 EPO and derivatives in animal models of neurological disorders (selected studies).

Study	References	Study design	Outcome/conclusion
Hypoxic-ischemic brain injury			
Aydin et al.	[5]	Ligation of the carotid artery and exposure to hypoxia in postnatal (P7) rat pups; intracerebroventricular injection of rhEPO or vehicle	Mean infarct volume significantly smaller in treatment group seven days after hypoxia
Demers et al.	[6]	Ligation of carotid artery and exposure to hypoxia in P7 rat pups; administration (s.c.) of rhEpo (2500 U kg^{-1} per day) for three or vehicle	No differences in gross brain injury between rhEpo and vehicle-treated animals, increased numbers of dopamine neurons in substantial nigra and ventral tegmental area in rhEpo-treated animals after four weeks. Improvement of behavioral measures in treatment group
Oxygen-induced brain injury			
Yiş et al.	[24]	Exposure of neonatal rat pups to 80% oxygen or 21% oxygen (controls). I. p. injection of rhEPO (1000 U kg^{-1} per day) for five days	Increased neuronal density in dentate gyrus in EPO-treated animals compared to untreated animals. Decreased numbers of apoptotic neurons in some hippocampal areas
Kaindl et al.	[25]	Exposure of neonatal C57Bl/6 mice or rats to normoxia (21%) or hyperoxia (80%). Treatment with i.p. injections of 20,000 IU kg^{-1} rhEpo or saline	Reduced apoptosis in the brains EPO-treated animals and prevention of hyperoxia-induced brain proteome changes
Stroke			
Bernaudin et al.	[26]	Permanent occlusion of the middle cerebral artery (MCAO) in mice, intracerebroventricular injection of rMoEPO (0.4 g kg^{-1}) or vehicle 24 h before MCAO	Significant reduction of infarct volume in EPO-treated animals

Table 15.1 (Continued).

Study	References	Study design	Outcome/conclusion
Brines et al.	[27]	Several animal models, for stroke: MCAO penumbra model, administration of 250–5000 U kg^{-1} rhEPO or saline i.p. 24 h before and 0, 3, 6, or 9 h after MCAO	Reduced brain infarct volume in rhEPO-treated animals before and up to 3 h after occlusion
Wang et al.	[28]	Rat MCAO model. I.v. administration of 50–5000 IU kg^{-1} rhEPO, CEPO 50 µg kg^{-1} or vehicle 6, 24, and 48 h after MCAO	Reduced infarct volume in rhEPO- and CEPO-treated animals
CNS traumatic injury			
Gorio et al.	[29]	Two traumatic spinal cord injury rat models. Administration (i.p) of rhEPO in various doses	Partial recovery of motor function in rhEPO-treated animals as compared to no recovery in saline-treated animals
Grasso et al.	[30]	Traumatic (cryogenic) brain injury. I.p. administration of 1000 IU/kg rhEPO 3 times per 24 h for up to eight days	Improvement in motor function, reduced brain edema, reduced injury volume in rhEPO-treated animals
Experimental allergic encephalomyelitis (EAE)			
Brines et al.	[27]	Active EAE, immunization of Lewis rats with guinea pig myelin basic protein (MBP) + CFA + M. tuberculosis Administration of rhEPO (5000 U kg^{-1}) or saline from 3 to 18 days	Delay of symptom onset and decrease in symptom severity in rhEPO-treated animals
Leist et al.	[17]	Several animal models including EAE: chronic EAE model in C57BL/6 mice, immunized with MOG peptide in IFA + M. tuberculosis Administration of CEPO i.p. 3 times per week (50 µg kg^{-1}) for eight weeks	Improvement in neurological function

(continued overleaf)

Table 15.1 (Continued).

Study	References	Study design	Outcome/conclusion
Peripheral neuropathy			
Campana and Myers	[31]	Crush of spinal L5 nerve in rats. S.c. injection of 1000–5000 U kg^{-1} rhEpo or vehicle daily	Reduced apoptosis of dorsal root ganglia neurons and improvement in pain behavioral measures in rhEPO-treated animals
Keswani et al.	[32]	Acrylamide neuropathy model in rats, Daily i.p. injection of 2500 IU kg^{-1} rhEPO or vehicle	Reduced axonal degeneration of sensory skin fibers, improvement in pain measures, less motor axonal degeneration, improved grip strength measures in rhEPO-treated animals

^{a}CEPO, carbamylated erythropoietin; CFA, complete Freund's adjuvants; EAE, experimental allergic encephalomyelitis; IFA, incomplete Freund's adjuvants; (I)U, (international) units; i.p., intraperitoneal; kg, kilograms; rhEPO, recombinant human erythropoietin; rMoEPO, recombinant mouse erythropoietin; and s.c., subcutaneous.

combination with insulin-like growth factor-1 (IGF-1) has been shown to reduce infarct size and improve neurological function in mice up to 90 days after MCAO [35]. In addition to functional and morphological outcomes, EPO treatment may also result in enhanced white matter reorganization as assessed by specific MRI techniques such as diffusion tensor imaging [37]. This observation suggests that EPO may enhance cortical and subcortical plasticity following a stroke.

Despite the limitations of rhEPO use for neurological conditions in humans, a safety study followed by a small, randomized clinical trial of rhEPO at a dose of 33 000 IU per day for three days was carried out in stroke patients [38]. In the safety study, the investigators were able to demonstrate that the systemic administration of rhEPO resulted in 60–100-fold increase in the EPO levels in the CSF, supporting feasibility of rhEPO in clinical use for CNS disorders. This single-center clinical trial was more recently followed by a large multicenter trial of rhEPO in stroke patients [39]. Unfortunately, rhEPO (40 000 IU per day for three doses) did not have an effect on the primary endpoint but there were significant protocol violations with some patients receiving thrombolytic therapy outside the accepted window of 3–6 h poststroke. This was more common in the rhEPO arm and resulted in higher incidence of hemorrhagic conversion of ischemic strokes and led to higher complication rates among the patients receiving rhEPO. Although it was unclear if there was any interaction between thrombolytic therapy and rhEPO, this study most probably precludes the use of rhEPO in stroke patients in the future.

15.4
EPO in Models of Traumatic Brain Injury and Spinal Cord Trauma

After CNS injury in adult mammals, functional recovery is very limited because of reduced intrinsic growth capacity and a nonpermissive environment for axonal regeneration. In spinal cord injury, a cascade of cell death mediated by an undesirable combination of necrosis, apoptosis, inflammation, and ischemia affect neurons and glial cells [40, 41]. In addition to its neuroprotective effects, EPO can also enhance axon regeneration, albeit in a limited manner [42]; thus, it is a potential therapeutic option to enhance the functional outcome after traumatic injury in the CNS.

The utility of EPO in preventing neuronal cell death has been studied in various models of traumatic injury in the brain and spinal cord. In models of traumatic brain injury, systemic EPO administration can exert beneficial effects when given immediately after the injurious event [29, 30, 43]. In a model of cryogenic injury in rats, EPO significantly improved motor recovery, and reduced brain edema and breakdown of the blood–brain barrier as assessed by Evans-blue staining [30]. Another study assessed the combination of heat acclimation and EPO treatment [43]. Heat acclimation had neuroprotective effects similar to administration of exogenous rhEPO but the combination did not confer asynergistic effect. Heat acclimation stimulated endogenous expression of EPO in areas of trauma, whereas treatment with an antibody against EPO blocked the effects of heat acclimation. Likewise, when traumatic brain injury was induced in mice that lack EPOR, mice showed more sensorimotor deficits and impaired cortical neurogenesis as compared to wild-type littermates [44].

In addition to traumatic brain injury, rhEPO administered systemically in doses as low as 350 units kg^{-1} has been shown to improve functional and histological outcome in models of transient spinal cord ischemia [12, 29] and traumatic spinal cord injury simulated by compression or contusion [45, 46]. A potential mechanism for these effects was proposed based on the observations that administration of rhEPO at the time of spinal cord injury led to preservation of the ventral white matter tracts and myelinating oligodendrocytes [47]. These encouraging results in models of spinal cord injury were recently questioned by a reassessment study initiated by an NIH "Facilities of Research – Spinal Cord Injury" contract. In this study, compression or contusion injury was applied to the thoracic spinal cord. Animals received intraperitoneal rhEPO of 1000 or 5000 IU kg^{-1} per day for seven days or as a single dose. However, none of the treatment regimens improved the behavioral measures or morphological parameters [48].

15.5
EPO in Experimental Autoimmune Encephalomyelitis

Experimental autoimmune (allergic) encephalomyelitis (EAE) is a widely used animal model that mimics typical clinical features of multiple sclerosis (MS) such as inflammation, demyelination, and axonal damage. EAE can be induced in different species by immunization with CNS tissue or CNS myelin in Freund's complete adjuvant [49]. Apart from this "actively-induced EAE," it can also be induced by passive transfer of T cells from actively immunized animals ("adoptive transfer EAE"), which highlights the crucial role of T cells in this animal model. EPO has been tested in actively-induced EAE and various studies have shown that the administration of EPO delays disease onset and attenuates disease severity [16, 50–54]. This clinical improvement is associated with a decrease in levels of TNF and IL-6 [16]. EPO and nonerythropoietic CEPO were also tested in a mouse model of EAE, characterized by a significantly longer disease course [17]. In this model, both treatments were effective in reducing disease symptoms and improving histological and behavioral outcomes. Likewise, mice that were immunized with myelin oligodendrocyte glycoprotein peptide (MOG35–55) as a model of chronic EAE, showed an attenuated disease course when treated with EPO or derivatives three days after immunization [52]. EPO was also effective when given at disease onset but to a lesser degree. These changes were coupled with a decrease in the inflammatory cytokine levels of TNF-α and IL-1 in the spinal cord. A more recent study investigated the role of EPO in modulating the pathogenic T cell response in EAE. Treatment of mice with EPO appears to promote T regulatory cells and Th17 cells, which inhibit proliferation of the antigen-presenting dendritic cell population [53].

On the basis of experience in stroke patients and relative safety of rhEPO, a small open-label clinical trial of rhEPO was carried out in patients with chronic progressive MS [55]. Patients receiving high-dose rhEPO (48 000 IU biweekly for 48 weeks) showed improvements in disability scores and performance in cognitive tests. This effect was not observed in the low-dose group (8000 IU biweekly). This study will need to be replicated in a larger double-blind placebo-controlled trial before any recommendations about clinical use can be made.

15.6
EPO in Models of Peripheral Neuropathy

Peripheral neuropathies are common and are associated with significant morbidity. The most common forms of peripheral neuropathies such as diabetic and human immunodeficiency virus -(HIV-) associated neuropathy are "dying back" axonopathies, characterized by degeneration of the most distal portions of axons, with centripetal progression [56]. Thus, prevention of axonal degeneration must be a major goal for therapeutic approaches to peripheral neuropathies. EPO has been shown to prevent apoptosis of DRG (dorsal root ganglia) sensory neurons [31] and axonal degeneration in various models of peripheral neuropathies [31, 32, 57–59]. These preclinical observations raise the possibility of use of rhEPO in peripheral neuropathies but the recent concern for vascular side effects hampered clinical trials. A clinical trial of rhEPO in ovarian cancer patients undergoing chemotherapy with paclitaxel and another clinical trial of rhEPO in HIV-associated sensory neuropathy were halted due to recruitment issues and concerns over the cardiovascular side effects that led to issuing of a black box warning by the Federal Drug Administration in 2007.

The mechanism of EPO's effect on peripheral axons is not fully understood, but both EPO and EPOR are expressed in peripheral neurons and Schwann cells [1, 32, 60]. Peripheral nerve injury leads to an increase of EPO mRNA levels in periaxonal and perineuronal Schwann cells and EPOR mRNA in lumbar DRG neurons [32]. Furthermore, after axonal transection in established dissociated DRG cultures, EPO immunostaining markedly increases in Schwann cells that are in close proximity to the transected axons as compared to those adjacent to uninjured axons. This observation was supported in studies on *in vitro* models of HIV-associated sensory neuropathies. Cotreatment of DRG neuron–Schwann cell cultures with rhEPO prevents axonal degeneration induced by the HIV envelope protein, gp120 or the neurotoxic HIV drug, zalcitabine [61]. In cultures exposed to gp120 or zalcitabine, up-regulation of endogenous EPO was seen only in periaxonal Schwann cells in response to an injury signal from axons because exposure of pure Schwann cells to gp120 or zalcitabine does not induce expression of EPO [32].

These observations raise the possibility of perineuronal and periaxonal Schwann cells providing endogenous neuroprotection mediated via EPO [62, 63]. Details of the mechanism of EPO up-regulation in periaxonal and perineuronal Schwann cells are unknown but this process most probably involves nitric oxide [32]. Exposure of DRG neurons to gp120 or zalcitabine causes up-regulation of nitric oxide production in neurons; this effect is blocked by specific inhibitors of neuronal nitric oxide synthase (nNOS). Nitric oxide donors induce EPO up-regulation in pure Schwann cell cultures and provide neuroprotection to DRG neurons. The mechanism by which nitric oxide induces EPO up-regulation in Schwann cells is unknown but activation of HIF-1a (hypoxia-inducible factor) in Schwann cells is likely to play an important role (Keswani and Höke, unpublished data).

In addition to the studies on the neuroprotective effects of EPO in models of peripheral neuropathies, a recent study investigated the neurotrophic effect of locally administered EPO in PNS regeneration. Administration of EPO via a conduit following sciatic crush injury increased the numbers of regenerating myelinated axons and prevented the retrograde degeneration of axons [64]. The effect accompanied improved behavioral measures in sensorimotor function. This effect was associated with increased expression of Akt and STAT3 (signal transducer and activator of transcription 3) in DRG neurons, suggesting that rhEPO up-regulated intracellular pathways that have been shown to be involved in actions of neurotrophic factors. Similarly, rhEPO was effective in improving functional recovery in a model of erectile dysfunction after prostatectomy [65]. In this model, crushed cavernous nerves showed increased axonal regeneration after daily administration of rhEPO and this observation correlated with earlier recovery of erectile function.

15.7
Summary

EPO and its nonerythropoietic derivative CEPO have proved to be protective in various preclinical models of diseases of the CNS and PNS. This large body of data indicates that especially in disease conditions in which a progressive degeneration of axons, rather than neuronal loss, is the predominant pathological change, EPO may have a therapeutic potential. As mentioned above, effect of rhEPO has already been evaluated in MS, ischemic stroke, and neonatal ischemia with limited success. The promising observations from small open-label studies extend to neuropsychiatric diseases such as schizophrenia [66].

However, enthusiasm for these studies has to be tempered because of the relatively limited success of rhEPO and the presence of many unanswered questions about the safety and effective dose range in humans. For example, these questions may be as follows: (i) What is the appropriate dose for a given neurological disorder? (ii) Does EPO interfere with other drugs such as immunosuppressive or immunomodulatory drugs used in MS? (iii) How can adverse events be avoided to enhance safety and tolerability, especially those related to the erythropoietic function? (iv) What is the long-term effect of EPO on the disease course in chronic disorders? Taken together, for almost all potential indications, further experimental and clinical studies are necessary to develop EPO into an optimized treatment modality.

References

1. Campana, W. and Myers, R. (2001) Erythropoietin and erythropoietin receptors in the peripheral nervous system: changes after nerve injury. FASEB J., 15, 1804–1806.
2. Kobayashi, T., Yanase, H., Iwanaga, T., Sasaki, R., and Nagao, M. (2002) Epididymis is a novel site of erythropoietin production in mouse reproductive organs. Biochem. Biophys. Res. Commun., 296, 145–151.
3. Marti, H., Wenger, R., Rivas, L. et al. (1996) Erythropoietin gene expression in human, monkey and murine brain. Eur. J. Neurosci., 8, 666–676.
4. Masuda, S., Kobayashi, T., Chikuma, M., Nagao, M., and Sasaki, R. (2000) The oviduct produces erythropoietin in an estrogen- and oxygen-dependent manner. Am. J. Physiol. Endocrinol. Metab., 278, E1038–E1044.
5. Aydin, A., Genç, K., Akhisaroglu, M., Yorukoglu, K., Gokmen, N., and Gonullu, E. (2003) Erythropoietin exerts neuroprotective effect in neonatal rat model of hypoxic-ischemic brain injury. Brain Dev., 25, 494–498.
6. Demers, E., McPherson, R., and Juul, S. (2005) Erythropoietin protects dopaminergic neurons and improves neurobehavioral outcomes in juvenile rats after neonatal hypoxia-ischemia. Pediatr. Res., 58, 297–301.
7. Jones, N. and Bergeron, M. (2001) Hypoxic preconditioning induces changes in HIF-1 target genes in neonatal rat brain. J. Cereb. Blood Flow Metab., 21, 1105–1114.
8. Kumral, A., Ozer, E., Yilmaz, O. et al. (2003) Neuroprotective effect of erythropoietin on hypoxic-ischemic brain injury in neonatal rats. Biol. Neonate, 83, 224–228.
9. Konishi, Y., Chui, D., Hirose, H., Kunishita, T., and Tabira, T. (1993) Trophic effect of erythropoietin and

other hematopoietic factors on central cholinergic neurons in vitro and in vivo. *Brain Res.*, **609**, 29–35.
10. Morishita, E., Masuda, S., Nagao, M., Yasuda, Y., and Sasaki, R. (1997) Erythropoietin receptor is expressed in rat hippocampal and cerebral cortical neurons, and erythropoietin prevents in vitro glutamate-induced neuronal death. *Neuroscience*, **76**, 105–116.
11. Campana, W., Misasi, R., and O'Brien, J. (1998) Identification of a neurotrophic sequence in erythropoietin. *Int. J. Mol. Med.*, **1**, 235–241.
12. Celik, M., Gökmen, N., Erbayraktar, S. et al. (2002) Erythropoietin prevents motor neuron apoptosis and neurologic disability in experimental spinal cord ischemic injury. *Proc. Natl. Acad. Sci. U.S.A.*, **99**, 2258–2263.
13. Silva, M., Benito, A., Sanz, C. et al. (1999) Erythropoietin can induce the expression of bcl-x(L) through Stat5 in erythropoietin-dependent progenitor cell lines. *J. Biol. Chem.*, **274**, 22165–22169.
14. Sirén, A., Fratelli, M., Brines, M. et al. (2001) Erythropoietin prevents neuronal apoptosis after cerebral ischemia and metabolic stress. *Proc. Natl. Acad. Sci. U.S.A.*, **98**, 4044–4049.
15. Villa, P., van Beek, J., Larsen, A. et al. (2007) Reduced functional deficits, neuroinflammation, and secondary tissue damage after treatment of stroke by nonerythropoietic erythropoietin derivatives. *J. Cereb. Blood Flow Metab.*, **27**, 552–563.
16. Agnello, D., Bigini, P., Villa, P. et al. (2002) Erythropoietin exerts an anti-inflammatory effect on the CNS in a model of experimental autoimmune encephalomyelitis. *Brain Res.*, **952**, 128–134.
17. Leist, M., Ghezzi, P., Grasso, G. et al. (2004) Derivatives of erythropoietin that are tissue protective but not erythropoietic. *Science*, **305**, 239–242.
18. Keller, M., Yang, J., Griesmaier, E. et al. (2006) Erythropoietin is neuroprotective against NMDA-receptor-mediated excitotoxic brain injury in newborn mice. *Neurobiol. Dis.*, **24**, 357–366.
19. Kumral, A., Baskin, H., Yesilirmak, D. et al. (2007) Erythropoietin attenuates lipopolysaccharide-induced white matter injury in the neonatal rat brain. *Neonatology*, **92**, 269–278.
20. McPherson, R., Demers, E., and Juul, S. (2007) Safety of high-dose recombinant erythropoietin in a neonatal rat model. *Neonatology*, **91**, 36–43.
21. Statler, P., McPherson, R., Bauer, L., Kellert, B., and Juul, S. (2007) Pharmacokinetics of high-dose recombinant erythropoietin in plasma and brain of neonatal rats. *Pediatr. Res.*, **61**, 671–675.
22. Kellert, B., McPherson, R., and Juul, S. (2007) A comparison of high-dose recombinant erythropoietin treatment regimens in brain-injured neonatal rats. *Pediatr. Res.*, **61**, 451–455.
23. Fauchère, J., Dame, C., and Vonthein, R. (2008) An approach to using recombinant erythropoietin for neuroprotection in very preterm infants. *Pediatrics*, **122**, 375–382.
24. Yiş, U., Kurul, S., Kumral, A. et al. (2008) Effect of erythropoietin on oxygen-induced brain injury in the newborn rat. *Neurosci. Lett.*, **448**, 245–249.
25. Kaindl, A., Sifringer, M., Koppelstaetter, A. et al. (2008) Erythropoietin protects the developing brain from hyperoxia-induced cell death and proteome changes. *Ann. Neurol.*, **64**, 523–534.
26. Bernaudin, M., Marti, H., Roussel, S. et al. (1999) A potential role for erythropoietin in focal permanent cerebral ischemia in mice. *J. Cereb. Blood Flow Metab.*, **19**, 643–651.
27. Brines, M., Ghezzi, P., Keenan, S. et al. (2000) Erythropoietin crosses the blood–brain barrier to protect against experimental brain injury. *Proc. Natl. Acad. Sci. U.S.A.*, **97**, 10526–10531.
28. Wang, Y., Zhang, Z., Rhodes, K. et al. (2007) Post-ischemic treatment with erythropoietin or carbamylated erythropoietin reduces infarction and improves neurological outcome in a rat model of focal cerebral ischemia. *Br. J. Pharmacol.*, **151**, 1377–1384.
29. Gorio, A., Gokmen, N., Erbayraktar, S. et al. (2002) Recombinant human erythropoietin counteracts secondary injury and markedly enhances neurological recovery from experimental spinal cord

trauma. *Proc. Natl. Acad. Sci. U.S.A.*, **99**, 9450–9455.

30. Grasso, G., Sfacteria, A., Meli, F., Fodale, V., Buemi, M., and Iacopino, D. (2007) Neuroprotection by erythropoietin administration after experimental traumatic brain injury. *Brain Res.*, **1182**, 99–105.

31. Campana, W., and Myers, R. (2003) Exogenous erythropoietin protects against dorsal root ganglion apoptosis and pain following peripheral nerve injury. *Eur. J. Neurosci.*, **18**, 1497–1506.

32. Keswani, S.C., Buldanlioglu, U., Fischer, A. *et al.* (2004) A novel endogenous erythropoietin-mediated pathway prevents axonal degeneration. *Ann. Neurol.*, **56**, 815–826.

33. Sadamoto, Y., Igase, K., Sakanaka, M. *et al.* (1998) Erythropoietin prevents place navigation disability and cortical infarction in rats with permanent occlusion of the middle cerebral artery. *Biochem. Biophys. Res. Commun.*, **253**, 26–32.

34. Wiessner, C., Allegrini, P., Ekatodramis, D., Jewell, U., Stallmach, T., and Gassmann, M. (2001) Increased cerebral infarct volumes in polyglobulic mice overexpressing erythropoietin. *J. Cereb. Blood Flow Metab.*, **21**, 857–864.

35. Fletcher, L., Kohli, S., Sprague, S. *et al.* (2009) Intranasal delivery of erythropoietin plus insulin-like growth factor-I for acute neuroprotection in stroke. *J. Neurosurg.*, **111**, 164–170.

36. Yu, Y., Xu, Q., Zhang, Q., Zhang, W., Zhang, L., and Wei, E. (2005) Intranasal recombinant human erythropoietin protects rats against focal cerebral ischemia. *Neurosci. Lett.*, **387**, 5–10.

37. Li, L., Jiang, Q., Ding, G. *et al.* (2009) MRI identification of white matter reorganization enhanced by erythropoietin treatment in a rat model of focal ischemia. *Stroke*, **40**, 936–941.

38. Ehrenreich, H., Hasselblatt, M., Dembowski, C. *et al.* (2002) Erythropoietin therapy for acute stroke is both safe and beneficial. *Mol. Med.*, **8**, 495–505.

39. Ehrenreich, H., Weissenborn, K., Prange, H. *et al.* (2009) Recombinant human erythropoietin in the treatment of acute ischemic stroke. *Stroke*, **40**, e647–e656.

40. Dumont, R., Okonkwo, D., Verma, S. *et al.* (2001) Acute spinal cord injury, part I: pathophysiologic mechanisms. *Clin. Neuropharmacol.*, **24**, 254–264.

41. Norenberg, M., Smith, J., and Marcillo, A. (2004) The pathology of human spinal cord injury: defining the problems. *J. Neurotrauma*, **21**, 429–440.

42. Toth, C., Martinez, J., Liu, W. *et al.* (2008) Local erythropoietin signaling enhances regeneration in peripheral axons. *Neuroscience*, **154**, 767–783.

43. Shein, N., Grigoriadis, N., Alexandrovich, A. *et al.* (2008) Differential neuroprotective properties of endogenous and exogenous erythropoietin in a mouse model of traumatic brain injury. *J. Neurotrauma*, **25**, 112–123.

44. Xiong, Y., Mahmood, A., Lu, D. *et al.* (2008) Histological and functional outcomes after traumatic brain injury in mice null for the erythropoietin receptor in the central nervous system. *Brain Res.*, **1230**, 247–257.

45. Gorio, A., Madaschi, L., Di Stefano, B. *et al.* (2005) Methylprednisolone neutralizes the beneficial effects of erythropoietin in experimental spinal cord injury. *Proc. Natl. Acad. Sci. U.S.A.*, **102**, 16379–16384.

46. Kaptanoglu, E., Solaroglu, I., Okutan, O., Surucu, H., Akbiyik, F., and Beskonakli, E. (2004) Erythropoietin exerts neuroprotection after acute spinal cord injury in rats: effect on lipid peroxidation and early ultrastructural findings. *Neurosurg. Rev.*, **27**, 113–120.

47. Vitellaro-Zuccarello, L., Mazzetti, S., Madaschi, L., Bosisio, P., Gorio, A., and De Biasi, S. (2007) Erythropoietin-mediated preservation of the white matter in rat spinal cord injury. *Neuroscience*, **144**, 865–877.

48. Pinzon, A., Marcillo, A., Pabon, D., Bramlett, H., Bunge, M., and Dietrich, W. (2008) A re-assessment of erythropoietin as a neuroprotective agent following rat spinal cord compression or contusion injury. *Exp. Neurol.*, **213**, 129–136.

49. Gold, R., Hartung, H.P., and Toyka, K.V. (2000) Animal models for autoimmune demyelinating disorders of the nervous system. *Mol. Med. Today*, **6**, 88–91.
50. Li, W., Maeda, Y., Yuan, R., Elkabes, S., Cook, S., and Dowling, P. (2004) Beneficial effect of erythropoietin on experimental allergic encephalomyelitis. *Ann. Neurol.*, **56**, 767–777.
51. Sättler, M., Merkler, D., Maier, K. et al. (2004) Neuroprotective effects and intracellular signaling pathways of erythropoietin in a rat model of multiple sclerosis. *Cell Death Differ.*, **11** (Suppl 2), S181–S192.
52. Savino, C., Pedotti, R., Baggi, F. et al. (2006) Delayed administration of erythropoietin and its non-erythropoietic derivatives ameliorates chronic murine autoimmune encephalomyelitis. *J. Neuroimmunol.*, **172**, 27–37.
53. Yuan, R., Maeda, Y., Li, W., Lu, W., Cook, S., and Dowling, P. (2008) Erythropoietin: a potent inducer of peripheral immuno/inflammatory modulation in autoimmune EAE. *PLoS One*, **3**, e1924.
54. Zhang, J., Li, Y., Cui, Y. et al. (2005) Erythropoietin treatment improves neurological functional recovery in EAE mice. *Brain Res.*, **1034**, 34–39.
55. Ehrenreich, H., Fischer, B., Norra, C. et al. (2007) Exploring recombinant human erythropoietin in chronic progressive multiple sclerosis. *Brain*, **130**, 2577–2588.
56. Keswani, S., Pardo, C., Cherry, C., Hoke, A., and McArthur, J. (2002) HIV-associated sensory neuropathies. *AIDS*, **16**, 2105–2117.
57. Bianchi, R., Gilardini, A., Rodriguez-Menendez, V. et al. (2007) Cisplatin-induced peripheral neuropathy: neuroprotection by erythropoietin without affecting tumour growth. *Eur. J. Cancer*, **43**, 710–717.
58. Keswani, S., Leitz, G., and Hoke, A. (2004) Erythropoietin is neuroprotective in models of HIV sensory neuropathy. *Neurosci. Lett.*, **371**, 102–105.
59. Melli, G., Jack, C., Lambrinos, G., Ringkamp, M., and Höke, A. (2006) Erythropoietin protects sensory axons against paclitaxel-induced distal degeneration. *Neurobiol. Dis.*, **24**, 525–530.
60. Digicaylioglu, M. and Lipton, S. (2001) Erythropoietin-mediated neuroprotection involves cross-talk between Jak2 and NF-kappaB signalling cascades. *Nature*, **412**, 641–647.
61. Keswani, S.C., Leitz, G.J., and Hoke, A. (2004) Erythropoietin is neuroprotective in models of HIV sensory neuropathy. *Neurosci. Lett.*, **371**, 102–105.
62. Hoke, A. (2006) Neuroprotection in the peripheral nervous system: rationale for more effective therapies. *Arch. Neurol.*, **63**, 1681–1685.
63. Hoke, A. and Keswani, S.C. (2005) Neuroprotection in the PNS: erythropoietin and immunophilin ligands. *Ann. N. Y. Acad. Sci.*, **1053**, 491–501.
64. Toth, C., Martinez, J.A., Liu, W.Q. et al. (2008) Local erythropoietin signaling enhances regeneration in peripheral axons. *Neuroscience*, **154**, 767–783.
65. Allaf, M.E., Hoke, A., and Burnett, A.L. (2005) Erythropoietin promotes the recovery of erectile function following cavernous nerve injury. *J. Urol.*, **174**, 2060–2064.
66. Ehrenreich, H., Hinze-Selch, D., Stawicki, S. et al. (2007) Improvement of cognitive functions in chronic schizophrenic patients by recombinant human erythropoietin. *Mol. Psychiatry*, **12**, 206–220.

Part IV
Hormones and Neurogenesis

16
Thyroid Hormone Actions on Glioma Cells
Min Zhou, Harold K. Kimelberg, Faith B. Davis, and Paul J. Davis

16.1
Introduction

The actions of thyroid hormone on glial cells in the brain have been reviewed recently [1–3]. Among these actions are morphological differentiation [4], modulation of the state of actin in the astrocyte [5], heightening of the antioxidative capacity of astrocytes [6], and secretion of growth factors such as fibroblast growth factor 2 (FGF2; basic fibroblast growth factor (bFGF)) [7]. Glioma cells are also affected by iodothyronines. There is evidence that thyroid hormone *in vitro* causes proliferation of rodent glioma cell lines [8] and human glioblastoma cells [9] via a nongenomic mechanism involving a newly recognized cell surface receptor for the hormone on integrin $\alpha v \beta 3$ [10, 11]. This site mediates the antiapoptotic action of the thyroid hormone [12]. Nongenomic actions of the hormone are those that do not require intranuclear binding of the thyroid hormone by its nuclear receptors (TRs) [11]. Specific blockade of the integrin receptor site on the plasma membrane inhibits the proliferative effect of the thyroid hormone on glioma cells and also appears to radiosensitize such cells [13]. Clinical data now indicate that chemical induction of mild hypothyroidism in glioblastoma patients importantly improves survival [14]. We propose that this clinical observation reflects decreased hormonal action at the integrin receptor. However, it is important to point out that abundance of one of the nuclear receptors for the thyroid hormone, $TR\beta 1$, is correlated with clinical aggressiveness of astrocytomas [15], and that mutations in nuclear $TR\beta$ can result in neoplasia in other, nonneural tissues [16].

In this chapter, we examine certain nongenomic actions of the thyroid hormone on glioma cells and emphasize the molecular basis for the effects where this is at least in part understood.

16.2
Origins of Glioma

Although the nomenclature of gliomas suggests a specific cell of origin, such as an astrocyte or committed astrocyte precursor for astrocytomas, it has not clearly

Hormones in Neurodegeneration, Neuroprotection, and Neurogenesis.
Edited by Achille G. Gravanis and Synthia H. Mellon
Copyright © 2011 WILEY-VCH Verlag GmbH & Co. KGaA, Weinheim
ISBN: 978-3-527-32627-3

been determined which cell population in the brain gives rise to tumors that develop several decades after birth in a tissue that is mitotically largely inactive. Gliomas are thought to arise from either preexisting neural stem cells in the adult brain or proliferating cells that are dedifferentiated from adult glial cell types, such as astrocytes or oligodendrocytes due to oncogenic mutations, but we still know very little about how abnormalities in the cell cycle or other cellular processes might trigger glial transformation for gliomagenesis [17]. Interestingly, when cell proliferation was examined by BrdU incorporation in mouse brain, 97% of BrdU-positive cells could also be labeled by the marker NG2 in the adult brain [18]. NG2 is a membrane proteoglycan specific for a subpopulation of glia showing a similar morphology to astrocytes but considered distinct from these cells [19–21]. These BrdU-positive cells are located outside of the subventricular zone (SVZ), the region that retains neurogenesis and gliogenesis in the adult brain. Importantly, the lineage relationship of NG2 to astrocytes and neurons remains to be elucidated. The fact that NG2 cells proliferate in the adult brain raises the question of whether these cells can directly transform themselves to oligodendrogliomas or astrocytomas by losing control of cell cycle regulation without dedifferentiation [22].

The new field of stem cell biology is also expected to shed light on the origin of glioma. Combined activation of Ras and AKT in neural progenitors results in the formation of glioblastomas in mice [23], providing a novel model for gliomagenesis. Similarly, autocrine stimulation by platelet-derived growth factor (PDGF) induces the formation of oligodendroglial tumors by neural progenitor cells [24]. These studies suggest that brain parenchymal stem cells can be a source of glioma formation in the adult mammalian brain.

16.3
Glioma Cell Biology

Invasiveness, migration, and angiogenesis are key biological features of malignant glioma cells, but glioma growth is physically restricted by the bony skull and vertebrae. Also, to metastasize they must migrate through the packed brain parenchyma and an extremely narrow and tortuous extracellular space. Therefore, to propagate and invade, glioma must make space available by actively eliminating surrounding healthy cells.

To achieve this, glioma cells release glutamate which causes excitotoxic death to surrounding neurons. The amount of glutamate synthesized and released by glioma cells is considerable; for example, cultured glioma cells can increase glutamate concentration in their media from 1 to 100 μM within 5–6 h. The released glutamate also explains peritumoral seizures which are a common symptom early on in the disease. Glutamate release occurs via a cystine–glutamate exchanger system X(c), which releases glutamate in exchange for cystine. The imported cystine in both astrocytes and gliomas is used for the synthesis of the cellular antioxidant GSH. GSH protects cells from endogenously produced reactive oxygen and nitrogen species and therefore, unfortunately, also endows tumors with an enhanced resistance to radiation and chemotherapy. Preclinical data demonstrate that pharmacological inhibition of system X(c) by 4-carboxyphenylglycine causes GSH depletion which slows tumor growth and curtails tumor invasion *in vivo* [25, 26].

Against this background, it is interesting to note that glutamate uptake by nonmalignant astrocytes is increased by the thyroid hormone (3,5,3′-triiodo-L-thyronine, T_3) [27]. This effect in the normal brain is seen to decrease the local excitotoxicity of glutamate and thus to be neuroprotective. We may assume that this action of the hormone is reduced or lost in glioma cells, but this has not been examined experimentally.

Glioma cells can migrate from one hemisphere to another so that gliomas can spread through the whole brain compromising vital nervous activities and leading to death. The mechanisms of glioma cell migration are several. First, glioma cells express a number of metalloproteinases which assist in breaking down the extracellular matrix (ECM) and establishing migration tunnels [28]. Second, glioma cells are able to undergo substantial shrinkage which helps them to attain an elongated shape and thus penetrate into narrow interstitial compartments [29]. The decrease in glioma cell volume is believed to be achieved through activation of several voltage- or osmotic-sensitive chloride channels [29, 30]. However, the inhibitor used to link such channel activity with invasiveness, chlorotoxin, also inhibits metalloproteinases, so its inhibition of cell invasiveness does not define shape changes as critical. There is a long literature on how cells move, involving intracellular contractile elements and different types of cell surface adhesion [28]. The thyroid hormone promotes the conversion of soluble actin to F-actin [1, 5] and thus contributes to the cytoskeletal support of cell movement. Animal cells do not maintain an intracellular pressure and, although water and therefore cytoplasm are incompressible, they can change their shape to squeeze through tissue spaces. Glioma migration might be also influenced by cytosolic Ca^{2+} oscillation which can be mediated by activation of Ca^{2+}-permeable AMPA receptors. Together, glioma cells utilize several mechanisms to expand within the CNS, giving rise to the high malignancy and rapid clinical progression of glioma.

Angiogenesis is another important biological feature of malignant glioma. In fact, microvascular hyperplasia is used as one of the criteria for distinguishing high- from low-grade gliomas. The clinical aggressiveness and likelihood of recurrence of malignant glioma appears to be closely related to the degrees of neovascularization. On the basis of this, antiangiogenic strategies with monoclonal antibodies or tyrosine kinase inhibitors have recently been proposed. It is generally believed that expansion of neoplasms causes hypoxia inside tumor tissues and this initiates angiogenesis through chronic activation of the hypoxia-inducible factor-1α (HIF-1α) pathway that produces vascular endothelial growth factor (VEGF) and bFGF (FGF2) [31]. Remarkably and independently of hypoxia, the thyroid hormone nongenomically regulates *HIF-1α* gene expression [9, 32]. By this mechanism and by its ability to induce bFGF release [8], the thyroid hormone is proangiogenic [10, 33]. Inhibition of the proangiogenic action of the thyroid hormones is reconsidered later in this chapter, as a contribution to antiglioma strategy.

16.4
Thyroid Hormone Analogs, Transport, and Metabolism

The thyroid hormones, L-thyroxine (T_4) and T_3, are diphenylether structures with an inner ring alanine side chain. The principal circulating form of the thyroid hormone is T_4. The half-life of T_4 in man is substantially longer than that of T_3, due to the higher affinity of a blood transport protein, thyroxine-binding globulin (TBG) for T_4. Another serum thyroid hormone transport protein, prealbumin (transthyretin), appears to be important to the import of T_4 into the brain [34]. The ratio of total concentration of T_4 to T_3 in blood is approximately 20 to 1. T_4 is converted to T_3 by cellular deiodinases. The thyroid hormone has genomic effects that require the interaction of nuclear T_3 with the nuclear thyroid receptor (TR) protein isoforms and results in specific gene transcription. Many of the genes transcribed are involved in maintenance of homeostasis. At the plasma membrane receptor on heterodimeric integrin $\alpha v \beta 3$ [10], cellular events are initiated by thyroid hormones that are defined as nongenomic; that is, they do not primarily require the interaction of nuclear T_3 with TRs. At this receptor, T_4 may be more active than T_3 [11, 35]. The cellular events that are initiated at the cell surface integrin include complex downstream actions that involve gene transcription such as tumor cell division or angiogenesis [9, 35]. Glioma/glioblastoma cell proliferation is regulated by the thyroid hormone at the plasma membrane integrin receptor [8, 9].

A thyroid hormone analog that blocks actions of T_4 and T_3 at the integrin receptor, but not in the cell nucleus, is tetraiodothyroacetic acid (tetrac) [10, 11, 35]. Tetrac is the deaminated analog of T_4. While it occurs naturally in small quantities, tetrac shows some promise therapeutically as an antiproliferative and antiangiogenic agent, because integrin $\alpha v \beta 3$ is primarily expressed on cancer cells and on endothelial and vascular smooth muscle cells that are already committed to angiogenesis. In the laboratory, tetrac has been covalently coupled to a nanoparticle [36] and acts exclusively at the cell surface integrin receptor for the thyroid hormone; nanoparticulate tetrac cannot enter the cell. Unmodified tetrac also acts at the integrin, but it is taken up by cells and is a low-grade thyromimetic in the cell nucleus.

16.5
Thyroid Hormones and Brain Development

Thyroid hormones are major physiological regulators of mammalian brain development [2]. Cell differentiation, cell migration, and gene expression are altered as a consequence of thyroid hormone deficiency or excess. The hormone is importantly involved in several phases of brain development; that is, cell migration, outgrowth of neuronal processes and acquisition of neuronal polarity, synaptogenesis, and myelin formation. These processes are slowed down in the setting of thyroid hormone deficiency (hypothyroidism), indicating that iodothyronines both directly and indirectly affect these central nervous system developmental processes. The actions of thyroid hormone can best be defined as ensuring the timed coordination of different developmental events, through specific effects on the rates of cell differentiation and gene expression. The critical contribution of the thyroid hormone to development and maintenance of brain vasculature has recently been demonstrated by Schlenker et al. [37].

The roles of thyroid hormone on glial cell differentiation and development have also been studied extensively. Astrocytes are currently suggested to be important mediators of thyroid hormone action during CNS development. These cells play a central role in thyroid hormone metabolism in the brain, being the principal transporters of thyroxine from blood into the brain and conducting much of the conversion of T_4 to T_3 in the brain. The hormone regulates production of ECM proteins and growth factors by glial cells, thus controlling neuronal growth and neuritogenesis [2]. Iodothyronines are instructive signals for the generation of oligodendrocytes and enhance the proliferation of the committed precursor oligodendrocyte cells. The thyroid hormone also regulates oligodendrocyte numbers in brain regions by directly promoting their differentiation. The hormone increases morphological and functional maturation of postmitotic oligodendrocytes by stimulating the expression of various myelin genes [38].

16.6
Nongenomic Actions of Thyroid Hormones

Nongenomic actions of thyroid hormones have been described in a variety of tissues or cellular models [8, 11, 39]. In the past several years, the molecular basis for some of these actions has been clarified. For example, a pool of nuclear TRβ1 has been described by Moeller et al. to reside in the cytoplasm of human skin fibroblasts, where the TR binds T_3 and forms a complex with the p85α subunit of signal transducing phosphatidylinositol 3-kinase (PI 3-K). This complex is the initiating step toward downstream transcription of a series of genes relevant to carbohydrate metabolism [40]. Among these genes is *HIF-1α*, whose gene product, mentioned above, is relevant to the angiogenesis that supports tumors. A truncated isoform of nuclear TRα1 may also reside in cytoplasm where it is essential to the actions of T_4 and 3,3′,5′-triiodothyronine (reverse T_3) on polymerization of actin [1]. T_3 is not involved in actin polymerization [41]. This nongenomic action of the thyroid hormone on the cytoskeleton has been noted above to occur in both astrocytes and neurons, and supports cell migration.

The identification of a plasma membrane receptor for thyroid hormones on integrin $\alpha v \beta 3$ in 2005 has provided a molecular basis for several actions of thyroid hormone analogs [10], including those pertaining to angiogenesis, cancer cell proliferation, shuttling of intracellular proteins between cytoplasm and nucleus

[42], and residence time of chemotherapeutic agents in cancer cells [43]. This receptor has been localized to the Arg-Gly-Asp (RGD) recognition site on the integrin [10, 44]. The RGD site is essential to the interaction of cells with ECM proteins such as vitronectin and fibronectin and with endogenous growth factors [45, 46]. The ECM–protein interactions of integrins provide important directions in the process of cell migration.

Mathematical modeling of the kinetics of the binding of thyroid hormones to the integrin receptors and the discrete effectiveness of certain inhibitors of hormone-binding at the RGD site have suggested that the hormone-binding domain includes two sites [9] (Figure 16.1). One site binds only T_3 (S1) and

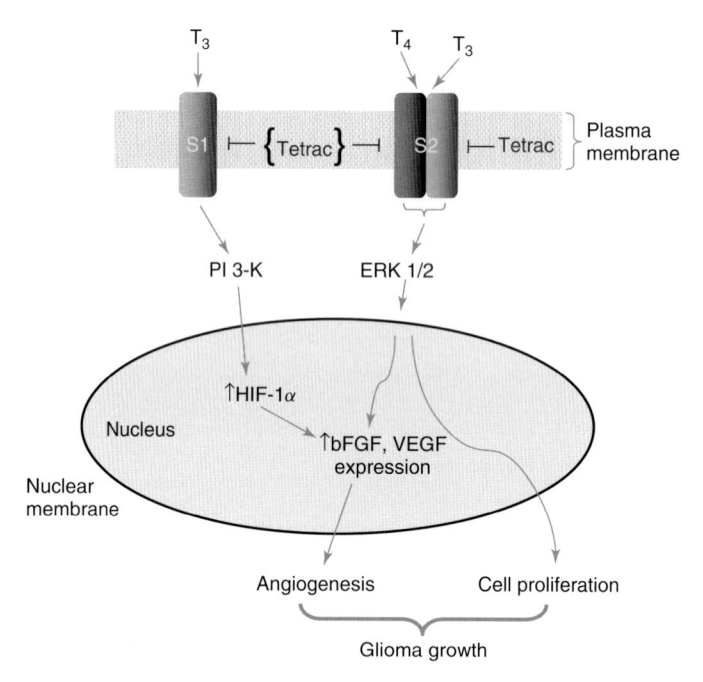

Figure 16.1 Molecular actions of the thyroid hormone on glioma cells that are initiated at the cell surface receptor for the hormone on heterodimeric integrin $\alpha v \beta 3$. The integrin receptor contains two binding sites: S1 binds T_3 exclusively and activates phosphatidyl-inositol 3-kinase (PI 3-K); the T_4 signal at S2 is transduced by mitogen-activated protein kinase (MAPK; extracellular signal-regulated kinase 1/2 (ERK1/2)) [9]. T_3 is also bound by S2. The proliferative effect of T_4 on glioma cells [8, 9] appears to be S2-mediated [9] and the angiogenic action of thyroid hormone on tumor-relevant blood supply [33, 47, 48] may be mediated by PI 3-K or ERK1/2. Hypoxia-inducible factor 1α (HIF 1α) gene expression may be directed from the integrin receptor by T_3 and the HIF 1α gene product in the cell nucleus may transcribe vascular growth factor genes (VEGF, bFGF) [31]. Tetraiodothyroacetic acid (tetrac) is a deaminated analog of T_4 that inhibits the effects of both T_4 and T_3 at the integrin receptor; tetrac is both antiproliferative and antiangiogenic [8–10]. Not shown is that from its integrin receptor on glioma cells, thyroid hormone is also antiapoptotic [12]. The antiapoptotic mechanism involves interference with activity of p. 53.

the hormone signal is transduced by PI 3-K, leading to cytoplasm-to-nucleus shuttling of TRα1 and transcription of the *HIF-1α* gene. The second site (S2) binds both T_4 and T_3, and uses the extracellular signal–regulated kinase 1/2 (ERK1/2) pathway to transduce the hormone signal into proliferation of tumor cells. This site also regulates cytoplasm-to-nucleus trafficking of TRβ1. Thus, there is distinctive regulation of the nuclear uptake of these two isoforms of TR. Importantly, tetrac blocks all the actions of T_3 and T_4 via both integrin hormone-binding domains, while the RGD peptide only inhibits T_3 action via S1 and PI-3-K. T_4 action via S2 is mediated by mitogen-activated protein kinase (MAPK; ERK1/2).

16.7
Hypothyroidism Suppresses Growth of Glioma in Patients

Glioblastoma multiforme (GBM) growth has been shown to be thyroid hormone dependent. Clinical observations by Hercbergs *et al.* [14] showed that, in 22 recurrent glioma patients, mild chemical hypothyroidism prolonged average survival duration threefold. A similar clinical trial conducted in Israel yielded comparable results. Three rodent glioma cell lines have been shown to proliferate in response to the thyroid hormone, an effect that is susceptible to inhibition with tetrac [8]. Addition of tetrac to GL261 rodent glioma cells *in vitro* has been reported recently to decrease the ability of glioma cells to recover from radiation-induced damage, indicating that inhibition of nongenomic thyroid hormone action increases radiosensitivity [13].

Integrin-linked kinase 1 (ILK1), a serine–threonine kinase, interacts with the cytoplasmic domains of the β subunits of integrins and is involved in growth factor signaling and organization of the intracellular actin cytoskeleton [49, 50]. A recent study has shown that the *ILK1* mRNA and ILK1 protein are increased 5.6-fold and 10.1-fold, respectively, in primary astrocytomas [51]. It will be useful to determine whether integrin-mediated actions of thyroid hormone in glial cells may also be influenced by ILK1.

16.8
Molecular Mechanisms of Hypothyroidism-Induced Clinical Suppression of Glioma Progression

To understand how mild hypothyroidism may influence the clinical course of GBM, extensive *in vitro* studies have been carried out on several rodent glioma cell lines – C6, F98, and GL261 [8] – and on human glioblastoma cells (U87 MG) [9]. The evidence available is consistent with the concept that the beneficial effect of clinical thyroid hormone withdrawal in GBM patients reflects decreased integrin receptor-mediated actions of endogenous thyroid hormone.

16.8.1
Thyroid Hormone and Proliferation of Tumor Cells

Integrin $\alpha v\beta 3$ is highly expressed by glioma cells. T_4 is a proliferation factor *in vitro* for several rodent glioma cell lines [8] and for human GBM cells [9]. This agonist effect of T_4 is inhibited by tetrac and by a proprietary RGD peptide specifically directed against integrin $\alpha v\beta 3$ [8]. At the same concentration, T_3 and T_4 both stimulate proliferating cell nuclear antigen (PCNA) accumulation in C6, F98, and GL261 cells but the physiologic concentration of T_3 is 50-fold lower than that of T_4. The S1 binding site in the iodothyronine receptor domain of integrin $\alpha v\beta 3$ binds T_3 exclusively and does not appear to support tumor cell proliferation *in vitro* [9]. From such observations, one may infer that T_4 is more important to glioma cell proliferation than T_3.

16.8.2
Angiogenic Action of Thyroid Hormones

Angiogenic action of thyroid hormones initiated at the integrin receptor has been demonstrated in the blood vessels of the chick chorioallantoic membrane (CAM) model and in a human dermal microvascular endothelial cell sprouting model [33, 47, 48]. Binding of T_3 to the integrin hormone receptor domain 1 (S1, Figure 16.1) causes *HIF-1α* gene expression [9] that, in turn, may lead to VEGF- and bFGF-supported angiogenesis. The angiogenic activity of thyroid hormone has been demonstrated in intact organs [52, 53], including the brain [37]. The blood supply of gliomas is generous *in vivo* and we postulate that the mechanism is in part attributable to the endogenous thyroid hormone.

16.8.3
Antiapoptotic Action of Thyroid Hormones

The thyroid hormone is antiapoptotic, an action which appears to be mediated by inhibition of one or more steps of the ERK-dependent activation of the oncogene suppressor protein, p53 (Figure 16.1) [12, 54]. This has been demonstrated in rodent glioma cells and in the presence of a potent experimental proapoptotic stilbene.

From the foregoing review and the favorable clinical impact of induced hypothyroidism on GBM [14], one may conclude that the thyroid hormone is an endogenous growth factor for glioma. The *in vitro* evidence that the proliferative effect of the thyroid hormone on glioma cells is expressed via the integrin receptor for the hormone predicted the effectiveness of tetrac as an antiproliferative agent in glioma cells.

16.8.4
Tumor Suppression Actions of Tetrac

As we have reported [9] and as shown in Figure 16.1, tetrac not only mimics the RGD peptide in terms of its antagonistic actions at receptor sites S1 and S2, but it has also been shown to have several other antitumor effects. Treating glioma cells with unmodified tetrac induces expression of proapoptotic *Bcl-Xs* and *caspase-2* genes [12]. A more comprehensive recent RNA microarray study has shown that exposing MDA-MB-231 cells to nanoparticulate tetrac up-regulates proapoptosis genes, down-regulates several antiapoptotic genes, and increases expression of *thrombospondin 1* [36]. Thrombospondin 1 is an antiangiogenic protein. This pattern of activities of nanoparticulate tetrac is a coherent antagonism of the cell survival program of these cancer cells. These studies need to be conducted in glioma cells.

16.9
Future Perspectives

Hypothyroidism appears to favorably affect the clinical courses of several cancers. These include glioblastoma [14], breast cancer [55], and head-and-neck cancers [56]. As discussed here, *in vitro* studies have shown that the thyroid hormone is a growth factor for multiple animal glioma cell lines and for human glioblastoma cells. Murine xenografts of human renal cell carcinoma [57], breast cancer [43], follicular thyroid carcinoma [58], and medullary carcinoma of the thyroid [59] respond favorably to administration of tetrac or nanoparticulate tetrac – agents that block actions of the thyroid hormone initiated at the recently recognized thyroid hormone receptor on cell surface integrin $\alpha v \beta 3$. This integrin is expressed primarily on cancer cells and on endothelial and vascular smooth muscle cells. It is not surprising, then, that in human cancer xenograft studies in nude mice, the vascular supply to the grafts is importantly and rapidly reduced by tetrac or nanoparticulate tetrac [57–59]. Xenograft studies of human glioblastoma cells are indicated. *In vitro* radiosensitization of rodent glioma cells by tetrac has also been shown [13] and these results encourage testing for the possibility of radiosensitization in tetrac-treated animals bearing glioblastoma xenografts. It appears that the thyroid hormone receptor on integrin $\alpha v \beta 3$ holds promise as a target for anticancer and antiangiogenic agents. In contrast to other single targets in cancer cells, the receptor coherently regulates cell proliferation, apoptosis, and angiogenesis by multiple mechanisms. Somewhat surprisingly, one of the mechanisms invoked by the cell surface receptor is up- or down-regulation of differentially regulated survival pathway genes in a pattern that promotes cell death [36].

References

1. Leonard, J.L. (2008) Non-genomic actions of thyroid hormone in brain development. *Steroids*, **73**, 1008–1012.
2. Trentin, A.G. (2006) Thyroid hormone and astrocyte morphogenesis. *J. Endocrinol.*, **189**, 189–197.
3. Bernal, J. (2005) Thyroid hormone and brain development. *Vitam. Horm.*, **71**, 95–122.
4. Ghosh, M., Gharami, K., Paul, S., and Das, S. (2005) Thyroid hormone-induced morphological differentiation and maturation of astrocytes involves activation of protein kinase A and ERK signaling pathways. *Eur. J. Neurosci.*, **22**, 1609–1617.
5. Farwell, A.P., Dubord-Tomasetti, S.A., Pietrzykowski, A.Z., Stachelek, S.J., and Leonard, J.L. (2005) Regulation of cerebellar neuronal migration and neurite outgrowth by thyroxine and 3,3′,5-triiodothyronine. *Brain Res. Dev. Brain Res.*, **154**, 121–135.
6. Dasgupta, A., Das, A., and Sarkar, P.K. (2007) Thyroid hormone promotes glutathione synthesis in astrocytes by up regulation of glutamate cysteine ligase through differential stimulation of its catalytic and modulator subunit mRNAs. *Free Radic. Biol. Med.*, **42**, 617–626.
7. Trentin, A.G., Alvarez-Silva, M., and Moura Neto, V. (2001) Thyroid hormone induces cerebellar astrocytes and C6 glioma cells to secrete mitogenic growth factors. *Am. J. Physiol. Endocrinol. Metab.*, **281**, E1088–E1094.
8. Davis, F.B., Tang, H.Y., Shih, A., Keating, T., Lansing, L., Hercbergs, A., Fenstermaker, R.A., Mousa, A., Mousa, S.A., Davis, P.J., and Lin, H.Y. (2006) Acting via a cell surface receptor, thyroid hormone is a growth factor for glioma cells. *Cancer Res.*, **64**, 7270–7275.
9. Lin, H.Y., Sun, M., Tang, H.Y., Lin, C., Luidens, M.K., Mousa, S.A., Incerpi, S., Drusano, G.L., Davis, F.B., and Davis, P.J. (2009) L-Thyroxine vs. 3,5,3′-triiodo-L-thyronine and cell proliferation: activation of mitogen-activated protein kinase and phosphatidylinositol 3-kinase. *Am. J. Physiol. Cell. Physiol.*, **296**, C980–C991.
10. Bergh, J.J., Lin, H.Y., Lansing, L., Mohamed, S.N., Davis, F.B., Mousa, S., and Davis, P.J. (2005) Integrin $\alpha V \beta 3$ contains a cell surface receptor for thyroid hormone that is linked to activation of mitogen-activated protein kinase and induction of angiogenesis. *Endocrinology*, **146**, 2864–2871.
11. Davis, P.J., Leonard, J.L., and Davis, F.B. (2008) Mechanisms of nongenomic actions of thyroid hormone. *Front. Neuroendocrinol.*, **29**, 211–218.
12. Lin, H.Y., Tang, H.Y., Keating, T., Wu, Y.H., Shih, A., Hammond, D., Sun, M., Hercbergs, A., Davis, F.B., and Davis, P.J. (2008) Resveratrol is pro-apoptotic and thyroid hormone is anti-apoptotic in glioma cells: both actions are integrin and ERK mediated. *Carcinogenesis*, **29**, 62–69.
13. Hercbergs, A.H., Davis, P.J., Davis, F.B., Ciesielski, M.J., and Leith, J.T. (2009) Radiosensitization of GL261 glioma cells by tetraiodothyroacetic acid (tetrac). *Cell Cycle*, **8**, 2586–2591.
14. Hercbergs, A.A., Goyal, L.K., Suh, J.H., Lee, S., Reddy, C.A., Cohen, B.H., Stevens, G.H., Reddy, S.K., Peereboom, D.M., Elson, P.J., Gupta, M.K., and Barnett, G.H. (2003) Propylthiouracil-induced hypothyroidism with high-dose tamoxifen prolongs survival in recurrent high grade glioma: a phase I/II study. *Anticancer Res.*, **23**, 617–626.
15. Huang, S.L., Lin, C.L., Lieu, A.S., Hwang, Y.F., Howng, S.L., Hong, Y.R., Chang, D.S., and Lee, K.S. (2008) The expression of thyroid hormone receptor isoforms in human astrocytomas. *Surg. Neurol.*, **70**(Suppl. 1), S1.4–S1.8.
16. Guigon, C.J. and Cheng, S.Y. (2009) Novel oncogenic actions of TRβ mutants in tumorigenesis. *IUBMB Life*, **61**, 528–536.
17. Ichimura, K., Schmidt, E.E., Goike, H.M., and Collins, V.P. (1996) Human glioblastomas with no alterations of the CDKN2A (p16INK4A, MTSI) and CDK4 genes have frequent mutations of the retinoblastoma gene. *Oncogene*, **13**, 1065–1072.

18. Dawson, M.R., Polito, A., Levine, J.M., and Reynolds, R. (2003) NG2-expressing glial progenitor cells: an abundant and widespread population of cycling cells in the adult rat CNS. *Mol. Cell. Neurosci.*, **24**, 476–488.
19. Butt, A.M., Hamilton, M., Hubbard, P., Pugh, M., and Ibrahim, M. (2005) Synantocytes: the fifth element. *J. Anat.*, **207**, 695–706.
20. Kimelberg, H.K. (2004) The problem of astrocyte identity. *Neurochem. Int.*, **45**, 191–202.
21. Nishiyama, A., Komitova, M., Suzuki, R., and Zhu, X. (2009) Polydendrocytes (NG2 cells): multifunctional cells with lineage plasticity. *Nat. Rev. Neurosci.*, **10**, 9–22.
22. Ge, W.P., Zhou, W., Luo, Q., Jan, L.Y., and Jan, Y.N. (2009) Dividing glial cells maintain differentiated properties including complex morphology and functional synapses. *Proc. Natl. Acad. Sci. U.S.A.*, **109**, 328–333.
23. Holland, E.C., Celestino, J., Dai, C., Schaefer, L., Sawaya, R.E., and Fuller, G.N. (2000) Combined activation of Ras and Akt in neural progenitors induces glioblastoma formation in mice. *Nat. Genet.*, **25**, 55–57.
24. Dai, C., Celestino, J.C., Okada, Y., Louis, D.N., Fuller, G.N., and Holland, E.C. (2001) PDGF autocrine stimulation dedifferentiates cultured astrocytes and induces oligodendrogliomas and oligoastrocytomas from neural progenitors and astrocytes in vivo. *Genes Dev.*, **15**, 1913–1925.
25. Sontheimer, H. (2008) A role for glutamate in growth and invasion of primary brain tumors. *J. Neurochem.*, **105**, 287–295.
26. Chung, W.J., Lyons, S.A., Nelson, G.M., Hamza, H., Gladson, C.L., Gillespie, G.Y., and Sontheimer, H. (2005) Inhibition of cystine uptake disrupts the growth of primary brain tumors. *J. Neurosci.*, **25**, 7101–7110.
27. Mendes-de-Aguitar, C.B., Alchini, R., Decker, H., Alvarez-Silva, M., Tasca, C.I., and Trentin, A.G. (2008) Thyroid hormone increases astrocyte glutamate uptake and protects astrocytes and neurons against glutamate toxicity. *J. Neurosci. Res.*, **86**, 3117–3125.
28. Nakada, M., Okada, Y., and Yamashita, J. (2003) The role of matrix metalloproteinases in glioma invasion. *Front. Biosci.*, **8**, 261–269.
29. Ransom, C., O'Neal, J.T., and Sontheimer, H. (2001) Volume-activated chloride currents contribute to the resting conductance and invasive migration of human glioma cells. *J. Neurosci.*, **21**, 7674–7683.
30. Kimelberg, H.K., Macvicar, B.A., and Sontheimer, H. (2006) Anion channels in astrocytes: biophysics, pharmacology, and function. *Glia*, **54**, 747–757.
31. Argyriou, A., Giannopoulou, E., and Kalofonos, H.P. (2009) Angiogenesis and anti-angiogenic molecularly targeted therapies in malignant gliomas. *Oncology*, **77**, 1–11.
32. Moeller, L.C., Dumitrescu, A.M., and Refetoff, S. (2005) Cytosolic action of thyroid hormone leads to induction of hypoxia-inducible factor-1α and glycolytic genes. *Mol. Endocrinol.*, **19**, 2955–2963.
33. Mousa, S.A., O'Connor, L., Davis, F.B., and Davis, P.J. (2006) Proangiogenesis action of the thyroid hormone analog 3,5-diiodothyropropionic acid (DITPA) is initiated at the cell surface and is integrin mediated. *Endocrinology*, **147**, 1602–1607.
34. Kassem, N.A. (2006) Role of transthyretin in thyroxine transfer from CSF to brain and choroid plexus. *Am. J. Physiol. Regul. Integr. Comp. Physiol.*, **291**, R1310–R1315.
35. Cheng, S.Y., Leonard, J.L., and Davis, P.J. (2009) Molecular mechanisms of thyroid hormone actions. *Endocr. Rev.*, **31**, 139–170.
36. Glinskii, A.B., Glinsky, G.V., Lin, H.Y., Tang, H.Y., Sun, M., Davis, F.B., Luidens, M.K., Mousa, S.A., Hercbergs, A.H., and Davis, P.J. (2009) Modification of survival pathway gene expression in human breast cancer cells by tetraiodothyroacetic acid (tetrac). *Cell Cycle*, **8**, 3562–3570.
37. Schlenker, E.H., Hora, M., Liu, Y., Redetzke, F.A., Morkin, E., and Gerdes, A.M. (2008) Effects of thyroidectomy,

37. T4, and DITPA replacement on brain blood vessel density in adult rats. *Am. J. Physiol. Regul. Integr. Comp. Physiol.*, **294**, R1504–R1509.
38. Rodriguez-Pena, A. (1999) Oligodendrocyte development and thyroid hormone. *J. Neurobiol.*, **40**, 497–512.
39. Davis, P.J., Davis, F.B., and Cody, V. (2005) Membrane receptors mediating thyroid hormone action. *Trends Endocrinol. Metab.*, **16**, 429–435.
40. Moeller, L.C., Cao, X., Dumitrescu, A.M., Seo, H., and Refetoff, S. (2006) Thyroid hormone mediated changes in gene expression can be initiated by cytosolic action of the thyroid hormone receptor β through the phosphatidylinositol 3-kinase pathway. *Nucl. Recept. Signal*, **4**, e020.
41. Leonard, J.L., and Farwell, A.P. (1997) Thyroid hormone-regulated actin polymerization in brain. *Thyroid*, **7**, 147–151.
42. Cao, H.J., Lin, H.Y., Luidens, M.K., Davis, F.B., and Davis, P.J. (2009) Cytoplasm-to-nucleus shuttling of thyroid hormone receptor-β (TRβ1) is directed from a plasma membrane integrin receptor by thyroid hormone. *Endocr. Res.*, **34**, 31–42.
43. Rebbaa, A., Chu, F., Davis, F.B., Davis, P.J., and Mousa, S.A. (2008) Novel function of the thyroid hormone analog tetraiodothyroacetic acid: a cancer chemosensitizing and anticancer agent. *Angiogenesis*, **11**, 269–276.
44. Cody, V., Davis, P.J., and Davis, F.B. (2007) Molecular modeling of the thyroid hormone interactions with $\alpha v\beta 3$ integrin. *Steroids*, **72**, 165–170.
45. Plow, E.F., Haas, T., Zhang, L., Loftus, J., and Smith, J.W. (2000) Ligand binding to integrins. *J. Biol. Chem.*, **275**, 21785–21788.
46. Arnaout, M.A., Mahalingam, B., and Xiong, J.P. (2005) Integrin structure, allostery, and bidirectional signaling. *Annu. Rev. Cell Dev. Biol.*, **21**, 381–410.
47. Mousa, S.A., Davis, F.B., Mohamed, S., Davis, P.J., and Feng, X. (2006) Pro-angiogenesis action of thyroid hormone and analogs in a three-dimensional in vitro microvascular endothelial sprouting model. *Int. Angiol.*, **25**, 407–413.
48. Davis, F.B., Mousa, S.A., O'Connor, L., Mohamed, S., Lin, H.Y., Cao, H.J., and Davis, P.J. (2004) Proangiogenic action of thyroid hormone is fibroblast growth factor-dependent and is initiated at the cell surface. *Circ. Res.*, **94**, 1500–1506.
49. Dedhar, S. (2000) Cell-substrate interactions and signaling through ILK. *Curr. Opin. Cell Biol.*, **12**, 250–256.
50. Sawai, H., Okada, Y., Funahashi, H., Matsuo, Y., Takahashi, H., Takcyama, H., and Manabe, T. (2006) Interleukin-1α enhances the aggressive behavior of pancreatic cancer cells by regulating the $\alpha_6\beta_1$ integrin and urokinase plaminogen activator receptor expression. *BMC Cell Biol.*, **7**, 8.
51. Li, J., Zhang, H., Wu, J., Guan, H., Yuan, J., Huang, Z., and Li, M. (2009) Prognostic significance of integrin-linked kinase 1 (ILK1) overexpression in astrocytoma. *Int. J. Cancer*, **126**, 1436–1444.
52. Tomanek, R.J., Zimmerman, M.B., Suvarna, P.R., Morkin, E., Pennock, G.D., and Goldman, S. (1998) A thyroid hormone analog stimulates angiogenesis in the post-infarcted rat heart. *J. Mol. Cell. Cardiol.*, **30**, 923–932.
53. Liu, Y., Wang, D., Redetzke, R.A., Sherer, B.A., and Gerdes, A.M. (2009) Thyroid hormone analog 3,5-diiodothyropropionic acid promotes healthy vasculature in the adult myocardium independent of thyroid effects on cardiac function. *Am. J. Physiol. Heart Circ. Physiol.*, **296**, H1551–H1557.
54. Lin, H.Y., Tang, H.Y., Shih, A., Keating, T., Cao, G., Davis, P.J., and Davis, F.B. (2007) Thyroid hormone is a MAPK-dependent growth factor for thyroid cancer cells and is anti-apoptotic. *Steroids*, **72**, 180–187.
55. Cristofanilli, M., Yamamura, Y., Kau, S.W., Bevers, T., Strom, S., Patangan, M., Hsu, L., Krishnamurthy, S., Theriault, R.L., and Hortobagyi, G.N. (2005) Thyroid hormone and breast carcinoma. Primary hypothyroidism is associated with a reduced incidence of

primary breast carcinoma. *Cancer*, **103**, 1122–1128.

56. Nelson, M., Hercbergs, A.H., Rybicki, L., and Strome, M. (2006) Association between development of hypothyroidism and improved survival in patients with head and neck cancer. *Arch. Otolaryngol. Head Neck Surg.*, **132**, 1041–1046.

57. Yalcin, M., Lansing, L., Dyskin, E., Bharali, D.J., Rebbaa, A., Mousa, S.S., Davis, F.B., Davis, P.J., Lin, H.Y., Hercbergs, A.H., and Mousa, S.A. (2009) Tetraiodothyroacetic acid (tetrac) and tetrac nanoparticles inhibit growth of human renal cell carcinoma. *Anticancer Res.*, **29**, 3825–3832.

58. Yalcin, M., Bharali, D.J., Dyskin, E., Dier, E., Lansing, L., Mousa, S.S., Davis, F.B., Davis, P.J., and Mousa, S.A. (2010) Tetraiodothyroacetic acid (tetrac) and tetrac nanoparticles effectively inhibit growth of human follicular thyroid carcinoma. *Thyroid*, **20**, 281–286.

59. Yalcin, M., Dyskin, E., Lansing, L., Bharali, D.J., Mousa, S.S., Bridoux, A., Hercbergs, A.H., Lin, H.Y., Davis, F.B., Glinsky, G.V., Glinskii, A., Ma, J., Davis, P.J., and Mousa, S.A. (2010) Tetraiodothyroacetic acid (tetrac) and nanoparticulate tetrac arrest growth of medullary carcinoma of the thyroid. *J. Clin. Endocrinol. Metab.*, **95**, 1972–1980.

17
Gonadal Hormones, Neurosteroids, and Clinical Progestins as Neurogenic Regenerative Agents: Therapeutic Implications

Lifei Liu and Roberta Diaz Brinton

17.1
Introduction

The regenerative capacity of the brain is most dramatically evident in the daily generation of a multitude of new neurons in the two proliferative zones of the brain, the subgranular zone (SGZ) and the subventricular zone (SVZ) [1]. During neural development, steroids are well-documented determinants of neural proliferation, differentiation, and organization of both the central and peripheral nervous systems [2–4]. The organizational properties of steroids can determine brain structure and function, for life [2–5]. The existence of proliferative zones within the adult mammalian nervous system [6] provides a mechanism whereby the organizational properties of steroid hormones can be sustained to provide a pathway for regenerating neural tissue throughout life and a strategy to repair the brain following damage [7–9].

Neurogenesis in the adult nervous system is now well documented in multiple species, including rodents [10–13], songbirds [14, 15], primates [16, 17], and humans [18]. Adult neurogenesis in the mammalian brain is restricted to the SVZ of the lateral ventricle [19, 20] and the SGZ of the dentate gyrus in the hippocampus [6, 12, 21]. In SGZ of the dentate gyrus, newly generated neurons contribute to multiple forms of hippocampal-dependent learning and memory [22–25], while SVZ neurogenesis enables the olfactory bulb circuit to adapt to novel sensory challenges [26].

While neurogenesis beyond these two proliferative zones is limited in the normal adult mammalian CNS, under certain pathological insults, newly generated neurons do occur in regions beyond the SGZ and SVZ [27–30]. Limited neuronal replacement occurred in the CNS [6] and newly generated neurons were functionally integrated into existing neuronal circuits [31–33].

The number of intrinsic and extrinsic factors regulating adult neurogenesis is rapidly growing [13, 21, 34]. It is clear that gonadal hormone function extends well beyond the reproductive system and are synthesized *de novo* in the central nervous system [35–37] to regulate neural function at multiple levels [2, 3], including neurogenesis [38–41]. When compared to other regulators of neurogenesis such

Hormones in Neurodegeneration, Neuroprotection, and Neurogenesis.
Edited by Achille G. Gravanis and Synthia H. Mellon
Copyright © 2011 WILEY-VCH Verlag GmbH & Co. KGaA, Weinheim
ISBN: 978-3-527-32627-3

as growth factors, steroid hormones possess unique advantages including small molecular weight and penetration of the blood–brain barrier [40], making them promising candidates as neuroregenerative therapeutics.

In this chapter, we first summarize the neurobiological basis of gonadal hormones and neurosteroids as neurogenic factors within the context of current understanding of neurogenesis in the central nervous system. We then focus our attention on gonadal hormone regulation of different stages of neurogenesis under both normal and disease status, with primary emphasis on neurosteroids, progestagens, estrogens, and androgens. Lastly, we discuss the challenges and strategies of using gonadal hormones as regenerative therapeutics for treating CNS disease by regulating endogenous adult neurogenesis in comparison to stem cell transplantation [42–44].

17.2
Gonadal Hormones, Neurosteroids, and Neurogenesis

Naturally occurring gonadal hormones, including androgens, estrogens, and progestins, are predominantly synthesized by the gonads and play significant regulatory roles in both reproductive and nonproductive behaviors throughout the life span [2–5]. In addition to their peripheral synthesis, steroid hormones can be *de novo* synthesized in both the central and the peripheral nervous systems where they modulate brain function by genomic and nongenomic mechanisms [36, 37, 45–47]. Neurosteroids include pregnenolone (PREG) and dehydroepiandrosterone (DHEA), their sulfated derivatives PREGS and DHEAS, as well as progesterone (P_4) and its 5α-reduced metabolites, 3α-hydroxy-5α-pregnan-20-one (allopregnanolone, APα) [36, 37, 48], and estradiol [49, 50]. Cholesterol, the parent molecule of all steroids including neurosteroids, is converted to PREG by mitochondrial enzyme P450scc (CYP11A1) and subsequently to DHEA by P450c17 (CYP17A1) (side-chain cleavage at carbon 17), or to P_4 by 3β-hydroxysteroid dehydrogenase (3β-HSD). DHEA can be further converted to testosterone via 17β-HSD and ultimately to estrogen by aromatase (CYP19A1). In parallel, P_4 can be converted to APα via 5α-reductase and, subsequently, 3α-hydroxysteroid dehydrogenase (3α-HSD) [45, 46]. Key neurosteroid-synthesizing enzymes, including P450c17 (CYP17A1), 3β-HSD, 3α-HSD, and 5α-reductase, are found in both the embryonic and the adult rodent central nervous systems, including hindbrain, mesencephalic nuclei, locus coeruleus [51], spinal cord [52], cortex, midbrain, and hindbrain [36, 37, 53].

Consistent with their localization within the brain, gonadal hormones play a significant regulatory role in the central nervous system [40, 41, 44], adding to their already broad effects on multiple targets including immune response [54, 55], cardio- and cerebrovascular function [56, 57], energy generation [41, 58, 59], lipid metabolism [60, 61], bone formation [62, 63], and tumorigenesis [64, 65]. Specifically, gonadal hormones modulate CNS functions including mood, locomotor activity, pain sensitivity, vulnerability to epilepsy, neurogenesis, attention mechanisms, and cognition, whereas deficiencies in hormone levels are closely associated with disease status and cognitive behavior [38, 40, 66].

Among the CNS modulatory effects mentioned above, the effects of gonadal hormones on adult neurogenesis are receiving increasing attention (Figure 17.1). *Adult neurogenesis*, as described earlier, is a term describing the process of production of new neurons in the adult brain, starting with the division of a precursor cell and culminating in the existence of a new functioning neuron [12, 19, 20]. More specifically, this process includes precursor cell proliferation, cell fate determination, survival, migration, maturation (dendrite/axon extension and synaptogenesis), and functional integration [6]. Adult neurogenesis is restricted to two regions within the adult brain: SGZ of the hippocampal dentate gyrus [12, 13, 31] and SVZ of the lateral ventricle [6, 19, 20, 33]. In both proliferative zones, GFAP-positive cells have been identified as neural stem cells [6, 20, 67]. In the SGZ area, neural

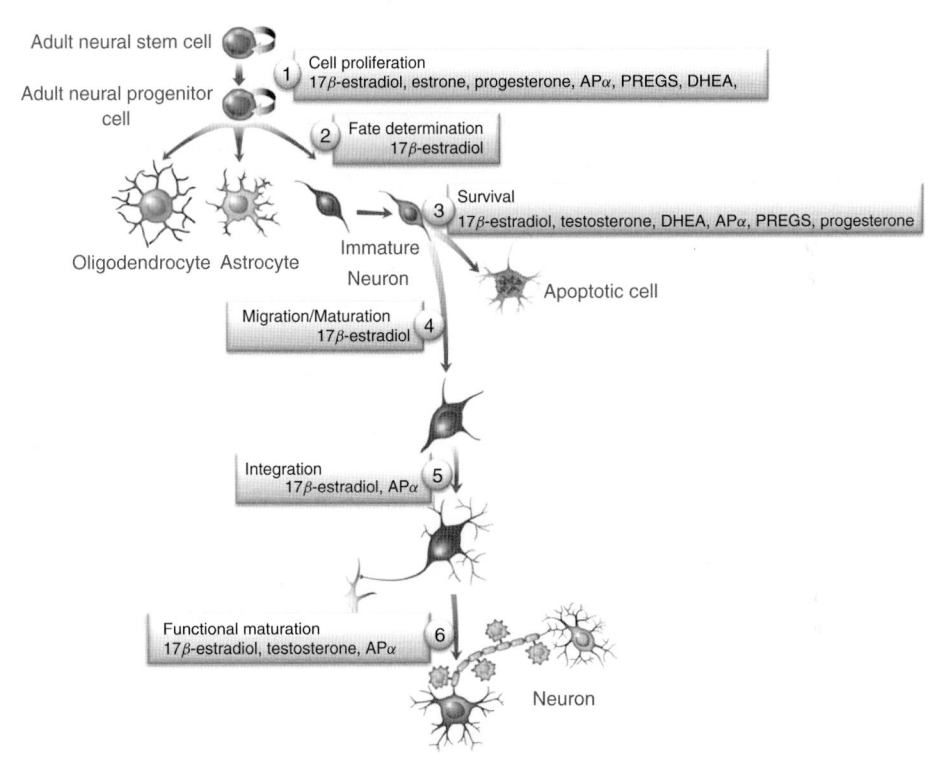

Figure 17.1 Summary of effects of gonadal hormone on adult neurogenesis. Adult neurogenesis includes six stages, which are numbered in the order of occurrence. Gonadal hormones, neurosteroids, and progestins differentially regulate stages within the neurogenic cascade, which is summarized in the corresponding boxes. DHEA, dehydroepiandrosterone; APα, allopregnanolone; PREGS, pregnanolone sulfate. Figure adapted from Lie *et al.* 2004 [6].

stem cells give rise to transit amplifying cells that differentiate into immature neurons, which then migrate into the granule cell layer of the dentate gyrus and differentiate into granule neurons, receiving inputs from the entorhinal cortex and extending projections into CA3 [6, 12, 13]. In the SVZ, neural progenitors give rise to transient amplifying cells that differentiate into immature neurons, which migrate along the rostral migratory stream (RMS) to the olfactory bulb. Immature neurons then differentiate into local interneurons within the granule cell layer and the periglomerular layer [19, 20, 26, 33]. In both cases, newly generated neurons are electrically active, capable of firing action potentials, and receiving synaptic inputs [22, 68–70].

17.2.1
Ovarian Hormone Regulation of Adult Neurogenesis

The first evidence indicating the potential relationship between gonadal hormones and neurogenesis arose from the observation that naturally occurring fluctuations in gonadal hormones could influence adult neurogenesis in multiple brain regions across different species [71, 72]. Seasonal sex differences in neurogenesis was observed in meadow voles, favoring females in hippocampal cell proliferation and survival rates, and was thought to be due to the large fluctuating levels of gonadal hormones across the seasons [71]. Similarly, female rats have higher rates of cell proliferation compared to males but only when females are in proestrus, when both estrogen and P_4 levels are highest [72]. A similar trend was observed in C57BL/6 mice where females in the proestrus showed increased levels of proliferation in the hippocampus, although the increase was not statistically significant [73].

Consistent with gonadal hormone regulation of neurogenesis, bilateral removal of ovaries differentially regulated neurogenesis, depending on the length of ovariectomy [72, 73]. Short-term ovariectomy significantly reduced cell proliferation [72], whereas long-term ovariectomy was without effect [73] suggesting a compensatory mechanism potentially involving extragonadal estrogen synthesis [74]. On the other hand, both short-term and long-term ovariectomy decreased cell survival in the dentate gyrus [72, 75, 76]. It is important to note that removal of ovaries leads to loss of both estrogen and P_4. Thus, changes in neurogenesis and neuron survival following ovariectomy could be the result of depletion of one or both of these hormones.

17.2.2
Estrogen Regulation of Adult Neurogenesis

While the number of factors that regulate neurogenesis grows, gonadal hormones, and in particular E_2, are among those factors for which there is substantial evidence documenting their neurogenic efficacy and functional relevance (Figure 17.1). *In vivo*, E_2 increased the proliferation of neural progenitor cells in the dentate gyrus

SGZ of ovariectomized rats and in the intact rat during the estrus cycle during proestrus, when ovarian hormone levels are highest, as compared to estrus and diestrus [72]. In rats, the neurogenic effect of E_2 is vulnerable to duration of ovarian hormone deprivation and regimen of hormone replacement. Prolonged absence of ovarian hormones was associated with a loss of neurogenic response to E_2, whether administered chronically or cyclically, and a decrease in the number of new cells expressing a neuronal phenotype [77]. These data suggest that prolonged deprivation of E_2 leads to diminished responsiveness to E_2 and a concomitant decline in neuron production [77]. These findings in a rodent model would suggest disturbing implications for neurogenesis in women deprived of ovarian hormones for an extended period of time. It remains to be determined whether the same extended window of estrogen responsivity in morphogenesis [78] applies to neurogenesis.

The question immediately arises as to the biological significance of generating new granule cells in the dentate gyrus. While the role of neurogenesis in learning and memory continues to be debated, increasing evidence, initially from Shors and colleagues [25], indicates that newly generated neurons in the dentate gyrus contribute to the association of stimuli that are separated in time, a function that is a hippocampal-dependent type of associative learning and memory [79–87]. As the dentate gyrus is the first region of the hippocampus that receives and integrates sensory information via the perforant path [40], expanding the temporal window of associating information from multiple inputs should result in greater integration of information over time. Consistent with this postulate, Aimone and Gage suggest that the neural circuits generated by neurogenesis lead to the formation of temporal clusters of long-term episodic memories [23].

Estrogen-induced neurogenesis is also potentially relevant to disease states. In an animal model of ischemic stroke, Wise and colleagues reported that E_2 enhanced neurogenesis in a mouse model of ischemia [88]. Interestingly, the neurogenic effect of E2 was limited to the dorsal region of the SVZ and was absent from the ventral SVZ. Both estrogen receptor α and β (ERα and ERβ) contributed to the generation of new neurons; however, ERα contributed to a greater extent than ERβ [88]. The greater contribution of ERα to E_2-induced neural proliferation observed in the ischemic rodents [88] is in contrast to the near-equal contribution in normal rats [89]. The ERα preference could be due to the higher level of ERα expression in the ischemia model, in which there is a two- to threefold increase in ERα, whereas there is a decrease in ERβ expression (reviewed in [90]). In a different disease model, E_2 restored neurogenesis in the SGZ and SVZ of the chronic diabetic mouse brain [91].

To pursue the relevance of E_2-induced neurogenesis in the human realm, we investigated E_2-regulated proliferation of human neural progenitor cells (hNPCs) [92]. E_2 induced a significant increase in hNPC proliferation in a time- and dose-dependent manner. E_2-induced hNPC DNA replication was paralleled by elevated cell cycle protein expression and centrosome amplification, which was associated with augmentation of total cell number [92]. To determine whether ERs and which ER subtype were required for E_2-induced hNPC proliferation, ER expression was first determined by real-time RT-PCR followed by Western blot analysis and subsequently verified pharmacologically using ERα- or ERβ-selective ligands. ERβ expression was predominant relative to ERα, which was barely detectable in hNPC. Activation of ERβ by the ERβ-selective ligand, diarylpropionitrile, led to an increase in pERK, centrosome amplification, and hNPC proliferation, which were blocked by the MEKK antagonist. The efficacy of E_2 as a neurogenic factor was comparable to that induced by bFGF plus heparin [8]. Our findings of ERβ-mediated proliferation of human cortical neural progenitor cells are relevant to findings in ERβ knockout mice. ERβ appears to play a major role in neurogenesis and brain development [92, 93]. In ERβ knockout mice, brain size was smaller and fewer neurons were observed [93]. Expression of ERβ in human embryonic brain cells suggests a comparable role of ERβ in human brain development. Estrogen-inducible neurogenesis and the role of ERα and ERβ in regulating neural progenitor proliferation, particularly in the nonhuman primate and human brain, remain to be fully characterized.

17.2.3
Progestagen Regulation of Adult Neurogenesis

All progestagens are produced in the first steps of the sterodogenic pathway and characterized by their basic pregnane skeleton (C21). Members of this group include, but are not limited to, PREG, P_4, 5α-dihydroprogesterone, APα, 17α-hydroxypregnenolone, and 17α-hydroxyprogesterone. As both PREG and its sulfated forms PREGS and APα are neurosteroids, these molecules are considered under the neurosteroid section of this review. Emerging data indicate that progestagens have multiple nonreproductive functions in the central nervous system to regulate cognition, mood, inflammation, mitochondrial function, neurogenesis and regeneration, myelination, and recovery from traumatic brain injury [41]. Although the exact mode of action via which progestagens regulate neurogenesis remains to be unequivocally demonstrated, it is well accepted now that progestagens could, either by themselves or in concert with estrogens, modulate the neurogenic process [41]. The outcome of exposure varies depending on multiple factors including species, age, gender, the regimen of treatment, and disease status [41]. A number of progestagens, including PREG, its derivatives PREGS, P_4, and APα, have been reported to play significant roles in adult neurogenesis [7–9, 94–96] and their effects will be discussed below in the order of their position in the steroidogenic pathway. Notably, brain levels of those progestagens are closely associated with age-related cognitive decline and neurodegenerative disease progression, including Alzheimer's disease (AD) [7–9, 97–99]. Age-related decline on a hippocampus-dependent task is positively correlated with levels of concentration of PREGS, P_4, and APα in the brain; such decline can be reversed by manipulations of these progestagens [41, 97], suggesting that they may play critical roles in cognition and could potentially serve as a regenerative therapeutic intervention for neurological diseases.

17.2.4
Progesterone Regulation of Adult Neurogenesis

P_4 can promote cell proliferation in the breast [100–103], and has both inhibitory and stimulatory effects in the uterus depending upon cell type, the regimen of treatment, whether PR-A or PR-B is expressed, the dose of 17β-estradiol (E_2) and P_4, and when in the cycle P_4 is administered [41]. As in the uterus, P_4 regulation of mitosis in the nervous system is complex. Results from multiple laboratories have indicated that P_4 regulates neural cell proliferation in both the peripheral and the central nervous systems [95, 104–107]. A 24-h exposure of embryonic day 18 (E18) rat hippocampal neurons to P_4 alone induced a 20% increase in [^{3}H] thymidine uptake [95]; a phenomenon similar to that was observed after APα exposure [8], indicating that P_4 alone may promote hippocampal progenitor cell proliferation and thereby act as a neurogenic agent (Figure 17.1). Further analyses indicated that P_4 significantly increased adult neural progenitor cell proliferation in a dose-dependent manner [95]. This effect was mediated by the progesterone receptor membrane component 1 (PGRMC1) and not dependent on its conversion to metabolites [95]. In contrast, P_4 induced an $\sim$20% increase in [^{3}H] thymidine incorporation in polysialylated form of the neural cell adhesion molecule (PSA-NCAM)-positive progenitors derived from rat brain [108] and this effect required the conversion from P_4 to APα.

Differences in P_4 modes of action may be due to a number of factors, including differences in experimental model (adult neural progenitors vs embryonic neural progenitors) or to different cell populations (granular cells vs oligodendrocyte precursor cells). The *in vivo* ability of P_4 to regulate adult neurogenesis remains to be further elucidated, and recent studies from our laboratory indicated that single injection of P_4 could also increase BrdU incorporation in the brain from three month old OVXed rats [109, 110]. Tanapat *et al.* have shown that exposure to P_4 antagonized E_2-induced enhancement of cell proliferation following an initial increase [77]. Independent of its influence on cell proliferation, recent data confirmed that P_4 promotes the survival of newborn neurons in the dentate gyrus of adult male mice [111]. Thus, changes in P_4 biosynthesis and its receptor expression may affect neurogenesis, including neuronal proliferation and survival; the potential effect of P_4 on neuronal cell migration and functional integration remains undetermined.

17.2.5
Clinical Progestin Regulation of Neurogenesis

As described above, P_4 promoted rat neural progenitor cell (rNPC) proliferation and the concomitant regulation of mitotic cell cycle genes via the PGRMC/ERK pathway [95]. On the basis of these findings, we investigated the efficacy of eight clinically relevant hormone therapies and contraceptive progestins to promote rNPC proliferation and hippocampal cell viability *in vitro* and *in vivo*. Results of these analyses indicated that P_4, norgestimate, Nestorone, norethynodrel, norethindrone, and levonorgestrel induced a significant increase in rNPC proliferation *in vitro*, whereas norethindrone acetate was without effect and medroxyprogesterone acetate suppressed rNPC proliferation [109, 110]. All rNPC proliferative progestins were neuroprotective *in vitro*. P_4, Nestorone, levonorgestrel, and medroxyprogesterone acetate were selected for analysis of hippocampal rNPC proliferation *in vivo* in three month old ovariectomized Sprague-Dawley rats. Nestorone significantly increased protein expression of PCNA and CDC2, two well-documented mitotic markers, while P_4 increased CDC2, whereas levonorgestrel and MPA had no significant effect. Nestorone and P_4 induced a significant increase in the total number of hippocampal BrdU-positive cells as determined by FACS. Complex V (ATP synthase) subunit α (CVα) expression was used as an indicator of *in vivo* cell viability. P_4, Nestorone, and levonorgestrel significantly increased CVα expression, while medroxyprogesterone acetate had no effect. These results indicate that select clinical progestins, in particular P4 and Nestorone, exert proliferative and neuroprotective effects in the brain that belie their antiproliferative action in the uterus. These basic science findings are relevant when considering the impact of the progestins used in hormone therapy and contraception on the regenerative function of the brain to sustain neurological function in premenopausal and postmenopausal women [109, 110].

17.2.6
Androgen Regulation of Adult Neurogenesis

Androgen control of neurogenesis was the first of the gonadal hormones discovered to promote neurogenesis in the brain. Decades of research on song learning in birds was conducted by Fernando Nottlebaum and his many colleagues; he was the first to demonstrate that neurogenesis occurs in the vertebrate brain, that generation of new neurons can be a determinant of new learning, and that failure to generate new neurons leads to a failure to learn song [19]. In contrast, androgen regulation of neurogenesis in the mammalian brain is not as well studied. Among the major androgens in mammalians – testosterone, dihydrotestosterone (DHT), and DHEA – testosterone and DHEA can be produced peripherally and centrally [35, 112–114], while DHT is converted from testosterone within peripheral tissues [115]. Despite differences in location of synthesis, all three androgens can exert effects on CNS function and with some evidence indicating a role in regulating adult neurogenesis [112, 116, 117]. Compared to other gonadal hormones, far fewer experiments have directly manipulated androgens to determine their effect on adult neurogenesis. Instead, much of the evidence that androgens may influence adult neurogenesis is inferred based on association of androgen level and neurogenesis level [94].

17.2.7
Testosterone and DHT Regulation of Adult Neurogenesis

Endogenous testosterone was first shown to influence adult neurogenesis when the association between seasonal testosterone level changes and seasonal neurogenesis changes was discovered [94]. Specifically, lower testosterone induced cell death within the avian high vocal center (HVC), providing vacancies for newly proliferated cells to grow into [116]. Meanwhile, increased testosterone level contributes to cell survival without affecting cell proliferation, as evidenced by the enhanced survival of ^{3}H-thymidine-labeled cells in the dentate gyrus of reproductively active male meadow voles (when testosterone levels were high) when compared with reproductively inactive male meadow voles (when testosterone levels are low) [118]. Indeed, males with higher testosterone levels are found with larger hippocampal volumes [71], probably a result of testosterone's effect on cell survival within the dentate gyrus.

Exogenous application of testosterone indicated that, consistent with the effect of endogenous testosterone, androgens enhance cell survival but have no significant effect on cell proliferation in the dentate gyrus [119]. Interestingly, testosterone implants enhanced cell proliferation within the amygdala in castrated male voles, a phenomenon dependent on the conversion of testosterone into estrogen [119]. When compared with intact male rats, castration in adult male rats causes a decrease in cell survival without affecting cell proliferation within the dentate gyrus, and this decrease could be reversed after 30 days of administering testosterone injections [94]. This phenomenon was shown to be solely dependent on the androgen system, as 30 days of injections of DHT but not estradiol resulted in a significant increase in hippocampal neurogenesis [120]. Further, studies in starlings and canaries confirmed that testosterone exerts similar effects on cell survival in the ventricular zone (SVZ) [121, 122], while having no significant effect on cell proliferation in the the same region, [123] in adult female canaries (Figure 17.1).

17.2.8
DHEA Regulation of Adult Neurogenesis

Unlike testosterone and DHT, DHEA has been reported to increase both cell proliferation and survival [112, 117]. As a neurosteroid [35], DHEA could promote cell survival in rat embryonic neural progenitor cells by activating the serine–threonine protein kinase Akt pathway, a well-documented cell survival signaling pathway [124]. Sulfation of DHEA to DHEAS abolished this prosurvival effect and increased apoptosis [124], suggesting that the balance between these two neurosteroids may play a critical role in CNS development and plasticity. Indeed, in embryonic cortical neurons, DHEA was shown to stimulate axon growth but not dendrite growth, while DHEAS-induced neurite growth showed a preference to dendrite over axon [125]. It is unclear, however, whether this effect is applicable to adult neural progenitor cells. DHEA could also promote cell proliferation and survival in cultures of human neural stem cells [112], an effect that requires the existence of both the epidermal growth factor and the leukemia inhibitory factor. In addition, this effect is mediated through the NMDA and sigma-1 receptor [112].

In vivo, DHEA implants given to intact male rats were found to increase cell proliferation and cell survival (28 days) in the dentate gyrus [117]. Furthermore,

DHEA prevented the corticosterone-induced suppressing effects on cell proliferation in adult male rats [117]. Interestingly, the synergy between DHEA and the antidepressant-selective serotonin-uptake inhibitor, fluoxetine, was found on proliferation of progenitor cells in the dentate gyrus of the adult male rat [126].

17.3
Neurosteroid Regulation of Adult Neurogenesis

17.3.1
Pregnenolone and Pregnenolone Sulfate Regulation of Adult Neurogenesis

While the identification of PREG in the central nervous system is well accepted, the classification of PREGS as a neurosteroid is still under debate [127]. Multiple functions for PREG and PREGS in the nervous system have been proposed based on both *in vitro* and *in vivo* studies [127]. PREG promotes neurite growth and this function is achieved via binding to microtubule-associated protein 2 (MAP2) to control microtubule formation [128, 129]. On the other hand, PREGS can increase memory retention and cognitive performance in rodents, an effect dependent upon its ability to potentiate and antagonize NMDA/AMPA receptors and $GABA_A$ receptors, respectively [127].

Their roles in neurogenesis, however, are less well studied. PREGS was reported to decrease cell proliferation of primary culture neural progenitor cells obtained from both E18 [8] and adult rat hippocampus [95], consistent with its function of promoting morphological differentiation by increasing neurite growth in both early neuronal development and adult neural plasticity [128, 129]. In contrast, recent studies found that PREGS promotes proliferation in neurospheres derived from SVZ in adult mice [130]. PREGS administration also increased cell proliferation in the dentate gyrus and SVZ in both young and aged male rats, and increased cell survival in the young adult rats; longer periods of survival were not assessed in aged animals [130] (Figure 17.1). Interestingly, although controversial, PREGS effects in both studies were shown to be dependent on its modulatory effect on the $GABA_A$ receptor [8, 130]. The discrepancies across studies could be explained by differences between the model systems (embryonic vs adult hippocampal neurons, SGZ vs SVZ), endpoint measurements (BrdU incorporation vs neurosphere diameter), and treatment regimens (24 h vs 48 h); alternatively, the effects of PREG may be due to the activation of a regulatory mechanism that normalizes transient imbalances in neurogenesis following introduction of an extrinsic factor [131]. It is proposed that two distinct populations of proliferative cells exist within the neurogenic niche: the progenitor steady-state population and a replenishing stem cell population [131]. Equilibrium exists between these two cell populations, and modifications in one will lead to a compensatory change in the other [131]. Hence, it is possible that a compensatory increase in the stem cell population occurred after an initial decrease in the progenitor population following PREGS administration. The exact effect of

PREGS on neurogenesis and its underlying mechanism, however, remains to be determined.

17.3.2
Allopregnanolone Regulation of Adult Neurogenesis

APα is synthesized from P_4 via two enzymatic reactions: 5α reduction by 5α reductase, followed by 3α reduction of the C3 ketone by 3α-HSD. Unlike P_4, APα mediates its effects through modulation of the $GABA_A$ receptor [41] and the nuclear pregnane-X receptor (PXR) [53, 127]. Very early findings from our group indicated that *in vitro*, APα induced regression of neurites from hippocampal neurons that have not yet made contact with other neurons or glia, but had no effect on neurites that had existing neural connectivity in culture [7]. This regression of neurites was later determined to be a prelude to mitosis. Our follow-up analyses demonstrated that APα increased cell proliferation in rat embryonic hippocampal progenitors, as well as in human embryonic progenitor cells (Figure 17.1). APα-induced proliferation was a dose-dependent process with concentrations within the 10^{-9} to mid-10^{-7} M range promoting proliferation, while concentrations in excess of 10^{-6} M significantly inhibiting neurogenesis. The biphasic effect of APα on neurogenesis is consistent with other studies in which nanomolar levels of APα increased while micromolar levels of APα inhibited the proliferation of PSA-NCAM-positive neural progenitors [132]. At high concentrations (i.e., micromolar), APα can be converted by 20α-HSD to allopregnanediol (5α-pregnan-3α,20α-diol) [133] and hence may increase the local concentration of allopregnanediol that inhibited DNA replication of rNPC, thereby inducing a biphasic dose response.

Mechanistically, APα-induced proliferation was mediated via $GABA_A$ receptor-activated voltage-gated L-type Ca^{2+} channels [8] leading to a rapid rise in intracellular calcium in neural progenitors [134]. In neural progenitor cells, the high intracellular chloride content leads to an *efflux* of chloride through the $GABA_A$ channels upon opening, which leads to depolarization of the membrane and influx of Ca^{2+} through voltage-dependent L-type Ca^{2+} channels and activation of the CREB transcription factor [96, 134, 135]. Through this pathway, APα stimulation of GABA-mediated excitation and CREB signaling can activate a key pathway in adult hippocampal neurogenesis, to promote the proliferation, survival, and differentiation of neural progenitor cells. While this APα-activated pathway is relevant to the induction of NPC proliferation, the mechanisms by which APα promotes survival of rNPCs, which could involve either delayed or prolonged actions of APα on gene expression and neuron survival mechanisms, remain to be determined.

Others have reported similar findings of APα stimulating cerebellar granule cell proliferation *in vitro* [136]. In addition, it is found that i.c.v. administration of APα in young adult male rats leads to a decreased cell proliferation in the dentate gyrus 44 h later [130]. The suppression of *in vivo* proliferation by APα could be explained by the dose-dependent relationship described by Wang *et al.* [8], as the high concentration of APα (6.3 nmol 7 µl^{-1}) used in the *in vivo* experiment falls into the dose range where inhibitory effects occurred. Indeed, results from recent studies have shown that i.v. injection of APα, when given at a dose leading to brain APα level comparable to that *in vitro*, actually increased cell proliferation in both SGZ and SVZ areas in a transgenic mouse model of AD [96, 109, 110].

On the basis of our findings *in vitro*, we investigated the efficacy of APα to promote neurogenesis in the hippocampal SGZ, to reverse learning and memory deficits in the three month old male triple transgenic mouse model of Alzheimer's (3xTgAD) and the correlation between APα-induced neural progenitor cell survival and memory function in 3xTgAD mice. Neural progenitor cell proliferation was determined by unbiased stereological analysis of BrdU incorporation and survival determined by FACS for BrdU$^+$ cells. Learning and memory function was assessed using the hippocampal-dependent trace eye-blink conditioning paradigm. At three months, basal level of BrdU$^+$ cells in the SGZ of 3xTgAD mice was significantly lower relative to non-Tg mice, despite the lack of evident AD pathology. APα significantly increased, in a dose-dependent manner, BrdU$^+$ cells in SGZ in 3xTgAD mice and restored SGZ proliferation to normal magnitude. As with the deficit in proliferation, 3xTgAD mice exhibited deficits in learning and memory. APα reversed the cognitive deficits to restore learning and memory performance to the level of normal non-Tg mice. In 3xTgAD mice, APα-induced survival of neural progenitors was significantly correlated with APα-induced memory performance (Figure 17.1). These results suggest that early neurogenic deficits, which were evident prior to immunodetectable Aβ, may contribute to the cognitive phenotype of AD. Importantly, these results demonstrate that APα reversed deficits in SGZ neurogenesis, learning, and memory in the 3xTgAD mouse model of AD to restore both regenerative and cognitive functions to that of the normal nontransgenic mouse. From a translational perspective, blood–brain penetrance of APα, efficacy of a single exposure, and the mechanism of APα action make it an ideal molecule and a promising regenerative therapeutic candidate for promoting neural regeneration and reversing cognitive deficits associated with the prodromal stage of AD.

17.4
Gonadal Steroids, Clinical Progestins, and Neurosteroids as Neuroregenerative Therapeutics: Challenges and Strategies

While the therapeutic potential of neural stem cells is great, so too are the associated challenges. Neural stem therapy for neurodegenerative disease would presumably target a neurogenic deficit or enhancement of normal neurogenesis to restore lost neural tissue. In either case, multiple variables beyond a neurogenic deficit will regulate and determine efficacy of neural stem therapy. For example, regeneration of brain in AD requires that neural stem/progenitor cell proliferation, migration, differentiation, integration into neural circuits, and ultimately functioning, occur in a brain that has typically undergone a protracted process of degeneration. Confronting the efficacy of a regenerative therapeutic is the degenerative environment, which is characterized by neuronal loss, physical plague and glial scar barriers, and inflammation. A regenerative therapeutic strategy must address the challenge of regenerating neural circuits in a brain that is in various states of degeneration. In the following paragraphs, AD will be used as the example to discuss the potentials and challenges of utilizing stem cell therapy as a potential treatment.

17.4.1
Targeting Neurogenesis as a Treatment for Neurodegenerative Disease

AD is characterized by a progressive development of pathology, which ultimately becomes widely distributed with neuronal death occurring in multiple brain regions. In addition to these sites, the late-stage neuron loss of cholinergic neurons most probably due to the loss of trophic survival factors retrogradely transported from the hippocampus to cholinergic neurons of the nucleus of Maynert [137, 138] and locus coeruleus [139], are well described. Adding to the spatial complexity of neuronal loss is the phenotypic diversity of neurons targeted for demise. The spatial and phenotypic diversity of degeneration in AD predicts that a multipotent neural stem or progenitor cell population will be required for regenerative therapeutic efficacy. A more recently described pathological change in the AD brain is aberrant entry into the cell cycle, which ultimately leads to apoptotic cell death [140–143]. Aberrant entry into the cell cycle has been found in four regions of the AD brain; the hippocampus, subiculum, locus coeruleus, and dorsal raphe nucleus [144].

In addition to the phenotype diversity of degenerating systems is the topographical landscape of the degenerating terrain. The degenerative phenotype of AD is characterized by an increased number of neuritic plaques and neurofibrillary tangles in the hippocampus and cerebral cortex [145]. The former (neuritic plaques) are composed of tortuous neuritic processes surrounding a central amyloid (Aβ) core. The latter is characterized by the abnormal hyperphosphorylation and accumulation of tau protein in neurons and, less commonly, in astrocytes. The neurofibrillary tangles, formed by abnormal hyperphosphorylated tau, are frequently seen in AD brains and accompanied by neuronal loss and gliosis [146]. Inflammation also plays an important role in pathogenesis of neurodegenerative disorders including AD [147]. In AD brains, compacted Aβ plaques are often associated with activated astrocytes and microglia and a variety of cytokines and other inflammatory proteins secreted by activated astrocytes or microglia, including C1q, C3, C9, C3d, and C4d, which are found in brains from human AD patients and mouse models of AD [148]. Thus, the regenerative stem cell population must survive and traverse a landscape riddled with degenerative debris and replete with a biochemical cauldron of inflammatory, cellular stress, and defense molecular signaling.

Further, while altered patterns of adult neurogenesis appear to be a feature shared among different neurodegenerative disorders, and neural stem cells in neurogenic areas can respond to injury [149, 150], it remains to be elucidated whether increased neurogenesis leads to regeneration of the comprised neuronal network [6]. Most disturbing of all, Herrup and colleagues found that cell cycle events precede neuronal death in the cortex and CA3 regions at all stages of AD, from MCI to late-stage AD and within AD mouse models [140]. Expression of the ectopic cell cycle proteins ultimately predicts the demise of these neurons [140]. Further support for the aberrant entry into the cell cycle and cell death in AD are findings indicating that mutants of APP known to cause familial AD also lead to apoptosis and DNA synthesis [151, 152]. These findings are especially challenging for therapeutics targeting the regenerative potential of the endogenous neural stem/progenitor populations, as an unintended side effect may be to promote ectopic entry of neurons into the cell cycle and thereby exacerbate neuron demise.

Last but far from least, a regenerative therapy that targets neurogenesis may not address the fundamental problems imposed by neurodegenerative diseases such as the pathology associated with Alzheimer's, Parkinson's, or multiple sclerosis to name but a few. For example, newly generated neurons in the adult brain are unlikely to alleviate the burden of pathology (e.g., $A\beta$ plaques/pTau tangles) but will instead most probably become a target of the degenerative process. As bleak as this situation appears, preclinical evidence indicates that single neurosteroid molecules can promote both increased neurogenesis and decreased AD pathology. In that regard, APα is a promising candidate for a neuroregenerative therapeutics that both promotes neurogenesis and reduces AD pathology.

17.4.2
APα as a Regenerative Factor to Promote Functional Neurogenesis and Diminish Alzheimer's Pathology

As discussed above, APα promotes neural progenitor proliferation ranged in human and rodent neural progenitor cells [8]. Results of our *in vivo* analyses demonstrated that APα significantly increased SGZ neurogenesis in 3xTgAD mice and restored SGZ proliferation to normal magnitude. APα reversed the cognitive deficits of 3xTgAD mice to restore learning and memory performance to the level of normal non-Tg mice. In 3xTgAD mice, APα-induced survival of neural progenitors was significantly correlated with APα-induced memory performance [96]. We then went on to determine the efficacy of APα in preventing or delaying progression of AD pathology; APα, $10\,\mathrm{mg}^{-1}\,\mathrm{kg}^{-1}\,\mathrm{week}^{-1}$, was initiated either prior to $A\beta$ appearance at three months of age or following robust intraneuronal $A\beta$ accumulation at six months of age [153]. APα induced greatest decline in $A\beta$ immunoreactivity in the CA1 region of the hippocampus and frontal cortex, with amygdala and subiculum exhibiting a lesser magnitude of decline. In parallel to the decline in $A\beta$ expression, β amyloid-binding alcohol dehydrogenase (ABAD), a mitochondrial $A\beta$-binding enzyme was reduced by APα treatment [153]. APα induced a modest reduction in phosphorylated tau. In addition, APα treatment significantly reduced microglia expression. Initial mechanistic analyses indicate that APα when administered prior to development of pathology, induces LXR, PXR, and HMG-CoA reductase expression, whereas administration subsequent to intraneuronal $A\beta$ results in decreased expression of both LXR and PXR and no effect on HmG-CoA reductase. Collectively, these data indicate that long-term once-a-week APα treatment prior to expression of $A\beta$ results in a significant reduction of AD pathology and a concomitant rise in cholesterol regulatory pathways. Treatment subsequent to early-stage AD pathology development results in a diminution of therapeutic efficacy. These findings together with the regenerative capacity provide preclinical evidence for APα as a disease-modifying therapeutic agent to delay and diminish AD pathology progression. It is unclear at this point whether the neurogenic and pathology-modifying efficacy of APα generalizes to other neurodegenerative diseases, but analysis of APα efficacy in Parkinson's disease model is underway.

To conclude, gonadal hormones, neurosteroids, and clinical progestins exert significant effects on the neurogenic and regenerative capacity of the developing and adult brains (Figure 17.1). Caveats to the efficacy of these molecules are their structure–function relationships, dose–response profiles, and the nature of the regenerative response itself. Activating and regulating the regenerative capacity of the brain requires an understanding of the cellular regenerative process, which requires a therapeutic strategy quite different from that which targets synaptic transmission. The multifactorial nature of select steroid hormones to promote cellular regeneration in the brain [40, 95, 96] while simultaneously modifying disease pathology [153–155] make them promising therapeutic candidates to promote neurogenesis and prevent or delay neurodegenerative disease.

References

1. Zhao, C., Deng, W., and Gage, F.H. (2008) Mechanisms and functional implications of adult neurogenesis. *Cell*, **132**, 645–660.
2. Arnold, A.P. and Breedlove, S.M. (1985) Organizational and activational effects of sex steroids on brain and behavior: a reanalysis. *Horm. Behav.*, **19**, 469–498.
3. McEwen, B.S., Davis, P.G., Parsons, B., and Pfaff, D.W. (1979) The brain as a target for steroid hormone action. *Annu. Rev. Neurosci.*, **2**, 65–112.
4. Pfaff, D.W. and McEwen, B.S. (1983) Actions of estrogens and progestins on nerve cells. *Science*, **219**, 808–814.
5. Nottebohm, F. (2005) The neural basis of birdsong. *PLoS Biol.*, **3**, e164.
6. Lie, D.C., Song, H., Colamarino, S.A., Ming, G.L., and Gage, F.H. (2004) Neurogenesis in the adult brain: new strategies for central nervous system diseases. *Annu. Rev. Pharmacol. Toxicol.*, **44**, 399–421.
7. Brinton, R.D. (1994) The neurosteroid 3 alpha-hydroxy-5 alpha-pregnan-20-one induces cytoarchitectural regression in cultured fetal hippocampal neurons. *J. Neurosci.*, **14**, 2763–2774.
8. Wang, J.M., Johnston, P.B., Ball, B.G., and Brinton, R.D. (2005) The neurosteroid allopregnanolone promotes proliferation of rodent and human neural progenitor cells and regulates cell-cycle gene and protein expression. *J. Neurosci.*, **25**, 4706–4718.
9. Wang, J.M., Liu, L., Irwin, R.W., Chen, S., and Brinton, R.D. (2008) Regenerative potential of allopregnanolone. *Brain Res. Rev.*, **57**, 398–409.
10. Altman, J. and Das, G.D. (1965) Post-natal origin of microneurones in the rat brain. *Nature*, **207**, 953–956.
11. Kaplan, M.S. and Hinds, J.W. (1977) Neurogenesis in the adult rat: electron microscopic analysis of light radioautographs. *Science*, **197**, 1092–1094.
12. Gage, F.H. (2000) Mammalian neural stem cells. *Science*, **287**, 1433–1438.
13. Kempermann, G. and Gage, F.H. (2000) Neurogenesis in the adult hippocampus. *Novartis Found. Symp.*, **231**, 220–235; discussion 235–241, 302–226.
14. Nottebohm, F. (1980) Testosterone triggers growth of brain vocal control nuclei in adult female canaries. *Brain Res.*, **189**, 429–436.

15. Alvarez-Buylla, A. and Nottebohm, F. (1988) Migration of young neurons in adult avian brain. *Nature*, **335**, 353–354.
16. Kornack, D.R. and Rakic, P. (2001) Cell proliferation without neurogenesis in adult primate neocortex. *Science*, **294**, 2127–2130.
17. Kornack, D.R. and Rakic, P. (2001) The generation, migration, and differentiation of olfactory neurons in the adult primate brain. *Proc. Natl. Acad. Sci. U.S.A.*, **98**, 4752–4757.
18. Eriksson, P.S., Perfilieva, E., Bjork-Eriksson, T., Alborn, A.M., Nordborg, C., Peterson, D.A., and Gage, F.H. (1998) Neurogenesis in the adult human hippocampus. *Nat. Med.*, **4**, 1313–1317.
19. Alvarez-Buylla, A., Theelen, M., and Nottebohm, F. (1990) Proliferation "hot spots" in adult avian ventricular zone reveal radial cell division. *Neuron*, **5**, 101–109.
20. Alvarez-Buylla, A., Garcia-Verdugo, J.M., and Tramontin, A.D. (2001) A unified hypothesis on the lineage of neural stem cells. *Nat. Rev. Neurosci.*, **2**, 287–293.
21. Ming, G.L. and Song, H. (2005) Adult neurogenesis in the mammalian central nervous system. *Annu. Rev. Neurosci.*, **28**, 223–250.
22. van Praag, H., Schinder, A.F., Christie, B.R., Toni, N., Palmer, T.D., and Gage, F.H. (2002) Functional neurogenesis in the adult hippocampus. *Nature*, **415**, 1030–1034.
23. Aimone, J.B., Wiles, J., and Gage, F.H. (2006) Potential role for adult neurogenesis in the encoding of time in new memories. *Nat. Neurosci.*, **9**, 723–727.
24. Shors, T.J., Mathew, J., Sisti, H.M., Edgecomb, C., Beckoff, S., and Dalla, C. (2007) Neurogenesis and helplessness are mediated by controllability in males but not in females. *Biol. Psychiatry*, **62**, 487–495.
25. Sisti, H.M., Glass, A.L., and Shors, T.J. (2007) Neurogenesis and the spacing effect: learning over time enhances memory and the survival of new neurons. *Learn. Mem.*, **14**, 368–375.
26. Lledo, P.M., Alonso, M., and Grubb, M.S. (2006) Adult neurogenesis and functional plasticity in neuronal circuits. *Nat. Rev. Neurosci.*, **7**, 179–193.
27. Parent, J.M. and Murphy, G.G. (2008) Mechanisms and functional significance of aberrant seizure-induced hippocampal neurogenesis. *Epilepsia*, **49** (Suppl 5), 19–25.
28. Banasr, M. and Duman, R.S. (2007) Regulation of neurogenesis and gliogenesis by stress and antidepressant treatment. *CNS Neurol. Disord. Drug Targets*, **6**, 311–320.
29. Lindvall, O., Kokaia, Z., and Martinez-Serrano, A. (2004) Stem cell therapy for human neurodegenerative disorders-how to make it work. *Nat. Med.*, **10**, S42–S50.
30. Steiner, B., Wolf, S., and Kempermann, G. (2006) Adult neurogenesis and neurodegenerative disease. *Regen. Med.*, **1**, 15–28.
31. Kempermann, G., Wiskott, L., and Gage, F.H. (2004) Functional significance of adult neurogenesis. *Curr. Opin. Neurobiol.*, **14**, 186–191.
32. Nissant, A., Bardy, C., Katagiri, H., Murray, K., and Lledo, P.M. (2009) Adult neurogenesis promotes synaptic plasticity in the olfactory bulb. *Nat. Neurosci.*, **12**, 728–730.
33. Lledo, P.M., Merkle, F.T., and Alvarez-Buylla, A. (2008) Origin and function of olfactory bulb interneuron diversity. *Trends Neurosci.*, **31**, 392–400.
34. Pozniak, C.D. and Pleasure, S.J. (2006) A tale of two signals: Wnt and Hedgehog in dentate neurogenesis. *Sci. STKE*, **2006**, e5.
35. Baulieu, E.E. (1997) Neurosteroids: of the nervous system, by the nervous system, for the nervous system. *Recent Prog. Horm. Res.*, **52**, 1–32.
36. Mellon, S.H., Griffin, L.D., and Compagnone, N.A. (2001) Biosynthesis and action of neurosteroids. *Brain Res. Brain Res. Rev.*, **37**, 3–12.
37. Mellon, S.H. and Vaudry, H. (2001) Biosynthesis of neurosteroids and regulation of their synthesis. *Int. Rev. Neurobiol.*, **46**, 33–78.

38. McEwen, B. (2002) Estrogen actions throughout the brain. *Recent Prog. Horm. Res.*, **57**, 357–384.
39. Gould, E., McEwen, B.S., Tanapat, P., Galea, L.A., and Fuchs, E. (1997) Neurogenesis in the dentate gyrus of the adult tree shrew is regulated by psychosocial stress and NMDA receptor activation. *J. Neurosci.*, **17**, 2492–2498.
40. Brinton, R.D. (2009) Estrogen-induced plasticity from cells to circuits: predictions for cognitive function. *Trends Pharmacol. Sci.*, **30**, 212–222.
41. Brinton, R.D., Thompson, R.F., Foy, M.R., Baudry, M., Wang, J., Finch, C.E., Morgan, T.E., Pike, C.J., Mack, W.J., Stanczyk, F.Z., and Nilsen, J. (2008) Progesterone receptors: form and function in brain. *Front Neuroendocrinol.*, **29**, 313–339.
42. Fricker-Gates, R.A., Lundberg, C., and Dunnett, S.B. (2001) Neural transplantation: restoring complex circuitry in the striatum. *Restor. Neurol. Neurosci.*, **19**, 119–138.
43. Bjorklund, A. and Lindvall, O. (2000) Cell replacement therapies for central nervous system disorders. *Nat. Neurosci.*, **3**, 537–544.
44. Lindvall, O. and Hagell, P. (2001) Cell therapy and transplantation in Parkinson's disease. *Clin. Chem. Lab. Med.*, **39**, 356–361.
45. Baulieu, E.E., Robel, P., and Schumacher, M. (2001) Neurosteroids: beginning of the story. *Int. Rev. Neurobiol.*, **46**, 1–32.
46. Belelli, D., Herd, M.B., Mitchell, E.A., Peden, D.R., Vardy, A.W., Gentet, L., and Lambert, J.J. (2006) Neuroactive steroids and inhibitory neurotransmission: mechanisms of action and physiological relevance. *Neuroscience*, **138**, 821–829.
47. Schumacher, M., Guennoun, R., Ghoumari, A., Massaad, C., Robert, F., El-Etr, M., Akwa, Y., Rajkowski, K., and Baulieu, E.E. (2007) Novel perspectives for progesterone in hormone replacement therapy, with special reference to the nervous system. *Endocr. Rev.*, **28**, 387–439.
48. Charalampopoulos, I., Remboutsika, E., Margioris, A.N., and Gravanis, A. (2008) Neurosteroids as modulators of neurogenesis and neuronal survival. *Trends Endocrinol. Metab.*, **19**, 300–307.
49. Rune, G.M., Lohse, C., Prange-Kiel, J., Fester, L., and Frotscher, M. (2006) Synaptic plasticity in the hippocampus: effects of estrogen from the gonads or hippocampus? *Neurochem. Res.*, **31**, 145–155.
50. von Schassen, C., Fester, L., Prange-Kiel, J., Lohse, C., Huber, C., Bottner, M., and Rune, G.M. (2006) Oestrogen synthesis in the hippocampus: role in axon outgrowth. *J. Neuroendocrinol.*, **18**, 847–856.
51. Compagnone, N.A., Bulfone, A., Rubenstein, J.L., and Mellon, S.H. (1995) Steroidogenic enzyme P450c17 is expressed in the embryonic central nervous system. *Endocrinology*, **136**, 5212–5223.
52. Kibaly, C., Patte-Mensah, C., and Mensah-Nyagan, A.G. (2005) Molecular and neurochemical evidence for the biosynthesis of dehydroepiandrosterone in the adult rat spinal cord. *J. Neurochem.*, **93**, 1220–1230.
53. Mellon, S.H. and Griffin, L.D. (2002) Neurosteroids: biochemistry and clinical significance. *Trends Endocrinol. Metab.*, **13**, 35–43.
54. Cutolo, M., Sulli, A., Capellino, S., Villaggio, B., Montagna, P., Seriolo, B., and Straub, R.H. (2004) Sex hormones influence on the immune system: basic and clinical aspects in autoimmunity. *Lupus*, **13**, 635–638.
55. Gilliver, S.C. and Ashcroft, G.S. (2007) Sex steroids and cutaneous wound healing: the contrasting influences of estrogens and androgens. *Climacteric*, **10**, 276–288.
56. Dubey, R.K., Oparil, S., Imthurn, B., and Jackson, E.K. (2002) Sex hormones and hypertension. *Cardiovasc. Res.*, **53**, 688–708.
57. Krause, D.N., Duckles, S.P., and Pelligrino, D.A. (2006) Influence of sex steroid hormones on cerebrovascular function. *J. Appl. Physiol.*, **101**, 1252–1261.

58. Gavrilova-Jordan, L.P. and Price, T.M. (2007) Actions of steroids in mitochondria. *Semin Reprod. Med.*, **25**, 154–164.
59. Brinton, R.D. (2008) The healthy cell bias of estrogen action: mitochondrial bioenergetics and neurological implications. *Trends Neurosci.*, **31**, 529–537.
60. Mayes, J.S. and Watson, G.H. (2004) Direct effects of sex steroid hormones on adipose tissues and obesity. *Obes. Rev.*, **5**, 197–216.
61. Ng, M.K., Jessup, W., and Celermajer, D.S. (2001) Sex-related differences in the regulation of macrophage cholesterol metabolism. *Curr. Opin. Lipidol.*, **12**, 505–510.
62. Bland, R. (2000) Steroid hormone receptor expression and action in bone. *Clin. Sci. (Lond)*, **98**, 217–240.
63. Balasch, J. (2003) Sex steroids and bone: current perspectives. *Hum. Reprod. Update*, **9**, 207–222.
64. Foster, P.A. (2008) Steroid metabolism in breast cancer. *Minerva Endocrinol.*, **33**, 27–37.
65. Zheng, H., Kavanagh, J.J., Hu, W., Liao, Q., and Fu, S. (2007) Hormonal therapy in ovarian cancer. *Int. J. Gynecol. Cancer*, **17**, 325–338.
66. Kim, S.B., Hill, M., Kwak, Y.T., Hampl, R., Jo, D.H., and Morfin, R. (2003) Neurosteroids: Cerebrospinal fluid levels for Alzheimer's disease and vascular dementia diagnostics. *J. Clin. Endocrinol. Metab.*, **88**, 5199–5206.
67. Kriegstein, A. and Alvarez-Buylla, A. (2009) The glial nature of embryonic and adult neural stem cells. *Annu. Rev. Neurosci.*, **32**, 149–184.
68. Carleton, A., Petreanu, L.T., Lansford, R., Alvarez-Buylla, A., and Lledo, P.M. (2003) Becoming a new neuron in the adult olfactory bulb. *Nat. Neurosci.*, **6**, 507–518.
69. Carlen, M., Cassidy, R.M., Brismar, H., Smith, G.A., Enquist, L.W., and Frisen, J. (2002) Functional integration of adult-born neurons. *Curr. Biol.*, **12**, 606–608.
70. Markakis, E.A. and Gage, F.H. (1999) Adult-generated neurons in the dentate gyrus send axonal projections to field CA3 and are surrounded by synaptic vesicles. *J. Comp. Neurol.*, **406**, 449–460.
71. Galea, L.A. and McEwen, B.S. (1999) Sex and seasonal differences in the rate of cell proliferation in the dentate gyrus of adult wild meadow voles. *Neuroscience*, **89**, 955–964.
72. Tanapat, P., Hastings, N.B., Reeves, A.J., and Gould, E. (1999) Estrogen stimulates a transient increase in the number of new neurons in the dentate gyrus of the adult female rat. *J. Neurosci.*, **19**, 5792–5801.
73. Lagace, D.C., Fischer, S.J., and Eisch, A.J. (2007) Gender and endogenous levels of estradiol do not influence adult hippocampal neurogenesis in mice. *Hippocampus*, **17**, 175–180.
74. Fester, L., Ribeiro-Gouveia, V., Prange-Kiel, J., von Schassen, C., Bottner, M., Jarry, H., and Rune, G.M. (2006) Proliferation and apoptosis of hippocampal granule cells require local oestrogen synthesis. *J. Neurochem.*, **97**, 1136–1144.
75. Barker, J.M. and Galea, L.A. (2008) Repeated estradiol administration alters different aspects of neurogenesis and cell death in the hippocampus of female, but not male, rats. *Neuroscience*, **152**, 888–902.
76. Green, A.D. and Galea, L.A. (2008) Adult hippocampal cell proliferation is suppressed with estrogen withdrawal after a hormone-simulated pregnancy. *Horm. Behav.*, **54**, 203–211.
77. Tanapat, P., Hastings, N.B., and Gould, E. (2005) Ovarian steroids influence cell proliferation in the dentate gyrus of the adult female rat in a dose- and time-dependent manner. *J. Comp. Neurol.*, **481**, 252–265.
78. Morrison, J.H., Brinton, R.D., Schmidt, P.J., and Gore, A.C. (2006) Estrogen, menopause, and the aging brain: how basic neuroscience can inform hormone therapy in women. *J. Neurosci.*, **26**, 10332–10348.
79. Shors, T.J. (2008) From stem cells to grandmother cells: how neurogenesis relates to learning and memory. *Cell Stem Cell*, **3**, 253–258.

80. Leuner, B., Gould, E., and Shors, T.J. (2006) Is there a link between adult neurogenesis and learning? *Hippocampus*, **16**, 216–224.
81. Shors, T.J. (2004) Memory traces of trace memories: neurogenesis, synaptogenesis and awareness. *Trends Neurosci.*, **27**, 250–256.
82. Shors, T.J., Townsend, D.A., Zhao, M., Kozorovitskiy, Y., and Gould, E. (2002) Neurogenesis may relate to some but not all types of hippocampal-dependent learning. *Hippocampus*, **12**, 578–584.
83. Shors, T.J., Miesegaes, G., Beylin, A., Zhao, M., Rydel, T., and Gould, E. (2001) Neurogenesis in the adult is involved in the formation of trace memories. *Nature*, **410**, 372–376.
84. Aimone, J.B., Wiles, J., and Gage, F.H. (2009) Computational influence of adult neurogenesis on memory encoding. *Neuron*, **61**, 187–202.
85. Clelland, C.D., Choi, M., Romberg, C., Clemenson, G.D. Jr., Fragniere, A., Tyers, P., Jessberger, S., Saksida, L.M., Barker, R.A., Gage, F.H., and Bussey, T.J. (2009) A functional role for adult hippocampal neurogenesis in spatial pattern separation. *Science*, **325**, 210–213.
86. Jessberger, S., Clark, R.E., Broadbent, N.J., Clemenson, G.D. Jr., Consiglio, A., Lie, D.C., Squire, L.R., and Gage, F.H. (2009) Dentate gyrus-specific knockdown of adult neurogenesis impairs spatial and object recognition memory in adult rats. *Learn. Mem.*, **16**, 147–154.
87. Thuret, S., Toni, N., Aigner, S., Yeo, G.W., and Gage, F.H. (2009) Hippocampus-dependent learning is associated with adult neurogenesis in MRL/MpJ mice. *Hippocampus*, **19**, 658–669.
88. Suzuki, S., Gerhold, L.M., Bottner, M., Rau, S.W., Dela Cruz, C., Yang, E., Zhu, H., Yu, J., Cashion, A.B., Kindy, M.S., Merchenthaler, I., Gage, F.H., and Wise, P.M. (2007) Estradiol enhances neurogenesis following ischemic stroke through estrogen receptors alpha and beta. *J. Comp. Neurol.*, **500**, 1064–1075.
89. Mazzucco, C.A., Lieblich, S.E., Bingham, B.I., Williamson, M.A., Viau, V., and Galea, L.A. (2006) Both estrogen receptor alpha and estrogen receptor beta agonists enhance cell proliferation in the dentate gyrus of adult female rats. *Neuroscience*, **141**, 1793–1800.
90. Wise, P.M. (2006) Estrogen therapy: does it help or hurt the adult and aging brain? Insights derived from animal models. *Neuroscience*, **138**, 831–835.
91. Saravia, F., Revsin, Y., Lux-Lantos, V., Beauquis, J., Homo-Delarche, F., and De Nicola, A.F. (2004) Oestradiol restores cell proliferation in dentate gyrus and subventricular zone of streptozotocin-diabetic mice. *J. Neuroendocrinol.*, **16**, 704–710.
92. Wang, J.M., Liu, L., and Brinton, R.D. (2008) Estradiol-17 beta-induced human neural progenitor cell proliferation is mediated by an estrogen receptor beta-phosphorylated extracellularly regulated kinase pathway. *Endocrinology*, **149**, 208–218.
93. Wang, L., Andersson, S., Warner, M., and Gustafsson, J.A. (2003) Estrogen receptor (ER)beta knockout mice reveal a role for ERbeta in migration of cortical neurons in the developing brain. *Proc. Natl. Acad. Sci. U.S.A.*, **100**, 703–708.
94. Galea, L.A., Spritzer, M.D., Barker, J.M., and Pawluski, J.L. (2006) Gonadal hormone modulation of hippocampal neurogenesis in the adult. *Hippocampus*, **16**, 225–232.
95. Liu, L., Wang, J., Zhao, L., Nilsen, J., McClure, K., Wong, K., and Brinton, R.D. (2009) Progesterone increases rat neural progenitor cell cycle gene expression and proliferation via ERK and Progesterone Receptor Membrane Components (PGRMC) 1 and 2. *Endocrinology*, **150**, 3186–3196.
96. Wang, J.M., Singh, C., Liu, L., Irwin, R.W., Chen, S., Chung, E.J., Thompson, R.F., and Brinton, R.D. (2010) Allopregnanolone reverses neurogenic and cognitive deficits in mouse model of Alzheimer's disease. *Proc. Natl. Acad. Sci. U.S.A.*, **107**, 6498–6503.

97. Vallee, M., Mayo, W., Darnaudery, M., Corpechot, C., Young, J., Koehl, M., Le Moal, M., Baulieu, E.E., Robel, P., and Simon, H. (1997) Neurosteroids: deficient cognitive performance in aged rats depends on low pregnenolone sulfate levels in the hippocampus. *Proc. Natl. Acad. Sci. U.S.A.*, **94**, 14865–14870.

98. Vallee, M., Mayo, W., and Le Moal, M. (2001) Role of pregnenolone, dehydroepiandrosterone and their sulfate esters on learning and memory in cognitive aging. *Brain Res. Brain Res. Rev.*, **37**, 301–312.

99. Weill-Engerer, S., David, J.P., Sazdovitch, V., Liere, P., Eychenne, B., Pianos, A., Schumacher, M., Delacourte, A., Baulieu, E.E., and Akwa, Y. (2002) Neurosteroid quantification in human brain regions: comparison between Alzheimer's and nondemented patients. *J. Clin. Endocrinol. Metab.*, **87**, 5138–5143.

100. Bernstein, L. (2002) Epidemiology of endocrine-related risk factors for breast cancer. *J. Mammary Gland Biol. Neoplasia*, **7**, 3–15.

101. Chlebowski, R.T., Hendrix, S.L., Langer, R.D., Stefanick, M.L., Gass, M., Lane, D., Rodabough, R.J., Gilligan, M.A., Cyr, M.G., Thomson, C.A., Khandekar, J., Petrovitch, H., and McTiernan, A. (2003) Influence of estrogen plus progestin on breast cancer and mammography in healthy postmenopausal women: the Women's Health Initiative randomized trial. *J. Am. Med. Assoc.*, **289**, 3243–3253.

102. Kiss, R., Paridaens, R.J., Heuson, J.C., and Danguy, A.J. (1986) Effect of progesterone on cell proliferation in the MXT mouse hormone-sensitive mammary neoplasm. *J. Natl. Cancer Inst.*, **77**, 173–178.

103. Pike, M.C. and Ross, R.K. (2000) Progestins and menopause: epidemiological studies of risks of endometrial and breast cancer. *Steroids*, **65**, 659–664.

104. Magnaghi, V., Ballabio, M., Roglio, I., and Melcangi, R.C. (2007) Progesterone derivatives increase expression of Krox-20 and Sox-10 in rat Schwann cells. *J. Mol. Neurosci.*, **31**, 149–157.

105. Svenningsen, A.F. and Kanje, M. (1999) Estrogen and progesterone stimulate Schwann cell proliferation in a sex- and age-dependent manner. *J. Neurosci. Res.*, **57**, 124–130.

106. Ghoumari, A.M., Baulieu, E.E., and Schumacher, M. (2005) Progesterone increases oligodendroglial cell proliferation in rat cerebellar slice cultures. *Neuroscience*, **135**, 47–58.

107. Marin-Husstege, M., Muggironi, M., Raban, D., Skoff, R.P., and Casaccia-Bonnefil, P. (2004) Oligodendrocyte progenitor proliferation and maturation is differentially regulated by male and female sex steroid hormones. *Dev. Neurosci.*, **26**, 245–254.

108. Gago, N., Akwa, Y., Sananes, N., Guennoun, R., Baulieu, E.E., El-Etr, M., and Schumacher, M. (2001) Progesterone and the oligodendroglial lineage: stage-dependent biosynthesis and metabolism. *Glia*, **36**, 295–308.

109. Liu, L., Zhao, L., She, H., Chen, S., Wang, J.M., Wong, C., McClure, K., Sitruk-Ware, R., and Brinton, R.D. (2010) Clinically revelent progestins regulate neurogenic and neuroprotective responses in vitro and in vivo. *Endocrinology* 2010 Oct 13. [Epub ahead of print].

110. Brinton, R.D., Wang, J., Liu, L., Mao, Z., Zhao, L., and Morgan, T.E. (2009) Progesterone and clinical progestins differentially regulate proliferation of rat neural progenitor cells both in vitro and in vivo. Program No. 499.4 2009 Neuroscience Meeting Planner, Society for Neuroscience, Chicago, IL. http://www.abstractsonline.com/plan/start.aspx?mkey={081F7976-E4CD-4F3D-A0AF-E8387992A658}.

111. Zhang, Z., Yang, R., Zhou, R., Li, L., Sokabe, M., and Chen, L. (2010) Progesterone promotes the survival of newborn neurons in the dentate gyrus of adult male mice. *Hippocampus*, **20**, 402–412.

112. Suzuki, M., Wright, L.S., Marwah, P., Lardy, H.A., and Svendsen, C.N. (2004) Mitotic and neurogenic effects of dehydroepiandrosterone (DHEA) on human

113. Corpechot, C., Robel, P., Axelson, M., Sjovall, J., and Baulieu, E.E. (1981) Characterization and measurement of dehydroepiandrosterone sulfate in rat brain. *Proc. Natl. Acad. Sci. U.S.A.*, **78**, 4704–4707.
114. Corpechot, C., Synguelakis, M., Talha, S., Axelson, M., Sjovall, J., Vihko, R., Baulieu, E.E., and Robel, P. (1983) Pregnenolone and its sulfate ester in the rat brain. *Brain Res.*, **270**, 119–125.
115. Hu, M.C., Hsu, N.C., El Hadj, N.B., Pai, C.I., Chu, H.P., Wang, C.K., and Chung, B.C. (2002) Steroid deficiency syndromes in mice with targeted disruption of Cyp11a1. *Mol. Endocrinol.*, **16**, 1943–1950.
116. Alvarez-Buylla, A. and Kirn, J.R. (1997) Birth, migration, incorporation, and death of vocal control neurons in adult songbirds. *J. Neurobiol.*, **33**, 585–601.
117. Karishma, K.K. and Herbert, J. (2002) Dehydroepiandrosterone (DHEA) stimulates neurogenesis in the hippocampus of the rat, promotes survival of newly formed neurons and prevents corticosterone-induced suppression. *Eur. J. Neurosci.*, **16**, 445–453.
118. Ormerod, B.K., Lee, T.T., and Galea, L.A. (2003) Estradiol initially enhances but subsequently suppresses (via adrenal steroids) granule cell proliferation in the dentate gyrus of adult female rats. *J. Neurobiol.*, **55**, 247–260.
119. Fowler, C.D., Freeman, M.E., and Wang, Z. (2003) Newly proliferated cells in the adult male amygdala are affected by gonadal steroid hormones. *J. Neurobiol.*, **57**, 257–269.
120. Spritzer, M.D. and Galea, L.A. (2007) Testosterone and dihydrotestosterone, but not estradiol, enhance survival of new hippocampal neurons in adult male rats. *Dev. Neurobiol.*, **67**, 1321–1333.
121. Absil, P., Pinxten, R., Balthazart, J., and Eens, M. (2003) Effect of age and testosterone on autumnal neurogenesis in male European starlings (Sturnus vulgaris). *Behav. Brain Res.*, **143**, 15–30.
122. Absil, P., Pinxten, R., Balthazart, J., and Eens, M. (2003) Effects of testosterone on Reelin expression in the brain of male European starlings. *Cell Tissue Res.*, **312**, 81–93.
123. Brown, S.D., Johnson, F., and Bottjer, S.W. (1993) Neurogenesis in adult canary telencephalon is independent of gonadal hormone levels. *J. Neurosci.*, **13**, 2024–2032.
124. Zhang, L., Li, B., Ma, W., Barker, J.L., Chang, Y.H., Zhao, W., and Rubinow, D.R. (2002) Dehydroepiandrosterone (DHEA) and its sulfated derivative (DHEAS) regulate apoptosis during neurogenesis by triggering the Akt signaling pathway in opposing ways. *Brain Res. Mol. Brain Res.*, **98**, 58–66.
125. Compagnone, N.A. and Mellon, S.H. (1998) Dehydroepiandrosterone: a potential signalling molecule for neocortical organization during development. *Proc. Natl. Acad. Sci, U.S.A.*, **95**, 4678–4683.
126. Pinnock, S.B., Lazic, S.E., Wong, H.T., Wong, I.H., and Herbert, J. (2009) Synergistic effects of dehydroepiandrosterone and fluoxetine on proliferation of progenitor cells in the dentate gyrus of the adult male rat. *Neuroscience*, **158**, 1644–1651.
127. Mellon, S.H. (2007) Neurosteroid regulation of central nervous system development. *Pharmacol. Ther.*, **116**, 107–124.
128. Fontaine-Lenoir, V., Chambraud, B., Fellous, A., David, S., Duchossoy, Y., Baulieu, E.E., and Robel, P. (2006) Microtubule-associated protein 2 (MAP2) is a neurosteroid receptor. *Proc. Natl. Acad. Sci. U.S.A.*, **103**, 4711–4716.
129. Murakami, K., Fellous, A., Baulieu, E.E., and Robel, P. (2000) Pregnenolone binds to microtubule-associated protein 2 and stimulates microtubule assembly. *Proc. Natl. Acad. Sci. U.S.A.*, **97**, 3579–3584.
130. Mayo, W., Lemaire, V., Malaterre, J., Rodriguez, J.J., Cayre, M., Stewart, M.G., Kharouby, M., Rougon, G., Le Moal, M., Piazza, P.V., and Abrous, D.N. (2005) Pregnenolone sulfate enhances neurogenesis and PSA-NCAM

131. Pawluski, J.L., Brummelte, S., Barha, C.K., Crozier, T.M., and Galea, L.A. (2009) Effects of steroid hormones on neurogenesis in the hippocampus of the adult female rodent during the estrous cycle, pregnancy, lactation and aging. *Front Neuroendocrinol.*, **30**, 343–357.
132. Gago, N., El-Etr, M., Sananes, N., Cadepond, F., Samuel, D., Avellana-Adalid, V., Baron-Van Evercooren, A., and Schumacher, M. (2004) $3\alpha,5\alpha$-Tetrahydroprogesterone (allopregnanolone) and gamma-aminobutyric acid: autocrine/paracrine interactions in the control of neonatal PSA-NCAM + progenitor proliferation. *J. Neurosci. Res.*, **78**, 770–783.
133. Wiebe, J.P. and Lewis, M.J. (2003) Activity and expression of progesterone metabolizing 5alpha-reductase, 20α-hydroxysteroid oxidoreductase and $3\alpha(\beta)$-hydroxysteroid oxidoreductases in tumorigenic (MCF-7, MDA-MB-231, T-47D) and nontumorigenic (MCF-10A) human breast cancer cells. *BMC Cancer*, **3**, 9.
134. Wang, J.M. and Brinton, R.D. (2008) Allopregnanolone-induced rise in intracellular calcium in embryonic hippocampal neurons parallels their proliferative potential. *BMC Neurosci.*, **9** (Suppl 2), S11.
135. Jagasia, R., Steib, K., Englberger, E., Herold, S., Faus-Kessler, T., Saxe, M., Gage, F.H., Song, H., and Lie, D.C. (2009) GABA-cAMP response element-binding protein signaling regulates maturation and survival of newly generated neurons in the adult hippocampus. *J. Neurosci.*, **29**, 7966–7977.
136. Keller, E.A., Zamparini, A., Borodinsky, L.N., Gravielle, M.C., and Fiszman, M.L. (2004) Role of allopregnanolone on cerebellar granule cells neurogenesis. *Brain Res. Dev. Brain Res.*, **153**, 13–17.
137. Phillips, H.S., Hains, J.M., Armanini, M., Laramee, G.R., Johnson, S.A., and Winslow, J.W. (1991) BDNF mRNA is decreased in the hippocampus of individuals with Alzheimer's disease. *Neuron*, **7**, 695–702.
138. Terry, R.D. and Davies, P. (1980) Dementia of the Alzheimer type. *Annu. Rev. Neurosci.*, **3**, 77–95.
139. Zarow, C., Lyness, S.A., Mortimer, J.A., and Chui, H.C. (2003) Neuronal loss is greater in the locus coeruleus than nucleus basalis and substantia nigra in Alzheimer and Parkinson diseases. *Arch. Neurol.*, **60**, 337–341.
140. Yang, Y., Mufson, E.J., and Herrup, K. (2003) Neuronal cell death is preceded by cell cycle events at all stages of Alzheimer's disease. *J. Neurosci.*, **23**, 2557–2563.
141. Bhaskar, K., Miller, M., Chludzinski, A., Herrup, K., Zagorski, M., and Lamb, B.T. (2009) The PI3K-Akt-mTOR pathway regulates Abeta oligomer induced neuronal cell cycle events. *Mol. Neurodegener.*, **4**, 14.
142. Varvel, N.H., Bhaskar, K., Patil, A.R., Pimplikar, S.W., Herrup, K., and Lamb, B.T. (2008) Abeta oligomers induce neuronal cell cycle events in Alzheimer's disease. *J. Neurosci.*, **28**, 10786–10793.
143. Herrup, K. and Yang, Y. (2007) Cell cycle regulation in the postmitotic neuron: oxymoron or new biology? *Nat. Rev. Neurosci.*, **8**, 368–378.
144. Busser, J., Geldmacher, D.S., and Herrup, K. (1998) Ectopic cell cycle proteins predict the sites of neuronal cell death in Alzheimer's disease brain. *J. Neurosci.*, **18**, 2801–2807.
145. Jellinger, K.A. (2006) Clinicopathological analysis of dementia disorders in the elderly – an update. *J. Alzheimers Dis.*, **9**, 61–70.
146. Iqbal, K. and Grundke-Iqbal, I. (2006) Discoveries of tau, abnormally hyperphosphorylated tau and others of neurofibrillary degeneration: a personal historical perspective. *J. Alzheimers Dis.*, **9**, 219–242.
147. Xie, Z., Wei, M., Morgan, T.E., Fabrizio, P., Han, D., Finch, C.E., and Longo, V.D. (2002) Peroxynitrite mediates neurotoxicity of amyloid beta-peptide 1-42- and

lipopolysaccharide-activated microglia. *J. Neurosci.*, **22**, 3484–3492.

148. Eikelenboom, P., Rozemuller, J.M., Kraal, G., Stam, F.C., McBride, P.A., Bruce, M.E., and Fraser, H. (1991) Cerebral amyloid plaques in Alzheimer's disease but not in scrapie-affected mice are closely associated with a local inflammatory process. *Virchows Arch., B, Cell Pathol. Incl. Mol. Pathol.*, **60**, 329–336.

149. Jin, K., Minami, M., Lan, J.Q., Mao, X.O., Batteur, S., Simon, R.P., and Greenberg, D.A. (2001) Neurogenesis in dentate subgranular zone and rostral subventricular zone after focal cerebral ischemia in the rat. *Proc. Natl. Acad. Sci. U.S.A.*, **98**, 4710–4715.

150. Parent, J.M., Valentin, V.V., and Lowenstein, D.H. (2002) Prolonged seizures increase proliferating neuroblasts in the adult rat subventricular zone-olfactory bulb pathway. *J. Neurosci.*, **22**, 3174–3188.

151. Chen, Y., Liu, W., McPhie, D.L., Hassinger, L., and Neve, R.L. (2003) APP-BP1 mediates APP-induced apoptosis and DNA synthesis and is increased in Alzheimer's disease brain. *J. Cell Biol.*, **163**, 27–33.

152. Chen, Y., Liu, W., Naumovski, L., and Neve, R.L. (2003) ASPP2 inhibits APP-BP1-mediated NEDD8 conjugation to cullin-1 and decreases APP-BP1-induced cell proliferation and neuronal apoptosis. *J. Neurochem.*, **85**, 801–809.

153. Chen, S., Wang, J., Irwin, R.W., Yao, J., Liu, L., Hamilton, R.T., Lemus, J., and Brinton, R.D. (2009) Allopregnanolone promotes myelin formation associated with decreased accumulation of Amyloid β in the triple transgenic mouse model of Alzheimer's disease. Program No. 543.2. 2009 Neuroscience Meeting Planner, Society for Neuroscience, Chicago, IL. http://www.abstractsonline.com/plan/start.aspx?mkey={081F7976-E4CD-4F3D-A0AF-E8387992A658}.

154. Zhao, L., Yao, J., Mao, Z., Chen, S., Wang, Y., and Brinton, R.D. 17 beta-Estradiol regulates insulin-degrading enzyme expression via an ER beta/PI3-K pathway in hippocampus: relevance to Alzheimer's prevention. *Neurobiol. Aging* 2010 Jan 4 [Epub ahead of print].

155. Carroll, J.C., Rosario, E.R., Chang, L., Stanczyk, F.Z., Oddo, S., LaFerla, F.M., and Pike, C.J. (2007) Progesterone and estrogen regulate Alzheimer-like neuropathology in female 3xTg-AD mice. *J. Neurosci.*, **27**, 13357–13365.

18
Gonadotropins and Progestogens: Obligatory Developmental Functions during Early Embryogenesis and their Role in Adult Neurogenesis, Neuroregeneration and Neurodegeneration

Craig S. Atwood and Sivan Vadakkadath Meethal

18.1
Introduction

Hormones of the hypothalamic-pituitary-gonadal (HPG) axis are well known for their roles in reproduction. However, sex hormones have roles far beyond the regulation and coordination of reproduction; they also control the entire molecular process of life from the conception of the embryo through its development into an adult and finally its demise during senescence [1]. This recent understanding of the "big picture," that is, of how HPG hormones regulate life processes, has gathered support from studies identifying the function of reproductive hormones in embryonic development [2–5], as well as from studies into the extrareproductive functions of HPG hormones in the postreproductive physiology of mammals (reviewed in [6]).

This chapter describes our current, albeit incomprehensively small, knowledge of how reproductive (pregnancy) hormones direct growth and development during embryogenesis. It also describes the requirement for these reproductive hormones during adult life, and how the age-related dysregulation in the signaling of these hormones drives neurodegeneration and cognitive decline during senescence.

18.2
Hormonal Regulation of Human Embryogenesis

18.2.1
The Missing Links

Embryogenesis is a complex coordinated series of molecular and cellular changes that takes place within a well-defined internal environment. Human embryogenesis is orchestrated by a complex array of endocrine signals that commences with conception, is followed by the growth and development of the zygote into a blastocyst, its implantation into the endometrium, and the subsequent growth and development of the blastocyst into the neonate. Following conception, the

Hormones in Neurodegeneration, Neuroprotection, and Neurogenesis.
Edited by Achille G. Gravanis and Synthia H. Mellon
Copyright © 2011 WILEY-VCH Verlag GmbH & Co. KGaA, Weinheim
ISBN: 978-3-527-32627-3

developing embryo (zygote/morula/blastocyst) has 7–14 days in which to produce sufficient human chorionic gonadotropin (hCG), and subsequently progesterone (P_4; from both embryonic and corpus luteal sources), to allow implantation and halt degradation and discharging of the endometrium (menstruation) [7]. The up-regulation of hCG and P_4 not only allows for the maintenance of the endometrium, blastocyst attachment, and synctiotrophoblast proliferation into the endometrium [8–10], but also prevents ovulation and prepares the immune, metabolic, and psychological systems of the mother for pregnancy.

P_4 production is an *absolute* requirement for the maintenance of pregnancy [8]. Indeed, administration of RU-486 (mifepristone), an antiprogesterone and antiglucocorticosteroid agent to humans, is used for the medical termination of pregnancies of up to 49 days gestation (up to 63 days gestation in Britain and Sweden), and in combination with prostaglandin E1, for termination of pregnancies between 13 and 24 weeks gestation [11]. Inhibition of P_4 signaling using RU-486, a P_4 receptor (PR) competitive antagonist, results in endometrial decidual degeneration, trophoblast detachment, and decreased syncytiotrophoblast production of hCG, which in turn decreases P_4 production by the corpus luteum. In addition, RU-486 induces cervical softening and dilatation, release of endogenous prostaglandins, and an increase in the sensitivity of the myometrium to the contractile effects of prostaglandins leading to the expulsion of the embryo/fetus [12].

Despite our understanding of the endocrinology of pregnancy, only recently have we been able to answer fundamental questions such as what endocrine/paracrine/juxtacrine/autocrine factors (i) regulate embryonic cell division, (ii) regulate cell migration, (iii) specify differentiation into particular lineages, and (iv) induce apoptosis, during early embryogenesis.

18.2.2
Trophoblastic Hormone Secretion

Zygotic division into a blastocyst establishes the extraembryonic tissues (trophoblast layer or outer cell mass) and the hypoblast (extraembryonic endoderm) that support the embryonic embryoblast (inner cell mass) early in embryogenesis [13]. Trophoblasts secrete an array of hormones [4, 14–17] including P_4, endorphins, hCG, 17β-estradiol (E_2), and gonadotropin-releasing hormone (GnRH). The dramatic elevation in the production of hCG by the trophoblastic layer of the blastocyst during the early embryonic stage (from 5 to ≥ 1000 mIU ml^{-1} in the maternal serum) [17, 18] signals both the corpus lutea and trophoblast [19] to synthesize and secrete P_4 [20–25]. Secretion of these hormones appears to occur during the migration of the blastocyst through the fallopian tube and its implantation into the endometrium and subsequently from the placental tissues during later stages of embryogenesis. This is most clearly demonstrated by the elevation in maternal hCG during the early embryonic period (Figure 18.1).

The secretion of P_4, endorphins, hCG, E_2, and GnRH by trophoblasts that lie adjacent to the embryoblast in the blastocyst suggests that these hormones

18.2 Hormonal Regulation of Human Embryogenesis | 307

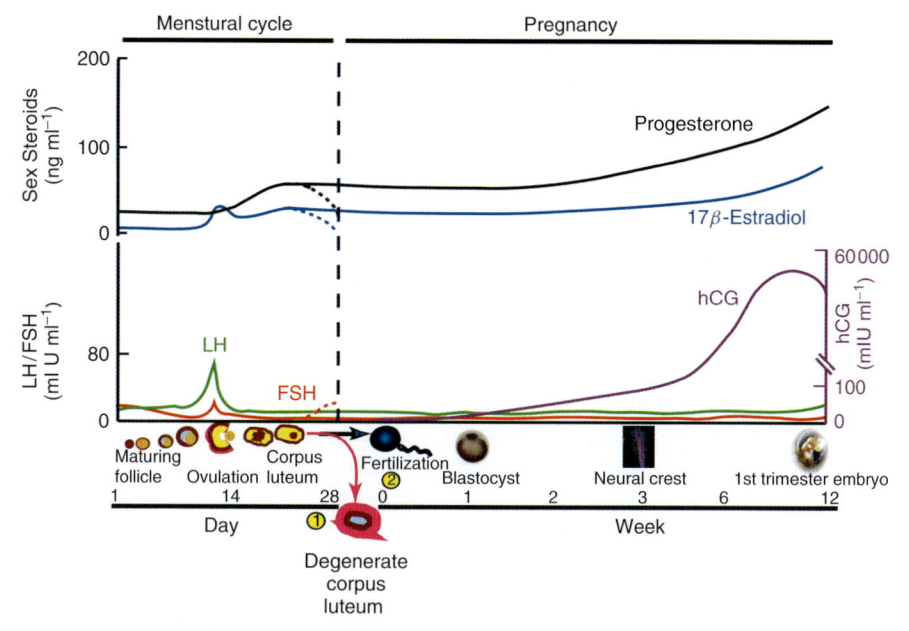

Figure 18.1 Hormonal regulation of the menstrual cycle and embryogenesis. The menstrual cycle is regulated by cyclical changes in the hypothalamic-pituitary-gonadal axis. The FSH-induced E_2 production increases the negative feedback on the hypothalamic-pituitary unit and slows the LH pulse frequency over the course of the follicular phase. The elevated GnRH levels result in an LH and FSH surge nearing the end of the follicular phase, which ultimately results in the LH/FSH-stimulated degradation of the follicular wall and subsequent release of the mature ovum into the ampulary duct. Large amounts of P_4 and some E_2 are secreted as the cells in the follicular wall undergo hypertrophy and hyperplasia. During the luteal phase, P_4 and E_2 negatively feedback on the hypothalamus, and the pituitary markedly slows the pulsitile release of LH. When the ova are fertilized, the pregnancy hormone hCG is secreted by the synctiotrophoblast, where levels rise in a logarithmic fashion until 8–10 weeks after the last menstrual period. After 10 weeks, a decline in hCG occurs until 18 weeks, whereupon levels remain constant until term. hCG induces the synthesis of P_4 from the corpus luteum and trophoblast for the maintenance of pregnancy. Patterns of plasma hormonal change for hCG, LH, FSH, E_2, and P_4 are taken from Braustein et al. [26], Reyes et al. [27], Tulchinsky et al. [28], Levitz et al. [29], and Kaplan et al. [30].

may *directly* signal the growth and development of the embryoblast. Evidence supporting this notion includes the presence of placental opioid-enhancing factor in amniotic fluid and placenta, and that the ingestion of placenta potentiates δ- and κ-opioid antinociception [16]. Likewise, trophoblastic and corpa luteal production of hCG/P_4 is markedly elevated postconception and is obligatory for the maintenance of pregnancy [8]. An autocrine/paracrine role for hCG secreted from invasive extravillous cytotrophoblasts [31] in the induction of neoangiogenesis during endometrial vascularization has previously been proposed [9]. hCG signaling via full-length luteinizing hormone/human chorionic gonadotropin receptors

(LHCGRs) on trophoblasts has been shown to modulate differentiation of the trophoblasts for subsequent villus projection and placentation.

Given the close spatial localization of trophoblasts to the embryoblast, and the availability of human embryonic stem cells (hESCs), that is, embryoblast-derived stem cells, it has become possible to explore trophoblastic hormone function in the development of the embryo.

18.2.3
Human Embryonic Stem Cells: A Complete Model System for Understanding the Cellular and Molecular Mechanisms Regulating Early Human Embryogenesis

hESCs derived from the inner cell mass of the blastocyst are a very useful model for the study of the molecular and cellular mechanisms of early embryogenesis [2–5, 32, 33]. hESCs can be differentiated into embryoid bodies (EBs), which resemble the early postimplantation embryo (blastocyst containing all three germ layers) [34]. hESCs also can be differentiated into columnar neuroectodermal cells [35] and mimics *in vivo* neuroectodermal development in terms of timing and morphology [35]. *In vitro*, hESCs differentiate into primitive neuroectodermal (or neural precursor) cells at around day 10 and then into neuroectodermal cells that exhibit neural tube-like rosettes in 14–17 days of differentiation in a chemically defined neural induction media [36]. These structures are predominantly composed of neuroectodermal cells akin to those that form the neural tube and are neural precursor cells (NPCs)/neural stem cells (NSCs) that can be further differentiated into different neural lineages. Considering hESCs are equivalent to a 5–6 day embryo, development of the neuroectoderm *in vitro* takes about 18–20 days, the time window when the neural tube forms in a human embryo [37, 38].

18.2.4
Progesterone, Human Chorionic Gonadotropin, and Early Human Embryogenesis

18.2.4.1 Regulation of Blastulation by Human Chorionic Gonadotropin and Progesterone

Ironically, the first evidence for a function of any trophoblastic hormone in the regulation of human embryogenesis was demonstrated by the finding that hCG induces the expression of the adhesion and neuritogenic protein amyloid-β precursor protein (AβPP; [3, 39]), a protein normally associated with neurodegeneration associated with Alzheimer's disease (AD) [40]. AβPP expression was detected at the transcriptional and translational levels, with mature and immature forms of AβPP as well as truncated variants (~53, 47, and 29 kDa) being detected using N- and C-terminal antibodies by immunoblot analyses [39]. These results indicated a critical molecular signaling link between the hormonal environment of pregnancy and the expression of AβPP in hESCs, which was suggestive of an important function for this protein during early human embryogenesis prior to the formation of NPCs.

That hCG could induce changes in epiblast protein expression was subsequently supported by the finding that hESCs express mRNA and protein for the full-length mature LHCGR (92 kDa) [2, 4]. That LHCGR is expressed on hESCs implicates hCG as an important signaling molecule in the growth and development of the embryo. LHCGR expression did not alter upon differentiation into EBs (structures that resemble early postimplantation embryos containing all three germ layers) [34], or into neuroectodermal rosettes, which consist of >90% columnar NPCs and are the *in vitro* equivalent of a rudimentary neural tube [2, 35]. The comparable level of LHCGR expression between the different cell lineages was suggestive of (i) the existence of a tight regulatory system for the maintenance of hCG signaling during embryonic stem cell division and differentiation, and (ii) a basal requirement for LH/hCG signaling during this early stage of embryogenesis. Indeed, hCG signaling via its hESC receptor was found to be essential for the proliferation of hESCs; inhibition of LHCGR signaling with P-antisense oligonucleotides suppressed hESC proliferation, as did a specific blocking antibody against the extracellular activation site of LHCGR, an effect that was reversed by treatment with hCG [2, 4]. These data are supported by the known proliferative properties of (hyperglycosylated) hCG, which has been demonstrated to act as an autocrine factor on extravillous invasive cytotrophoblast cells to initiate and control invasion as occurs at implantation of pregnancy and the establishment of hemochorial placentation, and malignancy as occurs in invasive hydatidiform mole and choriocarcinoma [41].

In addition to its cell cycle signaling activity, signaling of hCG via the hESC receptor rapidly up-regulates steroidogenic acute regulatory (StAR) protein-mediated cholesterol transport and the synthesis of P_4, a neurogenic steroid [42, 43]. StAR, a key rate-limiting step in the production of sex steroids in reproductive tissues, was detected in hESCs at both mRNA and protein (37, 30, and 20 kDa variants) levels. hCG treatment dose-dependently suppressed the expression of these StAR variants, while P_4 treatment decreased the truncation of the 37 kDa to the 30/32 kDa variants of StAR, indicative of decreased cholesterol transport across the mitochondrial membrane for steroidogenesis [44–49]. Importantly, hCG treatment markedly increases P_4 secretion 15-fold, indicating that embryoblast-derived hESCs already possess the machinery to transport cholesterol and synthesize sex steroids. Together, these findings indicate that negative feedback pathways exist for the regulation of hCG/LH signaling and mitochondrial cholesterol uptake for the synthesis of sex steroids in hESCs and differentiating lineages.

hESCs and EBs express PR-A [2, 50] suggesting that P_4 can signal hESC differentiation. That PR-A is functional in hESCs is supported by the fact that P_4 treatment decreases truncation of the 37 kDa StAR variant [2, 4]. These results indicate mechanisms at the level of both StAR expression and processing exist to regulate hESC steroidogenesis. The requirement for P_4 in the differentiation of hESCs into EBs was confirmed by the finding that RU-486 (mifepristone), a PR competitive antagonist [51], potently inhibits the differentiation of hESCs into EBs. RU-486-treated colonies failed to form normal cystic structures after 10 days in culture, and instead formed solid irregular spheres that were ∼20% the size of normal spheroidal EBs (Figure 18.2). Thus, the tight regulation of hCG signaling

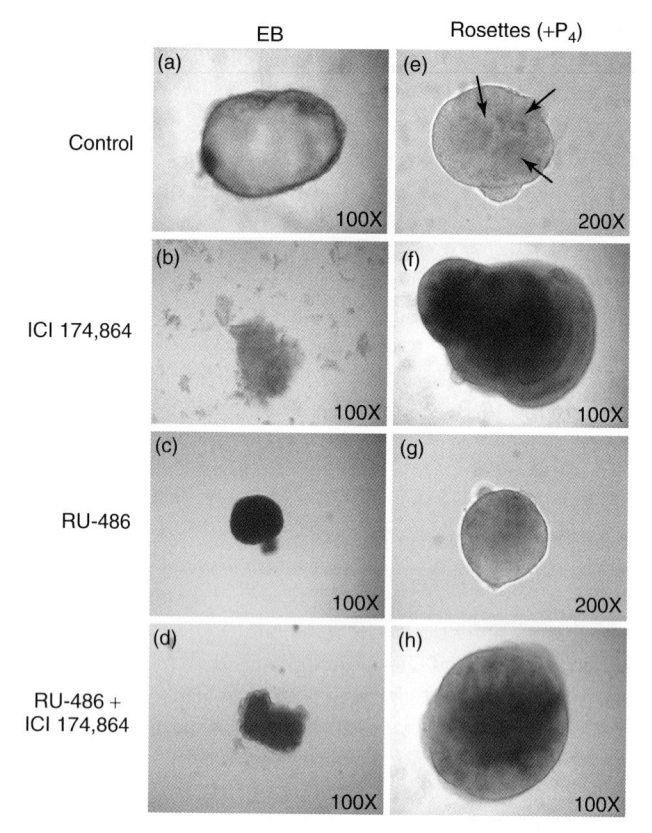

Figure 18.2 RU-486 and ICI 174,864 prevent embryoid body and neuroectodermal rosette formation. Embryoid body formation: H9 hESC colonies (day 4) cultured in the presence of (a) embryoid body media containing, (b) ICI 174,864 (0.1 mM), (c) RU-486 (20 μM), or (d) RU-486 (20 μM) + ICI 174,864 (0.1 mM). Colonies were followed for 10 days (when embryoid body formation occurs under normal conditions at day 14) and then examined morphologically. Rosette formation: H9 hESC colonies (day 4) were cultured in hESC growth media for four days prior to being placed in (e) neural induction media (+P_4) containing (f) ICI 174,864 (0.1 mM), (g) RU-486 (20 μM), or (h) RU-486 (20 μM) + ICI 174,864 (0.1 mM). Colonies were assessed morphologically at 17+ days (when rosette formation occurs). Control structures typically display a minimum of three rosettes within the neuroectodermal aggregate (arrows). Adapted from Gallego et al. [5].

and sex steroid synthesis and signaling is required to coordinate hESC proliferation and differentiation during gastrulation.

18.2.4.2 Regulation of Neurulation by Human Chorionic Gonadotropin and Progesterone

In addition to the obligatory signaling of P_4 for gastrulation, P_4 signaling also is required for the specification of NPCs from hESCs [2, 4, 5]. hCG treatment suppresses expression of the pluripotent marker Oct-3/4, suggesting that hCG, or steroid production initiated by hCG signaling, could direct lineage commitment [2, 4, 5]. In the presence of P_4, hESC colonies differentiate into spherical structures containing a minimum of three neuroectodermal rosettes inside of the cavity, while hESC colonies treated without P_4 or with RU-486 failed to form rosettes with columnar neuroectodermal cells after 17 days in culture (Figure 18.2; [2, 4, 5]). Morphological changes were more severe in the absence of P_4 than with RU-486. P_4, and to a lesser extent E_2, were found to increase the expression of nestin, an early marker of NPC formation, in hESCs. RU-486 completely suppressed nestin expression. Interestingly, "E_2 priming" is required for induction of PR expression in other tissues [52, 53]. Thus, the increase in nestin expression with E_2 treatment may reflect increased PR expression together with endogenous P_4 signaling, and explain the current requirement for serum priming of hESC colonies in the preparation of neuroectodermal rosettes.

Interestingly, hESCs default toward a primitive NSC fate if maintained for any length of time in culture [54]. Since hCG treatment induces nestin expression in hESCs, endogenous gonadotropin production by hESCs or trophoblastic cells [19] may be sufficient for NPC formation, thereby explaining the intrinsic hormonal signals regulating the "default pathway" of hESC differentiation into neuronal lineages [54].

These results suggest that trophoblastic hCG production adjacent to the embryoblast is required not only for trophoblast steroidogenesis and attachment of the blastocyst to the uterine wall [55] but also for signaling normal proliferation and differentiation of the epiblast. hCG-induced P_4 synthesis therefore has, in addition to its role in uterine decidualization for the implantation and maintenance of pregnancy, an obligatory role prior to the formation of NPCs, as well as an inductive role in the directed differentiation and specification of the first neuronal cell types (organogenesis) and the formation of the neural tube. While the structural importance of P_4 and alloprogesterone has previously been recognized by its early synthesis (by at least day 13) within the developing rat central nervous system [56], our results demonstrate an early (within the first seven days) and absolute requirement for P_4 during human blastulation and neurulation. In this respect, it has been shown that P_4 is necessary and sufficient (in neurobasal media) for the maintenance and differentiation of primary hippocampal/cortical/striatal neurons *in vitro* [42]. That P_4 is the hormone regulating these key events is perhaps not surprising given its location high in the synthetic pathway; P_4 is the first steroid synthesized from pregnenolone, the precursor to all other steroids.

Previous studies have demonstrated the importance of P_4 and related steroids as neurotrophic agents that promote adult neurogenesis, neuronal survival, and neuroprotection [42, 43, 57–63]. Clinical studies supporting the neurotrophic actions of P_4 administration are demonstrated by the decrease in mortality rate and improved outcome following acute traumatic brain injury (TBI) in humans [64].

Dependent upon the timing of administration during pregnancy, suppression of P_4 signaling with RU-486 [11] aside from intrauterine disruptive functions (decidual breakdown and trophoblast detachment) will also disrupt time-sensitive developmental processes. The requirement of P_4 during cavitation processes not only indicates the structural influence of these molecular pathways on the developing embryo within the first seven days, but also on the formation of the neural tube at around day 17–19, which will influence future neural connectivity. The relative binding affinity of RU-486 for the PR is twice that of P_4 [65], and is used at a dose of 200–600 mg for the termination of pregnancies (this equates to ~6–19 µM, equivalent to that used in Gallego et al. [2, 4, 5, 51]). Thus, the abortifacient effects of RU-486 in blocking PR signaling, therefore, also extends to blocking blastulation and neurulation and the normal growth and development of the embryo.

18.2.4.3 Regulation of Organogenesis by Human Chorionic Gonadotropin and Progesterone

Aside from the induction of blastulation and neurulation early in embryogenesis, hCG/LH and P_4 signaling may play a role in the development of other tissues. LHCGR and PR have been identified on numerous reproductive and nonreproductive tissues [66–69]. With regard hCG/LH, the free glycoprotein α-subunit of gonadotropins has been shown to stimulate differentiation of prolactin cells in the pituitary [70] and endometrial stromal cell decidualization in the placenta [71]. Although it has not been demonstrated if hCG/LH has a developmental function during organogenesis, hyperglycosylated hCGβ has potent cell growth and invasion properties as observed in early pregnancy, gestational choriocarcinoma, and testicular cancers [41]. It has been shown that hCG promotes angiogenesis by inducing vascular endothelial growth factor up-regulation [72–74]. Recent data demonstrate that physiological concentrations of hCG (10–400 IU mL^{-1}) significantly enhance pericyte sprouting and migration and give rise to the maturation and coverage of endothelial capillaries [75].

The potential for P_4 to regulate organogenesis has been reported during puberty and adulthood, where P_4 is obligatory for the development of the tertiary ducts on the mammary gland, and the physiological differentiation of the lobuloalveolar system from the lobular buds [52]. In the adult, P_4 also has been demonstrated to regulate bone formation [76], promote angiogenesis and arteriogenesis [77], and promote formation of the placenta. While our knowledge of hCG/LH and P_4 during organogenesis is rudimentary at this point, the above evidence indicates that these hormones probably play important functions in many tissues during organogenesis.

18.2.5
Opioid Signaling and Early Human Embryogenesis

Trophoblastic production of endorphins [14] also is crucial for embryogenesis [5]. Treatment of hESC colonies with the delta opioid receptor selective antagonist ICI 174,864 [78, 79] inhibits the formation of the EB cystic structure, and instead forms nonspherical structures ~40% the size of normal spheroidal EBs. The mechanism by which opioid signaling promotes blastulation is unclear; however, delta opioid antagonists may function to inhibit embryogenesis by regulating hCG release [80] required for P_4 production. Opioid signaling is also required for neuroectodermal rosette formation since ICI 174,864 [78, 79] inhibits normal neuroectodermal rosette formation and nestin expression [5]. Previous data has implicated P_4 as acting in the arcuate nucleus and anteroventral periventricular nucleus through β-endorphin and dynorphin B neurons to affect preoptic area GnRH neurons and gonadotropin secretion [81, 82]. Thus, delta opioid receptor signaling is required both for normal human blastulation and neurulation, but it remains to be determined if there is cross talk between opioid signaling and the regulation of GnRH/gonadotropin secretion.

18.3
Progesterone: an Essential Neurotrophic Hormone during All Phases of Life

The neurotrophic functions of P_4 during blastulation *and* neurulation in the human fit nicely with the known functions of sex steroids in the adult: an inductive or organizational role on the undifferentiated brain, together with an activational role on the differentiated brain that regulates cognition and expression of overt patterns of sexual behavior [83, 84]. In addition, sex steroids have a neuroregenerative (or neurotrophic) role that helps repair the brain following injury [85] as well as maintain normal brain health (reviewed in [86]). The developmental functions of P_4 during the fetal period and in adulthood are summarized by Brinton (Chapter 17). Briefly, it is known that sex steroids are primary hormonal signals that regulate neuronal growth and differentiation [87], promoting neurite development and migration that lead to changes in synaptogenesis [88–91]. As part of these differentiation processes, P_4, E_2, and testosterone are known to modulate the growth of dendrites and dendritic spine density, with the loss of sex steroids generally resulting in decreased spine density [92, 93]. More recently, allopregnanolone has been shown to significantly promote neurogenesis by increasing rat neuroprogenitor cell proliferation and human NSC proliferation [43]. Aside from these long-term changes, sex steroids also regulate synapse turnover in the CA1 region of the hippocampus during the four day estrous cycle of the female rat.

Over the last 15 years, a number of laboratories, and in particular, the group of Donald Stein, have reported the neurotrophic properties of P_4 in limiting tissue damage and improving functional outcome after blunt TBI, stroke, spinal cord injury, diabetic neuropathies, and other types of acute neuroinjuries in several species [64, 86]. P_4 administration after TBI improves short- and long-term behavioral recovery; protects and rebuilds the blood–brain barrier (BBB); and reduces inflammation, apoptosis, lesion volume, and cerebral edema in laboratory animals [63, 85, 86, 94–106]. These neurotrophic properties together with P_4's ability to regulate myelinization and microtubule assembly [57] are all plausible mechanisms for P_4's neuroprotective effects. A recent Phase IIa clinical trial [64] reported that four days of intravenous P_4 post-TBI reduces mortality by more than 50% in moderately to severely injured human patients and enhances functional outcomes at 30 days for the moderately injured. Thus, P_4 may be an effective clinical treatment for TBI in humans.

Given the differentiative properties of P_4 in human NSCs [43] and in hESCs [2, 4, 5], it is probable that P_4 also acts to drive neurogenesis from the resident totipotent NSC population in the brain following TBI. Thus, P_4-induced neurogenesis may be an integral component of the repair process following TBI leading to functional improvement. The existence of NSC's in the adult hippocampal dentate gyrus [107, 108] and their heightened proliferative response following injury [109–115] indicates neurogenesis as an important component of cognitive recovery following TBI. In addition, newly generated granule neurons are capable of extending projections along the hippocampal mossy fiber pathway in the acute posttraumatic period [115], and this extensive anatomical integration of new born dentate granule neurons occurs at the time when innate cognitive recovery is observed [116]. Finally, administration of human bone marrow stromal cells into the rat brain after TBI provides functional improvements, thought to be a result of neurogenesis and synaptogenesis in the lesioned area [117]. These data explain the substantive spontaneous cognitive recovery observed following TBI [118].

18.4
Age-Related Loss of Progesterone: Implications in the Pathophysiology of Neurodegenerative Diseases

The postreproductive changes in sex hormones are particularly important due to their association with disease [1]. The decline in sex steroid production by the gonads following menopause and during andropause leads to a loss of hypothalamic feedback inhibition that stimulates GnRH and gonadotropin production [119]. In addition, the decrease in gonadal inhibin production at this time [120] results in decreased activin receptor inhibition, and together with the increase in bioavailable activin [121] leads to a further increase in the secretion of GnRH and gonadotropins [122–124]. Thus, the lack of negative feedback from the ovary (P_4, E_2, and inhibin) is responsible for the unopposed and marked elevations in the secretion of GnRH and gonadotropins with ovarian and testicular senescence [6, 125–128].

The concentration of brain sex steroids, including P_4, is a mixture of peripherally derived sex steroids, converted peripheral steroids, and de novo synthesized neuro-sex steroids. The contribution of peripheral sex steroids to total brain sex steroids is unknown, but postreproductive declines in peripheral P_4 would be expected to impact brain P_4 concentrations. While elevations in GnRH and gonadotropin concentrations might promote brain neurosteroid production [129], including P_4, it is not know if this is sufficient to counter the loss of peripherally derived sex steroids. The consequences of these hormonal alterations, and in particular P_4, are discussed below in the context of AD and stroke.

18.4.1
Alzheimer's Disease

The age-related loss of P_4 is of particular importance given the differentiative properties of this steroid as described above. Also of importance are the age-related elevations in serum GnRH, FSH, and LH (the adult hCG homolog with 83% homology that binds the same receptor), especially given the known proliferative properties of hCG/LH [2, 4]. In this regard, hCG/LH is known to have powerful mitogenic properties in certain reproductive tissues [130–134], and is frequently expressed by tumor cells [135–137] (reviewed in [41]). Thus, these multiple changes in hormonal signaling with menopause/andropause, that is, increased mitogenic signaling but decreased differentiation signaling (*dyotic signaling*) might be expected to impact normal cell cycle dynamics. Indeed, accumulating evidence suggests there is a reactivation of the cell cycle with aging [1] as has been demonstrated for differentiated neurons of the brains of postreproductive individuals with AD [138, 139]. This data includes (i) the ectopic expression of cell cycle proteins in those regions of the brain affected by AD (e.g., cyclin B1, cdc2, PCNA, cdk4, Ki-67, p16), but not in areas unaffected by AD pathology or in control brains, (ii) chromosomal replication (endoreduplication) in differentiated AD neurons, demonstrating entry into the S-phase of the cell cycle, (iii) elevated cytoplasmic mitochondrial DNA and Cox-1 expression, suggestive of *de novo* mitochondrion synthesis, and (iv) up-regulated growth factor signal transduction pathways. Other parallels between embryonic neurogenesis and adult neurodegeneration include the expression of AβPP, secretases, and tau, together with the processing of AβPP either toward the amyloidogenic or nonamyloidogenic pathways, and the phosphorylation of tau ([39] and Atwood and Porayette, unpublished data). Similarly, the fetal brain has been reported to display a number of biochemical similarities to the AD brain, namely the presence of Aβ and AβPP [140, 141], presenilin-1 expression [142], and hyperphosphorylated tau [143]. The phosphorylation of tau is a mitogenic-associated event that normally occurs during metaphase of neuronal division, and is observed during differentiation of neurons in the fetal brain [143, 144]. These data suggest that AD neuropathology is a result of an imbalance in developmental/regenerative programs in the adult brain.

18.4.1.1 Amyloid-β Precursor Protein and Neurogenesis

That amyloidogenic pathways are involved in neurogenesis has recently been reported by a number of workers [144–147]. In this context, an increase in neurogenesis has been reported in young transgenic mice overexpressing human mutant APP [148, 149]. Moreover, the overexpression of wild-type or FAD mutant AβPP, which promotes Aβ generation [150], also has been shown to promote the reentry of primary neurons into the cell cycle, as demonstrated by the induction of DNA synthesis and cell cycle markers [151]. Not surprisingly, AβPP has structural similarity to growth factors [152] and modulates several important neurotrophic functions, including neuritogenesis, synaptogenesis, and synaptic plasticity [153].

18.4.1.2 Hormonal Regulation of Neurogenesis via Modulation of AβPP Metabolism

In hESCs, the differential processing of AβPP via secretase enzymes regulates the proliferation and differentiation of hESCs; processing toward the amyloidogenic pathway is associated with cell proliferation, and processing toward the nonamyloidogenic pathway is associated with cell specification and differentiation. Specifically, P_4 induces processing of AβPP toward the production of the sAβPPα in hESCs [39], which has known differentiative properties [154]. Similarly, E_2 has been shown to stimulate the processing of AβPP by the nonamyloidogenic α-secretory pathway and reduces cellular Aβ production in both nonneuronal [155] and neuronal cultures [156–158]. Conversely, the loss of sex steroids and elevation in gonadotropins following ovariectomy has been shown to increase Aβ generation in nontransgenic animals [159]. Importantly, we and others have demonstrated that LH promotes the processing of AβPP toward the amyloidogenic pathway *in vitro*, while suppression of serum LH in mice using GnRH agonists decreases the concentration of brain Aβ1-40 and Aβ1-42, the two major variants that deposit in the AD brain [160–163]. It is therefore plausible that the interaction of these hormonal pathways on the modulation of the processing of AβPP may regulate cell cycle events throughout life, with dyotic signaling by these hormones leading to the reactivation of the cell cycle in differentiated neurons of the AD brain [139]. This aberrant, albeit unsuccessful, reentry of neurons into the cell cycle leads to synapse contraction and neuron death (see [6, 138, 164] for reviews). In addition to the loss of neurons following the reactivation of the cell cycle in differentiated neurons, it is possible that dyotic signaling prevents normal neurogenesis from resident NSCs, thereby preventing replacement of neurons. Further studies are required to determine whether postreproductive levels of GnRH/gonadotropins are sufficient to induce neurosteroidogenesis in NSCs and neuronal cell types, and whether the level of postreproductive neurosteroid synthesis dictates normal or dyotic signaling in these cell types.

18.4.1.3 Progesterone in the Treatment of AD

The above studies, including the recent clinical trials for TBI, suggest the neuroprotective and neurotrophic functions of P_4 as a potential therapeutic for AD that might be expected to offset and delay the onset of AD. However, inconceivably, to date no clinical studies have assessed the effect of P_4 on cognitive function in aging or AD, although recent studies have assessed the effect of P_4 on AD neuropathology and cognitive function in normal and transgenic animals. Importantly, P_4 has been shown to enhance learning and memory of aged wild-type and progesterone receptor knockout mice [165], and following ovariectomy of young adult rats that are E_2 primed [166, 167]. Moreover, among mid-aged rats, forgetting in the water maze is attenuated by P_4 and/or E_2 after ovariectomy [168] at 14 months, while working memory in the radial arm maze is better among 14 month old rats administered with P_4 and/or E_2 following ovariectomy two months before [169].

P_4 does not, however, prevent ovariectomy-induced increases in Aβ accumulation as does E_2 in triple transgenic mice overexpressing mutant forms of AβPP, PS1, and tau, but does dramatically reduce tau phosphorylation when administered alone or in combination with E_2 [170]. Administration of P_4 in combination with E_2 in this model blocks E_2 attenuation of $A\beta$ accumulation, suggesting the influence of P_4 on AD neuropathology, and cognition may be modulated by E_2. Coadministration of E_2 and P_4 has been shown to rescue neuronal loss in rodent paradigms of neural injury [171, 172], while P_4 alone has been shown to attenuate seizures [173, 174]. Contrary to previous studies [168, 169] using normal mice, P_4 does not increase cognitive performance in ovariectomized triple transgenic mice [170]. Interpretation of these results is further complicated by findings that treatment with these sex steroids can markedly affect ER and PR expression [52, 53, 175]. In this respect, E_2-induced increases in hippocampal spine density in female rats is potentiated for 4 h following P_4 treatment, but is blocked after an 18 h P_4 treatment [92]. Thus, the neural beneficial effects of P_4 both directly and interactively with E_2 [176] may be optimized by delivery in a cyclic manner (which parallels natural fluctuations) rather than continuously. In this respect, Gibbs [177] found that activity of choline acetyltransferase in ovariectomized female rats was modestly increased by E_2, significantly elevated by E_2 paired with a cyclic P_4 regimen, but reduced by E_2 with continuous P_4 exposure. The influence of these hormones upon their actions is also demonstrated by the fact that estrogen inhibits P_4-induced proliferation during angiogenesis in ovariectomized mice [178]. Taken together, there is compelling evidence that P_4 is crucial for normal cognitive health, and may enhance cognitive performance in humans.

18.4.2
Stroke

Recent studies have highlighted the role of sex steroids including P_4 in the regulation of the BBB and impact on vascular integrity. Suppression of sex steroid production and elevation of gonadotropin production following ovariectomy has been shown to increase Evan's blue dye extravasation into the brain and a redistribution of the gap-junction protein connexin-43 (Cx43) along the extracellular microvascular endothelium [179]. Similarly, there is a significantly greater transfer of intravenously injected Evan's blue dye into the forebrain of acyclic (reproductive senescent) females compared to young adult females, indicating that BBB permeability is compromised in the brain following reproductive senescence [180]. Moreover, IgG expression is dramatically increased in the hippocampus and thalamus, but not the hypothalamus of reproductive senescent females compared to young adult females [181]. An increase in the expression of the gap protein Cx43 in the mouse brain following ovariectomy was suppressed in ovariectomized animals treated with leuprolide acetate, indicating serum gonadotropins in the absence of sex steroids may compromise gap junction function [179]. However, there also is support for P_4 as a major regulator of BBB integrity, since P_4 promotes the reconstitution of the BBB after TBI (reviewed in [86]). In this respect, the effects of P_4 in promoting angiogenesis and arteriogenesis have been well described [77, 178]. Likewise, E_2 has been shown to modulate the expression of occludin, transendothelial resistance and paracellular permeability in human vascular endothelial cells [182], and human cervical epithelial cells [183].

Together, these results suggest that sex steroids and gonadotropins can regulate the tissue-barrier function and that dyotic signaling by these hormones following endocrine dyscrasia with menopause/andropause may induce loss of selective permeability of the BBB at this time. Whether this constitutive loss of barrier integrity during aging is due to alterations in tight and gap junction function as a result of the selective loss of gap proteins or tight junction proteins in endothelial junctions; as a result of aberrant smooth muscle cells or endothelial cell division; as a result of the loss of endothelial cells; and/or as a result of the lack of endothelial cell replacement by endothelial totipotent stem cells, remains to be determined. Such changes in BBB permeability, however, may be responsible for the cerebropathophysiology of age-related neurodegenerative diseases such as stroke and AD, and of encephalitis and meningitis.

18.5
Conclusion

P_4 and hCG have obligatory developmental functions from embryogenesis through to the adult brain. The advent of the hESC era has allowed experimental determination of the physiological hormone requirements for early embryogenesis. In this respect, although progestagens are often considered primarily as reproductive hormones with maternal influences, it is now clear that progestogens are essential for the growth and development of the embryo as well as the normal health of both the female and male brain throughout life. Paracrine/juxtacrine signaling of hCG (and opioids) for mobilization of cholesterol for P_4 production by the epiblast/synctiotrophoblast following conception is essential for human blastulation and neurulation. This paracrine/juxtacrine signaling by extraembryonic tissues is the commencement of trophic support by placental tissues in the growth and development of the human embryo. While appropriate gonadotropin/GnRH and P_4 signaling is necessary for normal growth and development during embryogenesis, fetal life and childhood, and for the maintenance of brain health during adult reproductive life, the unopposed elevations in GnRH/gonadotropins with the loss of sex steroids following menopause/andropause might be expected to lead to dysregulation of cell cycle events given the proliferation and differentiation properties of GnRH/hCG/LH and sex steroids [1]. In this respect, accumulating evidence points toward reproductive hormone–induced cell cycle changes as being at the heart of age-related neurological diseases such as AD [6] and stroke [179]. As such, appropriate therapeutic strategies aimed at maintaining or returning HPG axis hormones to young adult levels would be expected to delay the onset of neurodegeneration and cognitive decline.

References

1. Bowen, R.L. and Atwood, C.S. (2004) Living and dying for sex. A theory of aging based on the modulation of cell cycle signaling by reproductive hormones. *Gerontology*, **50**, 265–290.
2. Gallego, M.J., Porayette, P., Kaltcheva, M.M., Bowen, R.L., Meethal, S.V., and Atwood, C.S. (2010) The pregnancy hormones human chorionic gonadotropin and progesterone induce human embryonic stem cell proliferation and differentiation into neuroectodermal rosettes. *Stem. Cell. Res. Ther.*, **1**, 28.
3. Porayette, P., Gallego, M.J., Kaltcheva, M.M., Meethal, S.V., and Atwood, C.S. (2007) Amyloid-beta precursor protein expression and modulation in human embryonic stem cells: a novel role for human chorionic gonadotropin. *Biochem. Biophys. Res. Commun.*, **364**, 522–527.
4. Gallego, M.J., Porayette, P., Kaltcheva, M., Bowen, R.L., Vadakkadath Meethal, S., and Atwood, C.S. (2008) Trophoblastic Hormones Direct Early Human Embryogenesis. Available from Nature Precedings, http://hdl.handle.net/10101/npre.2008.2671.1. (accessed 28.09.2010)
5. Gallego, M.J., Porayette, P., Kaltcheva, M.M., Meethal, S.V., and Atwood, C.S. (2009) Opioid and progesterone signaling is obligatory for early human embryogenesis. *Stem Cells Dev.*, **18**, 737–740.
6. Atwood, C.S., Meethal, S.V., Liu, T., Wilson, A.C., Gallego, M., Smith, M.A., and Bowen, R.L. (2005) Dysregulation of the hypothalamic-pituitary-gonadal axis with menopause and andropause promotes neurodegenerative senescence. *J. Neuropathol. Exp. Neurol.*, **64**, 93–103.
7. Gupta, A.K., Holzgreve, W., and Hahn, S. (2008) Decrease in lipid levels of syncytiotrophoblast micro-particles reduced their potential to inhibit endothelial cell proliferation. *Arch. Gynecol. Obstet.*, **277**, 115–119.
8. Larson, P., Kronenberg, H., Melmed, S., and Polonsky, K. (2002) *Williams Textbook of Endocrinology*, Saunders, Philadelphia, PA.
9. Licht, P., Russu, V., and Wildt, L. (2001) On the role of human chorionic gonadotropin (hCG) in the embryo-endometrial microenvironment: implications for differentiation and implantation. *Semin. Reprod. Med.*, **19**, 37–47.
10. Pepe, G.J. and Albrecht, E.D. (1995) Actions of placental and fetal adrenal steroid hormones in primate pregnancy. *Endocr. Rev.*, **16**, 608–648.
11. Fiala, C. and Gemzell-Danielsson, K. (2006) Review of medical abortion using mifepristone in combination with a prostaglandin analogue. *Contraception*, **74**, 66–86.
12. Gemzell-Danielsson, K., Bygdeman, M., and Aronsson, A. (2006) Studies on uterine contractility following mifepristone and various routes of misoprostol. *Contraception*, **74**, 31–35.
13. Gilbert, S. (2003) *Developmental Biology*, 7th edn, Sinauer Associates, Sunderland.
14. Zhuang, L.Z. and Li, R.H. (1991) Study on reproductive endocrinology of human placenta (II) – Hormone secreting activity of cytotrophoblast cells. *Sci. China B*, **34**, 1092–1097.
15. Cemerikic, B., Zamah, R., and Ahmed, M.S. (1994) Opioids regulation of human chorionic gonadotropin release from trophoblast tissue is mediated by gonadotropin releasing hormone. *J. Pharmacol. Exp. Ther.*, **268**, 971–977.
16. DiPirro, J.M. and Kristal, M.B. (2004) Placenta ingestion by rats enhances delta- and kappa-opioid antinociception, but suppresses mu-opioid antinociception. *Brain Res.*, **1014**, 22–33.
17. Pidoux, G., Gerbaud, P., Tsatsaris, V., Marpeau, O., Ferreira, F., Meduri, G., Guibourdenche, J., Badet, J., Evain-Brion, D., and Frendo, J.L. (2007) Biochemical characterization and modulation of LH/CG-receptor during human trophoblast differentiation. *J. Cell Physiol.*, **212**, 26–35.
18. Braunstein, G.D., Rasor, J., and Danzer, H. (1976) Serum human

chorionic gonadotropin levels throughout normal pregnancy. *Am. J. Obstet. Gynecol.*, **126**, 678.

19. Golos, T.G., Pollastrini, L.M., and Gerami-Naini, B. (2006) Human embryonic stem cells as a model for trophoblast differentiation. *Semin. Reprod. Med.*, **24**, 314–321.

20. Richardson, M.C. and Masson, G.M. (1985) Lack of direct inhibitory action of oxytocin on progesterone production by dispersed cells from human corpus luteum. *J. Endocrinol.*, **104**, 149–151.

21. Carr, B.R., MacDonald, P.C., and Simpson, E.R. (1982) The role of lipoproteins in the regulation of progesterone secretion by the human corpus luteum. *Fertil. Steril.*, **38**, 303–311.

22. Duncan, W.C., McNeilly, A.S., Fraser, H.M., and Illingworth, P.J. (1996) Luteinizing hormone receptor in the human corpus luteum: lack of down-regulation during maternal recognition of pregnancy. *Hum. Reprod.*, **11**, 2291–2297.

23. Strauss, J.F., Christenson, L.K., Devoto, L., and Martinez, F. III (2000) Providing progesterone for pregnancy: control of cholesterol flux to the side-chain cleavage system. *J. Reprod. Fertil. Suppl.*, **55**, 3–12.

24. Bukovsky, A., Caudle, M.R., Keenan, J.A., Wimalasena, J., Upadhyaya, N.B., and Van Meter, S.E. (1995) Is corpus luteum regression an immune-mediated event? Localization of immune system components and luteinizing hormone receptor in human corpora lutea. *Biol. Reprod.*, **53**, 1373–1384.

25. Casper, R.F. and Yen, S.S. (1979) Induction of luteolysis in the human with a long-acting analog of luteinizing hormone-releasing factor. *Science*, **205**, 408–410.

26. Braunstein, G.D., Rasor, J., Danzer, H., Adler, D., and Wade, M.E. (1976) Serum human chorionic gonadotropin levels throughout normal pregnancy. *Am. J. Obstet. Gynecol*, **126**, 678–681.

27. Reyes, F.I., Boroditsky, R.S., Winter, J.S., and Faiman, C. (1974) Studies on human sexual development. II. Fetal and maternal serum gonadotropin and sex steroid concentrations. *J. Clin. Endocrinol. Metab.*, **38**, 612–617.

28. Tulchinsky, D., Hobel, C.J., Yeager, E., and Marshall, J.R. (1972) Plasma estrone, estradiol, estriol, progesterone, and 17-hydroxyprogesterone in human pregnancy. I. Normal pregnancy. *Am. J. Obstet. Gynecol.*, **112**, 1095–1100.

29. Levitz, M. and Young, B.K. (1977) Estrogens in pregnancy. *Vitam. Horm.*, **35**, 109–147.

30. Kaplan, S.L., Grumbach, M.M., and Aubert, M.L. (1976) The ontogenesis of pituitary hormones and hypothalamic factors in the human fetus: maturation of central nervous system regulation of anterior pituitary function. *Recent Prog. Horm. Res.*, **32**, 161–243.

31. Handschuh, K., Guibourdenche, J., Tsatsaris, V., Guesnon, M., Laurendeau, I., Evain-Brion, D., and Fournier, T. (2007) Human chorionic gonadotropin expression in human trophoblasts from early placenta: comparative study between villous and extravillous trophoblastic cells. *Placenta*, **28**, 175–184.

32. Thomson, J.A., Itskovitz-Eldor, J., Shapiro, S.S., Waknitz, M.A., Swiergiel, J.J., Marshall, V.S., and Jones, J.M. (1998) Embryonic stem cell lines derived from human blastocysts. *Science*, **282**, 1145–1147.

33. Xia, X. and Zhang, S.C. (2009) Differentiation of neuroepithelia from human embryonic stem cells. *Methods Mol. Biol.*, **549**, 51–58.

34. O'Shea, K.S. (1999) Embryonic stem cell models of development. *Anat. Rec.*, **257**, 32–41.

35. Li, X. and Zhang, S. (2006) In vitro differentiation of neural precursors from human embryonic stem cells. *Methods Mol. Biol.*, **331**, 169–177.

36. Zhang, S.C., Wernig, M., Duncan, I.D., Brustle, O., and Thomson, J.A. (2001) In vitro differentiation of transplantable neural precursors from human embryonic stem cells. *Nat. Biotechnol.*, **19**, 1129–1133.

37. Muller, F. and O'Rahilly, R. (1985) The first appearance of the neural tube and

38. optic primordium in the human embryo at stage 10. *Anat. Embryol. (Berl.)*, **172**, 157–169.
38. Zhang, S.C. (2003) Embryonic stem cells for neural replacement therapy: prospects and challenges. *J. Hematother. Stem Cell Res.*, **12**, 625–634.
39. Porayette, P., Gallego, M.J., Kaltcheva, M.M., Bowen, R.L., Vadakkadath Meethal, S., and Atwood, C.S. (2009) Differential processing of amyloid-beta precursor protein directs human embryonic stem cell proliferation and differentiation into neuronal precursor cells. *J. Biol. Chem.*, **284**, 23806–23817.
40. Hardy, J. and Selkoe, D.J. (2002) The amyloid hypothesis of Alzheimer's disease: progress and problems on the road to therapeutics. *Science*, **297**, 353–356.
41. Cole, L.A. (2009) New discoveries on the biology and detection of human chorionic gonadotropin. *Reprod. Biol. Endocrinol.*, **7**, 8.
42. Brewer, G.J., Torricelli, J.R., Evege, E.K., and Price, P.J. (1993) Optimized survival of hippocampal neurons in B27-supplemented Neurobasal, a new serum-free medium combination. *J. Neurosci. Res.*, **35**, 567–576.
43. Wang, J.M., Johnston, P.B., Ball, B.G., and Brinton, R.D. (2005) The neurosteroid allopregnanolone promotes proliferation of rodent and human neural progenitor cells and regulates cell-cycle gene and protein expression. *J. Neurosci.*, **25**, 4706–4718.
44. Krueger, R.J. and Orme-Johnson, N.R. (1983) Acute adrenocorticotropic hormone stimulation of adrenal corticosteroidogenesis. Discovery of a rapidly induced protein. *J. Biol. Chem.*, **258**, 10159–10167.
45. Pon, L.A., Epstein, L.F., and Orme-Johnson, N.R. (1986) Acute cAMP stimulation in Leydig cells: rapid accumulation of a protein similar to that detected in adrenal cortex and corpus luteum. *Endocr. Res.*, **12**, 429–446.
46. Epstein, L.F. and Orme-Johnson, N.R. (1991) Regulation of steroid hormone biosynthesis. Identification of precursors of a phosphoprotein targeted to the mitochondrion in stimulated rat adrenal cortex cells. *J. Biol. Chem.*, **266**, 19739–19745.
47. Stocco, D.M. and Chen, W. (1991) Presence of identical mitochondrial proteins in unstimulated constitutive steroid-producing R2C rat Leydig tumor and stimulated nonconstitutive steroid-producing MA-10 mouse Leydig tumor cells. *Endocrinology*, **128**, 1918–1926.
48. Stocco, D.M. (2001) StAR protein and the regulation of steroid hormone biosynthesis. *Annu. Rev. Physiol.*, **63**, 193–213.
49. Yamazaki, T., Matsuoka, C., Gendou, M., Izumi, S., Zhao, D., Artemenko, I., Jefcoate, C.R., and Kominami, S. (2006) Mitochondrial processing of bovine adrenal steroidogenic acute regulatory protein. *Biochim. Biophys. Acta*, **1764**, 1561–1567.
50. Hong, S.H., Nah, H.Y., Lee, Y.J., Lee, J.W., Park, J.H., Kim, S.J., Lee, J.B., Yoon, H.S., and Kim, C.H. (2004) Expression of estrogen receptor-alpha and -beta, glucocorticoid receptor, and progesterone receptor genes in human embryonic stem cells and embryoid bodies. *Mol. Cells*, **18**, 320–325.
51. Fiala, C. (2006) Exelgyn, (February 2006). Mifegyne UK Summary of Product Characteristics (SPC) Retrieved on 9-9-08. http://emc.medicines.org.uk/emc/assets/c/html/displaydoc.asp?documentid=617. (accessed 28.09.2010).
52. Atwood, C.S., Hovey, R.C., Glover, J.P., Chepko, G., Ginsburg, E., Robison, W.G., and Vonderhaar, B.K. (2000) Progesterone induces side-branching of the ductal epithelium in the mammary glands of peripubertal mice. *J. Endocrinol.*, **167**, 39–52.
53. Mylonas, I., Jeschke, U., Shabani, N., Kuhn, C., Kunze, S., Dian, D., Friedl, C., Kupka, M.S., and Friese, K. (2007) Steroid receptors ERalpha, ERbeta, PR-A and PR-B are differentially expressed in normal and atrophic human endometrium. *Histol. Histopathol.*, **22**, 169–176.

54. Munoz-Sanjuan, I. and Brivanlou, A.H. (2002) Neural induction, the default model and embryonic stem cells. *Nat. Rev. Neurosci.*, **3**, 271–280.
55. Wahabi, H., Abed Althagafi, N., and Elawad, M. (2007) Progestogen for treating threatened miscarriage. *Cochrane Database Syst. Rev.*, **3**, CD005943.
56. Pomata, P.E., Colman-Lerner, A.A., Baranao, J.L., and Fiszman, M.L. (2000) In vivo evidences of early neurosteroid synthesis in the developing rat central nervous system and placenta. *Brain Res. Dev. Brain Res.*, **120**, 83–86.
57. Schumacher, M., Guennoun, R., Robert, F., Carelli, C., Gago, N., Ghoumari, A., Gonzalez Deniselle, M.C., Gonzalez, S.L., Ibanez, C., Labombarda, F., Coirini, H., Baulieu, E.E., and De Nicola, A.F. (2004) Local synthesis and dual actions of progesterone in the nervous system: neuroprotection and myelination. *Growth Horm. IGF Res.*, **14** (Suppl A), 18–33.
58. Schumacher, A., Arnhold, S., Addicks, K., and Doerfler, W. (2003) Staurosporine is a potent activator of neuronal, glial, and "CNS stem cell-like" neurosphere differentiation in murine embryonic stem cells. *Mol. Cell. Neurosci.*, **23**, 669–680.
59. Mauch, D.H., Nagler, K., Schumacher, S., Goritz, C., Muller, E.C., Otto, A., and Pfrieger, F.W. (2001) CNS synaptogenesis promoted by glia-derived cholesterol. *Science*, **294**, 1354–1357.
60. Ciriza, I., Carrero, P., Azcoitia, I., Lundeen, S.G., and Garcia-Segura, L.M. (2004) Selective estrogen receptor modulators protect hippocampal neurons from kainic acid excitotoxicity: differences with the effect of estradiol. *J. Neurobiol.*, **61**, 209–221.
61. VanLandingham, J.W., Cutler, S.M., Virmani, S., Hoffman, S.W., Covey, D.F., Krishnan, K., Hammes, S.R., Jamnongjit, M., and Stein, D.G. (2006) The enantiomer of progesterone acts as a molecular neuroprotectant after traumatic brain injury. *Neuropharmacology*, **51**, 1078–1085.
62. Cutler, S.M., VanLandingham, J.W., Murphy, A.Z., and Stein, D.G. (2006) Slow-release and injected progesterone treatments enhance acute recovery after traumatic brain injury. *Pharmacol. Biochem. Behav.*, **84**, 420–428.
63. Guo, Q., Sayeed, I., Baronne, L.M., Hoffman, S.W., Guennoun, R., and Stein, D.G. (2006) Progesterone administration modulates AQP4 expression and edema after traumatic brain injury in male rats. *Exp. Neurol.*, **198**, 469–478.
64. Wright, D.W., Kellermann, A.L., Hertzberg, V.S., Clark, P.L., Frankel, M., Goldstein, F.C., Salomone, J.P., Dent, L.L., Harris, O.A., Ander, D.S., Lowery, D.W., Patel, M.M., Denson, D.D., Gordon, A.B., Wald, M.M., Gupta, S., Hoffman, S.W., and Stein, D.G. (2007) ProTECT: a randomized clinical trial of progesterone for acute traumatic brain injury. *Ann. Emerg. Med.*, **49**, 391–402, e391–e392.
65. Heikinheimo, O., Kekkonen, R., and Lahteenmaki, P. (2003) The pharmacokinetics of mifepristone in humans reveal insights into differential mechanisms of antiprogestin action. *Contraception*, **68**, 421–426.
66. Bouchard, P. (1999) Progesterone and the progesterone receptor. *J. Reprod. Med.*, **44**, 153–157.
67. Ascoli, M., Fanelli, F., and Segaloff, D.L. (2002) The lutropin/choriogonadotropin receptor, a 2002 perspective. *Endocr. Rev.*, **23**, 141–174.
68. Mulac-Jericevic, B. and Conneely, O.M. (2004) Reproductive tissue selective actions of progesterone receptors. *Reproduction*, **128**, 139–146.
69. Bukovsky, A., Indrapichate, K., Fujiwara, H., Cekanova, M., Ayala, M.E., Dominguez, R., Caudle, M.R., Wimalsena, J., Elder, R.F., Copas, P., Foster, J.S., Fernando, R.I., Henley, D.C., and Upadhyaya, N.B. (2003) Multiple luteinizing hormone receptor (LHR) protein variants, interspecies reactivity of anti-LHR mAb clone 3B5, subcellular localization of LHR in human placenta, pelvic floor and brain,

and possible role for LHR in the development of abnormal pregnancy, pelvic floor disorders and Alzheimer's disease. *Reprod. Biol. Endocrinol.*, **1**, 46.

70. Avsian-Kretchmer, O. and Hsueh, A.J. (2004) Comparative genomic analysis of the eight-membered ring cystine knot-containing bone morphogenetic protein antagonists. *Mol. Endocrinol.*, **18**, 1–12.

71. Blithe, D.L., Richards, R.G., and Skarulis, M.C. (1991) Free alpha molecules from pregnancy stimulate secretion of prolactin from human decidual cells: a novel function for free alpha in pregnancy. *Endocrinology*, **129**, 2257–2259.

72. Berndt, S., Perrier d'Hauterive, S., Blacher, S., Pequeux, C., Lorquet, S., Munaut, C., Applanat, M., Herve, M.A., Lamande, N., Corvol, P., van den Brule, F., Frankenne, F., Poutanen, M., Huhtaniemi, I., Geenen, V., Noel, A., and Foidart, J.M. (2006) Angiogenic activity of human chorionic gonadotropin through LH receptor activation on endothelial and epithelial cells of the endometrium. *FASEB J.*, **20**, 2630–2632.

73. Licht, P., Russu, V., Lehmeyer, S., Moll, J., Siebzehnrubl, E., and Wildt, L. (2002) Intrauterine microdialysis reveals cycle-dependent regulation of endometrial insulin-like growth factor binding protein-1 secretion by human chorionic gonadotropin. *Fertil. Steril.*, **78**, 252–258.

74. Zygmunt, M., Herr, F., Keller-Schoenwetter, S., Kunzi-Rapp, K., Munstedt, K., Rao, C.V., Lang, U., and Preissner, K.T. (2002) Characterization of human chorionic gonadotropin as a novel angiogenic factor. *J. Clin. Endocrinol. Metab.*, **87**, 5290–5296.

75. Berndt, S., Blacher, S., Perrier d'Hauterive, S., Thiry, M., Tsampalas, M., Cruz, A., Pequeux, C., Lorquet, S., Munaut, C., Noel, A., and Foidart, J.M. (2009) Chorionic gonadotropin stimulation of angiogenesis and pericyte recruitment. *J. Clin. Endocrinol. Metab.*, **94**, 4567–4574.

76. Prior, J.C. (1990) Progesterone as a bone-trophic hormone. *Endocr. Rev.*, **11**, 386–398.

77. Rogers, P.A., Donoghue, J.F., Walter, L.M., and Girling, J.E. (2009) Endometrial angiogenesis, vascular maturation, and lymphangiogenesis. *Reprod. Sci.*, **16**, 147–151.

78. Corbett, A.D., Gillan, M.G., Kosterlitz, H.W., McKnight, A.T., Paterson, S.J., and Robson, L.E. (1984) Selectivities of opioid peptide analogues as agonists and antagonists at the delta-receptor. *Br. J. Pharmacol.*, **83**, 271–279.

79. Paterson, S.J., Corbett, A.D., Gillan, M.G., Kosterlitz, H.W., McKnight, A.T., and Robson, L.E. (1984) Radioligands for probing opioid receptors. *J. Recept. Res.*, **4**, 143–154.

80. Cemerikic, B., Schabbing, R., and Ahmed, M.S. (1992) Selectivity and potency of opioid peptides in regulating human chorionic gonadotropin release from term trophoblast tissue. *Peptides*, **13**, 897–903.

81. Gu, G. and Simerly, R.B. (1994) Hormonal regulation of opioid peptide neurons in the anteroventral periventricular nucleus. *Horm. Behav.*, **28**, 503–511.

82. Dufourny, L., Caraty, A., Clarke, I.J., Robinson, J.E., and Skinner, D.C. (2005) Progesterone-receptive beta-endorphin and dynorphin B neurons in the arcuate nucleus project to regions of high gonadotropin-releasing hormone neuron density in the ovine preoptic area. *Neuroendocrinology*, **81**, 139–149.

83. Harris, G.W. (1964) Sex hormones, brain development and brain function. *Endocrinology*, **75**, 627–648.

84. Young, W.C., Goy, R.W., and Phoenix, C.H. (1964) Hormones and sexual behavior. *Science*, **143**, 212–218.

85. Roof, R.L., Duvdevani, R., Braswell, L., and Stein, D.G. (1994) Progesterone facilitates cognitive recovery and reduces secondary neuronal loss caused by cortical contusion injury in male rats. *Exp. Neurol.*, **129**, 64–69.

86. Stein, D.G., Wright, D.W., and Kellermann, A.L. (2008) Does progesterone have neuroprotective properties? *Ann. Emerg. Med.*, **51**, 164–172.
87. Gould, E., Tanapat, P., Rydel, T., and Hastings, N. (2000) Regulation of hippocampal neurogenesis in adulthood. *Biol. Psychiatry*, **48**, 715–720.
88. McEwan, P.E., Lindop, G.B., and Kenyon, C.J. (1996) Control of cell proliferation in the rat adrenal gland in vivo by the renin-angiotensin system. *Am. J. Physiol.*, **271**, E192–E198.
89. Masumoto, A., Natori, S., Iwamoto, H., Uchida, E., Ohashi, M., Sakamoto, S., and Nawata, H. (1991) Effect of insulin, glucagon or dexamethasone on the production of insulin-like growth factor I in cultured rat hepatocytes. *Fukuoka Igaku Zasshi*, **82**, 136–141.
90. Leranth, C., Shanabrough, M., and Redmond, D.E. Jr. (2002) Gonadal hormones are responsible for maintaining the integrity of spine synapses in the CA1 hippocampal subfield of female nonhuman primates. *J. Comp. Neurol.*, **447**, 34–42.
91. Simerly, R.B. (2002) Wired for reproduction: organization and development of sexually dimorphic circuits in the mammalian forebrain. *Annu. Rev. Neurosci.*, **25**, 507–536.
92. Woolley, C.S. and McEwen, B.S. (1993) Roles of estradiol and progesterone in regulation of hippocampal dendritic spine density during the estrous cycle in the rat. *J. Comp. Neurol.*, **336**, 293–306.
93. Leranth, C., Petnehazy, O., and MacLusky, N.J. (2003) Gonadal hormones affect spine synaptic density in the CA1 hippocampal subfield of male rats. *J. Neurosci.*, **23**, 1588–1592.
94. Roof, R.L., Hoffman, S.W., and Stein, D.G. (1997) Progesterone protects against lipid peroxidation following traumatic brain injury in rats. *Mol. Chem. Neuropathol.*, **31**, 1–11.
95. Attella, M.J., Nattinville, A., and Stein, D.G. (1987) Hormonal state affects recovery from frontal cortex lesions in adult female rats. *Behav. Neural. Biol.*, **48**, 352–367.
96. Asbury, E.T., Fritts, M.E., Horton, J.E., and Isaac, W.L. (1998) Progesterone facilitates the acquisition of avoidance learning and protects against subcortical neuronal death following prefrontal cortex ablation in the rat. *Behav. Brain Res.*, **97**, 99–106.
97. Chang, L.L., Lo, M.J., Kan, S.F., Huang, W.J., Chen, J.J., Kau, M.M., Wang, J.L., Lin, H., Tsai, S.C., Chiao, Y.C., Yeh, J.Y., Wun, W.S., and Wang, P.S. (1999) Direct effects of prolactin on corticosterone release by zona fasciculata-reticularis cells from male rats. *J. Cell. Biochem.*, **73**, 563–572.
98. Kumon, Y., Kim, S.C., Tompkins, P., Stevens, A., Sakaki, S., and Loftus, C.M. (2000) Neuroprotective effect of postischemic administration of progesterone in spontaneously hypertensive rats with focal cerebral ischemia. *J. Neurosurg.*, **92**, 848–852.
99. Galani, R., Hoffman, S.W., and Stein, D.G. (2001) Effects of the duration of progesterone treatment on the resolution of cerebral edema induced by cortical contusions in rats. *Restor. Neurol. Neurosci.*, **18**, 161–166.
100. Wright, D.W., Bauer, M.E., Hoffman, S.W., and Stein, D.G. (2001) Serum progesterone levels correlate with decreased cerebral edema after traumatic brain injury in male rats. *J. Neurotrauma.*, **18**, 901–909.
101. Grossman, K.J., Goss, C.W., and Stein, D.G. (2004) Effects of progesterone on the inflammatory response to brain injury in the rat. *Brain Res.*, **1008**, 29–39.
102. He, J., Evans, C.O., Hoffman, S.W., Oyesiku, N.M., and Stein, D.G. (2004) Progesterone and allopregnanolone reduce inflammatory cytokines after traumatic brain injury. *Exp. Neurol.*, **189**, 404–412.
103. O'Connor, C.A., Cernak, I., and Vink, R. (2005) Both estrogen and progesterone attenuate edema formation following diffuse traumatic brain injury in rats. *Brain Res.*, **1062**, 171–174.
104. Pettus, E.H., Wright, D.W., Stein, D.G., and Hoffman, S.W. (2005) Progesterone treatment inhibits the inflammatory agents that accompany

traumatic brain injury. *Brain Res.*, **1049**, 112–119.
105. Stein, D.G. (2005) The case for progesterone. *Ann. N. Y. Acad. Sci.*, **1052**, 152–169.
106. Leonelli, E., Bianchi, R., Cavaletti, G., Caruso, D., Crippa, D., Garcia-Segura, L.M., Lauria, G., Magnaghi, V., Roglio, I., and Melcangi, R.C. (2007) Progesterone and its derivatives are neuroprotective agents in experimental diabetic neuropathy: a multimodal analysis. *Neuroscience*, **144**, 1293–1304.
107. Lois, C. and Alvarez-Buylla, A. (1993) Proliferating subventricular zone cells in the adult mammalian forebrain can differentiate into neurons and glia. *Proc. Natl. Acad. Sci. U.S.A.*, **90**, 2074–2077.
108. Kempermann, G. and Gage, F.H. (2000) Neurogenesis in the adult hippocampus. *Novartis Found. Symp.*, **231**, 220–235; discussion 235–241, 302–226.
109. Chirumamilla, S., Sun, D., Bullock, M.R., and Colello, R.J. (2002) Traumatic brain injury induced cell proliferation in the adult mammalian central nervous system. *J. Neurotrauma*, **19**, 693–703.
110. Rice, A.C., Khaldi, A., Harvey, H.B., Salman, N.J., White, F., Fillmore, H., and Bullock, M.R. (2003) Proliferation and neuronal differentiation of mitotically active cells following traumatic brain injury. *Exp. Neurol.*, **183**, 406–417.
111. Sun, D., Colello, R.J., Daugherty, W.P., Kwon, T.H., McGinn, M.J., Harvey, H.B., and Bullock, M.R. (2005) Cell proliferation and neuronal differentiation in the dentate gyrus in juvenile and adult rats following traumatic brain injury. *J. Neurotrauma*, **22**, 95–105.
112. Dash, P.K., Mach, S.A., and Moore, A.N. (2001) Enhanced neurogenesis in the rodent hippocampus following traumatic brain injury. *J. Neurosci. Res.*, **63**, 313–319.
113. Yoshimura, S., Teramoto, T., Whalen, M.J., Irizarry, M.C., Takagi, Y., Qiu, J., Harada, J., Waeber, C., Breakefield, X.O., and Moskowitz, M.A. (2003) FGF-2 regulates neurogenesis and degeneration in the dentate gyrus after traumatic brain injury in mice. *J. Clin. Invest.*, **112**, 1202–1210.
114. Kleindienst, A., McGinn, M.J., Harvey, H.B., Colello, R.J., Hamm, R.J., and Bullock, M.R. (2005) Enhanced hippocampal neurogenesis by intraventricular S100B infusion is associated with improved cognitive recovery after traumatic brain injury. *J. Neurotrauma*, **22**, 645–655.
115. Emery, C.A. (2005) Injury prevention and future research. *Med. Sport Sci.*, **49**, 170–191.
116. Sun, D., McGinn, M.J., Zhou, Z., Harvey, H.B., Bullock, M.R., and Colello, R.J. (2007) Anatomical integration of newly generated dentate granule neurons following traumatic brain injury in adult rats and its association to cognitive recovery. *Exp. Neurol.*, **204**, 264–272.
117. Mahmood, A., Lu, D., Qu, C., Goussev, A., and Chopp, M. (2005) Human marrow stromal cell treatment provides long-lasting benefit after traumatic brain injury in rats. *Neurosurgery*, **57**, 1026–1031; discussion 1026–1031.
118. Schmidt, R.H., Scholten, K.J., and Maughan, P.H. (1999) Time course for recovery of water maze performance and central cholinergic innervation after fluid percussion injury. *J. Neurotrauma*, **16**, 1139–1147.
119. Carr, B.R. (1998) in *Williams Textbook of Endocrinology* (eds J.D. Wilson, H.M. Kronenberg, and P.R. Larsen), WB Saunders Co., Philadelphia, PA, pp. 751–817.
120. Reichlin, S. (1998) in *Williams Textbook of Endocrinology*, 10 edn (eds J.D. Wilson, D.W. Foster, H.M. Kronenberg, and P.R. Larsen), WB Saunders Co., Philadelphia, PA, pp. 165–248.
121. Gray, P.C., Bilizikjian, L.M., and Vale, W. (2002) Antagonism of activin by inhibin and inhibin receptors: a functional role for betaglycan. *Mol. Cell. Endocrinol.*, **188**, 254–260.
122. Schwall, R.H., Szonyi, E., Mason, A.J., and Nikolics, K. (1988) Activin stimulates secretion of follicle-stimulating

hormone from pituitary cells desensitized to gonadotropin-releasing hormone. *Biochem. Biophys. Res. Commun.*, **151**, 1099–1104.
123. MacConell, L.A., Lawson, M.A., Mellon, P.L., and Roberts, V.J. (1999) Activin A regulation of gonadotropin-releasing hormone synthesis and release in vitro. *Neuroendocrinology*, **70**, 246–254.
124. Weiss, J., Crowley, W.F., Halvorson, L.M., and Jameson, J.L. Jr. (1993) Perifusion of rat pituitary cells with gonadotropin-releasing hormone, activin, and inhibin reveals distinct effects on gonadotropin gene expression and secretion. *Endocrinology*, **132**, 2307–2311.
125. Chakravarti, S., Collins, W.P., Forecast, J.D., Newton, J.R., Oram, D.H., and Studd, J.W. (1976) Hormonal profiles after the menopause. *Br. Med. J.*, **2**, 784–787.
126. Neaves, W.B., Johnson, L., Porter, J.C., Parker, C.R., and Petty, C.S. Jr. (1984) Leydig cell numbers, daily sperm production, and serum gonadotropin levels in aging men. *J. Clin. Endocrinol. Metab.*, **59**, 756–763.
127. Reame, N.E., Kelche, R.P., Beitins, I.Z., Yu, M.Y., Zawacki, C.M., and Padmanabhan, V. (1996) Age effects of follicle-stimulating hormone and pulsatile luteinizing hormone secretion across the menstrual cycle of premenopausal women. *J. Clin. Endocrinol. Metab.*, **81**, 1512–1518.
128. Schmidt, P.J., Gindoff, P.R., Baron, D.A., and Rubinow, D.R. (1996) Basal and stimulated gonadotropin levels in the perimenopause. *Am. J. Obstet. Gynecol.*, **175**, 643–650.
129. Meethal, S.V., Liu, T., Chan, H.W., Ginsburg, E., Wilson, A.C., Gray, D.N., Bowen, R.L., Vonderhaar, B.K., and Atwood, C.S. (2009) Identification of a regulatory loop for the synthesis of neurosteroids: a steroidogenic acute regulatory protein-dependent mechanism involving hypothalamic-pituitary-gonadal axis receptors. *J. Neurochem.*, **110**, 1014–1027.
130. Sriraman, V., Rao, V.S., Rajesh, N., Vasan, S.S., and Rao, A.J. (2001) A preliminary study on the possible role for luteinizing hormone in androgen independent growth of prostate. *Reprod. Biomed. Online*, **3**, 6–13.
131. Harris, D., Bonfil, D., Chuderland, D., Kraus, S., Seger, R., and Naor, Z. (2002) Activation of MAPK cascades by GnRH: ERK and Jun N-terminal kinase are involved in basal and GnRH-stimulated activity of the glycoprotein hormone LHbeta-subunit promoter. *Endocrinology*, **143**, 1018–1025.
132. Horiuchi, A., Nikaido, T., Yoshizawa, T., Itoh, K., Kobayashi, Y., Toki, T., Konishi, I., and Fujii, S. (2000) HCG promotes proliferation of uterine leiomyomal cells more strongly than that of myometrial smooth muscle cells in vitro. *Mol. Hum. Reprod.*, **6**, 523–528.
133. Davies, B.R., Finnigan, D.S., Smith, S.K., and Ponder, B.A., and Aogsponmose, G.E. (1999) Administration of gonadotropins stimulates proliferation of normal mouse ovarian surface epithelium. *Gynecol. Endocrinol.*, **13**, 75–81.
134. Webber, R.J. and Sokoloff, L. (1981) In vitro culture of rabbit growth plate chondrocytes. 1. Age-dependence of response to fibroblast growth factor and "chondrocyte growth factor". *Growth*, **45**, 252–268.
135. Whitfield, G.K. and Kourides, I.A. (1985) Expression of chorionic gonadotropin alpha- and beta-genes in normal and neoplastic human tissues: relationship to deoxyribonucleic acid structure. *Endocrinology*, **117**, 231–236.
136. Yokotani, T., Koizumi, T., Taniguchi, R., Nakagawa, T., Isobe, T., Yoshimura, M., Tsubota, N., Hasegawa, K., Ohsawa, N., Baba, S., Yasui, H., and Nishimura, R. (1997) Expression of alpha and beta genes of human chorionic gonadotropin in lung cancer. *Int. J. Cancer*, **71**, 539–544.
137. Krichevsky, A., Campbell-Acevedo, E.A., Tong, J.Y., and Acevedo, H.F. (1995) Immunological detection of membrane-associated human luteinizing hormone correlates with gene

expression in cultured human cancer and fetal cells. *Endocrinology*, **136**, 1034–1039.
138. Raina, A.K., Zhu, X., Rottkamp, C.A., Monteiro, M., Takeda, A., and Smith, M.A. (2000) Cyclin' toward dementia: cell cycle abnormalities and abortive oncogenesis in Alzheimer disease. *J. Neurosci. Res.*, **61**, 128–133.
139. Herrup, K. and Yang, Y. (2007) Cell cycle regulation in the postmitotic neuron: oxymoron or new biology? *Nat. Rev. Neurosci.*, **8**, 368–378.
140. Takashima, S., Kuruta, H., Mito, T., Nishizawa, M., Kunishita, T., and Tabira, T. (1990) Developmental and aging changes in the expression patterns of beta-amyloid in the brains of normal and Down syndrome cases. *Brain Dev.*, **12**, 367–371.
141. Arai, Y., Suzuki, A., Mizuguchi, M., and Takashima, S. (1997) Developmental and aging changes in the expression of amyloid precursor protein in Down syndrome brains. *Brain Dev.*, **19**, 290–294.
142. Berezovska, O., Xia, M.Q., Page, K., Wasco, W., Tanzi, R.E., and Hyman, B.T. (1997) Developmental regulation of presenilin mRNA expression parallels notch expression. *J. Neuropathol. Exp. Neurol.*, **56**, 40–44.
143. Goedert, M., Jakes, R., Crowther, R.A., Six, J., Lubke, U., Vandermeeren, M., Cras, P., Trojanowski, J.Q., and Lee, V.M. (1993) The abnormal phosphorylation of tau protein at Ser-202 in Alzheimer disease recapitulates phosphorylation during development. *Proc. Natl. Acad. Sci. U.S.A.*, **90**, 5066–5070.
144. Liu, T., Perry, G., Chan, H.W., Verdile, G., Martins, R.N., Smith, M.A., and Atwood, C.S. (2004) Amyloid-beta-induced toxicity of primary neurons is dependent upon differentiation-associated increases in tau and cyclin-dependent kinase 5 expression. *J. Neurochem.*, **88**, 554–563.
145. Lopez-Toledano, M.A. and Shelanski, M.L. (2004) Neurogenic effect of beta-amyloid peptide in the development of neural stem cells. *J. Neurosci.*, **24**, 5439–5444.
146. Calafiore, M., Battaglia, G., Zappala, A., Trovato-Salinaro, E., Caraci, F., Caruso, M., Vancheri, C., Sortino, M.A., Nicoletti, F., and Copani, A. (2006) Progenitor cells from the adult mouse brain acquire a neuronal phenotype in response to beta-amyloid. *Neurobiol. Aging*, **27**, 606–613.
147. Heo, C., Chang, K.A., Choi, H.S., Kim, H.S., Kim, S., Liew, H., Kim, J.A., Yu, E., Ma, J., and Suh, Y.H. (2007) Effects of the monomeric, oligomeric, and fibrillar Abeta42 peptides on the proliferation and differentiation of adult neural stem cells from subventricular zone. *J. Neurochem.*, **102**, 493–500.
148. Jin, K., Galvan, V., Xie, L., Mao, X.O., Gorostiza, O.F., Bredesen, D.E., and Greenberg, D.A. (2004) Enhanced neurogenesis in Alzheimer's disease transgenic (PDGF-APPSw,Ind) mice. *Proc. Natl. Acad. Sci. U.S.A.*, **101**, 13363–13367.
149. Lopez-Toledano, M.A. and Shelanski, M.L. (2007) Increased neurogenesis in young transgenic mice overexpressing human APP_(Sw, Ind). *J. Alzheimers Dis.*, **12**, 229–240.
150. Citron, M., Westaway, D., Xia, W., Carlson, G., Diehl, T., Levesque, G., Johnson-Wood, K., Lee, M., Seubert, P., Davis, A., Kholodenko, D., Motter, R., Sherrington, R., Perry, B., Yao, H., Strome, R., Lieberburg, I., Rommens, J., Kim, S., Schenk, D., Fraser, P., St George Hyslop, P., and Selkoe, D.J. (1997) Mutant presenilins of Alzheimer's disease increase production of 42-residue amyloid beta-protein in both transfected cells and transgenic mice. *Nat. Med.*, **3**, 67–72.
151. McPhie, D.L., Coopersmith, R., Hines-Peralta, A., Chen, Y., Ivins, K.J., Manly, S.P., Kozlowski, M.R., Neve, K.A., and Neve, R.L. (2003) DNA synthesis and neuronal apoptosis caused by familial Alzheimer disease mutants of the amyloid precursor protein are mediated by the p21 activated kinase PAK3. *J. Neurosci.*, **23**, 6914–6927.
152. Trapp, B.D. and Hauer, P.E. (1994) Amyloid precursor protein is enriched

in radial glia: implications for neuronal development. *J. Neurosci. Res.*, **37**, 538–550.

153. Gralle, M. and Ferreira, S.T. (2007) Structure and functions of the human amyloid precursor protein: the whole is more than the sum of its parts. *Prog. Neurobiol.*, **82**, 11–32.

154. Milward, E.A., Papadopoulos, R., Fuller, S.J., Moir, R.D., Small, D., Beyreuther, K., and Masters, C.L. (1992) The amyloid protein precursor of Alzheimer's disease is a mediator of the effects of nerve growth factor on neurite outgrowth. *Neuron*, **9**, 129–137.

155. Jaffe, A.B., Toran-Allerand, C.D., Greengard, P., and Gandy, S.E. (1994) Estrogen regulates metabolism of Alzheimer amyloid beta precursor protein. *J. Biol. Chem.*, **269**, 13065–13068.

156. Xu, H., Gouras, G.K., Greenfield, J.P., Vincent, B., Naslund, J., Mazzarelli, L., Fried, G., Jovanovic, J.N., Seeger, M., Relkin, N.R., Liao, F., Checler, F., Buxbaum, J.D., Chait, B.T., Thinakaran, G., Sisodia, S.S., Wang, R., Greengard, P., and Gandy, S. (1998) Estrogen reduces neuronal generation of Alzheimer beta-amyloid peptides. *Nat. Med.*, **4**, 447–451.

157. Greenfield, J.P., Leung, L.W., Cai, D., Kaasik, K., Gross, R.S., Rodriguez-Boulan, E., Greengard, P., and Xu, H. (2002) Estrogen lowers Alzheimer beta-amyloid generation by stimulating trans-Golgi network vesicle biogenesis. *J. Biol. Chem.*, **277**, 12128–12136.

158. Manthey, D., Heck, S., Engert, S., and Behl, C. (2001) Estrogen induces a rapid secretion of amyloid beta precursor protein via the mitogen-activated protein kinase pathway. *Eur. J. Biochem.*, **268**, 4285–4291.

159. Petanceska, S.S., Nagy, V., Frail, D., and Gandy, S. (2000) Ovariectomy and 17β-estradiol modulate the levels of Alzheimer's amyloid beta peptides in brain. *Neurology*, **54**, 2212–2217.

160. Bowen, R.L., Verdile, G., Liu, T., Parlow, A.F., Perry, G., Smith, M.A., Martins, R.N., and Atwood, C.S. (2004) Luteinizing hormone, a reproductive regulator that modulates the processing of amyloid-beta precursor protein and amyloid-beta deposition. *J. Biol. Chem.*, **279**, 20539–20545.

161. Casadesus, G., Webber, K.M., Atwood, C.S., Pappolla, M.A., Perry, G., Bowen, R.L., and Smith, M.A. (2006) Luteinizing hormone modulates cognition and amyloid-beta deposition in Alzheimer APP transgenic mice. *Biochim. Biophys. Acta*, **1762**, 447–452.

162. Berry, A., Tomidokoro, Y., Ghiso, J., and Thornton, J. (2008) Human chorionic gonadotropin (a luteinizing hormone homologue) decreases spatial memory and increases brain amyloid-beta levels in female rats. *Horm. Behav.*, **54**, 143–152.

163. Lin, J., Li, X., Yuan, F., Lin, L., Cook, C.L., Rao, Ch V., and Lei, Z. (2010) Genetic ablation of luteinizing hormone receptor improves the amyloid pathology in a mouse model of Alzheimer disease. *J. Neuropathol. Exp. Neurol.*, **69**, 253–261.

164. Herrup, K., Neve, R., Ackerman, S.L., and Copani, A. (2004) Divide and die: cell cycle events as triggers of nerve cell death. *J. Neurosci.*, **24**, 9232–9239.

165. Frye, C.A. and Walf, A.A. (2010) Progesterone enhances learning and memory of aged wildtype and progestin receptor knockout mice. *Neurosci. Lett.* **472**, 38–42.

166. Markham, J.A., Pych, J.C., and Juraska, J.M. (2002) Ovarian hormone replacement to aged ovariectomized female rats benefits acquisition of the Morris water maze. *Horm. Behav.*, **42**, 284–293.

167. Sandstrom, N.J. and Williams, C.L. (2001) Memory retention is modulated by acute estradiol and progesterone replacement. *Behav. Neurosci.*, **115**, 384–393.

168. Gibbs, R.B. (2000) Long-term treatment with estrogen and progesterone enhances acquisition of a spatial memory task by ovariectomized aged rats. *Neurobiol. Aging*, **21**, 107–116.

169. Sato, T., Tanaka, K., Ohnishi, Y., Teramoto, T., Irifune, M., and Nishikawa, T. (2004) Effects of estradiol

and progesterone on radial maze performance in middle-aged female rats fed a low-calcium diet. *Behav. Brain Res.*, **150**, 33–42.

170. Carroll, J.C., Rosario, E.R., Chang, L., Stanczyk, F.Z., Oddo, S., LaFerla, F.M., and Pike, C.J. (2007) Progesterone and estrogen regulate Alzheimer-like neuropathology in female 3xTg-AD mice. *J. Neurosci.*, **27**, 13357–13365.

171. Azcoitia, I., Fernandez-Galaz, C., Sierra, A., and Garcia-Segura, L.M. (1999) Gonadal hormones affect neuronal vulnerability to excitotoxin-induced degeneration. *J. Neurocytol.*, **28**, 699–710.

172. Toung, T.J., Chen, T.Y., Littleton-Kearney, M.T., Hurn, P.D., and Murphy, S.J. (2004) Effects of combined estrogen and progesterone on brain infarction in reproductively senescent female rats. *J. Cereb. Blood Flow Metab.*, **24**, 1160–1166.

173. Frye, C.A. and Scalise, T.J. (2000) Anti-seizure effects of progesterone and 3α,5α-THP in kainic acid and perforant pathway models of epilepsy. *Psychoneuroendocrinology*, **25**, 407–420.

174. Rosario, E.R., Ramsden, M., and Pike, C.J. (2006) Progestins inhibit the neuroprotective effects of estrogen in rat hippocampus. *Brain Res.*, **1099**, 206–210.

175. Jayaraman, A. and Pike, C.J. (2009) Progesterone attenuates oestrogen neuroprotection via downregulation of oestrogen receptor expression in cultured neurons. *J. Neuroendocrinol.*, **21**, 77–81.

176. Brinton, R.D., Thompson, R.F., Foy, M.R., Baudry, M., Wang, J., Finch, C.E., Morgan, T.E., Pike, C.J., Mack, W.J., Stanczyk, F.Z., and Nilsen, J. (2008) Progesterone receptors: form and function in brain. *Front Neuroendocrinol.*, **29**, 313–339.

177. Gibbs, R.B. (2000) Effects of gonadal hormone replacement on measures of basal forebrain cholinergic function. *Neuroscience*, **101**, 931–938.

178. Walter, L.M., Rogers, P.A., and Girling, J.E. (2005) The role of progesterone in endometrial angiogenesis in pregnant and ovariectomised mice. *Reproduction*, **129**, 765–777.

179. Wilson, A.C., Clemente, L., Liu, T., Bowen, R.L., Meethal, S.V., and Atwood, C.S. (2008) Reproductive hormones regulate the selective permeability of the blood-brain barrier. *Biochim. Biophys. Acta*, **1782**, 401–407.

180. Bake, S. and Sohrabji, F. (2004) 17β-estradiol differentially regulates blood-brain barrier permeability in young and aging female rats. *Endocrinology*, **145**, 5471–5475.

181. Bake, S., Friedman, J.A., and Sohrabji, F. (2009) Reproductive age-related changes in the blood brain barrier: Expression of IgG and tight junction proteins. *Microvasc. Res.*, **78**, 413–424.

182. Ye, L., Martin, T.A., Parr, C., Harrison, G.M., Mansel, R.E., and Jiang, W.G. (2003) Biphasic effects of 17-beta-estradiol on expression of occludin and transendothelial resistance and paracellular permeability in human vascular endothelial cells. *J. Cell. Physiol.*, **196**, 362–369.

183. Zeng, R., Li, X., and Gorodeski, G.I. (2004) Estrogen abrogates transcervical tight junctional resistance by acceleration of occludin modulation. *J. Clin. Endocrinol. Metab.*, **89**, 5145–5155.

19
Human Neural Progenitor Cells: Mitotic and Neurogenic Effects of Growth Factors, Neurosteroids, and Excitatory Amino Acids

Masatoshi Suzuki, Jacalyn McHugh, and Narisorn Kitiyanant

19.1
Introduction

Over the last few decades, the discovery of stem cells has raised hopes with reference to assisting the natural healing processes of damaged tissues and prevention of cell or tissue loss due to degenerative diseases. For neuroscience, neural stem cells (NSCs) hold great promise for neurodegenerative diseases including Alzheimer's disease, Parkinson's disease, amyotrophic lateral sclerosis, and spinal cord injury. NSCs have been isolated at all stages of development, from the early developing embryo to the adult organism. These cells are unspecialized ones that can replenish themselves indefinitely and can be directed to generate specialized cells of the body. Under appropriate conditions, NSCs can be expanded to large numbers in culture and differentiated into restricted cell types of brain cells: neurons and glial cells. The differentiated neurons and glial cells can be used for developmental or disease modeling *in vitro* or cell-based therapy to transplant into patients [1, 2].

We have used human neural stem/progenitor cell (hNPCs) cultures as an ideal and innovative model to understand the developmental and molecular mechanisms of growth factors and other possible molecules such as neurosteroids and excitatory amino acids in the human central nervous system (CNS). This review summarizes our recent observations using hNPCs to elucidate mitotic and neurogenic effects of these factors and discusses how these cultures reflect human cortical development *in vivo*.

19.2
Neural Stem/Progenitor Cells as a Model of Human Cortical Development

Stem cells have been found to reside in all embryonic and adult tissues investigated so far, including fast and slowly renewing tissues (such as brain) and throughout the life of an individual. *NSCs* are defined as clonogenic cells capable of self-renewal and multipotent differentiation into both neuronal and glial lineages [3, 4]. These cells can be isolated from different sources such as embryonic stem (ES) cells, fetal and adult tissues, and also induced pluripotent stem (iPS) cells – a new technique to establish pluripotent stem cells from adult somatic cells by overexpressing a set of pluripotent genes [5, 6]. These cells can be used for the direct examination of proliferation, differentiation, and migration in the culture dish [3, 4]. These cells are regionally specified based on the brain region from which they were isolated [7, 8] and also differ between species [9]. NSCs can be cultured using either of two principal schemes: (i) the generation of floating spherical aggregates termed *neurospheres* [10] or (ii) the generation of monolayer cultures on substrate-coated tissue culture plates [3, 11]. NSCs are responsive to the mitogens epidermal growth factor (EGF) and fibroblast growth factor-2 (FGF-2) [12].

Human neurospheres can be easily isolated from human fetal tissues and maintained in culture. We have grown them from the human fetal cortex for extended periods of time in EGF, FGF-2, and leukemia inhibitory factor (LIF) [10, 13]. Human neurosphere lines in our lab have been established by dissection of the developing human fetal brain between 90 and 120 days of gestation and expanded as neurospheres in the presence of EGF, FGF-2, and heparin for two to four weeks after isolation (Figure 19.1). For passaging of neurospheres, we used a chopping method that does not require trypsin or mechanical dissociation, and cell/cell contact was continuously maintained [10]. The neurospheres are then cultured in a defined medium containing only EGF for 15–30 weeks. LIF is added in late passages to increase cell proliferation [13]. These cultures have shown that they represent a fairly homogeneous population of cells, which are immunoreactive for the NSC markers nestin and glial fibrillary acidic protein (GFAP) [14, 15]. Although we have previously designated them as long-term human NSCs [16], we recently found that proliferation reduces after 50 population doublings and neuronal production declines (Figure 19.1) [15]. Our cultures may represent a transient amplifying population of hNPCs. These cells maintain the developmental timeline of neural/astrocyte production in human cortex *in vivo*.

19.3
Mitotic and Neurogenic Effects of a Neurosteroid: Dehydroepiandrosterone (DHEA)

The development of the cerebral cortex takes place through coordinated cell division, migration, and differentiation of progenitor cells. Both extrinsic and intrinsic factors have been suggested to play a role in the regulation of cortical neurogenesis – the birth of new neurons from progenitor cells. hNPCs may represent an ideal and innovative model to understand the developmental and molecular mechanisms of these factors in human CNS. Using hNPCs cultures, we have demonstrated the potential roles of a neurosteroid, dehydroepiandrosterone (DHEA), in cell proliferation and neurogenesis [17].

DHEA and its sulfate are the most abundant steroids in the blood of young adult humans. Levels of DHEA peak at 20 years of age and then decline to reach values of 20–30% at 70–80 years of age [18]. Significant interest has arisen from the hypothesis that declining DHEA concentrations in adults may serve as an indicator

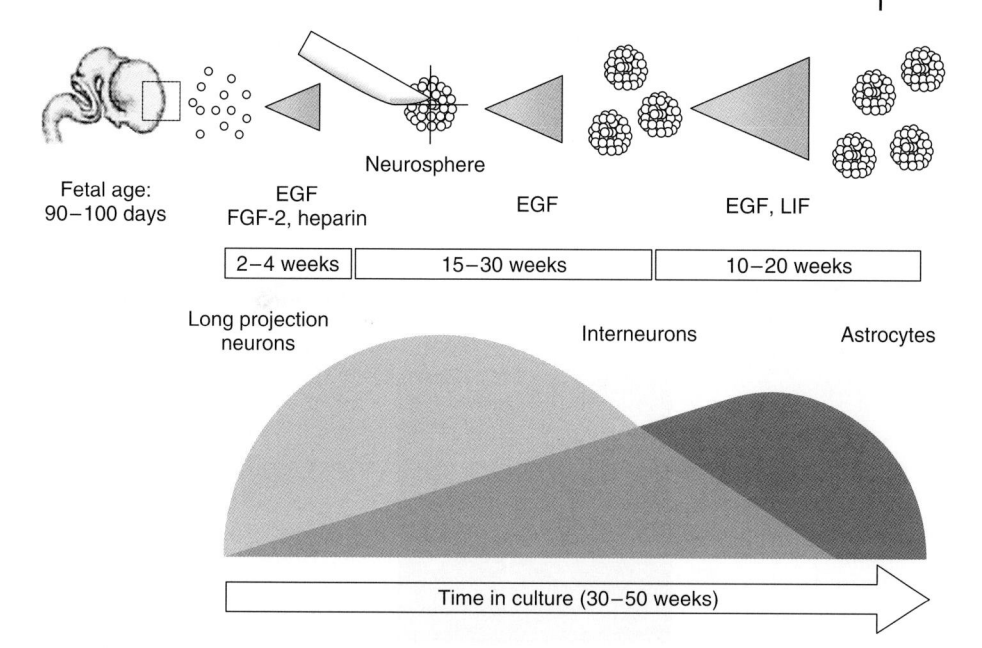

Figure 19.1 Schematic timeline of human neural progenitor cell (hNPCs) culture. Human NPC lines have been established by dissection from the developing human fetal brain between 90 and 120 days of gestation and expanded as neurospheres in the presence of EGF, FGF-2, and heparin for two to four weeks after isolation. The cells are then cultured in a defined medium containing only EGF for 15–30 weeks. LIF is added in late passages for 10–20 weeks to increase cell proliferation. These cultures have shown that they represent a fairly homogeneous population of cells immunoreactive for neural progenitor cell markers such as nestin and glial fibrillary acidic protein (GFAP). We recently found that proliferation reduces after 50 population doublings and neuronal production declines, indicating that our cultures may represent a transient amplifying population of hNPCs. These cells maintain the developmental timeline of neural/astrocyte production in human cortex *in vivo*.

of a number of conditions including the loss of insulin sensitivity, obesity, diabetes, cardiovascular diseases, stress, and aging [19]. In clinical studies, levels of DHEA were found to decline in mental illnesses such as a major depressive disorder or in systemic diseases that respond to DHEA supplementation [20, 21]. While DHEA represents the most abundant steroid product of the adrenal cortex, it has also been identified as a "neurosteroid." Neurosteroids are synthesized *de novo* in the CNS independent, at least in part, of peripheral organ activity [22, 23]. Neurosteroids possess the ability to affect neurons through activation of N-methyl-D-aspartate (NMDA), γ-aminobutyric acid (GABA)$_A$, and the opiate sigma receptor [22, 23], although the exact role of each with regard to DHEA is not well established. DHEA has been known to show multiple effects such as enhancing memory and neuroprotective effects in rodent studies [24, 25]. DHEA has also been shown to be a potential signaling molecule in neuronal differentiation during development [26] and has recently been shown to increase neurogenesis in the adult rodent hippocampus [27]. However, most studies on DHEA have been performed in rodents, and there is little direct evidence for biological effects on the human CNS. Furthermore, levels of DHEA in the adult rat brain are very low in contrast to the human brain [23, 28]. Thus, there are some questions as to the relevance of the effect of DHEA on rodents and how this relates to the human situation. To answer these questions, we recently used hNPCs cultures as an innovative model to

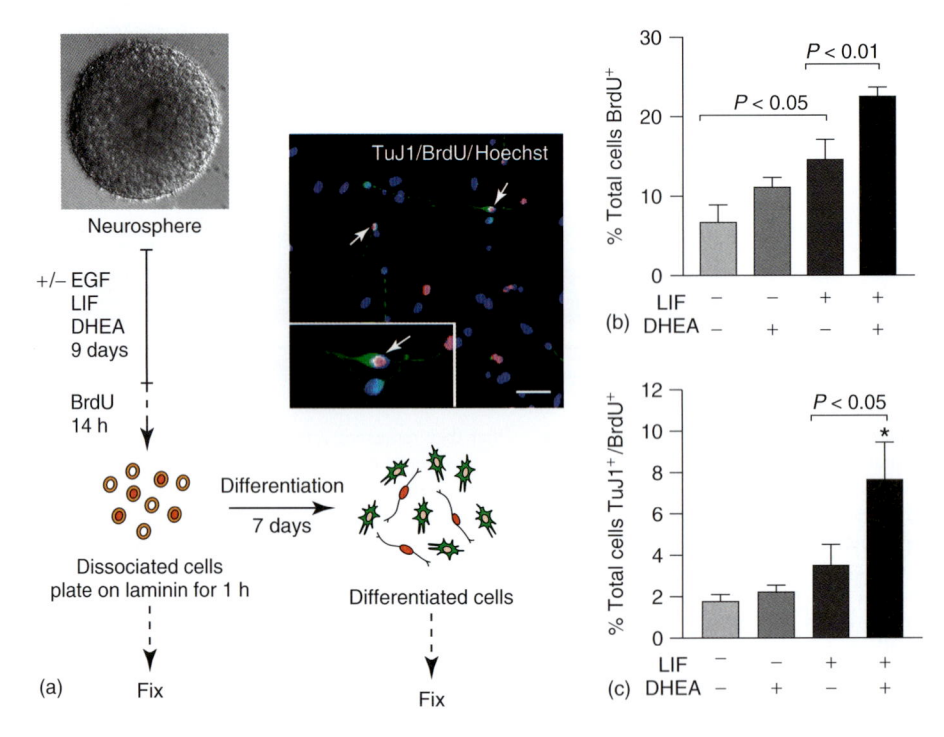

Figure 19.2 Mitotic and neurogenic effects of dehydroepiandrosterone (DHEA) in hNPCs. (a) Schematic describing culture schedules and treatments for hNPCs. Neurospheres were cultured with basal medium in the presence or absence of EGF, LIF, and DHEA. After nine days, neurospheres were pulsed with BrdU for 14 h. The cultures were then dissociated into a single-cell suspension with enzyme and acutely plated onto coverslips. The fixed cells were immunostained with BrdU or other cell markers. For differentiation studies, the plated cells were differentiated for seven days under serum-free conditions and then visualized with antibodies against a neuronal marker β-tubulin-III (TuJ1) and BrdU. Many TuJ1 (green) and BrdU (red) double-positive cells (designated by arrows) were found after seven days of differentiation (Scale bar, 40 µm). (b) DHEA increases the number of BrdU cells in hNPCs in the presence of EGF and LIF, although no significant effect was observed with DHEA in the absence of LIF. (c) hNPCs grown in the presence of DHEA generated more maturing neurons. Neurospheres were cultured in the presence or absence of EGF, LIF, and DHEA for nine days, pulsed with BrdU, plated, and differentiated for seven days in differentiation medium. The numbers of TuJ1$^+$/BrdU$^+$ double-positive cells were not affected by either DHEA or LIF alone. However, TuJ1$^+$/BrdU$^+$ double-positive cells were increased by LIF/DHEA. *$P < 0.01$ versus LIF or DHEA alone. (Reproduced with permission from Suzuki et al. 2004 [17].)

understand the developmental and molecular mechanisms of neurosteroid actions in the human CNS.

Neurospheres were generated from fetal cortex samples, initially grown for four weeks in EGF and FGF-2 (Figure 19.1). The cultures were then switched to medium supplemented with EGF alone. After >10 weeks of growth, the cultures were supplemented with EGF and LIF and maintained for a further four weeks. Thirty to forty neurospheres were cultured with basal medium in the presence or absence of EGF, LIF, and DHEA (Figure 19.2a). After nine days, neurospheres were pulsed with 5-bromo-2′-deoxyuridine (BrdU, 0.2 µM) for 14 h and then dissociated into a single-cell suspension with enzyme. Dissociated cells were plated onto coverslips for 1 h. To check the number of cells immunostained with BrdU or other cell markers, cells were fixed for 1 h after plating. For differentiation studies, the plated cells were differentiated for seven days under serum-free conditions and then visualized with primary antibodies to a neural marker β-tubulin-III (TuJ1) and an astrocyte marker glial fibrillary acidic protein (GFAP).

We first asked what role DHEA may play in the growth of hNPCs cultures. To establish whether DHEA could enhance growth rates in the absence of EGF or LIF, we designed a withdrawal experiment (Figure 19.2a). EGF, LIF, or both factors were withdrawn from the medium in the presence or absence of DHEA for nine days [17]. In the complete absence of EGF, the numbers of BrdU-positive (BrdU$^+$) cells tended to increase in the presence of either LIF or DHEA, although this did not reach significant proportions. This increase may have been due to trace amounts of EGF remaining associated with membranes. In the presence of EGF, LIF significantly increased BrdU incorporation as we have shown previously [13], whereas DHEA again tended to increase BrdU incorporation but did not reach significant levels (Figure 19.2b). However, in the presence of EGF/LIF, DHEA significantly increased the number of BrdU$^+$ cells (Figure 19.2b). Furthermore, a precursor of DHEA such as pregnenolone or six of its major metabolites, had no significant effect on proliferation rates [17]. These data show that DHEA selectively increases the division of the EGF/LIF-responsive cell within the neurospheres.

We next hypothesized that DHEA treatment may alter the potential to generate neurons in hNPCs cultures. To test the possibility, the DHEA-treated neurospheres were labeled with BrdU, dissociated, plated down on coverslips, and allowed to differentiate for seven days (Figure 19.2a). When we immunostained these cells using TuJ1 and BrdU, we found that DHEA treatment significantly increased the number of neurons after differentiation (Figure 19.2c). These results indicate that DHEA can increase proliferation of hNPCs in the presence of EGF/LIF and positively regulate the number of neurons produced by these cultures.

We next asked which types of cells are proliferating in response to DHEA in hNPCs and found that the majority of cells dividing within the neurospheres were GFAP-positive. Our previous results indicated that the expression of GFAP was positively regulated by LIF in hNPCs [13], presumably through the gp130 receptor system and the Janus kinase–signal transducer and activators and transcription (JAK–STAT) signaling pathway [29, 30]. We confirmed these findings and showed that DHEA also showed powerful effects on both GFAP expression and the number of GFAP-positive cells in response to LIF [17]. Interestingly, NSCs capable of forming large numbers of neurons in adult mammals have been identified as astrocyte-resembling cells that are GFAP- and nestin-positive [31, 32]. Given that the DHEA cultures showed increased GFAP production and neurogenesis, our results lend further support to the idea that an astrocyte-like cell may act as a stem cell in the human CNS [32–35].

It has been unclear which receptors mediated the effect of DHEA in human cells. Neurosteroids have been known to modulate NMDA receptor functions through sigma receptors, a type of opiate receptor [26, 36]. We could confirm that the functional subunits of NMDA receptor, NR1 and NR2B, were detected in neurospheres [17]. Interestingly, NR1 mRNA and proteins were strongly induced by LIF and LIF/DHEA treatments. Thus, functional subunits of NMDA receptors on cells within hNPCs could be induced by LIF and LIF/DHEA. We also confirmed that the increased proliferation induced by LIF/DHEA based on BrdU$^+$ cell number could be completely blocked by a NMDA receptor antagonist MK801.

Sigma receptors (a type of opiate receptor) may affect cellular response to neurosteroids by modulating NMDA receptor signaling [37–39]. To test this possibility, we first checked sigma 1 receptor expression in hNPCs by using RT-PCR. Sigma 1 receptor mRNA was detected in all neurosphere cultures and was not seen to be affected by either LIF or DHEA [17]. To establish whether these receptors were important for the effects of DHEA, neurospheres were cultured for nine days in basal medium with EGF/LIF or EGF/LIF/DHEA with or without 3 μM BD1063 or haloperidol, which specifically antagonize sigma 1 receptor function [37]. These experiments suggested that NMDA receptor signaling after sigma 1 receptor activation might involve DHEA actions on these cells. Together, a neurosteroid DHEA has powerful effects on cell proliferation of hNPCs and this effect was modulated through NMDA receptor signaling.

19.4
Glutamate Enhances Proliferation and Neurogenesis in hNPCs

Our previous observation using DHEA suggested that glutamate receptor activation with glutamate may also be able to enhance proliferation of cells and neurogenesis in hNPCs [17]. Excitatory amino acids such as glutamate are the major excitatory neurotransmitters in the CNS [40] but at high concentrations, can also lead to excitotoxic pathology [41, 42]. Numerous studies indicated that glutamate also plays an important role in neural development by influencing the migration, survival, differentiation, and neuritogenesis of new neurons [43–46]. Furthermore, recent reports suggested that glutamate modulates proliferation of rodent forebrain neuronal precursors during neural development [47–49], although there is little direct evidence for biological effects of glutamate on human neural precursor cells during CNS development.

We continued to extend our idea and hypothesized that glutamate treatment with LIF might be involved with cell growth and neurogenesis in hNPCs (Figure 19.3)

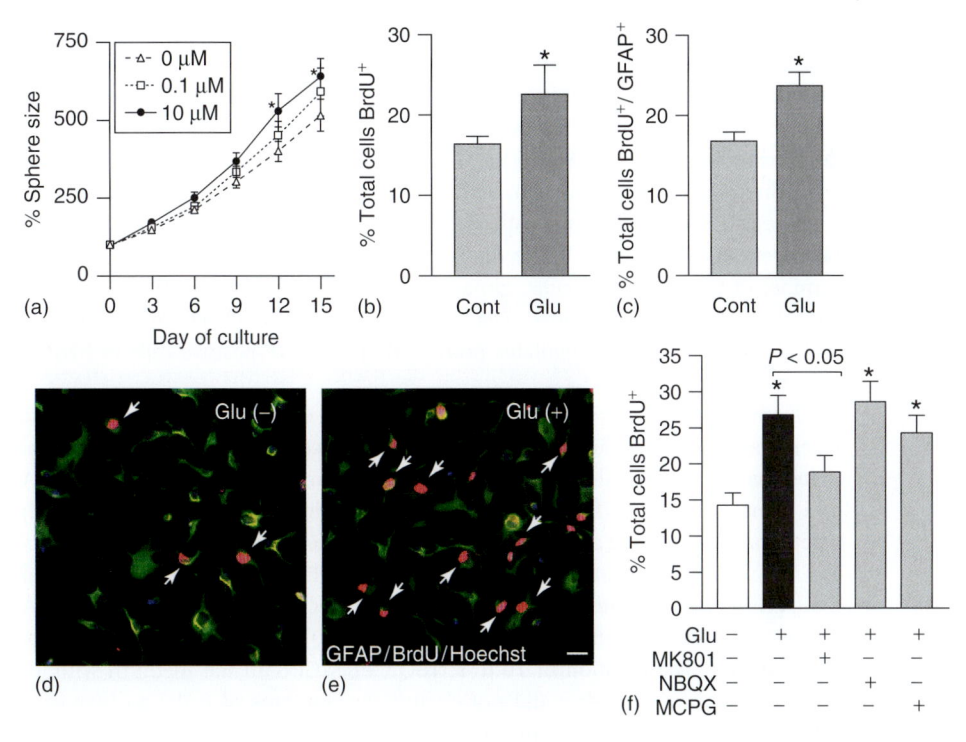

Figure 19.3 Glutamate has a significant effect on proliferation in hNPCs. (a) Glutamate increases growth rate of hNPCs. Neurospheres were cultured in medium supplemented with glutamate for 18 days and sphere volume measurements were taken every three days and compared with control. $^*P < 0.05$ versus control (Control: 0 μM). (b) Glutamate increases the number of BrdU-positive cells in hNPCs in the presence of EGF and LIF. Neurospheres were cultured in the continuous presence or absence of glutamate (10 μM) for nine days, pulse-labeled with BrdU, dissociated, plated for 1 h, and stained with BrdU antibody. $^*P < 0.05$ versus control. (c) The percentages of double-staining with GFAP and BrdU antibody after nine days culture with glutamate. $^*P < 0.05$ versus control. (d and e) Photomicrographs of cells cultured in the absence (d) and presence (e) of glutamate for nine days show a difference in the number of GFAP$^+$/BrdU$^+$ cells (Scale bar, 20 μM) (f) NMDA receptor signaling is involved in the effect of glutamate. Specific NMDA receptor antagonist MK-801 blocks the effects of glutamate, but not Kainate/AMPA receptor antagonist NBQX or metabotropic receptor antagonist MCPG. $^*P < 0.05$ versus control. Reproduced with permission from Suzuki et al. 2006 [50].

[50]. We first checked the effects of adding glutamate to the sphere volume of hNPCs. Neurospheres were cultured in medium supplemented with glutamate for 18 days and sphere volume measurements were taken every 3 days. The addition of 10 μM glutamate significantly increased growth rate of neurospheres when compared to the control (Figure 19.3a). We next asked whether this effect of adding glutamate was not due to protection but rather, due to proliferation of hNPCs. Neurospheres were cultured in the continuous presence of glutamate with the neurobasal medium containing EGF/LIF for nine days. The cultures were then

pulse-labeled with BrdU, dissociated, plated for 1 h, and immunostained with anti-BrdU antibody. The addition of glutamate (10 µM) significantly increased the number of BrdU-positive cells from 16–23% ($P < 0.05$; Figure 19.3b). To further determine whether newly dividing cells express GFAP, we performed double labeling with GFAP and BrdU antibodies, so as to identify cell types that were directly responsive to glutamate in the neurospheres during this specific time period [17]. In keeping with our previous results [13, 17], 80–90% of the cells were GFAP-positive. While the addition of glutamate did not alter the total number of GFAP-positive cells, glutamate significantly increased the number of $BrdU^+/GFAP^+$ double-positive cells by 40% (Figure 19.3c–e). These data suggest that glutamate positively regulates proliferation of GFAP-positive cells in hNPCs cultures.

Glutamate is known to act through both ligand-gated ion channels (ionotropic receptors) and G-protein-coupled (metabotropic) receptors. The ionotropic glutamate receptors are multimeric assemblies of subunits, subdivided into NMDA and kainate/AMPA ((RS)-α-amino-3-hydroxy-5-methyl-4-isoxazolepropionic acid) receptors. We designed a receptor blockade experiment using specific antagonists against glutamate receptors. Neurospheres were cultured in medium containing EGF/LIF/glutamate with or without 10 µM of specific blockers against NMDA receptors (MK801), kainate/AMPA receptors (NBQX), or metabotrophic receptors (MCPG). The addition of 10 µM NBQX or MCPG in the medium with 10 µM glutamate for nine days did not change the numbers of BrdU-positive cells (Figure 19.3f). However, the increased proliferation induced by glutamate based on $BrdU^+$ cell number could be blocked by the NMDA receptor antagonist MK801. These results show that glutamate action on these cells is mediated through the NMDA receptor. Furthermore, we hypothesized that glutamate treatment may alter neurogenic levels in the culture system and found that a subset of progenitor cells divided in response to glutamate and subsequently generated neurons following differentiation [51].

We next asked which types of cells in hNPCs were responsive to glutamate. For this purpose, we checked NMDA receptor expression in neurospheres according to immunohistochemistry. Neurospheres were dissociated, plated, fixed, and immunostained for anti-NR1 antibody. Interestingly, the NR1 subunit of the NMDA receptor was detectable in elongated, bipolar/unipolar cells with small cell bodies in undifferentiated hNPCs [50]. These NR1-positive cells were colocalized with GFAP immunoreactivity.

19.5
Increased Neurogenic "Radial Glial"-like Cells within Human Neurosphere Cultures

Radial glial cells are classically defined as cells with a radial morphology and glial characteristics typical of astrocytes, such as GFAP expression in primates [52]. In the developing brain, radial glial cells have long been known to produce cortical astrocytes, but recent data indicate that radial glial cells also possess neurogenic potential and divide asymmetrically to produce cortical neurons that migrate along their own processes to appropriate cortical layers [53, 54]. Similar radial glial cells have also been found in the human fetal cortex although no real-time examination of their division and migration has been possible [55–57]. In some cases, these neurogenic radial glial cells isolated from rodents may also be stimulated to divide *in vitro* within cultures derived from the developing cortex [58]. However, the exact composition of neurospheres grown in culture from the developing human cortex is currently poorly understood.

We recently found that there are different types of cells in proliferating hNPCs by using double labeling with GFAP and nestin following acute plating [50, 59]. We found that in undifferentiated progenitor cells there are three different morphologies in hNPCs: (i) small and flat, (ii) elongated and radial, or (iii) large and flat cells (Figure 19.4e) [59]. During the process of analyzing glutamate effects on hNPCs, we hypothesized that DHEA and glutamate treatment, which increased neural production in culture, may mainly affect the elongated progenitor cells possessing morphological characteristics like those of radial glial cells. To test this possibility, we compared the three morphologically different cell types in proliferating hNPCs using GFAP staining (Figure 19.4a–d) [50]. The addition of glutamate significantly increased the number of radial GFAP$^+$ cells from ≈11 to 22% ($P < 0.05$; Figure 19.4b,d). Furthermore, we checked the number of radial GFAP$^+$ cells in neurospheres cultured with DHEA and found that DHEA also increased the proportion of radial cells ($P < 0.05$; Figure 19.4c,d). These data suggest that glutamate and DHEA increased the number of radial cells growing within neurospheres. Thus, the "radial glial"-like progenitor cells may represent intermediate progenitor cells or some type of radial cells with neurogenic potential.

This possibility was also confirmed in another paper using culture with a high concentration of EGF, which also increased neural production in hNPCs [59]. Traditional methods have used 20 ng ml^{-1} EGF to drive the proliferation of neurospheres. We recently showed that 100 ng ml^{-1} EGF can significantly increase

Figure 19.4 Enhancement of an elongated cell population in hNPCs is enhanced in response to DHEA, glutamate, or high EGF treatment (a–d). Glutamate and DHEA altered the cellular population in hNPCs. Neurospheres in the continual presence or absence of glutamate were cultured in medium containing EGF for nine days. After dissociation and plating for 1 h, the cells were dissociated, fixed, and stained with GFAP antibody. Photomicrographs represent GFAP immunoreactivity (green) in the acutely plated cells grown in the absence (a), presence of glutamate (b), or presence of DHEA (c), (Scale bar, 20 μM). Many elongated bipolar/unipolar cells with a small cell body (designated by arrows) were found in the cells grown in the presence of glutamate or DHEA. (d) The numbers of elongated cells were significantly increased by glutamate or DHEA. *$P < 0.05$ versus control. (e) Examples of the different cell types (i.e., large/flat, short/flat, and elongated) are shown stained with GFAP$^+$/Nestin$^+$. The values listed are the ranges of nuclear areas of cells that fall within these categories. (f) A significant increase in elongated cell morphologies in the high EGF treated cultures (100 ng ml^{-1}) with a reduction in the large, flat astrocytic cell morphologies was observed. *$P < 0.05$ versus 20 ng ml^{-1} EGF (Reproduced with permission from Suzuki *et al.* 2006 [50] and Nelson *et al.* 2008 [59].)

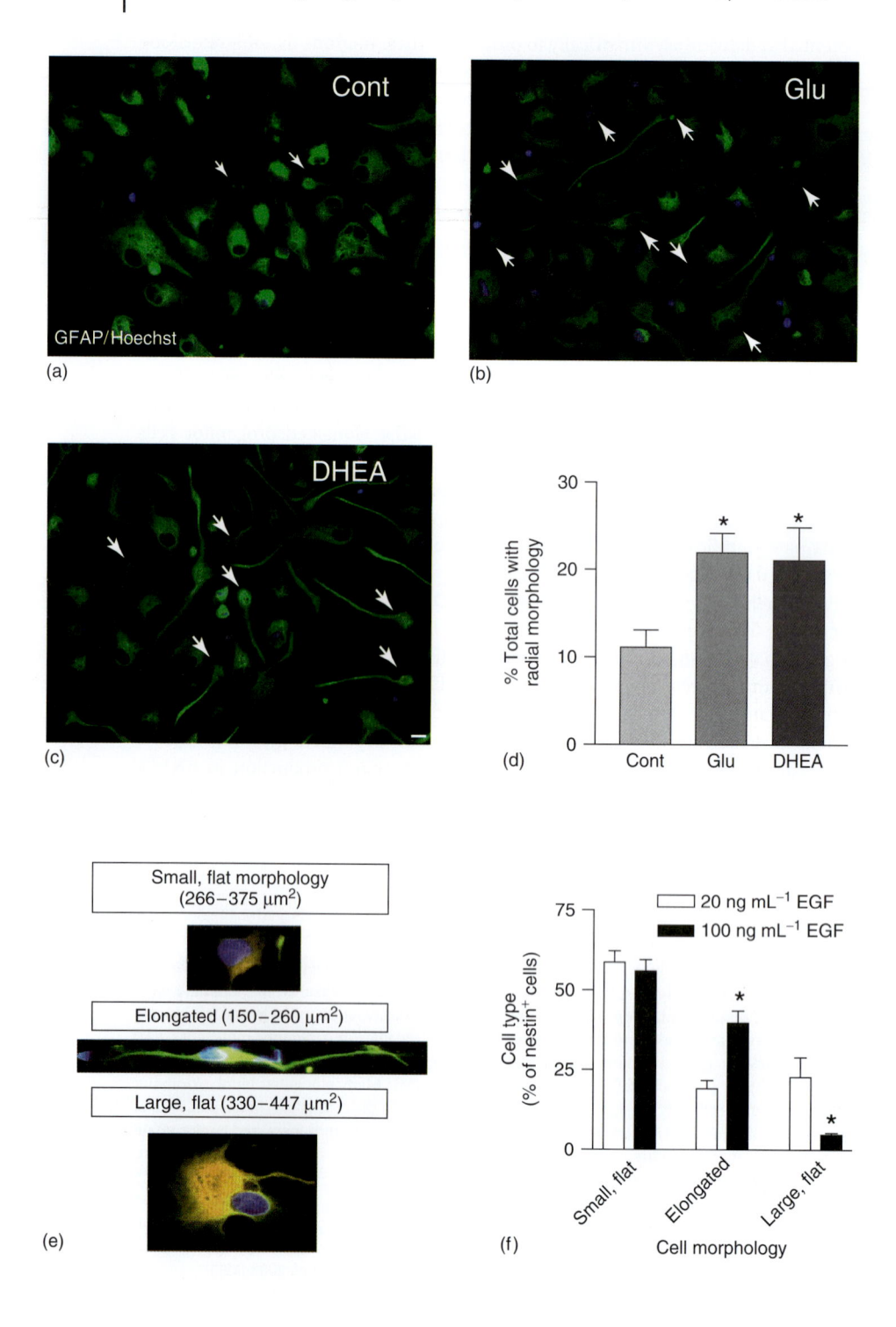

Figure (a) GFAP/Hoechst — Cont
Figure (b) Glu
Figure (c) DHEA
Figure (d) % Total cells with radial morphology — Cont, Glu*, DHEA*
Figure (e) Small, flat morphology (266–375 µm²); Elongated (150–260 µm²); Large, flat (330–447 µm²)
Figure (f) Cell type (% of nestin⁺ cells) vs. Cell morphology (Small, flat; Elongated; Large, flat) for 20 ng mL⁻¹ EGF and 100 ng mL⁻¹ EGF

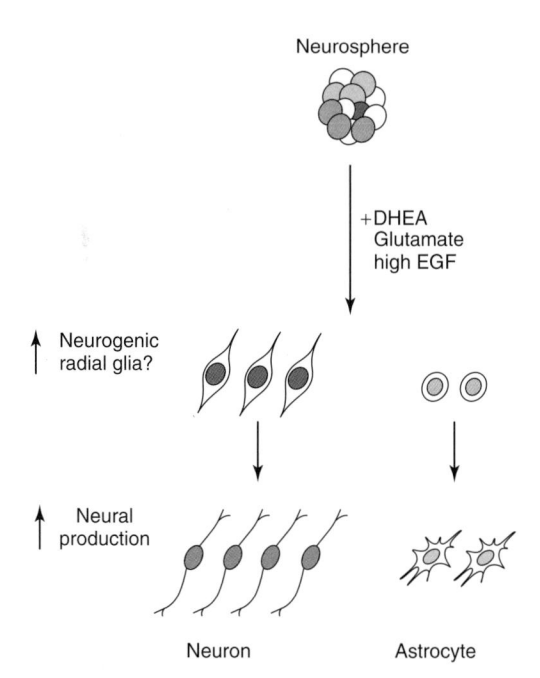

Figure 19.5 Possible involvement of the "radial glial"-like progenitor cells in proliferating hNPCs cultures. Supplementation with glutamate, DHEA, or high EGF in the media can selectively increase elongated progenitor cells with neurogenic potential, which leads to a subsequent increase in the rate of neurogenesis after differentiation.

growth rate of hNPCs at later passages [59]. This was through increased survival of dividing cells rather than increased proliferation. High EGF also resulted in a larger population of elongated "radial glial"-like cells within the growing neurospheres (Figure 19.4f) and increased expression of radial glial markers [59]. The number of new neurons generated from high EGF cultures was significantly higher when compared to the cultures with 20 ng ml^{-1} of EGF [59].

19.6 Conclusions

We have shown that a neurosteroid, DHEA, can increase proliferation of hNPCs in the presence of EGF/LIF and positively regulate the number of neurons produced by these cultures. This effect was modulated through NMDA receptor signaling. It is important to understand whether these direct effects of DHEA (which is currently available as a health supplement in the United States) on human cells warn against its long-term use until the consequences of its action. Although it is impossible to test the effects of DHEA on the adult human brain *in vivo*, our results show an effect in hNPCs, and this may explain some of the beneficial effects of DHEA seen in clinical studies [20, 21].

We have also demonstrated that glutamate itself can increase cell proliferation of hNPCs *in vitro*, and that this is modulated through NMDA receptors. The importance of glutamate receptor activity has recently been highlighted in the pathogenesis of many chronic neurological illnesses such as Huntington's disease,

Alzheimer's disease, and amyotrophic lateral sclerosis [60, 61]. Our results show that in addition to being toxic, glutamate enhances the proliferation of human progenitor cells. This may contribute to normal development and has implications for repairing the damaged brain. Furthermore, it is interesting that selective increases in elongated progenitor cells by glutamate and DHEA correlated to a subsequent increase in the rate of neurogenesis. These observations support our hypothesis that the "radial glial"-like progenitor cells may represent intermediate progenitor cells of some type of radial cells with neurogenic potential (Figure 19.5). Additional studies using time-lapse and immunohistochemical analysis with specific markers against human radial glial cells might be helpful in answering these possibilities.

Currently, the influence of growth factors and other extrinsic molecules on proliferation and differentiation of human progenitor/stem cells is poorly understood. However, human neural progenitor cells offer an excellent model to investigate the effects and gain an understanding of molecular interactions on normal CNS development and neurodegenerative diseases.

Acknowledgments

This work was supported by funds from the University of Wisconsin Foundation, Japan Society for the Promotion of Science, NIH/NINDS 1R21NS06104, and the Les Turner ALS foundation.

References

1. Jakel, R.J., Schneider, B.L., and Svendsen, C.N. (2004) Using human neural stem cells to model neurological disease. *Nat. Rev. Genet.*, **5**, 136–144.
2. Schneider, B.L., Seehus, C.R., Capowski, E.E., Aebischer, P., Zhang, S.C., and Svendsen, C.N. (2007) Over-expression of alpha-synuclein in human neural progenitors leads to specific changes in fate and differentiation. *Hum. Mol. Genet.*, **16**, 651–666.
3. McKay, R. (1997) Stem cells in the central nervous system. *Science*, **276**, 66–71.
4. Gage, F.H., Kempermann, G., Palmer, T.D., Peterson, D.A., and Ray, J. (1998) Multipotent progenitor cells in the adult dentate gyrus. *J. Neurobiol.*, **36**, 249–266.
5. Lederer, C.W. and Santama, N. (2008) Neural stem cells: mechanisms of fate specification and nuclear reprogramming in regenerative medicine. *Biotechnol. J.*, **3**, 1521–1538.
6. Svendsen, C.N. and Smith, A.G. (1999) New prospects for human stem-cell therapy in the nervous system. *Trends Neurosci.*, **22**, 357–364.
7. Hitoshi, S., Tropepe, V., Ekker, M., and van der Kooy, D. (2002) Neural stem cell lineages are regionally specified, but not committed, within distinct compartments of the developing brain. *Development*, **129**, 233–244.
8. Ostenfeld, T., Joly, E., Tai, Y.T., Peters, A., Caldwell, M., Jauniaux, E., and Svendsen, C.N. (2002) Regional specification of rodent and human neurospheres. *Brain Res. Dev. Brain Res.*, **134**, 43–55.
9. Svendsen, C.N., Skepper, J., Rosser, A.E., ter Borg, M.G., Tyres, P., and Ryken, T. (1997) Restricted growth

potential of rat neural precursors as compared to mouse. *Brain Res. Dev. Brain Res.*, **99**, 253–258.

10. Svendsen, C.N., ter Borg, M.G., Armstrong, R.J., Rosser, A.E., Chandran, S., Ostenfeld, T., and Caldwell, M.A. (1998) A new method for the rapid and long-term growth of human neural precursor cells. *J. Neurosci. Methods*, **85**, 141–152.

11. Ray, J., Peterson, D.A., Schinstine, M., and Gage, F.H. (1993) Proliferation, differentiation, and long-term culture of primary hippocampal neurons. *Proc. Natl. Acad. Sci. U.S.A.*, **90**, 3602–3606.

12. Ciccolini, F. and Svendsen, C.N. (1998) Fibroblast growth factor 2 (FGF-2) promotes acquisition of epidermal growth factor (EGF) responsiveness in mouse striatal precursor cells: identification of neural precursors responding to both EGF and FGF-2. *J. Neurosci.*, **18**, 7869–7880.

13. Wright, L.S., Li, J., Caldwell, M.A., Wallace, K., Johnson, J.A., and Svendsen, C.N. (2003) Gene expression in human neural stem cells: effects of leukemia inhibitory factor. *J. Neurochem.*, **86**, 179–195.

14. Lendahl, U., Zimmerman, L.B., and McKay, R.D. (1990) CNS stem cells express a new class of intermediate filament protein. *Cell*, **60**, 585–595.

15. Wright, L.S., Prowse, K.R., Wallace, K., Linskens, M.H., and Svendsen, C.N. (2006) Human progenitor cells isolated from the developing cortex undergo decreased neurogenesis and eventual senescence following expansion in vitro. *Exp. Cell Res.*, **312**, 2107–2120.

16. Svendsen, C.N., Caldwell, M.A., and Ostenfeld, T. (1999) Human neural stem cells: isolation, expansion and transplantation. *Brain Pathol.*, **9**, 499–513.

17. Suzuki, M., Wright, L.S., Marwah, P., Lardy, H.A., and Svendsen, C.N. (2004) Mitotic and neurogenic effects of dehydroepiandrosterone (DHEA) on human neural stem cell cultures derived from the fetal cortex. *Proc. Natl. Acad. Sci. U.S.A.*, **101**, 3202–3207.

18. Orentreich, N., Brind, J.L., Vogelman, J.H., Andres, R., and Baldwin, H. (1992) Long-term longitudinal measurements of plasma dehydroepiandrosterone sulfate in normal men. *J. Clin. Endocrinol. Metab.*, **75**, 1002–1004.

19. Celec, P. and Starka, L. (2003) Dehydroepiandrosterone – is the fountain of youth drying out? *Physiol. Res.*, **52**, 397–407.

20. Bloch, M., Schmidt, P.J., Danaceau, M.A., Adams, L.F., and Rubinow, D.R. (1999) Dehydroepiandrosterone treatment of midlife dysthymia. *Biol. Psychiatry*, **45**, 1533–1541.

21. Roshan, S., Nader, S., and Orlander, P. (1999) Review: ageing and hormones. *Eur. J. Clin. Invest.*, **29**, 210–213.

22. Baulieu, E.E. (1997) Neurosteroids: of the nervous system, by the nervous system, for the nervous system. *Recent Prog. Horm. Res.*, **52**, 1–32.

23. Baulieu, E.E. (1998) Neurosteroids: a novel function of the brain. *Psychoneuroendocrinology*, **23**, 963–987.

24. Bastianetto, S., Ramassamy, C., Poirier, J., and Quirion, R. (1999) Dehydroepiandrosterone (DHEA) protects hippocampal cells from oxidative stress-induced damage. *Brain Res. Mol. Brain Res.*, **66**, 35–41.

25. Kimonides, V.G., Khatibi, N.H., Svendsen, C.N., Sofroniew, M.V., and Herbert, J. (1998) Dehydroepiandrosterone (DHEA) and DHEA-sulfate (DHEAS) protect hippocampal neurons against excitatory amino acid-induced neurotoxicity. *Proc. Natl. Acad. Sci. U.S.A.*, **95**, 1852–1857.

26. Compagnone, N.A. and Mellon, S.H. (1998) Dehydroepiandrosterone: a potential signalling molecule for neocortical organization during development. *Proc. Natl. Acad. Sci. U.S.A.*, **95**, 4678–4683.

27. Karishma, K.K. and Herbert, J. (2002) Dehydroepiandrosterone (DHEA) stimulates neurogenesis in the hippocampus of the rat, promotes survival of newly formed neurons and prevents corticosterone-induced suppression. *Eur. J. Neurosci.*, **16**, 445–453.

28. Corpechot, C., Robel, P., Axelson, M., Sjovall, J., and Baulieu, E.E. (1981) Characterization and measurement of dehydroepiandrosterone sulfate in rat brain. *Proc. Natl. Acad. Sci. U.S.A.*, **78**, 4704–4707.

29. Bonni, A., Sun, Y., Nadal-Vicens, M., Bhatt, A., Frank, D.A., Rozovsky, I., Stahl, N., Yancopoulos, G.D., and Greenberg, M.E. (1997) Regulation of gliogenesis in the central nervous system by the JAK-STAT signaling pathway. *Science*, **278**, 477–483.

30. Nakashima, K., Wiese, S., Yanagisawa, M., Arakawa, H., Kimura, N., Hisatsune, T., Yoshida, K., Kishimoto, T., Sendtner, M., and Taga, T. (1999) Developmental requirement of gp130 signaling in neuronal survival and astrocyte differentiation. *J. Neurosci.*, **19**, 5429–5434.

31. Garcia-Verdugo, J.M., Doetsch, F., Wichterle, H., Lim, D.A., and varez-Buylla, A. (1998) Architecture and cell types of the adult subventricular zone: in search of the stem cells. *J. Neurobiol.*, **36**, 234–248.

32. Doetsch, F., Caille, I., Lim, D.A., Garcia-Verdugo, J.M., and varez-Buylla, A. (1999) Subventricular zone astrocytes are neural stem cells in the adult mammalian brain. *Cell*, **97**, 703–716.

33. Laywell, E.D., Rakic, P., Kukekov, V.G., Holland, E.C., and Steindler, D.A. (2000) Identification of a multipotent astrocytic stem cell in the immature and adult mouse brain. *Proc. Natl. Acad. Sci. U.S.A.*, **97**, 13883–13888.

34. Skogh, C., Eriksson, C., Kokaia, M., Meijer, X.C., Wahlberg, L.U., Wictorin, K., and Campbell, K. (2001) Generation of regionally specified neurons in expanded glial cultures derived from the mouse and human lateral ganglionic eminence. *Mol. Cell Neurosci.*, **17**, 811–820.

35. Imura, T., Kornblum, H.I., and Sofroniew, M.V. (2003) The predominant neural stem cell isolated from postnatal and adult forebrain but not early embryonic forebrain expresses GFAP. *J. Neurosci.*, **23**, 2824–2832.

36. Wu, F.S., Gibbs, T.T., and Farb, D.H. (1991) Pregnenolone sulfate: a positive allosteric modulator at the N-methyl-D-aspartate receptor. *Mol. Pharmacol.*, **40**, 333–336.

37. Monnet, F.P., Mahe, V., Robel, P., and Baulieu, E.E. (1995) Neurosteroids, via sigma receptors, modulate the [3H]norepinephrine release evoked by N-methyl-D-aspartate in the rat hippocampus. *Proc. Natl. Acad. Sci. U.S.A.*, **92**, 3774–3778.

38. Maurice, T., Phan, V.L., Urani, A., Kamei, H., Noda, Y., and Nabeshima, T. (1999) Neuroactive neurosteroids as endogenous effectors for the sigma 1 (sigma 1) receptor: pharmacological evidence and therapeutic opportunities. *Jpn. J. Pharmacol.*, **81**, 125–155.

39. Monnet, F.P. and Maurice, T. (2006) The sigma1 protein as a target for the non-genomic effects of neuro(active)steroids: molecular, physiological, and behavioral aspects. *J. Pharmacol. Sci.*, **100**, 93–118.

40. Watkins, J.C. (2000) l-glutamate as a central neurotransmitter: looking back. *Biochem. Soc. Trans.*, **28**, 297–309.

41. Olney, J.W. (1982) The toxic effects of glutamate and related compounds in the retina and the brain. *Retina*, **2**, 341–359.

42. Choi, D.W. (1988) Glutamate neurotoxicity and diseases of the nervous system. *Neuron*, **1**, 623–634.

43. Mattson, M.P. and Kater, S.B. (1987) Calcium regulation of neurite elongation and growth cone motility. *J. Neurosci.*, **7**, 4034–4043.

44. Rakic, P. and Komuro, H. (1995) The role of receptor/channel activity in neuronal cell migration. *J. Neurobiol.*, **26**, 299–315.

45. Simon, D.K., Prusky, G.T., O'Leary, D.D., and Constantine-Paton, M. (1992) N-methyl-D-aspartate receptor antagonists disrupt the formation of a mammalian neural map. *Proc. Natl. Acad. Sci. U.S.A.*, **89**, 10593–10597.

46. Rossi, D.J. and Slater, N.T. (1993) The developmental onset of NMDA receptor-channel activity during neuronal migration. *Neuropharmacology*, **32**, 1239–1248.

47. LoTurco, J.J., Owens, D.F., Heath, M.J., Davis, M.B., and Kriegstein, A.R. (1995) GABA and glutamate depolarize cortical progenitor cells and inhibit DNA synthesis. *Neuron*, **15**, 1287–1298.

48. Haydar, T.F., Wang, F., Schwartz, M.L., and Rakic, P. (2000) Differential modulation of proliferation in the neocortical

ventricular and subventricular zones. *J. Neurosci.*, **20**, 5764–5774.
49. Luk, K.C., Kennedy, T.E., and Sadikot, A.F. (2003) Glutamate promotes proliferation of striatal neuronal progenitors by an NMDA receptor-mediated mechanism. *J. Neurosc.*, **23**, 2239–2250.
50. Suzuki, M., Nelson, A.D., Eickstaedt, J.B., Wallace, K., Wright, L.S., and Svendsen, C.N. (2006) Glutamate enhances proliferation and neurogenesis in human neural progenitor cell cultures derived from the fetal cortex. *Eur. J. Neurosci.*, **24**, 645–653.
51. Noctor, S.C., Flint, A.C., Weissman, T.A., Wong, W.S., Clinton, B.K., and Kriegstein, A.R. (2002) Dividing precursor cells of the embryonic cortical ventricular zone have morphological and molecular characteristics of radial glia. *J. Neurosci.*, **22**, 3161–3173.
52. Levitt, P. and Rakic, P. (1980) Immunoperoxidase localization of glial fibrillary acidic protein in radial glial cells and astrocytes of the developing rhesus monkey brain. *J. Comp. Neurol.*, **193**, 815–840.
53. Noctor, S.C., Flint, A.C., Weissman, T.A., Dammerman, R.S., and Kriegstein, A.R. (2001) Neurons derived from radial glial cells establish radial units in neocortex. *Nature*, **409**, 714–720.
54. Malatesta, P., Hartfuss, E., and Gotz, M. (2000) Isolation of radial glial cells by fluorescent-activated cell sorting reveals a neuronal lineage. *Development*, **127**, 5253–5263.
55. Rakic, P. (2003) Developmental and evolutionary adaptations of cortical radial glia. *Cereb. Cortex*, **13**, 541–549.
56. Zecevic, N. (2004) Specific characteristic of radial glia in the human fetal telencephalon. *Glia*, **48**, 27–35.
57. Howard, B., Chen, Y., and Zecevic, N. (2006) Cortical progenitor cells in the developing human telencephalon. *Glia*, **53**, 57–66.
58. Gregg, C. and Weiss, S. (2003) Generation of functional radial glial cells by embryonic and adult forebrain neural stem cells. *J. Neurosci.*, **23**, 11587–11601.
59. Nelson, A.D., Suzuki, M., and Svendsen, C.N. (2008) A high concentration of epidermal growth factor increases the growth and survival of neurogenic radial glial cells within human neurosphere cultures. *Stem Cells*, **26**, 348–355.
60. Lipton, S.A. and Rosenberg, P.A. (1994) Excitatory amino acids as a final common pathway for neurologic disorders. *N. Engl. J. Med.*, **330**, 613–622.
61. Rao, S.D. and Weiss, J.H. (2004) Excitotoxic and oxidative cross talk between motor neurons and glia in ALS pathogenesis. *Trends Neurosci.*, **27**, 17–23.

20
Corticosterone, Dehydroepiandrosterone, and Neurogenesis in the Adult Hippocampus
Joe Herbert and Scarlet Bella Pinnock

20.1
Background

The phenomenon of continuing neurogenesis in the adult brain is remarkable. It is astonishing that this was not discovered until 1965 [1], and even more astonishing that the clear evidence of its existence [2] was not accepted until the 1990s. These facts alone invite reflection on the nature of scientific discovery and the difficulty of overcoming accepted ideas even in those dedicated to uncovering new ones. It is remarkable that a developmental process can continue in the environment of the adult brain, one that is generally hostile to mitosis and differentiation of neuronal progenitors. It is intriguing that adult neurogenesis seems to occur only in two areas of the brain, such as the dentate gyrus of the hippocampus and the subventricular zone of the forebrain (though claims have been made for other parts), raising questions about the continuing need for such a process in these regions but not others. Finally, in the case of the dentate gyrus, it has become clear that neurogenesis does not occur at a constant rate, but is highly labile. Hormones have turned out to be some of the most powerful regulators of the rate of progenitor proliferation in the hippocampus, and these are the subjects of this chapter.

Because hormones themselves are highly sensitive to environmental and adventitious events, they form a link between the animals' external world and the internal neurogenic process in the adult brain. In this regard, corticosterone and dehydroepiandrosterone (DHEA) are quite different [3]. Corticosterone (cortisol in man) is highly reactive to environmental events, particularly those involving either physiological or psychological demand ("stress"). Glucocorticoid secretion, and its sensitivity to events in the environment, is a consistent feature of mammalian life. DHEA, on the other hand, is secreted from a different zone of the adrenal (zona reticularis) and only in appreciable quantities in Old World primates, including man [4]. Unlike corticoids, DHEA shows a marked developmental trajectory with levels altering dramatically during intrauterine life, postnatally, and with increasing age in adulthood [5]. Furthermore, though DHEA levels do react to external events, the sensitivity and degree of this response is much less than for corticoids. Finally, the two sets of steroids interact with DHEA moderating the action of corticoids [6].

Hormones in Neurodegeneration, Neuroprotection, and Neurogenesis.
Edited by Achille G. Gravanis and Synthia H. Mellon
Copyright © 2011 WILEY-VCH Verlag GmbH & Co. KGaA, Weinheim
ISBN: 978-3-527-32627-3

The upshot is that these two steroids provide two very different control systems on adult neurogenesis, despite their interaction. Corticoids moderate neurogenesis largely as a result of adventitious events in the animal's physical or social environment (though there may be changes with age as well [7]), whereas DHEA has a much more predictable role that is largely related to the progression of the individual's life span [8]. However, it is extremely important to note that the basal levels and response to external events of both these hormones are highly variable individually, which adds another component to the way they control neurogenesis, and one that may have particular significance for such variable events as aging or the onset of certain illnesses [9].

20.2
Glucocorticoids and Neurogenesis in the Adult Hippocampus

20.2.1
Regulation by Corticoid Levels

The inner layer of the dentate gyrus in the adult rat contains a population of cells that have the phenotypic features of progenitor or stem cells. That is, they stain for nestin, sox-2, and other markers of stem cells. They are more accurately referred to as *progenitor cells* because in their natural environment at least, a high proportion of these cells (around 80%) will differentiate into mature neurons rather than other cell types (e.g., glia). At any one time, however, only a small proportion of these progenitor cells undergo mitosis, and alterations in this process offer one important control point. One of the earliest findings in the rather recent literature on adult neurogenesis was that adrenal corticoids regulate the mitosis rates of these progenitor cells [10]. Removal of the adrenals, and hence the major source of these steroids, results in a marked increase in proliferation rate, and giving excess corticoids reduces it well below basal levels (Figure 20.1) [11–13]. The first finding is intriguing, since it suggests that the basal levels of neurogenesis are held below their potential ceiling by endogenous glucocorticoids. The second shows that there exists a potential for individual or environmentally-driven differences in corticoids to alter rates of neurogenesis very significantly indeed. For example, the differences in basal mitosis rates between two strains of rat seem entirely dependent on corresponding differences in corticosterone levels [14]. It is important to note the range over which corticoids can alter progenitor mitosis rates: we find a threefold difference between adrenalectomized rats and those given a large dose (20 mg kg^{-1} per day) of corticosterone [15].

The cellular mechanisms on which this action of corticoids depends are only partly understood [13]. There are high concentrations of glucocorticoid receptors (GRs) in the dentate gyrus (Figure 20.1). Blockade of the GR (and/or mineralocorticoid receptors: MRs) prevents corticoid-induced reduction in proliferation [16, 17]. However, attempts to demonstrate these receptors in progenitor cells have not been successful [18]. A more likely scenario is that corticoids act on other cells

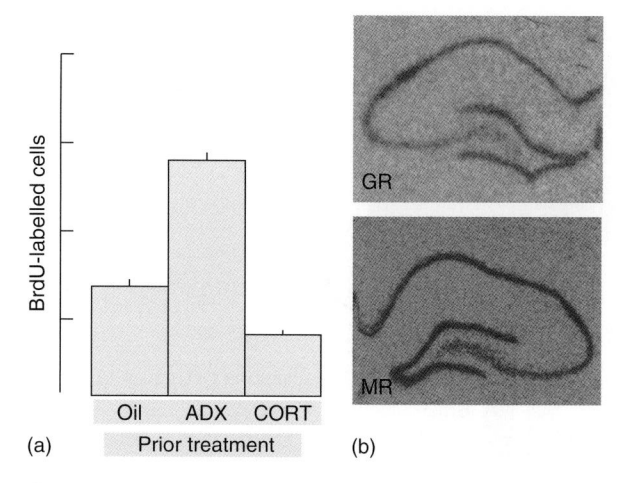

Figure 20.1 (a) Adrenalectomy (ADX) increases the rate of proliferation of progenitor cells in the rat, whereas excess corticosterone reduces it. (b) Both glucocorticoid (GR) and mineralocorticoid (MR) receptor mRNAs are highly expressed in the hippocampus.

forming part of the neurogenetic niche (e.g., glia) though their identity remains obscure. Downstream mechanisms may include the NMDA (N-methyl-D-aspartate) receptor, since activation of this receptor stimulates progenitor cell division [19] and blockade prevents the action of glucocorticoids [20, 21]. Nitric oxide (NO) may play a role. Inhibiting nitric oxide synthase (NOS) stimulates progenitor proliferation [22]. iNOS is markedly increased by corticosterone, and inhibiting it counteracts the suppressive action of excess corticosterone [23] (Figure 20.3). The dentate has a plentiful serotonin innervation (see below); depleting serotonin pharmacologically prevents the suppressive action of excess corticosterone Figure 20.2, suggesting that serotonin may interact with corticoids. This conclusion is reinforced by the interaction between serotonin-acting drugs and the daily corticosterone rhythm (discussed in more detail below). It is clear that BDNF (brain-derived neurotrophic factor) has an essential part in regulating adult neurogenesis (as it does during development), and excess corticosterone or stress reduce BDNF expression in the dentate gyrus [24]. This may, then, be one target of systemic corticoids, though the *BDNF* gene does not contain a glucocorticoid response element (GRE), so the relation between the two may be indirect [25]. The role of BDNF is discussed further below.

But neurogenesis is more than proliferating progenitor cells. The progenitors have to pass through a series of intermediate stages, identified by the expression of markers such as PCA-NCAM and double-cortin (DCX) before attaining the status of mature neurons (NeuN and β-tubulin positive). There is much less information on whether corticoids regulate this maturation process. We showed that corticosterone treatment given for a period of 19–27 days after mitosis [prelabeled (BrdU, 5-bromo-2′-deoxyuridine) progenitor cells] reduced their capacity to survive and

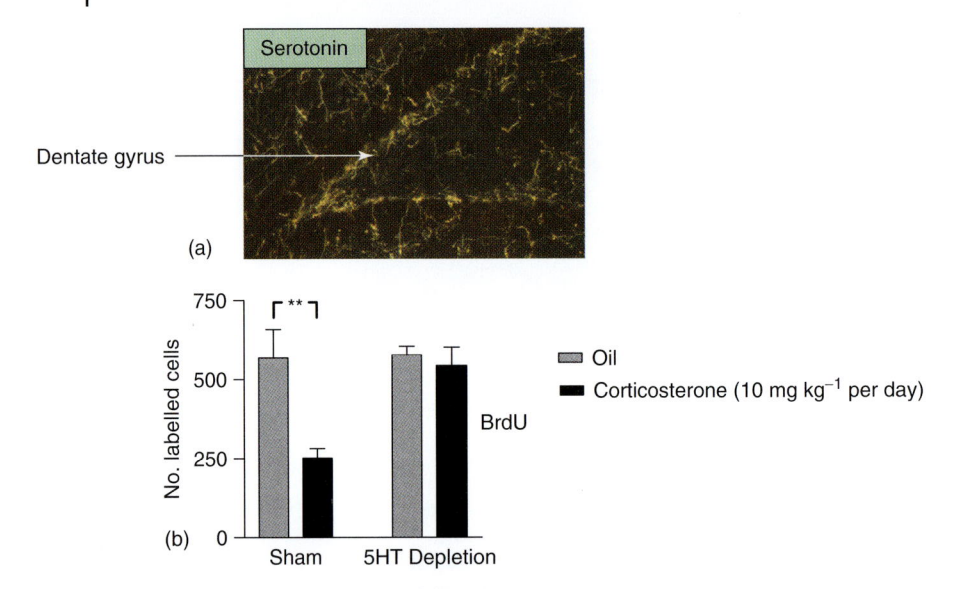

Figure 20.2 (a) There is a rich serotonin innervation of the dentate gyrus (immunohistochemistry). (b) Reducing serotonin by giving the drug PCPA prevents the suppressive action of excess corticosterone kin the rat.

become mature neurons [26] (Figure 20.3). So the combined effect of increased corticoid levels is to reduce the number of new mature neurons in the dentate gyrus by both suppressing progenitor proliferation and discouraging maturation. The final stage in the neurogenic process is the acquisition of afferent input from the entorhinal cortex, and the growth of projections to the CA3 region of the hippocampal pyramidal cell layer. Whether corticoids regulate this part of neurogenesis is currently unknown.

20.2.2
Regulation by the Corticoid Diurnal Rhythm

Corticoid levels in both rats and man show a marked diurnal (nyctohemeral) rhythm. In man, the rhythm appearing in the blood may be damped by the large amount of corticoid-binding globulin (CBG). In the saliva, a better measure of levels in the CSF values may change from six- to eightfold during the day [27, 28]. The progenitor cells are so sensitive to corticoids that their daily mitosis rate tracks that in the corticosterone rhythm of the rat. But a more important fact is that the presence of this rhythm is a requirement for other control systems to access the progenitor cell.

The dentate gyrus has a rich serotonin nerve supply. There is plentiful evidence that this plays a major role in the control of neurogenesis. Giving either drugs (such as fluoxetine) that act to inhibit the serotonin transporter (reuptake) or those that act directly on 5HT1A receptors (e.g., 5-OH DPAT) have a markedly stimulating

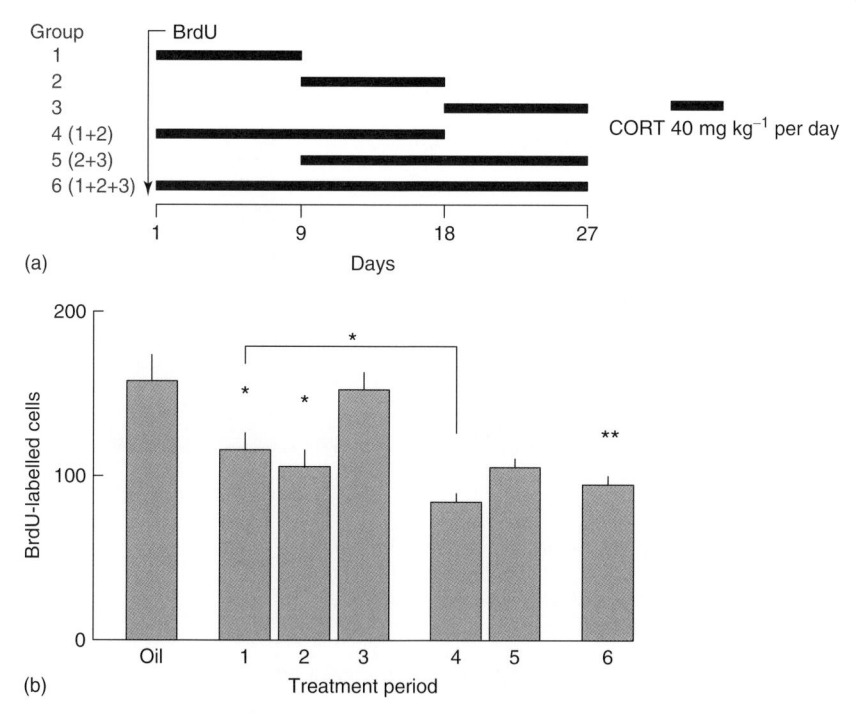

Figure 20.3 Rats were given BrdU which labels the dividing progenitor cells, and then treated with corticosterone for varying periods during the next 28 days. Treatment during the first or second (but not the third) seven day period (a) reduced the number of labeled cells surviving at the end of this period (b).

action on progenitor mitosis rates [12, 29]. Interestingly, fluoxetine, a selective serotonin reuptake inhibitor (SSRI), takes around 14 days to act, despite the more immediate inhibition on the 5HT (5-hydroxytryptamine, serotonin) transporter, a situation recalling the similar latent period preceding therapeutic efficacy in the treatment of depression (see below).

Fluoxetine needs an intact diurnal corticosterone rhythm to be effective on progenitor mitosis (Figure 20.4a). Flattening this rhythm by implanting a subcutaneous pellet of corticosterone prevents fluoxetine from increasing the release of 5HT from the forebrain [30]. It also prevents the stimulating action of fluoxetine on progenitor proliferation, which is restored by an additional injection of corticosterone at the beginning of the dark phase; this also restores the diurnal rhythm [15]. An intact corticosterone rhythm is also required for the action of drugs altering NO. L-nitro-arginine-methyl ester (L-NAME) a general inhibitor of NOS and aminoguanidine, a more specific inhibitor of iNOS, increases mitosis rates in intact rats but not in the presence of a flattened corticosterone rhythm (Figure 20.4b) [23, 31].

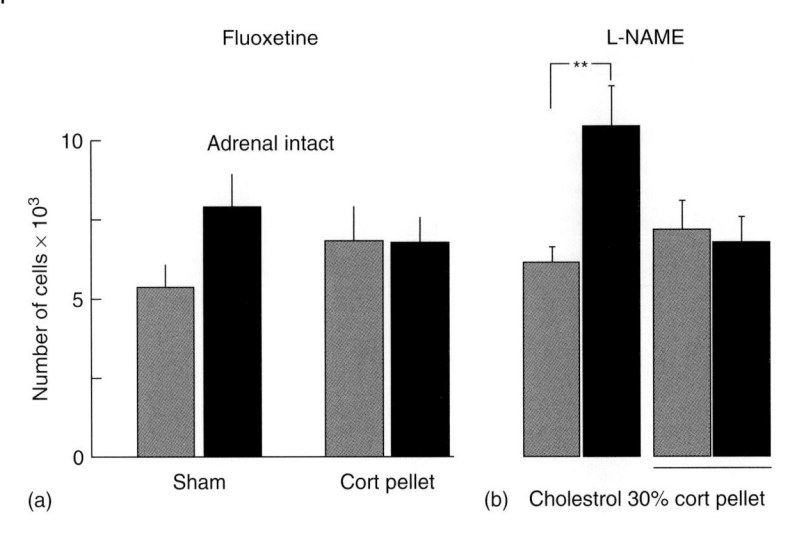

Figure 20.4 Flattening the diurnal corticosterone rhythm by a subcutaneous implant of this steroid prevented both the action of fluoxetine (a) or l-NAMR (b) on the mitotic rate of progenitor cells in the dentate gyrus.

Fluoxetine becomes inoperative if rats are also given a Trk (neurotrophin tyrosine kinase receptor) antagonist (Pinnock et al., submitted), further reinforcing the notion that BDNF is essential for its action and maybe, for other serotonin-dependent mechanisms as well (TrkB is one of the two BDNF receptors). If BDNF is a common downstream regulator of the progenitor cells, responsible for the actions of both fluoxetine and NOS inhibitors, then the question becomes this: is the inactivation of BDNF by these drugs dependent upon an intact corticoid rhythm (i.e., is the rhythm acting as a "gate" to BDNF) or does BDNF itself depend on the presence of a corticoid rhythm for its activity on proliferation? These propositions were tested by infusing BDNF directly into the lateral ventricles of either intact rats or those in which the diurnal corticosterone rhythm had been flattened [31]. BDNF infusions increased mitosis rates on the infused side, as expected [32, 33]. However, they had no significant effect in rats with flattened corticosterone rhythms (Figure 20.5). This clearly supports the second of the hypotheses offered above, and begs the following question: which downstream action of BDNF (e.g., CREB (cAMP-response element binding), Wnt3a) requires an intact rhythm and why?

But is there evidence for diurnal rhythms in the dentate gyrus, which would point to a local control mechanism that is sensitive to the diurnal rhythm in glucocorticoids? Per-1 is an essential component of the cellular machinery underlying the generation of circadian rhythms in the suprachiasmatic nucleus (SCN), which in turn, enables daily variations in both physiological and behavioral activities to be locked into phase with the light–dark cycle [34]. Using a transgenic rat in which the per-1 gene is linked to luciferase, we have shown that the dentate gyrus also expresses per-1, and has a daily rhythm that is in phase with that in the SCN.

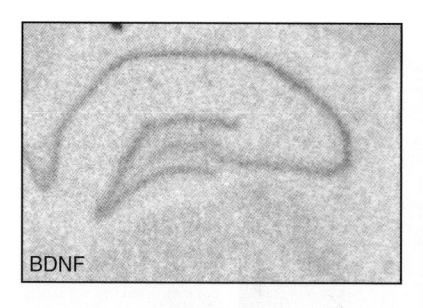

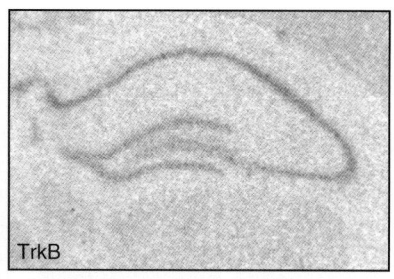

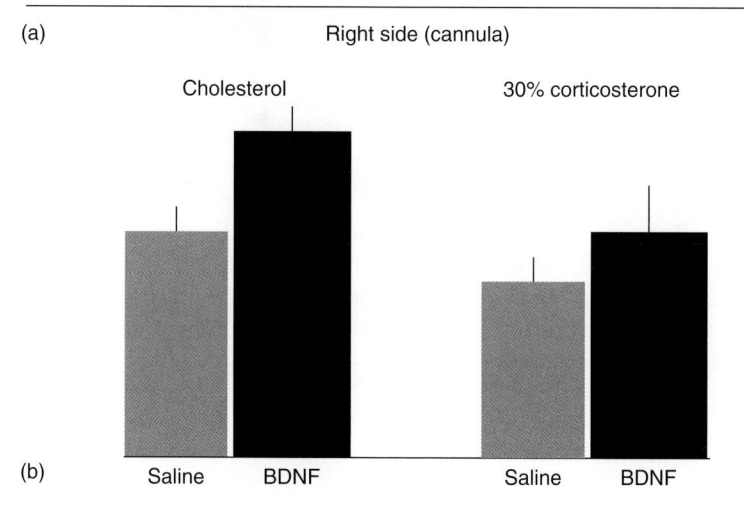

Figure 20.5 (a) High expression of both BDNF and TrkB mRNAs in the hippocampus of the rat. (b) Flattening the diurnal corticosterone rhythm attenuated the stimulating action of icv BDNF (seven days) on mitosis of progenitor cells in the dentate gyrus.

Whereas flattening the daily corticosterone rhythm has only a minor effect on the per-1 rhythm in the SCN, it abolishes that in the dentate gyrus (Figure 20.6, Gilhooley *et al*; submitted). This suggests that there is a "clock" system in the dentate gyrus. This may be driven by the daily corticoid rhythm, itself a product of the "master" oscillator in the SCN. It remains to be determined why the presence of this rhythm within the dentate gyrus seems so critical for the access of other regulatory systems to the progenitor cells, and whether per-1 is simply an index of rhythmic activity or an intrinsic part of the "gate" in the dentate gyrus regulating access of other controlling agents.

The powerful actions of glucocorticoids on hippocampal neurogenesis are thus multiple: absolute levels regulate overall amounts of mitotic activity of the progenitor cells, and the daily rhythm gates access of several other regulating factors (serotonin, NO, maybe BDNF) to the progenitor cells, by a mechanism which is still to be fully understood. This not only has theoretical interest, relevant both to

CT06

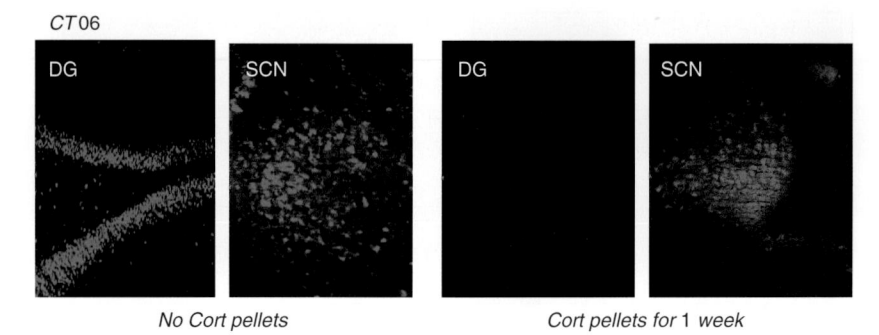

No Cort pellets   Cort pellets for 1 week

Figure 20.6 The expression of per1-luciferase in the suprachiasmatic nucleus SCN ir dentate gyrus (DG) in a transgenic rat. Both show a diurnal rhythm in control s (not shown) but the rhythm is highly attenuated in the DG, but not the SCN at CT6 by flattening the corticosterone rhythm.

the function of the corticoid circadian rhythm and the method by which progenitor cell mitosis rates – the essential first step in hippocampal neurogenesis – are controlled, but also in a more clinical context since disturbance of the daily cortisol rhythm is a prominent feature of major depression [35, 36]. This is discussed in greater detail below.

20.2.3
Dehydroepiandrosterone (DHEA)

Although DHEA is a steroid like the corticoids and is produced by cells that are neighbors (zona reticularis) to those secreting glucocorticoids (zona fasciculata), it is a very different hormone. In adult humans, DHEA circulates both as the free steroid and the sulfated form (DHEAS, dehydroepiandrosterone sulfate). While there have been speculations about the latter being an inactive form, interconversion in both the adrenal and other tissues is easy due to the plentiful supply of both sulfatases and sulfotransferases. Plasma levels of DHEAS easily exceed those of any other steroid in man and are about 10 times that of cortisol, the next abundant steroid. DHEA in the blood in humans is about 10% of DHEAS levels. However, the ratio in the CSF is different, since DHEA penetrates the blood–brain barrier more easily.

The human fetal adrenal produces large amount of DHEA, and this acts as the precursor to many other steroids, which are largely manufactured in the placenta. After birth, DHEA(S) levels fall dramatically, and remain very low until round about the eighth year, when they start to climb quite sharply. This event, adrenarche, has little obvious effect, though the subsequent growth of pubic hair is said to be DHEA-dependent. DHEA(S) levels continue to rise during puberty and adolescence, reaching a peak (a little higher in males than females) around the 20th year. One action is to alter the activity of sebaceous glands, so that adolescent acne may be a consequence of increasing DHEA. From that point, the trajectory is downwards [4]. There is a progressive, but individually variable, decline of DHEA(S) such that levels in the six to seventh decade may only be 20–30% of those at their peak [37]. There has been much debate about the possible consequences of this age-related decline: whether it is an index of the aging process, and whether supplementing DHEA in later life might be beneficial [38, 39]. In particular, there has been speculation that some of the mental deterioration occurring with age (e.g., in memory) might be counteracted by DHEA [40] but there is, as yet, no evidence for this or for a beneficial effect of additional DHEA on any cognitive process in man [41]. Addison's disease (failure of the adrenals) offers a natural example of lack of DHEA, but supplementation over 6–12 months had little effect on cognitive function or mood, though benefiting fatigue and loss of bone density [42, 43]. However, longer-term prospective studies are badly needed, particularly in vulnerable groups.

There is a daily rhythm in salivary DHEA in man, though it is less pronounced than that in cortisol. This means that the relative amounts of cortisol and DHEA (the cortisol/DHEA ratio) will alter in favor of DHEA as the day progresses. DHEA is also lowered in severe illness [44, 45] including major depression [46] and has been tried as an antidepressant [47–49].

The adult rat like other rodents (and carnivores) secretes hardly any DHEA(S). This is important, for it means that any experiments involving giving rats DHEA are essentially pharmacological, rather than (as in the case of glucocorticoids) physiological. Unlike corticosterone, on which a great deal of information is available about intracellular receptors and downstream actions on target genes, that on DHEA is sparse. Though DHEA may antagonize some of the actions of corticoids [50], no DHEA receptor has yet been identified.

DHEA increases the proliferation rate of progenitor cells in the dentate gyrus of the adult male rat. It also antagonizes, to some extent at least, the suppressive actions of corticosterone [6]. Given in amounts which do not, by themselves, increase dentate gyrus progenitor cell mitosis, DHEA can synergize with ineffective doses of fluoxetine (2.5 mg kg^{-1} per day) so that progenitor cell activity is increased [51] (Figure 20.7). Rather surprisingly, this was not accompanied by the expected increase in the expression of BDNF mRNA. Like other actions of DHEA, there is little information on the mechanisms by which it alters progenitor cell mitosis. For example, we do not yet know whether DHEA acts directly on the progenitor cell or on some other type of cell in the neurogenic niche. DHEA was not able to restore the effectiveness of a competent dose of fluoxetine (10 mg kg^{-1} per day) in rats with flattened corticosterone rhythms (see above), so despite the evidence that DHEA can counteract some of the actions of corticoids, it could not overcome the blocking action of a flattened rhythm. This is another indicator that absolute levels and rhythmic secretion of corticoids are separable parameters with distinct roles to play.

The rate of hippocampal neurogenesis declines with age in the rat and marmoset [52, 53], as it presumably does in man (though data on human neurogenesis is limited). Since there is a parallel reduction in DHEA levels, it might be tempting to promote DHEA supplementation as one way of increasing neurogenesis in the

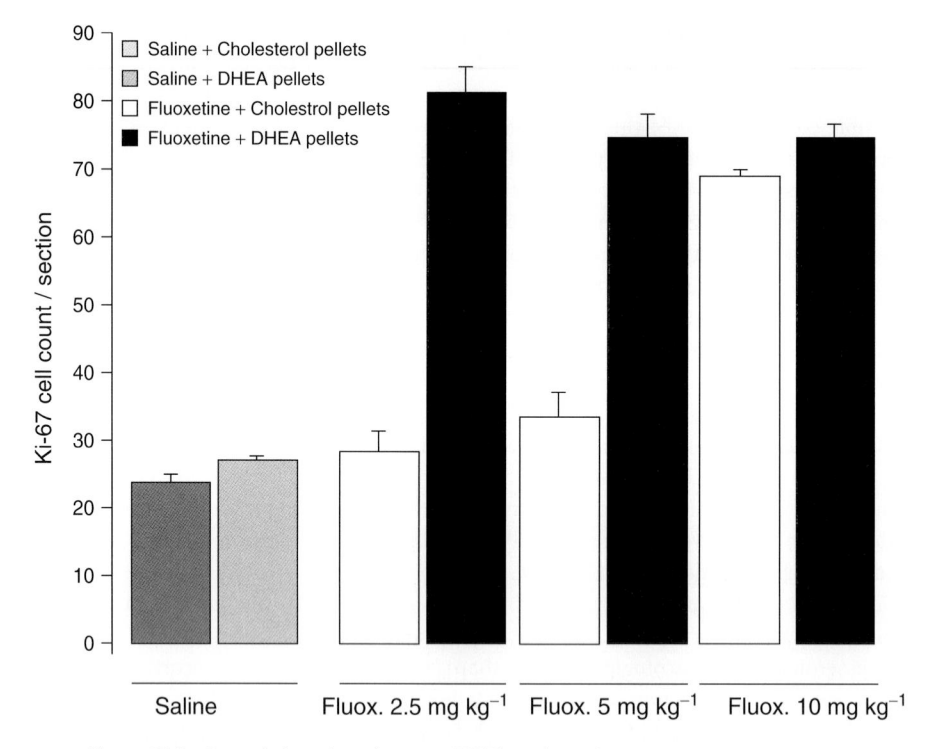

Figure 20.7 Synergistic actions between DHEA and a suboptimal dose of fluoxetine (2.5 mg kg^{-1} per day). Whereas neither DHEA (at this does) or fluoxetine increase the proliferation of progenitor cells in the dentate gurus, both have a similar effect to a much higher dose of fluoxetine alone (10 mg kg^{-1} per day).

aging brain – assuming this to be beneficial. While this may be true, it should not be forgotten that rodents, as pointed out earlier, have little DHEA, so the age-related reduction in neurogenesis cannot be attributed to DHEA in these species at least. However, corticoid levels tend to increase with age in both rats and humans [54] and this may be more generally important in age-related alterations in neurogenesis, accentuated perhaps in man by the associated decline in DHEA, thus attenuating the latter's restraining action on glucocorticoids.

These actions of DHEA on neurogenesis should be considered together with the other evidence that it has protective and positive actions on the brain. It reduces staurosporine-, NMDA-, and glutamate-induced neuronal damage [55–57] and increases the rate of neural stem cell growth [58]. It may do this by activating the serine–threonine kinase Akt [59] which is known to be involved in cell survival signaling. There is also evidence that it alters the nuclear localization of GR [60]; this is one example of its corticoid antagonistic properties. It also prevents oxidative damage induced by 3-nitropropionic acid in synaptosomes [38, 61]. It remains to be seen whether any of these findings apply directly to the way that DHEA regulates neurogenesis in the adult hippocampus.

20.2.4
Downstream Actions: pCREB and Wnt3a

CREB and Wnt3a represent two different intracellular signaling systems and there is evidence that both are involved in neurogenesis. CREB is phosphorylated (pCREB) following activation of cAMP by an extracellular signal. Change in the expression of pCREB is thus one index of the relative activity of this system. BDNF seems to be an essential link between factors regulating neurogenesis and the rate of progenitor mitosis. Blockade of trk receptors (including trkB, a major BDNF receptor) prevents fluoxetine from increasing progenitor proliferation (Pinnock et al., unpublished data). The relation between BDNF and pCREB is complex. Activation of neurotrophin receptors (including trkB) by BDNF results in increased expression of pCREB; that is, BDNF is upstream of pCREB. However, increased pCREB also results in increased BDNF (i.e., a downstream effect). It may be that these actions occur in different populations of cells in the neurogenic niche, but information on this point is still lacking.

Fluoxetine increases the expression of both pCREB and Wnt3a in the dentate gyrus (Figure 20.8). However, blockade of trk receptors prevents that in pCREB but not Wnt3a. This suggests that, in this context, pCREB is "downstream" of BDNF, and that up-regulation is essential for the action of this serotonin-acting drug. pCREB is colocalized with NeuN, a marker of mature neurons, but not

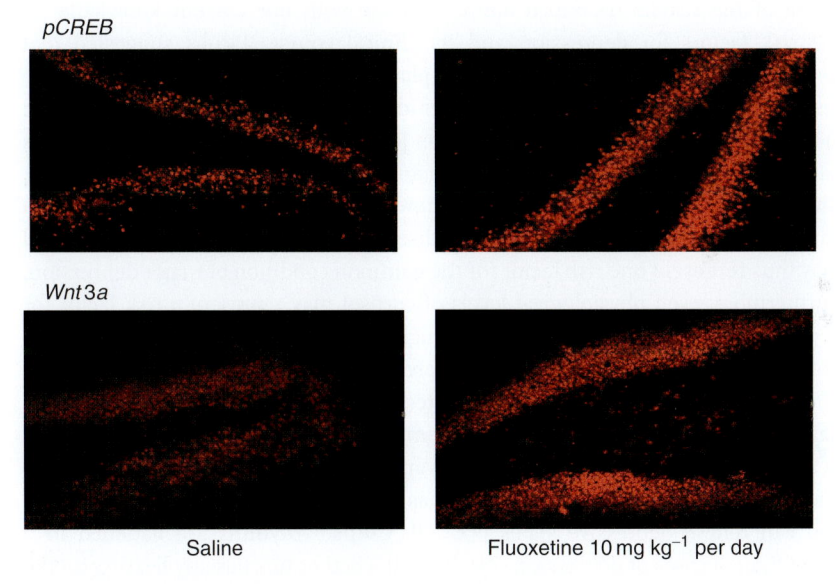

Figure 20.8 Fluoxetine increases the expression of both pCREB and Wnt3a in the dentate gyrus.

Ki-67, a marker of dividing progenitor cells, further suggesting that the action of fluoxetine on these cells is indirect. Adding DHEA to an ineffective dose of fluoxetine (discussed earlier) also increases both pCREB and Wnt3a, and blockade of trk receptors blocks this synergistic effect on both progenitor mitosis and pCREB, but not Wnt3a. Does this mean that Wnt3a is not necessary?

Unilateral infusions of corticotrophin-releasing factor (CRF) reduce progenitor mitosis rates to below control levels and also prevent the stimulating action of fluoxetine (Gibson et al., unpublished data). This shows that the HPA (hypothalamo–pituitary–adrenal) axis has multiple ways of regulating neurogenesis. It also indicates a role for CRF, which has anxiogenic properties, in neurogenesis, separable from that of corticoids (CRF was given in an i.c.v. dose that did not alter plasma corticosterone). But this action of CRF on progenitor cell mitosis was associated with inhibition of Wnt3a expression, not that of pCREB. This suggests that both pCREB and Wnt3a may be required. Whether the latter involves the canonical β-catenin (and *frz* receptors) pathway, which is distinct from pCREB, remains to be determined. If it does, then we will need to dissect the contribution of each pathway to the overall control of neurogenesis. These results underscore our lack of knowledge of the way that hormones and other regulators act on different cells or pathways in the neurogenetic niche to control the rate of formation of new neurons, their maturation, survival, and connectivity.

20.2.5
Relevance to Depression

Some of the results described above resonate with our current knowledge of the risk factors for depression and its clinical course. Jacobs suggested that neurogenesis in the adult hippocampus played a part in either depression or the response to antidepressant drugs [62]. This idea has been enthusiastically promoted [63], sometimes without sufficient caution [64]. There are some striking parallels between factors controlling neurogenesis and those implicated in depression [65]. These include cortisol, which is now known to induce depression if present in pathological amounts, as in Cushing's disease [66]. Increasing levels of cortisol may thus represent one risk factor for the common condition of major depression. Furthermore, even physiological levels of cortisol may represent a risk factor for depression. Individuals with relatively higher levels of cortisol, well within the normal range, are at increased risk for depression, suggesting that the vulnerability of the brain to this disorder (and to antecedent adversity) is increased [67–69]. Salivary cortisol is around 20% higher in females [28], which might account for some of the higher prevalence of depression in women. Stress decreases neurogenesis [70] and predisposes to depression [71], whereas exercise both stimulates neurogenesis and can relieve depressive states [72, 73]. Cortisol rhythms are flattened in a proportion of cases of depression [74], though whether this has any bearing on the response to antidepressants – which might be a credible question in view of the findings described above – is not known.

However, there are serious deficiencies in the contribution that the experimental literature has made to the proposed link between hippocampal neurogenesis and depression. The first comes from the behavioral tests that are supposed to indicate "depressive" states. A number of claims have been made that abolishing neurogenesis (by either genetic manipulation or X-irradiation) prevents the "antidepressant" behavioral actions of drugs such as fluoxetine in the laboratory, though these are not consistent [75, 76]. Some (e.g., novelty-suppressed feeding [76]) are really tests of anxiety, a very different emotional state, though they are often presented as indicating a "depressive" state. Anxiety is a natural, normal, and an essential emotional state, without which survival would not be possible. It becomes pathological when it occurs in inappropriate contexts, and is thus disabling. It commonly occurs in association with major depression in humans, and there is a large literature on this topic. Depression, since it is always disabling, is pathological in itself. So to confuse tests of anxiety with those supposedly measuring "depression" is a major problem of definition. Another set of tests feature so-called *learned helplessness*, which is actually a label put on the behavior of animals exposed to inescapable stress (e.g., foot shocks, suspension by the tail, or immersion in warm water). Cessation of struggling (or increased latency to escape) is labeled as "depression." This lacks face validity: the behavior has no similarity to a depressed state, in contrast to "illness" behavior, which does [77]. It also lacks clinical validity: depression does not occur a few minutes or hours after a stress (an adverse "life event"), but after several weeks [71]. Finally, the rationale for labeling these tests as indicators of depression lacks pharmacological validity. The argument is that, since antidepressants (such as fluoxetine) reduce or delay the onset of these behaviors, this supports the notion that they are experimental parallels of the human depressive state. But the term *antidepressant* is not a pharmacological definition: SSRIs act on serotonin, which is active in many (most) areas of the brain. There is a wealth of experimental data showing that serotonin can influence many categories of behavior. The term *antidepressant* thus reflects clinical usage (though SSRIs are also used for disorders other than depression) and not pharmacological or neurobiological specificity. Furthermore, there is a lack of clinical evidence linking hippocampal neurogenesis either to the risk of developing depression or to the response to treatment [78]. Such evidence is only obtainable, with the present limitations of imaging technology, from postmortem examination since current imaging techniques lack the resolution to reveal neurogenesis in the living adult brain [79].

SSRIs are widely used to treat depression, and typically take around 14–28 days to become effective. This has been related to the time taken for new neurons to mature in the dentate gyrus (28 days). However, as also noted above, there is a delay following the onset of SSRI (fluoxetine) treatment of around 14 days before the mitosis rates of the progenitor cells respond and this has to be taken into account. It seems unlikely that increased mitosis would itself have any discernible therapeutic effect. Currently, there are no adequate explanations for this latent period either in experimental or clinical settings. However, BDNF has been implicated in depression, since a common polymorphism (Val/66/Met) alters the risk for this disorder [80]. Moreover, as discussed earlier, there are interactions between serotonin and corticoid in the control of neurogenesis and in the risk for depression [81].

So all the evidence linking hippocampal neurogenesis and clinical depression is parallel and associative: there is neither direct evidence that neurogenesis is altered in depression or is related to treatment outcome nor credible animal models of the depressive state with which to test the hypothesis. Finally, in contrast to that of the orbital or anterior cingulate cortex, or the amygdala, the role of the hippocampus in depression is yet to be established. Nevertheless, the possibility that understanding more about hippocampal neurogenesis may lead us into a new neurobiology of depression should not be dismissed [64].

20.3
Conclusion

It has to be recognized that whatever its true significance for the function of the hippocampus, the proliferation rate and subsequent maturation of progenitor cells in the neurogenic layer of the dentate gyrus is highly variable, both within and between individuals. As well as its lability, one of the remarkable facts about neurogenesis is the large number of factors that seem able to influence its rate. Amongst these are glucocorticoids and DHEA. These two steroids represent distinct signaling systems, linking the rate of neurogenesis to different events in an animal's (and human's) life. Glucocorticoids are more related to external events, particularly those that represent an unusual or urgent demand, while DHEA is more to be associated with time-dependent maturation processes in the body as a whole. However, this distinction is not absolute, and the two sets of steroids interact in many ways with each other and with the other major controlling agents of neurogenesis, amongst which serotonin is prominent. The full significance of these powerful regulatory agents will become apparent only when we understand more about why active neurogenesis continues throughout life (though at a diminishing pace) in the dentate gyrus [82].

Acknowledgments

The work of our laboratory is supported by grants from the Wellcome Trust and UK Medical Research Council. We thank our laboratory colleagues Samaher AlAhmed, GuoJen Huang, Edmund Wong, Helen Shiers, Jayn Wright, Sarah Cleary, and our clinical colleagues Ian Goodyer, Valerie Dunn, Tirril Harris, and George Brown for allowing us work with them, and for endless stimulation.

References

1. Altman, J. and Das, G.D. (1965) Autoradiographic and histological evidence of postnatal hippocampal neurogenesis in rats. *J. Comp. Neurol.*, **124**, 319–335.

2. Kaplan, M.S. and Hinds, J.W. (1977) Neurogenesis in the adult rat: electron microscopic analysis of light radioautographs. *Science*, **197**, 1092–1094.
3. Hucklebridge, F., Hussain, T., Evans, P., and Clow, A. (2005) The diurnal patterns of the adrenal steroids cortisol and dehydroepiandrosterone (DHEA) in relation to awakening. *Psychoneuroendocrinology*, **30**, 51–57.
4. Orentreich, N., Brind, J.L., Rizer, R.L., and Vogelman, J.H. (1984) Age changes and sex differences in serum dehydroepiandrosterone sulfate concentrations throughout adulthood. *J. Clin. Endocrinol. Metab.*, **59**, 551–555.
5. Sapolsky, R.M., Vogelman, J.H., Orentreich, N., and Altmann, J. (1993) Senescent decline in serum dehydroepiandrosterone sulfate concentrations in a population of wild baboons. *J. Gerontol.*, **48**, B196–B200.
6. Karishma, K.K. and Herbert, J. (2002) Dehydroepiandrosterone (DHEA) stimulates neurogenesis in the hippocampus of the rat, promotes survival of newly formed neurons and prevents corticosterone-induced suppression. *Eur. J. Neurosci.*, **16**, 445–453.
7. Montaron, M.F., Drapeau, E., Dupret, D., Kitchener, P., Aurousseau, C., Le Moal, M. *et al.* (2006) Lifelong corticosterone level determines age-related decline in neurogenesis and memory. *Neurobiol. Aging*, **27**, 645–654.
8. Barrett-Connor, E., Khaw, K.T., and Yen, S.S. (1986) A prospective study of dehydroepiandrosterone sulfate, mortality, and cardiovascular disease. *N. Engl. J. Med.*, **315**, 1519–1524.
9. Herbert, J., Goodyer, I.M., Grossman, A.B., Hastings, M.H., de Kloet, E.R., Lightman, S.L. *et al.* (2006) Do corticosteroids damage the brain? *J. Neuroendocrinol.*, **18**, 393–411.
10. Cameron, H.A. and Gould, E. (1996) Distinct populations of cells in the adult dentate gyrus undergo mitosis or apoptosis in response to adrenalectomy. *J. Comp. Neurol.*, **369**, 56–63.
11. Gould, E., Woolley, C.S., Cameron, H.A., Daniels, D.C., and McEwen, B.S. (1991) Adrenal steroids regulate postnatal development of the rat dentate gyrus: II. Effects of glucocorticoids and mineralocorticoids on cell birth. *J. Comp. Neurol.*, **313**, 486–493.
12. Huang, G.J. and Herbert, J. (2005b) The role of 5-HT1A receptors in the proliferation and survival of progenitor cells in the dentate gyrus of the adult hippocampus and their regulation by corticoids. *Neuroscience*, **135**, 803–813.
13. Joels, M., Karst, H., Krugers, H.J., and Lucassen, P.J. (2007) Chronic stress: implications for neuronal morphology, function and neurogenesis. *Front. Neuroendocrinol.*, **28**, 72–96.
14. Alahmed, S. and Herbert, J. (2008) Strain differences in proliferation of progenitor cells in the dentate gyrus of the adult rat and the response to fluoxetine are dependent on corticosterone. *Neuroscience*, **157**, 677–682.
15. Huang, G.J. and Herbert, J. (2005a) Serotonin modulates the suppressive effects of corticosterone on proliferating progenitor cells in the dentate gyrus of the hippocampus in the adult rat. *Neuropsychopharmacology*, **30**, 231–241.
16. Mayer, J.L., Klumpers, L., Maslam, S., de Kloet, E.R., Joels, M., and Lucassen, P.J. (2006) Brief treatment with the glucocorticoid receptor antagonist mifepristone normalises the corticosterone-induced reduction of adult hippocampal neurogenesis. *J. Neuroendocrinol.*, **18**, 629–631.
17. Wong, E.Y. and Herbert, J. (2005) Roles of mineralocorticoid and glucocorticoid receptors in the regulation of progenitor proliferation in the adult hippocampus. *Eur. J. Neurosci.*, **22**, 785–792.
18. Cameron, H.A., Woolley, C.S., and Gould, E. (1993) Adrenal steroid receptor immunoreactivity in cells born in the adult rat dentate gyrus. *Brain Res.*, **611**, 342–346.
19. Joo, J.Y., Kim, B.W., Lee, J.S., Park, J.Y., Kim, S., Yun, Y.J. *et al.* (2007) Activation of NMDA receptors increases proliferation and differentiation of hippocampal neural progenitor cells. *J. Cell Sci.*, **120**, 1358–1370.
20. Cameron, H.A., McEwen, B.S., and Gould, E. (1995) Regulation of adult neurogenesis by excitatory input and

NMDA receptor activation in the dentate gyrus. *J. Neurosci.*, **15**, 4687–4692.
21. Nacher, J., Varea, E., Miguel Blasco-Ibanez, J., Gomez-Climent, M.A., Castillo-Gomez, E., Crespo, C. *et al.* (2007) N-methyl-d-aspartate receptor expression during adult neurogenesis in the rat dentate gyrus. *Neuroscience*, **144**, 855–864.
22. Packer, M.A., Stasiv, Y., Benraiss, A., Chmielnicki, E., Grinberg, A., Westphal, H. *et al.* (2003) Nitric oxide negatively regulates mammalian adult neurogenesis. *Proc. Natl. Acad. Sci. U.S.A.*, **100**, 9566–9571.
23. Pinnock, S.B., Balendra, R., Chan, M., Hunt, L.T., Turner-Stokes, T., and Herbert, J. (2007) Interactions between nitric oxide and corticosterone in the regulation of progenitor cell proliferation in the dentate gyrus of the adult rat. *Neuropsychopharmacology*, **32**, 493–504.
24. Cavus, I. and Duman, R.S. (2003) Influence of estradiol, stress, and 5-HT2A agonist treatment on brain-derived neurotrophic factor expression in female rats. *Biol. Psychiatry*, **54**, 59–69.
25. Hansson, A.C., Sommer, W., Rimondini, R., Andbjer, B., Stromberg, I., and Fuxe, K. (2003) c-fos reduces corticosterone-mediated effects on neurotrophic factor expression in the rat hippocampal CA1 region. *J. Neurosci.*, **23**, 6013–6022.
26. Wong, E.Y. and Herbert, J. (2006) Raised circulating corticosterone inhibits neuronal differentiation of progenitor cells in the adult hippocampus. *Neuroscience*, **137**, 83–92.
27. Guazzo, E.P., Kirkpatrick, P.J., Goodyer, I.M., Shiers, H.M., and Herbert, J. (1996) Cortisol, dehydroepiandrosterone (DHEA), and DHEA sulfate in the cerebrospinal fluid of man: relation to blood levels and the effects of age. *J. Clin. Endocrinol. Metab.*, **81**, 3951–3960.
28. Netherton, C., Goodyer, I., Tamplin, A., and Herbert, J. (2004) Salivary cortisol and dehydroepiandrosterone in relation to puberty and gender. *Psychoneuroendocrinology*, **29**, 125–140.
29. Duman, R.S., Malberg, J., and Nakagawa, S. (2001) Regulation of adult neurogenesis by psychotropic drugs and stress. *J. Pharmacol. Exp. Ther.*, **299**, 401–407.
30. Gartside, S.E., Leitch, M.M., and Young, A.H. (2003) Altered glucocorticoid rhythm attenuates the ability of a chronic SSRI to elevate forebrain 5-HT: implications for the treatment of depression. *Neuropsychopharmacology*, **28**, 1572–1578.
31. Pinnock, S.B. and Herbert, J. (2008) Brain-derived neurotropic factor and neurogenesis in the adult rat dentate gyrus: interactions with corticosterone. *Eur. J. Neurosci.*, **27**, 2493–2500.
32. Schmidt, H.D. and Duman, R.S. (2007) The role of neurotrophic factors in adult hippocampal neurogenesis, antidepressant treatments and animal models of depressive-like behavior. *Behav. Pharmacol.*, **18**, 391–418.
33. Li, T., Jiang, L., Zhang, X., and Chen, H. (2009) In-vitro effects of brain-derived neurotrophic factor on neural progenitor/stem cells from rat hippocampus. *Neuroreport*, **20**, 295–300.
34. Hastings, M.H., Maywood, E.S., and Reddy, A.B. (2008) Two decades of circadian time. *J. Neuroendocrinol.*, **20**, 812–819.
35. Keller, J., Flores, B., Gomez, R.G., Solvason, H.B., Kenna, H., Williams, G.H. *et al.* (2006) Cortisol circadian rhythm alterations in psychotic major depression. *Biol. Psychiatry*, **60**, 275–281.
36. Sachar, E.J., Hellman, L., Roffwarg, H.P., Halpern, F.S., Fukushima, D.K., and Gallagher, T.F. (1973) Disrupted 24-hour patterns of cortisol secretion in psychotic depression. *Arch. Gen. Psychiatry*, **28**, 19–24.
37. Vermeulen, A. (1995) Dehydroepiandrosterone sulfate and aging. *Ann. N. Y. Acad. Sci.*, **774**, 121–127.
38. Baulieu, E.E. (1995) Studies on dehydroepiandrosterone (DHEA) and its sulphate during aging. *C. R. Acad. Sci. III*, **318**, 7–11.
39. Ebeling, P. and Koivisto, V.A. (1994) Physiological importance of dehydroepiandrosterone. *Lancet*, **343**, 1479–1481.

40. Morales, A.J., Haubrich, R.H., Hwang, J.Y., Asakura, H., and Yen, S.S. (1998) The effect of six months treatment with a 100 mg daily dose of dehydroepiandrosterone (DHEA) on circulating sex steroids, body composition and muscle strength in age-advanced men and women. *Clin. Endocrinol. (Oxf)*, **49**, 421–432.
41. Huppert, F.A. and Van Niekerk, J.K. (2001) Dehydroepiandrosterone (DHEA) supplementation for cognitive function. *Cochrane Database Syst. Rev.*, (2) CD000304.
42. Gurnell, E.M., Hunt, P.J., Curran, S.E., Conway, C.L., Pullenayegum, E.M., Huppert, F.A. *et al.* (2008) Long-term DHEA replacement in primary adrenal insufficiency: a randomized, controlled trial. *J. Clin. Endocrinol. Metab.*, **93**, 400–409.
43. Hunt, P.J., Gurnell, E.M., Huppert, F.A., Richards, C., Prevost, A.T., Wass, J.A. *et al.* (2000) Improvement in mood and fatigue after dehydroepiandrosterone replacement in Addison's disease in a randomized, double-blind trial. *J. Clin. Endocrinol. Metab.*, **85**, 4650–4656.
44. Chen, C.C. and Parker, C.R. Jr. (2004) Adrenal androgens and the immune system. *Semin. Reprod. Med.*, **22**, 369–377.
45. Beishuizen, A., Thijs, L.G., and Vermes, I. (2002) Decreased levels of dehydroepiandrosterone sulphate in severe critical illness: a sign of exhausted adrenal reserve? *Crit. Care*, **6**, 434–438.
46. Michael, A., Jenaway, A., Paykel, E.S., and Herbert, J. (2000) Altered salivary dehydroepiandrosterone levels in major depression in adults. *Biol. Psychiatry*, **48**, 989–995.
47. Rabkin, J.G., McElhiney, M.C., Rabkin, R., McGrath, P.J., and Ferrando, S.J. (2006) Placebo-controlled trial of dehydroepiandrosterone (DHEA) for treatment of nonmajor depression in patients with HIV/AIDS. *Am. J. Psychiatry*, **163**, 59–66.
48. Schmidt, P.J., Daly, R.C., Bloch, M., Smith, M.J., Danaceau, M.A., St Clair, L.S. *et al.* (2005) Dehydroepiandrosterone monotherapy in midlife-onset major and minor depression. *Arch. Gen Psychiatry*, **62**, 154–162.
49. Wolkowitz, O.M., Reus, V.I., Keebler, A., Nelson, N., Friedland, M., Brizendine, L. *et al.* (1999) Double-blind treatment of major depression with dehydroepiandrosterone. *Am. J. Psychiatry*, **156**, 646–649.
50. Hechter, O., Grossman, A., and Chatterton, R.T. Jr. (1997) Relationship of dehydroepiandrosterone and cortisol in disease. *Med. Hypotheses*, **49**, 85–91.
51. Pinnock, S.B., Lazic, S.E., Wong, H.T., Wong, I.H., and Herbert, J. (2009) Synergistic effects of dehydroepiandrosterone and fluoxetine on proliferation of progenitor cells in the dentate gyrus of the adult male rat. *Neuroscience*, **158**, 1644–1651.
52. Leuner, B., Kozorovitskiy, Y., Gross, C.G., and Gould, E. (2007) Diminished adult neurogenesis in the marmoset brain precedes old age. *Proc. Natl. Acad. Sci. U.S.A.*, **104**, 17169–17173.
53. Mirochnic, S., Wolf, S., Staufenbiel, M., and Kempermann, G. (2009) Age effects on the regulation of adult hippocampal neurogenesis by physical activity and environmental enrichment in the APP23 mouse model of Alzheimer disease. *Hippocampus*, **19**, 1008–1018.
54. Lupien, S.J., Nair, N.P., Briere, S., Maheu, F., Tu, M.T., Lemay, M. *et al.* (1999) Increased cortisol levels and impaired cognition in human aging: implication for depression and dementia in later life. *Rev. Neurosci.*, **10**, 117–139.
55. Kimonides, V.G., Khatibi, N.H., Svendsen, C.N., Sofroniew, M.V., and Herbert, J. (1998) Dehydroepiandrosterone (DHEA) and DHEA-sulfate (DHEAS) protect hippocampal neurons against excitatory amino acid-induced neurotoxicity. *Proc. Natl. Acad. Sci. U.S.A.*, **95**, 1852–1857.
56. Kimonides, V.G., Spillantini, M.G., Sofroniew, M.V., Fawcett, J.W., and Herbert, J. (1999) Dehydroepiandrosterone antagonizes the neurotoxic effects of corticosterone and translocation of stress-activated protein kinase 3 in hippocampal primary cultures. *Neuroscience*, **89**, 429–436.
57. Leskiewicz, M., Regulska, M., Budziszewska, B., Jantas, D.,

Jaworska-Feil, L., Basta-Kaim, A. *et al.* (2008) Effects of neurosteroids on hydrogen peroxide- and staurosporine-induced damage of human neuroblastoma SH-SY5Y cells. *J. Neurosci. Res.*, **86**, 1361–1370.

58. Suzuki, M., Nelson, A.D., Eickstaedt, J.B., Wallace, K., Wright, L.S., and Svendsen, C.N. (2006) Glutamate enhances proliferation and neurogenesis in human neural progenitor cell cultures derived from the fetal cortex. *Eur. J. Neurosci.*, **24**, 645–653.

59. Tunez, I., Munoz, M.C., and Montilla, P. (2005) Treatment with dehydroepiandrosterone prevents oxidative stress induced by 3-nitropropionic acid in synaptosomes. *Pharmacology*, **74**, 113–118.

60. Cardounel, A., Regelson, W., and Kalimi, M. (1999) Dehydroepiandrosterone protects hippocampal neurons against neurotoxin-induced cell death: mechanism of action. *Proc. Soc. Exp. Biol. Med.*, **222**, 145–149.

61. Zhang, L., Li, B., Ma, W., Barker, J.L., Chang, Y.H., Zhao, W. *et al.* (2002) Dehydroepiandrosterone (DHEA) and its sulfated derivative (DHEAS) regulate apoptosis during neurogenesis by triggering the Akt signaling pathway in opposing ways. *Brain Res. Mol. Brain Res.*, **98**, 58–66.

62. Jacobs, B.L. (2002) Adult brain neurogenesis and depression. *Brain Behav. Immun.*, **16**, 602–609.

63. Pittenger, C. and Duman, R.S. (2008) Stress, depression, and neuroplasticity: a convergence of mechanisms. *Neuropsychopharmacology*, **33**, 88–109.

64. Herbert, J. (2008) Neurogenesis and depression: breakthrough or blind alley? *J. Neuroendocrinol.*, **20**, 413–414.

65. Thomas, R.M. and Peterson, D.A. (2008) Even neural stem cells get the blues: evidence for a molecular link between modulation of adult neurogenesis and depression. *Gene Expr.*, **14**, 183–193.

66. Jeffcoate, W.J., Silverstone, J.T., Edwards, C.R., and Besser, G.M. (1979) Psychiatric manifestations of Cushing's syndrome: response to lowering of plasma cortisol. *Q. J. Med.*, **48**, 465–472.

67. Goodyer, I.M., Herbert, J., Tamplin, A., and Altham, P.M. (2000) First-episode major depression in adolescents: affective, cognitive and endocrine characteristics of risk status and predictors of onset. *Br. J. Psychiatry*, **176**, 142–149.

68. Goodyer, I.M., Park, R.J., Netherton, C.M., and Herbert, J. (2001) Possible role of cortisol and dehydroepiandrosterone in human development and psychopathology. *Br. J. Psychiatry*, **179**, 243–249.

69. Harris, T.O., Borsanyi, S., Messari, S., Stanford, K., Cleary, S.E., Shiers, H.M. *et al.* (2000) Morning cortisol as a risk factor for subsequent major depressive disorder in adult women. *Br. J. Psychiatry*, **177**, 505–510.

70. Mitra, R., Sundlass, K., Parker, K.J., Schatzberg, A.F., and Lyons, D.M. (2006) Social stress-related behavior affects hippocampal cell proliferation in mice. *Physiol. Behav.*, **89**, 123–127.

71. Brown, G.W. (1993) Life events and affective disorder: replications and limitations. *Psychosom. Med.*, **55**, 248–259.

72. Mead, G.E., Morley, W., Campbell, P., Greig, C.A., McMurdo, M., and Lawlor, D.A. (2009) Exercise for depression. *Cochrane Database. Syst. Rev.*, **8**(3), CD004366.

73. van Praag, H., Christie, B.R., Sejnowski, T.J., and Gage, F.H. (1999) Running enhances neurogenesis, learning, and long-term potentiation in mice. *Proc. Natl. Acad. Sci. U.S.A.*, **96**, 13427–13431.

74. Vreeburg, S.A., Hoogendijk, W.J., van Pelt, J., Derijk, R.H., Verhagen, J.C., and van Dyck, R. *et al.* (2009) Major depressive disorder and hypothalamic-pituitary-adrenal axis activity: results from a large cohort study. *Arch. Gen. Psychiatry*, **66**, 617–626.

75. Holick, K.A., Lee, D.C., Hen, R., and Dulawa, S.C. (2008) Behavioral effects of chronic fluoxetine in BALB/cJ mice do not require adult hippocampal neurogenesis or the serotonin 1A receptor. *Neuropsychopharmacology*, **33**, 406–417.

76. Santarelli, L., Saxe, M., Gross, C., Surget, A., Battaglia, F., Dulawa, S. *et al.* (2003) Requirement of hippocampal neurogenesis for the behavioral

effects of antidepressants. *Science*, **301**, 805–809.

77. Dantzer, R. (2009) Cytokine, sickness behavior, and depression. *Immunol. Allergy. Clin. North Am.*, **29**, 247–264.

78. Drew, M.R. and Hen, R. (2007) Adult hippocampal neurogenesis as target for the treatment of depression. *CNS Neurol. Disord. Drug Targets*, **6**, 205–218.

79. Reif, A., Fritzen, S., Finger, M., Strobel, A., Lauer, M., Schmitt, A. *et al.* (2006) Neural stem cell proliferation is decreased in schizophrenia, but not in depression. *Mol. Psychiatry*, **11**, 514–522.

80. Frodl, T., Schule, C., Schmitt, G., Born, C., Baghai, T., Zill, P. *et al.* (2007) Association of the brain-derived neurotrophic factor Val66Met polymorphism with reduced hippocampal volumes in major depression. *Arch. Gen. Psychiatry*, **64**, 410–416.

81. Goodyer, I.M., Bacon, A., Ban, M., Croudace, T., and Herbert, J. (2009) Serotonin transporter genotype, morning cortisol and subsequent depression in adolescents. *Br. J. Psychiatry*, **195**, 39–45.

82. Clelland, C.D., Choi, M., Romberg, C., Clemenson, G.D. Jr., Fragniere, A., Tyers, P. *et al.* (2009) A functional role for adult hippocampal neurogenesis in spatial pattern separation. *Science*, **325**, 210–213.

Index

a

acquired peripheral neuropathy   121
2-adamantyl-estrogens   22
adamantyl groups   17
Addison's disease   355
adrenalectomy   111
adrenal hypertrophy   74
adrenocorticotrophic hormone (ACTH)   104
AG490   194
aging process   3–4, 121
– implications of cross talk between ER and IGF-I receptors   6–8
– neuroactive steroids, effect of   126–127
allopregnanolone   32, 34, 41–42, 156, 313
– apoptosis reduction, role in   48
– approaches to treatment   45
– calcium influx, role in   46
– cellular oxidative stress reduction, role in   48
– in GABA$_A$ receptor complex   46–47, 50
– neurogenesis, regulation of   291–292
– P19 neurons, effect on   49
– pregnane-X receptors (PXRs), effect on   48
– therapy for Niemann–Pick type C (NP-C) disease   45–47
– voltage-gated calcium channels, effect on   46, 49–50
Alzheimer's disease (AD)   6, 13, 21, 29, 32, 173, 178–179, 196, 240, 286, 308, 315–317, 331
amacrine cells (ACs)   206
amino acid (RS)-α-amino-3-hydroxy-5-methyl-4-isoxazolepropionic acid hydrobromide (AMPA)   213
α-amino-3-hydroxy-5-methyl-4-isoxazolepropionic acid (AMPA)   123
4-aminopyridine   196

AMP-activated protein kinase (AMPK)   197
amyloid β-peptide–induced toxicity   31–32, 196
amyloid-β precursor protein (AβPP)   308, 315–316
5α-androstane-3α   122
androstenedione   158
angiogenesis   269
antalarmin   239
antenatal dexamethasone   74
antidepressant (AD) treatment   110–112, 359
antiglucocorticoid therapy   111–115
APα   286–287, 294–295
Arg-Gly-Asp (RGD) recognition site   272
A-ring derivatives   15–16
astrocytes   137, 269, 271
astrogliosis   114
ataxia–telangiectasia   173
atrophy of the immune system   104
attention-deficit/hyperactivity disorder   66
axonal damage   122
axonal neuropathies   122

b

Bax   49
Bcl-2   6
Bcl-Xs   275
17-benzoate group   19
benzodiazepine   112
birth weight, association with neuropsychiatric disorders   66
blood–brain barrier (BBB)   72, 142, 175, 190–192, 252, 257, 282, 313
blood clots   13
blood pressure   70, 79–80, 239
B- or C-ring derivatives   16

bovine serum albumin (BSA)-conjugated, iodinated progesterone 31
brain
– aging process of 3–4
– under neurodegenerative conditions 3
brain-derived neurotrophic factor (BDNF) 31, 110, 242, 352–353
BrdU-positive cells 268
breast cancer 13
bromodeoxyuridine (BrdU)-positive cell clusters 109

c
CaBP-non-IR cells 209
CA3 dendritic morphology 111–112
Caenorhabditis elegans 174
calbindin 209
cAMP response element-binding protein (CREB) 241
– phosphorylated 357–358
carbamylated erythropoietin (CEPO) 253, 258
carboxylic acid group 15
cardiovascular disease 13, 65
caspase-2 275
catechol-*O*-methyltransferase gene 66
β-catenin 3, 6
central nervous system (CNS) 155, 331
– developmental programming of 66, 76–78
– insulin/IGF axis activity 174–175
– insulin/IGF system, role of 173–175
– themolecular targets of insulin in 175
cerebral edema 32
cerebral ischemia 13–14, 143
cfos/c-Jun 21
CH275 211
Charcot–Marie–Tooth (CMT) disease 121
CHOL 157, 159
cholesterol trafficking, consequences of 43–44
choline acetyl transferase (ChAT) 36
classical steroid receptors 123
C3 methoxy ether analogs 15
conjugated equine estrogens (CEEs) 36
coronary artery atherosclerosis 36
corticosterone rhythm 349–355
corticotrophin-releasing factor (CRF) 358. see also corticotropin-releasing hormone (CRH)
corticotropin-releasing hormone-binding protein (CRH-BP) 238, 241

corticotropin-releasing hormone (CRH) 104
– and molecular signal transduction 238–239
– and neurodegenerative conditions 240
– neuroprotective properties of 241, 244–245
– overview 237–238
– pathophysiology 239–240
– protective activities of 240–244
– protein family 238
corticotropin-releasing hormone receptor 1 (CRH-R1) 238–239
cortistatin (CST) 206
cross talk, between ER and IGF-I receptors 6–8
Cushing's disease 106, 111, 358
cyclooxygenase-2 (COX2)mRNA 143
cyp17-iCRE construct 42
cytochrome c 49

d
Darwinian concept 65
db/db mice 192–193, 196
degenerative disorders 65
dehydroepiandrosterone (DHEA) 3, 44, 137, 156, 162–164, 282, 347
– and autoimmune neurodegenerative processes 142
– brain shaping and maintenance, role in 146
– development of cerebral cortex 332–336
– hippocampal neurons, effect on 138–140
– immunoregulatory actions of 142
– interaction with NGF and TrkA receptors 144–145
– interaction with NGF receptors 144
– levels during human aging 138
– neuroprotective and neurogenic effects of 138–143
– nigrostriatal dopaminergic system, effect on 140–141
– and oxidative stress levels 143
– and protection from ischemia or trauma 142–143
– regulation of neurogenesis 289–290
– signaling pathways of 144–146
– spatial learning performance, impact in 143
– synthetic analogs 146–147
– therapeutic potential of 146–147
– T lymphocytes, activation and proliferation of 142
– treatment of MPTP toxicity 141

dehydroepiandrosterone sulfate ester
    (DHEAS)   137
Déjérine–Sottas syndrome (DSS)   123
dementia   13, 196
demyelination   44, 121–122
dendritic spines   44
dendritic trees   44
dentate gyrus (DG)   350
– granular neurons   105
deoxycorticosterone (DOC)   159
developmental programming   63. *see also*
    glucocorticoids
– associations with adult disorders   65
– biological "purpose" of behavioral
    programming   65
– birth weight and neuropsychiatric
    disorders   66
– brain, impact on   64
– CNS   66, 76–78
– early life physiological "programming"
    64
– environmental challenges, effects of   64
– fetal tissue development   66–68
– of glucocorticoid metabolizing enzymes
    81–82
– HPA axis   74–75
– HPA function and behavior, impact on   64
– human epidemiological studies   65–66
DHEA replacement therapy   146
diabetic neuropathy   128
diabetic retinopathy (DR)   212
dihydroprogesterone (DHP)   47, 122, 159
dihydrotestosterone (DHT)   42, 122, 159,
    288
dihydroxyphenylacetic acid (DOPAC)   141
17β-diol (3α-diol)   122
disease diathesis   64
displaced amacrine cells (DACs)   208
distal symmetric sensory-motor
    polyneuropathy   128
docetaxel   128–129
D-ring derivatives
– *in vitro* assessment of neuroprotective
    activity of   17
Drosophila melanogaster   174
25-Dx   31
dyslipidemia   64

e

effective concentration (EC_{50}) values,
    neuroprotective   15–16
– MPA   35
– for TBARs inhibition   17
Egr-3   125

electroconvulsive seizures   112
embryogenesis, hormonal regulation during
– blastulation regulation   308–310
– cell cycle signaling activity   309
– human chorionic gonadotropin (hCG)
    306–308, 311
– molecular and cellular mechanisms of
    early   308
– neurulation regulation   310–312
– opioid signaling and   312–313
– organogenesis regulation   312
– progesterone, role of   313–314
– trophoblastic secretions of hormones
    306–308
– up-regulation of hCG and P4   306
endocannabinoid receptor 1 (CB1)   242
endometrial hyperplasia   111
endothelin-1 (ET-1)   180
epiallopregnanolone   156
epidermal growth factor (EGF)   110
ER antagonist ICI   182, 780, 4
ERK1/2 phosphorylation   22
erythropoietin (EPO)
– animal model studies of   254–256
– experimental autoimmune (allergic)
    encephalomyelitis (EAE), role in
    257–258
– as a neuroprotective agent in
    hypoxic-ischemic brain injury   251–252
– pharmacokinetics in neonatal CNS injury
    252
– pharmacokinetics of   252
– in preventing neuronal cell death   257
– in preventing peripheral neuropathies
    258–259
– in preventing traumatic injuries   257
– in protection against ischemic stroke
    253–256
esterification   17
estradiol   3–5
– impact of MPA on   35
– molecular interactions   5
– regulation of IGF-I receptor signaling
    5–6
– role in neuroprotection   4–6
– *in vitro* assessment of neuroprotective
    activity of   15
17α-estradiol (αE2)   14
17β-estradiol (β-E2)   14–15
estratrienes, protective activity of   15–17
– lipophilicity   20

estrogen receptor (ER) α 3, 5–6
estrogen receptor (ER) binding 17–18
– correlation with TBAR inhibition 18
estrogens 13
experimental autoimmune neuritis (EAN) 142
extracellular signal–regulated kinase 1/2 (ERK1/2) pathway 273
extracellular signal–related kinase (ERK) pathway 191, 227

f

fa/fa Zucker rats 192–193
Fe(II)/Fe(III) 5 : 1 ratio 19
$FeSO_4^{-}$ 32
fetal programming, role of glucocorticoids
– brith weight, impacts 66
– CNS 66, 76–78
– overexposure effects 75–76, 79–80
– during prenatal stress 68–72
– psychological development 66
– tissue development 66–68
– tissue sensitivity 72–74
fluasterone 143, 146
fluoxetine 351, 357, 359
FosB 125

g

$GABA_A$ receptor complex 32, 34, 41, 123, 143, 160
– mechanism of allopregnanolone in 46–47
GABA-induced chloride conductance 32, 34
gamma-aminobutyric acid (GABA) 156
ganaxolone 46
ganglion cell layer (GCL) 206
gastrointestinal ulceration 104
glaucoma 212
glial fibrillary acidic protein (GFAP) 33, 332
glioblastoma multiforme (GBM) growth 273
gliomas
– cell biology 268–270
– origin 267–268
glucocorticoid response element (GRE) 68
glucocorticoids (GCs)
– binding property of 68
– carbohydrate and lipid metabolism, impact on 104
– cerebral impacts 104
– clinical use of 78–79
– feedback regulation, role in 104–105
– in fetal brain, impact 72–73
– fetal programming during prenatal stress, role in 68–72
– fetal tissue development, effects upon 66–68
– fetal tissue sensitivity of 72–74
– future perspectives and therapeutic opportunities 82–83
– HPA axis, role in activation of 67, 104
– 11β-HSD2 as a barrier to maternal 69, 71
– metabolic and physiologic processes, effects on 67–68
– metabolizing enzymes, programming of 81–82
– and neurogenesis in adult hippocampus 348–360
– overexposure effects 75–76, 79–80
– and posttraumatic stress disorder (PTSD) 80–81
– prenatal administration, consequences 74
– and stress 74–76
– transgenerational effects 71
glucose deprivation–induced toxicity 32
glucose transporters (GLUTs) 176
glutamate 14, 156
glutamate-induced toxicity 13, 15, 32, 194
– and ERK1/2 phosphorylation 22
glutamate receptor 1 (GluR1) 123
glycogen synthase kinase 3β (GSK3β) 3, 6, 179, 241
gonadal hormones 282–284
G-protein-coupled receptor 31
G-proteins 239
growth factor receptor-bound protein 2 (GRB2) 191

h

HIF-1α gene expression 269, 271, 273–274
hippocampal neurogenesis 4, 107–110
– and regulation of glucocorticoids 348–360
– stress impact 110–111
homovanillic acid (HVA) 141
hormones, in aging process 3
hormone therapy (HT) 13, 29
– effectiveness of 14
– progesterone 316–317
– *in vitro* assessment of neuroprotective activity of estrogens 14–23
hot flashes 111
HPA axis programming
– alterations in adult behavior 75–76

– impacts 74–75
– sex-specific effects 75
5HT1A receptors 350
HT-22 cells 15, 19, 22
5-HT reuptake modulators 111
human embryonic stem cells (hESCs) 308–309
human epididymal protein 1 (HE1) 43
human immunodeficiency virus -(HIV-) associated neuropathy 258
human neural stem/progenitor cell (hNPC) cultures 331. *see also* neural stem cells
– DHEA, role of 332–336
– glutamate enhances proliferation and neurogenesis in 336–338
– as a model of human cortical development 331–332
– radial glial cells in 338–341
Huntington's disease (HD) 173, 197
25-hydroxycholesterol 17
5-hydroxy-1,5-dimethyl-hexyl group 19
6-hydroxydopamine (6-OHDA) 4, 197
17α-hydroxylase/17,20 lyase 158
17β-hydroxyl groups 17
2-hydroxy-1-methyl-ethyl group 19
3-hydroxy-3-methylglutaryl-coenzyme A (HMG CoA) reductase 43
3α-hydroxysteroid dehydrogenase (3α-HSD) 122
3β-hydroxysteroid dehydrogenase (3β-HSD) 44, 156–158, 160, 282
11β-hydroxysteroid dehydrogenase type 2 (11β-HSD2) 69–72
3α-hydroxysteroid oxidoreductase (3α-HSOR) 156, 159, 161
hypercorticism 113
hyperglycemia 104, 115, 179
hyperinsulinemia 179
hypertension 64–65, 180
hypothalamic corticotrophin-releasing hormone (CRH) mRNA 73
hypothalamic-pituitary-adrenocortical (HPA) axis 4, 64, 67, 103–104, 237, 239. *see also* HPA axis programming
hypothalamic-pituitary-gonadal (HPG) axis 305
hypothyroidism 273–274
hypoxia ischemia 195

i

ICI 174,864 312–313
immunoglobulin gene superfamily (IgCAM) 123
inner nuclear layer (INL) 206

inner plexiform layer (IPL) 206
insulin deficiency 115
insulin degrading enzyme (IDE) 179
insulin/IGF axis 173–174
– in brain 176
– and neuroprotection 176–178
– neurovascular health, effect on 180
– signaling pathway 175–176
insulin-like growth factor binding proteins (IGFBPs) 174, 182
insulin-like growth factor-I (IGF-I) 3, 110
– estradiol regulation of 5–6
– molecular interactions 5
– regulation of ER transcriptional activity 6
– role in neuroprotection 4–5
– and spatial learning abilities 4
insulin receptor substrate (IRS) proteins 175
integrin $\alpha v \beta 3$ 274
integrin-linked kinase 1 (ILK1) 273
interleukin-6 (IL-6) 252
intracellular signaling pathways, of estrogens 20–23
iodoacetic acid (IAA) 14
iodothyronines 271
ischemic heart disease 64
ischemic neuronal death 194

j

JAK2/STAT3 pathway 199
Janus tyrosine kinase 2 (JAK2) 190–191

k

K252a 145
kainate receptor 123
kainic acid 4
ketoconazole 110, 160, 164
ketogenic diet 196
Krox-20 125
Krox-24 125

l

L-779 976, 213
learned helplessness 359
leptin
– functions of 189–190
– level and cognitive loss 196
– mutation of 192
– neuroprotective effect of 193–199
– neurotrophic role of 193
– receptor (ObR or LEPR) 190–191
– signaling pathways 198–199
– source and structure 189
– transport across BBB 191–192

levodopa 179
levonorgestrel 288
Lewy body diseases 21
lipid antioxidants 19
lipophilic estrogens 16
luteinizing hormone/human chorionic gonadotropin receptors (LHCGRs) 307–309

m

Mac-1 immunoreactive cells 253
MAPK phosphorylation 21
maternal undernutrition/malnutrition, impacts 66, 71, 74
medroxyprogesterone acetate (MPA) 29, 34–36
– antagonistic effects of 36
– ineffectiveness as a neuroprotectant 35–36
MEK/ERK signaling kinase pathway 144
α-melanocyte-stimulating hormone (α-MSH) 176
memory disorders 4
metabolic stress 113
metabolic syndrome 64
13α-methyl groups 17
1-methyl-4-phenyl-1,2,3,6-tetrahydropyridine (MPTP) toxicity 5, 141
metyrapone 110
middle cerebral artery occlusion (MCAO) 32, 212, 253
mifepristone 110–111, 114–115
mineralocorticoid receptors (MRs) 68
mitochondrial swelling 50
mitogen-activated protein kinase (MAPK) 241
mitogen-activated protein kinase (MEK) pathway 191
MK678 211
MK801 336
mononeuritis multiplex 128
mononeuropathy 128
motor and intellectual function, progressive loss of 43
MrgX2 206
multidrug resistance (Mdr)1A 72
multiple sclerosis (MS) 141–142, 257
muscle wasting 104
myelination 125–127, 286
myocardial ischemia 36

n

"natural" progestin. see progesterone
Neel's thrifty genotype hypothesis 65
negative sensory symptoms 121
nerve conduction velocity (NCV) 128
nerve growth factor (NGF) 110
nestorone 287–288
neural stem cells (NSCs) 331. see also human neural stem/progenitor cell (hNPC) cultures
– isolation of 331–332
neuroactive steroids 122
– myelin abnormalities 126–127
– as protective agents in PNS 126–128
– protective and regenerative effects of 127
– Schwann cell response to 123–126
– sexually dimorphic changes of 126
– synthesis and metabolism of 122
neuroblastoma cells 6
neurodegenerative diseases 147, 173, 197. see also Niemann–Pick type C (NP-C) disease
– neural stem therapy for 292–293
– regenerative therapy 292–294
neurofibrillary tangles (NFTs) 196
neurogenesis, adult 42, 281–284, 347
– allopregnanolone regulation of 291–292
– androgen regulation of 288
– clinical progestin regulation of 287–288
– and depression 358–360
– DHEA regulation of 289–290
– estrogen-induced 285–286
– estrogen regulation of 284–286
– glutamate receptor activation in 336–338
– gonadal hormone regulation of 284
– in hippocampus 348–360
– in learning and memory 285
– ovarian hormone regulation of 284
– P_4-induced 314
– and AβPP metabolism 315–316
– pregnenolone and pregnenolone sulfate regulation of 290–291
– progestagen regulation of 286
– progesterone regulation of 287
– seasonal sex differences in 284
– testosterone and DHT regulation of 288–289
neuronal loss 196
neuropeptide Y (NPY) 176
neuroprotection
– estradiol, role in 4–5
– insulin/IGF axis activity 176–178
– insulin-like growth factor-I (IGF-I), role in 4–5
– of nonfeminizing estrogens in intracellular signaling 20–23
– in vitro assessment of estrogens 14–20

neurospheres 332, 335, 338–341
neurosteroidogenesis 155
neurosteroids 282–284, 333, 336
– definition 41
– endogenous production of 157–162
– general background 155–156
– in pain management 162–164
– pregnenolone 41
– 5α-reduced 164
– role in neuronal function and differentiation 42
– synthesis of 41
neurovascular degeneration 180–182
Niemann–Pick type C (NP-C) disease 43–44
– abnormal cholesterol trafficking in 43
– allopregnanolone therapy 45–47
– fibroblasts 48
– steroidogenesis and neurosteroidogenesis in 44–45
nitric oxide (NO) 180
N-methyl-d-aspartate (NMDA) 123, 194, 356
nonclassical steroid receptors 123
nonfeminizing estrogens 20–22
norethindrone 287
norethynodrel 287
norgestimate 287
NPC1 gene 43
NPC2 gene 43
nuclear factor kappa beta (NFκB) 33, 49, 142

o

ob/ob mice 192–193
octadecaneuropeptide 122
oligodendrocytes 137
ovariectomy 284

p

P4 287, 310–311, 313–314
p53 274
p85 5
pain
– antinociceptive and analgesic action 163
– aspects of 156–157
– defined 156
– endogenous neurosteroids in management of 162–164
paraventricular nucleus (PVN) 67, 104
Parkinson's disease (PD) 4, 13, 140, 173, 179, 197, 244, 331
P4-BSA 31

P450c17, cytochrome 42, 137, 145, 158–160, 282
P450c17 gene 42
P450c17-GFP · transgenic mice 42
P450c17-immunostaining 159
pentylenetetrazole 196
peripheral benzodiazepine receptor (PBR) 122
peripheral myelin protein 22 (PMP22) 123
peripheral nerves 121
– classical and nonclassical steroid receptors 123
peripheral nervous system (PNS) 156
peripheral neuropathy 121
– acquired 121
– chemotherapy-induced 128–129
– mechanism of EPO 258–259
P-glycoprotein (Pgp) 72
phenolic A-ring 15
phenolic hydroxyl group 15
phosphatidylinositol 3-kinase/Akt pathway 34, 144, 175, 178
phosphatidylinositol 3-kinase (PI3K) 3, 5, 20
phospholipase C (PLC) 227
PI3K/Akt/GSK3β pathway 6
pituitary adenylate cyclase-activating polypeptide (PACAP) 227
– cell differentiation, role in 231–233
– cell migration, role in 229–231
– cell proliferation, role in 229
– cell survival, role in 231
– cerebellum, effect on 229, 231
– endogenous 233
– expression in developing cerebellum 227–229
– functional relevance 233–234
– molecular characterization of receptors 229
– neurotrophic effects of 231–232
– promitotic or antimitotic activities 229, 233
Pleurodeles waltl 208
plexopathy 128
pluripotent stem cells 332
p75^{NTR} 144
polysialylated form of the neural cell adhesion molecule (PSA-NCAM)-positive progenitors 287
positive sensory symptoms 121
posttraumatic stress disorder (PTSD) 66, 80–81
PR-A 309
Prader–Willi syndrome 173

pregnenolone (PREG)   122, 156, 282, 290–291
premenstrual dysphoric disorder   41
prenatal stress   4
PR isoforms   123
progenitor cells   348
progestagen   286
progesterone (PROG)   156, 158–160, 287, 313–314
– in animal models of stroke, protective effects   33
– biology of   30–31
– in central nervous system, protective effects   33
– induced-protection   32–34
– influence on GABA$_A$ receptor complex   34
– membrane-associated reeptors of   31–32
– re-myelination, role in   33
– synthetic. *see* medroxyprogesterone acetate (MPA)
– in traumatic brain injury (TBI) model, protective effects   32–33
progesterone receptor membrane component 1 (PGRMC1)   287
proopiomelanocortin (POMC)   176
prostaglandin (PGE2)   143
protein zero (P0)   123–124
P450scc, cytochrome   42, 122, 156–157, 159, 282
psychosis   110
psychotic depression   111
Purkinje cell degeneration   44
Purkinje cell layer   230–231

r

R121919   239
radial glial cells   338–341
Ras/MAPK pathways   175–176
Ras/MAPK signaling pathway   6
reactive oxygen species (ROS)   48
recombinant human erythropoietin (rhEPO)   252–253, 256
redox iron cycling   19
5α-reductase (5α-R)   122, 156, 158–160
5α-reductase type I, II   42
retina, role of somatostatin in   206–207
– function in retinal circuitry   209–211
– localization of retinal neurons   207–209
– retinal ischemia   212
ROP   180–182
rotarod test   32
RU-486   124, 306, 309, 311–312

S

schizophrenia   66
Schwann cell proliferation   121–122
– EPO up-regulation in   259
– MCS80 line of   125
– responses to neuroactive steroids   123–126
seizures   43
selective estrogen receptor modulators (SERMs)   14
serotonin type 3 (5-HT3)   123
sex hormones, changes in aging process   3
SH2 domain-containing protein-tyrosine phosphatase (SHP2)   191
SH-SY5Y cell cultures   193
somatostatin
– dopamine release, activation of   210
– function in retinal circuitry   209–211
– glutamate release, activation of   210
– localization of retinal neurons   207–209
– neuroprotective ability of   212–213
– nitric oxide/GMP, effects on   210–211
– receptors, activation of   205
– and related peptides   206
– retina, role in   206–207
– signaling of   206
– therapeutic potential of   216–217
somatotropin release inhibitory factor (SRIF)   205
– anti-ischemic actions of   212–213
– neuroprotective role of   213–216
– SRIF-14   206
– SRIF-28   206
Sox-10   125
spatial learning impairments   4
spinal cord neural crest–derived tissues   137
spinogenesis   113
SSRIs   359
sst$_1$ immunoreactivity (IR)   208
sst$_2$ receptor   208, 213
sst$_3$ receptor   209
sst$_4$ receptor   209
sst$_5$ receptor   209
3β-stereoisomer   46
steroidogenic acute regulatory protein   122
steroidogenic enzymes   41–42
steroid receptor coactivator-1(SRC-1)   125
steroids, synthesis from cholesterol   41
sterol regulatory element binding protein cleavage-activating protein (SCAP)   43
streptozotocin (STZ)   126
stress response

- adult hippocampal neurogenesis, effect on 110–111
- and axonal changes 107
- cell proliferation, impact on 110
- and changes in dendritic reorganization 107
- and depression 105–106
- functional alterations 107
- hippocampal volume changes 106–107
- of hypothalamo-pituitary-adrenal (HPA) system 103–104
- inhibitory effects of 109–110
- in mammals 103
- neurotrophic factors, impact on 110
- treatment with GR antagonists 111–115
stroke 13, 194, 317–318
STZ-induced diabetes 128
subgranular zone (SGZ) 113
substance P 156
- antagonists 111
substantia nigra pars compacta (SNc) 141
supranuclear ophthalmoplegia 43
synaptogenesis 271

t

T cell factor (TCF)/lymphoid enhancer binding factor-1 (LEF-1)-mediated transcription 6
testosterone (T) 122
tetrac 270, 275
2,3,7,8-tetrachlorodibenzo-*p*-dioxin 31
tetrahydrodeoxycorticosterone (THDOC) 159–160
tetrahydroprogesterone (THP) 122, 124, 156
thrombospondin 1, 275
thyroid hormone, actions on glial cells 270
- angiogenic action 274
- antiapoptotic action 274–275
- and brain development 270–271

- hypothyroidism 273–274
- nongenomic actions 271–273
- tumor cell proliferation 274
thyroxine-binding globulin (TBG) 270
transcranial magnetic stimulation 112
transcription factors (TFs) 121
translocator protein-18 kDa (TSPO) 122
trophoblastic secretions of hormones 306–308
tumor cell proliferation 274
tumor necrosis factor-α (TNF-α) 252
type 1 diabetes 113
type 2 diabetes 65, 180
type-I IGF tyrosine–kinase membrane receptor (IGF-R) 174
Tyr985 191
Tyr1027 191
Tyr1138 191
tyrosine kinase (TrkA) receptor 144

u

urocortins, neuroprotective effects of 241, 244–245
urocortin 1 (UCN1) 238, 243–244
urocortin 2 (UCN2) 238
urocortin 3 (UCN3) 238

v

vascular dementia 173, 179–180
vascular endothelial growth factor (VEGF) 110, 181

w

Werner syndrome 173
Wnt3a 6
Women's Health Initiative-Memory Study (WHIMS) 29

z

zalcitabine 259